Emerging Contaminants Handbook

Emerging Contaminants Handbook

Edited by

*Caitlin H. Bell, Margaret Gentile,
Erica Kalve, Ian Ross,
John Horst, and Suthan Suthersan
With foreword by Thomas K.G. Mohr*

CRC Press
Taylor & Francis Group
Boca Raton London New York

CRC Press is an imprint of the
Taylor & Francis Group, an **informa** business

CRC Press
Taylor & Francis Group
6000 Broken Sound Parkway NW, Suite 300
Boca Raton, FL 33487-2742

© 2019 by Taylor & Francis Group, LLC

CRC Press is an imprint of Taylor & Francis Group, an Informa business

No claim to original U.S. Government works

Printed on acid-free paper

International Standard Book Number-13: 978-1-138-06294-8 (Hardback)

This book contains information obtained from authentic and highly regarded sources. Reasonable efforts have been made to publish reliable data and information, but the author and publisher cannot assume responsibility for the validity of all materials or the consequences of their use. The authors and publishers have attempted to trace the copyright holders of all material reproduced in this publication and apologize to copyright holders if permission to publish in this form has not been obtained. If any copyright material has not been acknowledged, please write and let us know so we may rectify in any future reprint.

Except as permitted under U.S. Copyright Law, no part of this book may be reprinted, reproduced, transmitted, or utilized in any form by any electronic, mechanical, or other means, now known or hereafter invented, including photocopying, microfilming, and recording, or in any information storage or retrieval system, without written permission from the publishers.

For permission to photocopy or use material electronically from this work, please access www.copyright.com (http://www.copyright.com/) or contact the Copyright Clearance Center, Inc. (CCC), 222 Rosewood Drive, Danvers, MA 01923, 978-750-8400. CCC is a not-for-profit organization that provides licenses and registration for a variety of users. For organizations that have been granted a photocopy license by the CCC, a separate system of payment has been arranged.

Trademark Notice: Product or corporate names may be trademarks or registered trademarks, and are used only for identification and explanation without intent to infringe.

Library of Congress Cataloging-in-Publication Data

Names: Bell, Caitlin H., author.
Title: Emerging contaminants handbook / Caitlin H. Bell, Erica Kalve, Ian Ross, John Horst and Margaret Gentile.
Description: Boca Raton : Taylor & Francis, a CRC title, part of the Taylor & Francis imprint, a member of the Taylor & Francis Group, the academic division of T&F Informa, plc, 2018. | Includes bibliographical references.
Identifiers: LCCN 2018036093 | ISBN 9781138062948 (hardback : acid-free paper)
Subjects: LCSH: Pollutants—Handbooks, manuals, etc.
Classification: LCC TD176.4 .B45 2018 | DDC 363.738—dc23
LC record available at https://lccn.loc.gov/2018036093

Visit the Taylor & Francis Web site at
http://www.taylorandfrancis.com

and the CRC Press Web site at
http://www.crcpress.com

Printed in Canada

In Memoriam

SUTHAN SUTHERSAN, PhD, PE

The idea for this book came from Dr. Suthan Suthersan, who sadly and unexpectedly passed away before the project could be completed. Throughout his 30-year career, Dr. Suthersan was an inspirational leader and trusted friend to many in the field of environmental remediation. His groundbreaking work and vision for the future showed us time after time why he was acknowledged as one of the most knowledgeable, energetic, and creative people in the field of remediation.

Dr. Suthersan grew up in numerous rural villages across Sri Lanka because of his father's frequent transfers working for the railways. Despite significant financial pressure, his parents unselfishly supported their children, both encouraging them and setting very high expectations for their educational progress. Dr. Suthersan's engineering journey began at the Jaffna College and then the University of Sri Lanka, Peradeniya, where he obtained a degree in Civil Engineering. The political climate of Sri Lanka was tumultuous, and he endured multiple closings of the university during his four *years in Peradeniya. After a brief time at the National Water Supply and Drainage Board, Dr. Suthersan obtained scholarships to the Asian Institute of Technology in Bangkok, Thailand, and then the University of Toronto where he obtained his Masters and PhD degrees in Environmental Engineering and Waste Water Engineering. After graduation, Dr. Suthersan secured a short-term research and development position at the University of Wyoming. During his stay in Wyoming, Dr. Suthersan married his wonderful wife, Sumathy.*

In 1986, Dr. Suthersan became a consulting engineer with Groundwater Technology Inc. (GTI) at their corporate office in Norwood, Massachusetts. It was a special place at that time where many of the leading remediation practitioners were trained during those early years of the industry. In 1990, Dr. Suthersan joined the much-respected consulting company Geraghty and Miller (which later became Arcadis) where he came under the direct tutelage of David Miller, the founder of the company. In Dr. Suthersan's own words, this move "began the most productive, enjoyable and rewarding portion of my engineering journey." Until his passing in early 2017, Dr. Suthersan was Chief Technical Officer and Executive Vice President at Arcadis, where he tirelessly worked to make groundbreaking contributions toward technology development, and developing best practices and knowledge-sharing platforms. Over the years, Dr. Suthersan was awarded more than 20 technology patents and he authored four books: Remediation Engineering *(1996),* Natural and Enhanced Remediation Systems *(2001),* In Situ Remediation Engineering *(2004), and the second edition of* Remediation Engineering *(2017). Dr. Suthersan also contributed to the groundwater remediation community through a regularly featured column in the journal of* Ground water Monitoring and Remediation, *which continues today. Throughout all of his success, Suthan maintained a passion for promoting and empowering young staff to "see beyond others"; mentoring them to embrace change and show courage and curiosity; and constantly reminding them that collaboration is the key to success.*

This is Dr. Suthersan's final book. His motivation for starting the project came from a recognition of how the Information Age has had a profound effect on the speed with which new contaminants make it into the public consciousness, such that the dialogue outpaces the science. He also recognized that the topic of emerging contaminants is vast, complex, and fraught with sensitivity. Consequently, rather than attempting something comprehensive, he envisioned a handbook that addresses a set of contaminants in the mainstream dialogue with information on contaminant sources, the status of toxicological research and regulations, and technologies for characterization and (mainly) groundwater treatment.

We dedicate this book to him. His legacy will continue to inspire and guide engineers, scientists, regulators, students, and researchers around the world to advance our shared mission of improving the quality of life—today, tomorrow, and long into the future. He will be dearly missed.

Table of Contents

List of Figures xv
List of Tables xix
Foreword xxiii
Acknowledgments xxv
Editors xxvii
Contributors xxix

Chapter 1   Introduction to Emerging Contaminants 1

    1.1   Introduction 1
    1.2   Who Identifies Emerging Contaminants? 3
        1.2.1   United States Environmental Protection Agency 4
        1.2.2   United States Department of Defense 6
        1.2.3   United States Geologic Survey 7
        1.2.4   State Agencies in the United States 7
        1.2.5   Stockholm Convention on Persistent Organic Pollutants 8
        1.2.6   European Union 9
        1.2.7   Australian National Environment Protection Council 10
    1.3   What is the Life Cycle of an Emerging Contaminant? 11
    1.4   What are the Key Challenges Associated with Emerging Contaminants? 15
    1.5   The Need for Balance 17
    1.6   This Book 19
        Acronyms 25

Chapter 2   1,4-Dioxane 27

    2.1   Introduction 27
    2.2   Basic Information 27
    2.3   Toxicity and Risk Assessment 32
        2.3.1   Potential Noncancer Effects 33
        2.3.2   Potential Cancer Effects 34
    2.4   Regulatory Status 35
    2.5   Site Characterization 38
        2.5.1   Investigation Approaches 39
        2.5.2   Analytical Methods 41
        2.5.3   Advanced Investigation Techniques 48
    2.6   Soil Treatment 50
    2.7   Groundwater Treatment 51
        2.7.1   In Situ Treatment 51
            2.7.1.1   In Situ Chemical Oxidation 51
            2.7.1.2   Bioremediation 54

		2.7.1.3	Phytoremediation	59
		2.7.1.4	Thermal Treatment	60
	2.7.2	\multicolumn{2}{l}{Ex Situ Treatment and Dynamic Groundwater Recirculation}	61
	2.7.3	\multicolumn{2}{l}{Natural Attenuation}	62

- 2.8 Drinking Water and Wastewater Treatment 65
  - 2.8.1 Point-of-Use and Point-of-Entry Treatment 66
- 2.9 1,4-Dioxane Treatment Technologies for Drinking Water Treatment and Ex Situ Groundwater Remediation 66
  - 2.9.1 Advanced Oxidation Processes 69
  - 2.9.2 Bioreactors 71
  - 2.9.3 Granular Activated Carbon and Other Sorbent Media 73
  - 2.9.4 Electrochemical Oxidation 75
- 2.10 Conclusion 76
- Acronyms 83

Chapter 3 Per- and Polyfluoroalkyl Substances 85

- 3.1 Introduction 85
- 3.2 PFASs Chemistry 88
  - 3.2.1 Ionic State 91
  - 3.2.2 Linear and Branched Isomers 91
  - 3.2.3 Perfluoroalkyl Substances 92
    - 3.2.3.1 Perfluoroalkyl Sulfonic Acids 95
    - 3.2.3.2 Perfluoroalkyl Carboxylic Acids 96
    - 3.2.3.3 Perfluoroalkyl Phosphonic and Phosphinic Acids 96
    - 3.2.3.4 Perfluoroalkyl Ether Carboxylates and Perfluoroalkyl Ether Sulfonates 97
  - 3.2.4 Polyfluoroalkyl Substances 98
    - 3.2.4.1 ECF-Derived Polyfluoroalkyl Substances 98
    - 3.2.4.2 Fluorotelomerization-Derived Polyfluoroalkyl Substances 100
  - 3.2.5 Long- and Short-Chain PFASs 105
  - 3.2.6 Polymeric PFASs 106
  - 3.2.7 Replacement PFASs 107
  - 3.2.8 Chemistry of PFASs in Class B Firefighting Foams 107
- 3.3 Physical, Chemical, and Biological Properties 110
  - 3.3.1 Biological Activity Towards PFASs 112
  - 3.3.2 Transformation of Polyfluoroalkyl Substances 114
    - 3.3.2.1 Abiotic Transformation 115
    - 3.3.2.2 Biotic Transformation 115
- 3.4 PFASs Production and Use 119
  - 3.4.1 Manufacturing Processes and Uses 120

	3.4.2	Electrochemical Fluorination	120
	3.4.3	Fluorotelomerization	122
	3.4.4	Oligomerization	122
	3.4.5	Uses	123
		3.4.5.1 Use as Surfactants	123
		3.4.5.2 Use as Surface Coatings	123
		3.4.5.3 Other Uses	123
3.5	Sampling and Analysis		125
	3.5.1	General Sampling Guidelines	125
		3.5.1.1 Soil and Sediment Sampling	125
		3.5.1.2 Surface Water and Groundwater Sampling	125
		3.5.1.3 Storage and Hold Times	128
	3.5.2	Chemical Analysis Methods	128
		3.5.2.1 Overview of Standard Methods	129
		3.5.2.2 Advanced Analytical Techniques	129
3.6	Health Considerations		139
	3.6.1	Exposure Routes	139
	3.6.2	Distribution in Tissue	139
	3.6.3	Bioaccumulation	140
	3.6.4	Elimination	142
	3.6.5	Toxicologic and Epidemiological Studies	144
		3.6.5.1 Acute Toxicity	144
		3.6.5.2 (Sub)Chronic Toxicity	144
		3.6.5.3 Epidemiological Studies	144
		3.6.5.4 Polyfluoroalkyl Substance Toxicity	145
		3.6.5.5 Derivation of Reference Doses/Tolerable Daily Intakes	146
		3.6.5.6 Carcinogenic Effects	146
3.7	Regulation		150
	3.7.1	Regulation of PFASs	150
		3.7.1.1 Global Treaties and Conventions	150
		3.7.1.2 United States of America	152
		3.7.1.3 Europe	153
		3.7.1.4 Australia	153
	3.7.2	Regulation of Perfluoroalkyl Ethers	154
3.8	Fate and Transport		154
	3.8.1	PFAS Distribution in Environmental Matrices	155
		3.8.1.1 PFASs in Soils	156
		3.8.1.2 Leaching	156
		3.8.1.3 Transport and Retardation in Groundwater	160
		3.8.1.4 Surface Waters and Sediments	165
		3.8.1.5 Vapor Migration	166
		3.8.1.6 Atmospheric Deposition	166
	3.8.2	Detections and Background Levels in the Environment	167

                3.8.3   Sites of Concern 180
                         3.8.3.1   CSM for Industrial Facilities 181
                         3.8.3.2   CSM for Fire Training Areas and Class
                                    B Fire Response Areas 181
                         3.8.3.3   CSM for WWTPs and Biosolid
                                    Application Areas 186
                         3.8.3.4   CSM for Landfills 187
        3.9   PFAS-Relevant Treatment Technologies 188
                3.9.1   Biological Treatment 191
                3.9.2   Soil and Sediment Treatment 191
                         3.9.2.1   Incineration 192
                         3.9.2.2   Stabilization/Solidification 193
                         3.9.2.3   Vapor Energy Generator Technology 194
                         3.9.2.4   Soil/Sediment Washing 196
                         3.9.2.5   High-Energy Electron Beam 196
                         3.9.2.6   Mechanochemical Destruction 197
                3.9.3   Water Treatment 198
                         3.9.3.1   Mature Water Treatment Technologies 198
                         3.9.3.2   Developing Treatment Technologies 210
                         3.9.3.3   Experimental Treatment Technologies 212
        3.10   Conclusions.. 217
                Acronyms.. 257

Chapter 4 Hexavalent Chromium 263

        4.1   Basic Information 263
                4.1.1   Geochemistry of Chromium 263
                4.1.2   Fate and Transport... 265
                4.1.3   Sources of Cr(VI).. 267
        4.2   Toxicity and Risk Assessment 267
        4.3   Regulatory Status 268
                4.3.1   U.S. Federal Regulations 269
                4.3.2   U.S. State Regulations 269
                         4.3.2.1   California... 269
                         4.3.2.2   North Carolina 270
                         4.3.2.3   New Jersey 270
                4.3.3   Other Countries 270
        4.4   Occurrence of Cr(VI).. 271
                4.4.1   Naturally Occurring (Background) Cr(VI)
                         in Groundwater.. 271
                4.4.2   Cr(VI) in Drinking Water 272
        4.5   Site Characterization 273
                4.5.1   Investigation of Cr(VI) in Groundwater................... 273
                4.5.2   Analytical Methods 274
                4.5.3   Advanced Investigation Techniques 277
                         4.5.3.1   Chromium Isotopes 277
                         4.5.3.2   Mineralogical Analyses 277

	4.6	Groundwater Treatment..279
		4.6.1 In Situ Reduction..279
		4.6.1.1 In Situ Chemical Reduction......................280
		4.6.1.2 In Situ Biological Reduction....................284
		4.6.1.3 Permeable Reactive Barriers 288
		4.6.1.4 Reoxidation of Cr(III) Formed by In Situ Reduction 290
		4.6.2 Ex Situ Treatment..290
		4.6.3 Dynamic Groundwater Recirculation 291
		4.6.4 Natural Attenuation..291
		4.6.4.1 Tier I 291
		4.6.4.2 Tier II...292
		4.6.4.3 Tier III..292
		4.6.4.4 Tier IV 292
	4.7	Drinking Water Treatment 292
		4.7.1 Point-of-Entry and Point-of-Use Treatment 293
	4.8	Cr(VI) Treatment Technologies for Drinking Water Treatment and Ex Situ Groundwater Remediation................294
		4.8.1 Reduction/Coagulation/Filtration with Ferrous Iron 294
		4.8.2 Ion Exchange...297
		4.8.2.1 Weak Base Anion Resins..........................297
		4.8.2.2 Strong Base Anion Resins 298
		4.8.3 Reverse Osmosis...300
		4.8.4 Bioreactors..300
		4.8.4.1 Phytostabilization 302
		4.8.4.2 Iron Media 304
		4.8.5 Reduction/Filtration via Stannous Chloride (RF-Sn[II])..304
	4.9	Conclusions..305
		Acronyms...312
Chapter 5	1,2,3-Trichloropropane 315
	5.1	Basic Information 315
	5.2	Toxicity and Risk Assessment...317
	5.3	Regulatory Status 321
		5.3.1 U.S. Federal Regulations..321
		5.3.2 U.S. State Regulations..322
		5.3.3 International Guidance...324
	5.4	Site Characterization 325
		5.4.1 Investigation 325
		5.4.2 Analytical Methods 325
		5.4.3 Advanced Investigation Techniques..........................327
	5.5	Groundwater Remediation Technologies 328
		5.5.1 In Situ Treatment..328
		5.5.1.1 In Situ Hydrolysis 328

		5.5.1.2	In Situ Biological Treatment.................... 331
		5.5.1.3	In Situ Chemical Reduction..................... 333
		5.5.1.4	In Situ Chemical Oxidation 336
		5.5.2	Ex Situ Treatment... 337
	5.6	Water Treatment 338
	5.7	TCP Treatment Technologies for Drinking Water Treatment and Ex Situ Groundwater Remediation............... 338
		5.7.1	Granular Activated Carbon 338
		5.7.2	Advanced Oxidation Processes 343
		5.7.3	Air Stripping 344
		5.7.4	Other Processes 345
	5.8	Conclusions... 345
		Acronyms.. 351

Chapter 6 Considerations for Future Contaminants of Emerging Concern..........355

	6.1	Introduction 355
	6.2	Categorizing Future Emerging Contaminants 355
	6.3	The Challenges Posed in Emerging Contaminant Management 359
		6.3.1	Challenges Associated with Release to the Environment... 359
		6.3.2	Challenges Associated with Assessing Toxicological Risk... 360
		6.3.3	Challenges Associated with Regulation................ 363
		6.3.4	Challenges Associated with Characterization and Analysis... 364
		6.3.5	Challenges Associated with Treatment 366
	6.4	The Future of Emerging Contaminants................................ 367
		Acronyms.. 372

Appendices

Appendix A	USEPA Candidate Contaminant List... 373
Appendix B	REACH Candidate List 377
Appendix C	Emerging Contaminants and Their Physical and Chemical Properties... 385
Appendix D	NGI Preliminary List of Substances That Could Be Considered to Meet the PMT or vPvM Criteria 393
Appendix E.1	Summary of PFAS Environmental Standards: Soil..................... 401
Appendix E.2	Summary of PFAS Environmental Standards: Groundwater 408
Appendix E.3	Summary of PFAS Environmental Standards: Surface Water 415
Appendix E.4	Summary of PFAS Environmental Standards: Drinking Water..... 419
Appendix E.5	Notes 426
Index	 430

List of Figures

Figure 1.1 Example definitions of emerging contaminants 2
Figure 1.2 Emerging contaminants life cycle 12
Figure 2.1 Chemical structure of 1,4-dioxane 27
Figure 2.2 1,4-Dioxane groundwater and drinking water standards and guidelines, by state 37
Figure 2.3 Visual representation of 1,4-dioxane high resolution investigation 42
Figure 2.4 Propane biosparge mixing system 57
Figure 2.5 1,4-Dioxane results from propane biosparge treatment at Vandenberg Air Force Base in California 58
Figure 2.6 Stable isotope probing results from propane biosparge treatment at Vandenberg Air Force Base in California 59
Figure 2.7 1,4-Dioxane breakthrough in various types of GAC 74
Figure 3.1 PFAS families and subgroups 89
Figure 3.2 Structure of 6:2 FTS and PFOS 90
Figure 3.3 Examples of branched and linear PFOS isomers 91
Figure 3.4 Naming convention for PFCAs and PFSAs 92
Figure 3.5 Chemical structure of the PFOS linear isomer 95
Figure 3.6 Chemical structure of linear PFOA 96
Figure 3.7 Chemical structure of PFHxPA, PFOPA, and C_8/C_8 PFPiA 97
Figure 3.8 Chemical structure of ADONA, HFPO-DA, and HFPO-TA 97
Figure 3.9 Naming conventions for FASA-based and fluorotelomer-based PFASs 99
Figure 3.10 Examples of long- and short-chain PFAAs 105
Figure 3.11 Published biotransformation pathways for fluorotelomers 117
Figure 3.12 Published biotransformation pathways for polyfluoroalkyl sulfonamide-based PFSA precursors 120
Figure 3.13 Three major PFAS manufacturing processes, starting material, intermediates, end products, and major uses 121

Figure 3.14	PFAS uses and associated manufacturing processes as identified in literature and patent information	124
Figure 3.15	Example water samples from a recent C6 fluorotelomer foam spill to surface water and analytical results using the pre-TOP and post-TOP Assay digest	130
Figure 3.16	PFAAs formed by TOP Assay of common precursors	131
Figure 3.17	Interpretation of TOP Assay results in five groundwater samples	134
Figure 3.18	Branching in PFHxSaAm is retained in PFHxA after TOP Assay digest	137
Figure 3.19	Comparison of AOF and TOP Assay using AFFF-impacted groundwater	138
Figure 3.20	Conceptual diagram of PFAS exposure to humans and the environment	140
Figure 3.21	Evolution of regulatory standards and awareness of PFASs over time.	151
Figure 3.22	TOP Assay distribution of PFASs in soils impacted by PFASs by AFFF	160
Figure 3.23	CSM for industrial sites	182
Figure 3.24	CSM for evolution of PFAAs and their precursors from FTA source areas	183
Figure 3.25	CSM of class B fire response, fire training, nozzle testing, and storage areas	185
Figure 3.26	CSM of waste management processes	187
Figure 3.27	PFAS treatment technologies for soil, range of practicality, and stage of development	189
Figure 3.28	PFAS treatment technologies for water, range of practicality, and stage of development	189
Figure 4.1	pe-pH stability diagram for the chromium-water system	264
Figure 4.2	Solubility of Cr(III)	267
Figure 4.3	Redox potentials of common redox couples relevant to Cr(VI) reduction	280
Figure 4.4	Biogeochemical oxidation-reduction reactions relevant to biologically mediated Cr(VI) reduction, simplified	284
Figure 4.5	Maximum dissolved iron concentrations collected from dose response wells versus the distance from the injection well	286

Figure 4.6	In situ biological reduction case study results	288
Figure 4.7	Example reduction filtration with Fe(II) process flow diagram	297
Figure 4.8	Example WBA process flow diagram	298
Figure 4.9	Example regenerable SBA process flow diagram	299
Figure 4.10	Example single-pass SBA process flow diagram	299
Figure 4.11	Example RO process flow diagram	300
Figure 4.12	Example submerged bioreactor process flow diagram	301
Figure 4.13	Example membrane bioreactor process flow diagram	302
Figure 4.14	Irrigation system used for treatment of Cr(VI) at the PG&E Hinkley Compressor Station in Hinkley, California	303
Figure 4.15	Example RF-Sn(II) process flow diagram	304
Figure 5.1	1,2,3-Trichloropropane chemical structure	316
Figure 5.2	TCP detections by state in the US UCMR3	316
Figure 5.3	Map of United States TCP standards	322
Figure 5.4	Likely degradation pathways and intermediates for TCP	328
Figure 5.5	TCP hydrolysis half-lives as a function of pH and temperature	329
Figure 5.6	Mass normalized rate constants (k_M) and surface-area normalized rate constants (k_{SA}) for the reduction of TCP and other chlorinated compounds	334
Figure 5.7	Change in TCP breakthrough time between Rapid Small Scale Column Tests (RSSCT) with 0.3 mg/L TOC and 1.3 mg/L TOC for various GAC types	341
Figure 5.8	Difference in mass transfer zone between a typical VOC influent and a TCP-containing influent	341
Figure 5.9	Bed volumes before 30 percent breakthrough for VOCs and various GACs	342
Figure 5.10	VOC breakthrough from pilot-scale adsorber	343
Figure 6.1	Example routes of entry to the environment for emerging contaminants	360

List of Tables

Table 1.1	Contributing factors for selecting featured emerging contaminants	20
Table 2.1	Common sources of 1,4-dioxane and environmental relevance	28
Table 2.2	Comparison of relative physical and chemical properties of 1,4-dioxane with other chemicals	30
Table 2.3	Summary of toxicological reference values developed by USEPA and ATSDR for use in evaluating chronic oral and inhalation non-carcinogenic exposures to 1,4-dioxane along with the basis of their derivation	33
Table 2.4	Summary of USEPA's oral cancer slope factor and inhalation unit risks for 1,4-dioxane and their derivation process	35
Table 2.5	Federal guidelines and health standards for 1,4-dioxane	36
Table 2.6	Summary of 1,4-dioxane groundwater analytical methods	43
Table 2.7	Detailed review of 1,4-dioxane groundwater analytical methods	44
Table 2.8	Summary of chemical oxidant efficacy on 1,4-dioxane and common chlorinated compounds	53
Table 2.9	Considerations for bioremediation of 1,4-dioxane	56
Table 2.10	Water treatment technologies and their application to 1,4-dioxane	67
Table 2.11	Comparison between commercially available ex situ AOPs	70
Table 3.1	Common perfluoroalkyl acids	93
Table 3.2	Selection of known polyfluoroalkyl substances	101
Table 3.3a	Concentrations of PFASs in electrochemical fluorination foams	108
Table 3.3b	Concentrations of PFASs in fluorotelomer foams	109
Table 3.4	Degradation rates of PFAA precursors	116
Table 3.5	Examples of biotransformation pathways	119
Table 3.6a	Summary of acceptable sampling equipment and materials items for PFAS sampling	126
Table 3.6b	Summary of sampling equipment and materials not recommended for PFAS site investigations	127
Table 3.7	Molar yields of perfluoroalkyl carboxylates from precursors subjected to the TOP assay	133
Table 3.8	Sample TOP Assay data output	135

Table 3.9	Bioconcentration, bioaccumulation, and biomagnification information for PFOS	141
Table 3.10	Bioconcentration, bioaccumulation, and biomagnification information for PFOA	142
Table 3.11	Elimination half-lives of PFASs	142
Table 3.12	Reported NOAELs and LOAELs	147
Table 3.13	The evolution of TDIs over time	149
Table 3.14	Summary statistics for select PFASs measured by matrix	157
Table 3.15	Select properties of PFASs	162
Table 3.16	Retardation factors of Select PFAAs	164
Table 3.17a	Summary of detections and background concentrations of PFAAs and total PFASs measured in the environment	168
Table 3.17b	Summary of detections and background concentrations of other PFASs measured in air samples	176
Table 3.17c	Summary of detections and background concentrations of other PFASs measured in surf samples	177
Table 3.18	Technology screening table for water	190
Table 3.19	Isotherm parameters for various types of GAC	200
Table 3.20	Freundlich isotherm constants for resins	205
Table 4.1	Estimated Range of Kd values for Cr(VI) as a function of pH, extractable iron, and soluble sulfate	266
Table 4.2	Summary of Cr(T) and Cr(VI) regulatory standards	268
Table 4.3	Summary of naturally occurring Cr(VI) in groundwater in association with chromium rich deposits	271
Table 4.4	Summary of peak levels of Cr(VI) in drinking water sources detected under the California UCMR (CSWQCB 2014)	273
Table 4.5	Summary of analytical methods to speciate chromium in soil and groundwater	275
Table 4.6	Summary of advanced mineralogical analyses relevant to Cr(VI) groundwater remediation	278
Table 4.7	Summary of example in situ chemical reduction reagents	281
Table 4.8	Summary of example in situ biological reduction reagents	285
Table 4.9	Cr(VI) treatment selection considerations	295

List of Tables

Table 5.1	Summary of occurrence of TCP in drinking water sources detected under the California UCMR	317
Table 5.2	Comparison of relative physical and chemical properties of TCP with other volatile organic compounds	318
Table 5.3	Summary of toxicological reference values for chronic oral and inhalation noncarcinogenic exposures to TCP	319
Table 5.4	Summary of cancer slope factors and inhalation unit risks for TCP	320
Table 5.5	Summary of TCP regulatory standards and guidance levels for the United States of America and several U.S. states	321
Table 5.6	Summary of analytical methods for TCP	326
Table 5.7	ZVZ properties	334
Table 5.8	PRB residence times for TCP concentration reduction	335
Table 5.9	Summary of oxidants for chemical oxidation of TCP	337
Table 5.10	Summary of case studies from California for TCP drinking water treatment	340
Table 5.11	Compounds in TCP GAC pilot influent	343
Table 6.1	Example classes of emerging contaminants	357
Table 6.2	Comparison of emerging contaminant management strategies	369

Foreword

You hold in your hands the key to navigating the complex, transient, unsettled, and at times intimidating topic of emerging contaminants. In the face of conflicting information, wide-ranging and changing regulatory response, and absence of a regulatory requirement, it is understandable that you might be inclined to wait and see, rather than ride the roller coaster of changing and sometimes unreliable information on emerging contaminants. It is tempting to let others sort it out and defer action on emerging contaminants until there is greater certainty regarding contaminant origins and occurrence, toxicity, remedial technology, and regulatory policy. However, as this volume clearly demonstrates, there can be significant consequences to delaying action, including failure to contain a growing plume, failure to limit and manage your client's liability exposure, and worst of all, failure to prevent exposure to toxic contaminants in drinking water.

Taking the plunge into the murky waters of emerging contaminants has been an uninviting prospect, owing to the noisy nature of the information on this topic, and the significant challenge of acquiring actionable intelligence. When a "new" contaminant such as per- and polyfluoroalkyl substances or 1,4-dioxane is found at a site that has long been undergoing successful remediation for the target contaminants under an approved regulatory agreement, projects are derailed, and project managers may find themselves stymied to even identify the right questions to ask.

In the *Emerging Contaminants Handbook*, the authors have faithfully carried the torch of Suthan Suthersan's systematic and disciplined analysis, reflecting the organized thinking that unifies Suthersan's previous definitive texts on remediation engineering. Rather than provide an encyclopedic treatment of an ever-expanding topic, the *Emerging Contaminants Handbook* examines a few of the most relevant emerging contaminants that are attracting attention in 2019 in detail, and most importantly, presents a template for the consideration and management of other emerging contaminants.

The first chapter presents a clearly examined framework and background on the regulatory context of emerging contaminants in general. While the authors are based in North America, their work often includes overseas challenges, and so this book includes a great deal of information on how other nations and groups of nations are responding to emerging contaminants. This chapter is especially useful to remediation engineers, hydrogeologists, and others whose daily work rarely examines regulatory policy.

The contaminant-specific chapters provide a succinct review of the most important features of each contaminant or class of contaminants. You will find the well-organized data and information tables to be especially useful in your work on emerging contaminants, as the authors have completed a tremendous amount of research and analysis, succinctly summarized in an accessible format. Another highly useful feature of the contaminant specific chapters is the case studies. The authors have selected a few examples from their extensive site investigation and remediation experience to illustrate and demonstrate the key characteristics of each profiled contaminant.

The final chapter looks at the horizon to consider what's next in the field of emerging contaminants, and expands the topic to include related emerging issues such as endocrine disrupting compounds, nanomaterials, cyanotoxins, and others. The authors round out the topic of emerging contaminants in the first and last chapters by discussing source management in the marketplace for new materials and chemicals. The future of emerging contaminants is profiled in the context of contaminant categorization, challenges and new developments in toxicity and risk assessment, challenges with developing a consistent framework for establishing and implementing regulation, challenges and solutions for laboratory analysis and site characterization, and a holistic approach to remediation engineering that anticipates and preempts the treatment challenge presented when a new contaminant requiring different treatment methods is detected at a site.

The clear language, highly useful "super-summary" tables, and great case studies, as well as many insightful observations on emerging contaminants make the *Emerging Contaminants Handbook* a great addition to your bookshelf.

Thomas Mohr
Author of *Environmental Investigation and Remediation: 1,4-Dioxane and Other Solvent Stabilizers*

Retired Hydrogeologist with the Santa Clara Valley Water District, California

Past President of the Groundwater Resources Association of California

Acknowledgments

Caitlin H. Bell would like to acknowledge Danielle Pfeiffer, Anthony Bonner, Erin Osborn, Joe Darby, James Collins, Richard Murphy, Olivia Marshall, Ellyn Gates, Jennifer Martin Tilton, Thomas Jensen, Anika Hall, Andy Lewandowski, and Dennis Capria for their contributions to various chapters. Particular thanks to Zachary S. Wahl and Ushna Nasir for all of their help in the last days.

Margaret Gentile would like to thank Mike Hay for contributions, Frank Lenzo for review of the hexavalent chromium chapter, and Scott Murphy for contributions to the trichloropropane chapter. She would also like to thank Courtney Lenzo for her assistance with technical editing of the manuscript.

Ian Ross would like to thank Donald, Veronica, and David for their unwavering support and encouragement over the years.

Erica Kalve would like to send a heartfelt thank you to her family, Joe, Morgan, and Oskar, as this could not have happened without their loving support. Also a big thank you to her mother and sisters for always being there when she needed an extra set of hands. Finally, Erica and Ian would like to thank all those who helped complete the PFAS chapter. There are too many to list out here including Jeff Burdick, Joe Quinnan, Thomas Held, and Erika Houtz for their support and input throughout the process; to Jean Zodrow, Meredith Frenchmeyer, and Allie Folcik for their support with the health considerations section; to Yousof Aly, Katie Barry, Mimi Sarkar, Julia Vidonish, and Courtney Lenzo for other technical support; to Andy Lewandowski and Rachel Stevens for graphical support, and to Caitlin Bell for the additional coordination she provided throughout the process. We could not have done this without the help of all these incredibly talented individuals, as well as many other people helped out behind the scenes.

John Horst would like to thank Caitlin Bell, Margy Gentile and Erica Kalve for their inspiring enthusiasm and commitment to completing this project; and to Dr. Paul Anderson for his thoughtful contributions and guidance on some of the more challenging topics. There have been a great many people involved, a big thank you to all those that have helped along the way. Finally, to Sumathy Suthersan – your husband was the inspiration that started the whole thing. God bless.

Editors

Caitlin H. Bell, PE, is a principal engineer and 1,4-Dioxane lead for Arcadis North America. Ms. Bell focuses on subsurface treatment of soil and groundwater using in situ techniques. Specifically, she focuses on in situ bioremediation applications for a variety of chemicals of concern, including emerging contaminants. She serves as a technical resource to clients on topics such as molecular biology tools, bioaugmentation, compound specific isotope analysis, and challenging bioremediation approaches for compounds like 1,4-dioxane. Ms. Bell was a member of the team that authored the Interstate Technology & Regulatory Council's Environmental Molecular Diagnostics technical guidance document.

Margaret E. Gentile, PE PhD, is an associate vice president at Arcadis. In her work as an environmental engineer, Dr. Gentile leads teams in the design and implementation of remediation systems. She specializes in biological treatment processes, large-scale remediation and remediation of metals, including hexavalent chromium. Dr. Gentile has published a number of papers on the microbial ecology of engineered biological treatment and metals remediation. She leads the In Situ Reactive Treatment technical group and is an active participant in the emerging contaminants team within Arcadis.

Erica Kalve, PG, is a principal geologist with 18 years of experience in the environmental field conducting site investigation and remediation projects. She is the North American Emerging Contaminants Focus Group leader and team coordinator for the Global PFAS Team at Arcadis. In her current role Ms. Kalve focuses on emerging contaminant issues. Within Arcadis U.S., she leads the Technical Knowledge and Innovation group on emerging contaminants in the development of best practices to facilitate project excellence.

John Horst, PE, is the executive director of Technical Knowledge and Innovation for North America at Arcadis. Mr. Horst is an expert in the development, application, and optimization of new and innovative environmental restoration technologies, and integrating multiple disciplines to address restoration challenges with significant scale and complexity. He studied Engineering at Drexel University, where he participated in a National Science Foundation pilot to produce more business-savvy graduates. Mr. Horst holds one technology patent and is corresponding author of *Advances in Remediation Solutions* published quarterly in Ground Water Monitoring and Remediation. He is also a co-author of *Remediation Engineering: Design Concepts* (2nd *Edition*).

Ian Ross, PhD, is a senior technical director and Global, In Situ Remediation technical lead/Global PFAS lead at Arcadis from Leeds, West Yorkshire, UK. His focus since 2014 has been solely on PFAS (per- and polyfluoroalkyl substances) after initially working on options for PFOS management in 2005. He was part of the team authoring and reviewing the CONCAWE PFAS guidance document and has published several articles on PFAS analysis, site investigation, and remediation. Dr. Ross has focused on the bioremediation of xenobiotics for over 26 years as a result of three applied industrially sponsored academic research projects. At Arcadis he has worked to design and implement innovative chemical, physical, and biological remediation technologies.

Contributors

Caitlin H. Bell
Arcadis U.S., Inc.
Seattle, Washington, USA

Erica Kalve
Arcadis U.S., Inc.
San Rafael, California, USA

Margaret Gentile
Arcadis U.S., Inc.
San Francisco, California, USA

Suthan Suthersan
Arcadis U.S., Inc.
Newtown, Pennsylvania, USA

John Horst
Arcadis U.S., Inc.
Newtown, Pennsylvania, USA

Shandra Justicia-Leon
Arcadis U.S., Inc.
Guaynabo, Puerto Rico

Sarah Page
City of Ann Arbor
Ann Arbor, Michigan, USA

Greg Imamura
Arcadis U.S., Inc.
Los Angeles, California, USA

Brent Alspach
Arcadis U.S., Inc.
Oceanside, California, USA

Paul Anderson
Arcadis U.S., Inc.
Chelmsford, Massachusetts, USA

Ian Ross
Arcadis U.K.
Frodsham, United Kingdom

Jonathan A. L. Miles
Arcadis U.K.
Leeds, United Kingdom

Jeff McDonough
Arcadis U.S., Inc.
Portland, Maine, USA

Tessa Pancras
Arcadis NL.
Arnhem, Netherlands

Jake Hurst
Arcadis U.K.
Leeds, United Kingdom

Norman Forsberg
Arcadis U.S., Inc.
Clifton Park, New York, USA

1 Introduction to Emerging Contaminants

Caitlin H. Bell
(Arcadis U.S., Inc., caitlin.bell@arcadis.com)

John Horst
(Arcadis U.S., Inc., john.horst@arcadis.com)

Paul Anderson
(Arcadis U.S., Inc., paul.anderson@arcadis.com)

1.1 INTRODUCTION

The development of new technologies and the underlying science that makes them possible is the foundation of human progress. Technology has allowed for huge leaps forward in our quality of life. It has brought clean water, advanced medicines, improved techniques for agriculture, renewable sources of energy, more durable materials for housing, and an increased standard of living for nearly everyone around the globe. In this way, technology has increased our lifespans and has led us to ever more efficient ways of learning, communicating, traveling, and conducting other daily activities.

These advancements often come with peripheral consequences that are not always immediately understood but must be managed as they emerge. For the purposes of this book, we focus on just one of these peripheral consequences: new chemicals or compounds that end up in the environment as a result of technology manufacturing or human use. Where unregulated compounds (or classes of compounds) are discovered to have been released to the environment and come to be seen as a potential threat to human health and the environment, they are often referred to as "emerging contaminants." Alternatively, the term "constituents/chemicals/compounds/contaminants of emerging concern" (CECs) is also used. Selection of the terms "constituent" or "chemical" or "compound" or "contaminant" to precede "emerging concern" varies among stakeholders. Often, the terms constituent, chemical, or compound are preferred when the potential toxicity of the compounds is still being evaluated and there is a desire to avoid passing judgment on whether something is a contaminant until there is more information. For example, the state of California (discussed later in the chapter) has used either "constituent" or "chemical" instead of "contaminant" in its studies related to drinking water and recycled water (Anderson et al., 2010; Southern California Coastal Water Research Project, 2012; Drewes et al., 2018). By comparison, the terms "emerging contaminants" and "contaminants of emerging concern" are used among groundwater remediation stakeholders, as occurrence in media requiring treatment is being evaluated. Given that this book is focused mainly

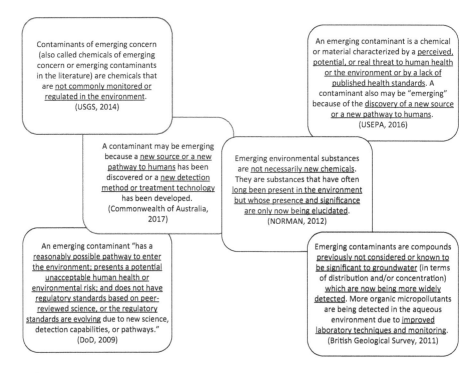

FIGURE 1.1 Example definitions of emerging contaminants: Examples of definitions of emerging contaminants from various entities.

on groundwater, it employs the latter two terms: "emerging contaminants" and "contaminants of emerging concern." This diction is simply a term of art—whether and to what degree a compound poses a risk to humans or the environment depends on each specific compound and the situation being evaluated.

There are as many definitions for "emerging contaminants" as there are groups that are focused on them (see examples in Figure 1.1). One thing that is common among them is that the term "emerging" reflects the overall developing state in our understanding of this class of compounds. Areas of developing understanding may include determining whether compound(s) have the potential to be harmful to humans or the environment and consequently to what extent they should be regulated, their occurrence in the environment, and techniques for treating environmental media to remove or transform them. These aspects are all interrelated and can take significant time frames to resolve.

- An accurate understanding of the toxicology associated with emerging contaminants can take many years (decades, in fact) to develop.
- Environmental occurrence may be poorly understood because chemical analysis is challenging and techniques for accurate detection and measurement are still developing. Or occurrence may not be understood because monitoring was not previously conducted or required and will take time (years) to complete.

Introduction to Emerging Contaminants 3

- Treatment can present challenges for water (drinking water, wastewater, groundwater, surface water) and other media depending on the nature of the compound, with effective technology requiring time and investment to develop.

The concern surrounding various emerging contaminants can often outpace the amount of time required to fully understand and manage them. With this in mind, this handbook is intended to serve as a resource to anyone who wishes to become familiar with the topic of emerging contaminants, and provides an emphasis on several contaminants that are currently very relevant and are at different points within and beyond the emerging contaminant life cycle. For these, the handbook is not intended to be a comprehensive reference, but rather a primer to provide an understanding of the current state of the science at the date of this publication regarding the properties, behavior, toxicity, regulatory status, and ability to analyze, characterize, and treat emerging contaminants. We hope this information can serve along with other sources to support the management of such contaminants and challenges they pose.

For the balance of this introductory chapter, we explore the following three questions:

- Who identifies emerging contaminants?
- What is the life cycle of an emerging contaminant?
- What are the key challenges associated with emerging contaminants?

Finally, we provide insight on the need for balance and the focus and purpose of this handbook with rationale for the specific compounds that are highlighted herein.

1.2 WHO IDENTIFIES EMERGING CONTAMINANTS?

There are a number of ways that emerging contaminants are described (see Figure 1.1 for examples), identified, and prioritized.

Various stakeholder groups around the world are working on identifying potential emerging contaminants and have developed systems for screening the potential candidates, prioritizing, and ultimately generating regulations to protect human health and the environment. Government environmental agencies, such as the United States Environmental Protection Agency (USEPA), the Australian National Environment Protection Council (Australian NEPC), and the German Environment Agency (Umweltbundesamt [UBA]) are active in this process. Multinational bodies such as the European Union and United Nations and international treaties such as the Stockholm Convention also advance awareness and actions around chemical manufacturing and emerging contaminants in the environment. Additional government and private organizations also contribute to the identification of emerging contaminants, such as scientific organizations (e.g., U.S. Geological Survey [USGS], Norwegian Geotechnical Institute [NGI], a private foundation) and the military (e.g., U.S. Department of Defense [DoD]).

The criteria for identifying and evaluating candidate compounds vary among these groups, and consequently various lists of potential contaminants are being developed. For example, Appendices A and B provide two example lists: the USEPA

Contaminant Candidate List (CCL) and the lists of compounds identified under the European Union Registration, Evaluation, Authorisation, and Restriction of Chemicals (REACH) regulations. Not every compound on these lists is an emerging contaminant. Some may not be considered emerging contaminants because they have not been released to the environment to cause significant exposure. Others are not considered emerging contaminants because of long-term awareness and programs/requirements to address them. For newer compounds, individual organizations have developed different methods for assessing them, determining which should have restricted manufacturing and use, and which are occurring in the environment and may deserve attention as emerging contaminants, as detailed in the following subsections. Appendix C provides a list of chemicals that are considered emerging contaminants at the time of this book's publication, compiled from the lists of various organizations discussed in this section, and considering the current state of awareness and response. This list was considered relevant for the discussion in this book, but a different list with another focus could easily have been developed from the same (or different) starting point.

These governmental and international organizations are often informed through the research of others (academics, industry groups, etc.), as well as through their own research. They then contribute to shaping the understanding of whether the compounds targeted deserve attention as an emerging concern and ultimately whether they qualify as contaminants that should be regulated and to what levels. The identification of new compounds for consideration as emerging contaminants is intertwined with the further and ongoing refinement of toxicological effects and the development of reliable and appropriate characterization and treatment methods. We explore the interplay between these various lines of development when we discuss the life cycle of an emerging contaminant later in this chapter, but for the balance of this discussion we focus on the various governmental and quasi-governmental organizations that play leading roles in the dialogue.

1.2.1 UNITED STATES ENVIRONMENTAL PROTECTION AGENCY

In the United States, the USEPA has a huge influence on identification, and later regulation, of emerging contaminants. The USEPA actively seeks to catalog unregulated compounds that may occur in public water supplies through the maintenance of the CCL. To further understand the frequency of occurrence and likelihood of public exposure, the USEPA specifies sampling of public water supplies in the Unregulated Contaminants Monitoring Rule (UCMR). In part, the UCMR and CCL are a catalog of emerging contaminants that could face future regulation. Occurrence data collected in the UCMR ultimately support regulatory determinations. Each of these elements of the USEPA process is discussed in further detail here.

Developing the CCL is the USEPA's first step in evaluating unregulated drinking water contaminants. The CCL is a list of contaminants that:

- Are not regulated by the National Primary Drinking Water Regulations.
- Are known or anticipated to occur at public water systems.
- May warrant regulation under the U.S. Safe Drinking Water Act (SDWA).

The list is compiled through consultation with technical experts, for example, representatives of the scientific community, the National Drinking Water Advisory Council, the National Academy of Sciences or the Science Advisory Board, as well as through solicitation of public input. In November 2016, the USEPA published the final CCL-4, the new drinking water priority contaminant list for regulatory decision-making and information collection. As summarized in Appendix A, the current CCL (CCL-4) contains 97 chemicals or chemical groups and 12 microbial contaminants and has been built over time from the historical CCLs (i.e., CCL-1 through CCL-3). Notable updates to the list were the removal of perchlorate and strontium, which are scheduled for regulation (so are no longer considered emerging).

Under the SDWA, USEPA has implemented its UCMR since 2001 to evaluate the frequency and occurrence of emerging contaminants in public water supplies. USEPA's selection of contaminants for a particular UCMR cycle is largely based on a review of the CCL. The UCMR program was developed in coordination with the CCL. Occurrence data are collected through UCMR to support the USEPA's determination of whether to regulate particular contaminants in the interest of protecting public health. This program includes:

- Monitoring no more than 30 contaminants every five years.
- Monitoring large systems serving more than 10,000 people and a representative sample of small public water systems serving fewer than 10,000 people.
- Storing analytical results in a National Contaminant Occurrence Database.

Notable examples of compounds included in past UCMR programs include perchlorate, methyl tertiary butyl ether (MTBE), flame retardants, and pesticides. More recent additions include 1,4-dioxane, 1,2,3-trichloropropane (TCP), per- and polyfluoroalkyl substances (PFASs), hexavalent chromium, and drinking water disinfection by-products—all of which persist in the environment. Increased attention on the identification, characterization, and remediation of these contaminants has served to reopen a review of surface water and groundwater used for public drinking purposes.

Finally, the SDWA requires the USEPA to make a regulatory determination for five CCL contaminants with sufficient health and occurrence data every five years. When making a "determination" to regulate a contaminant in drinking water, the law requires that USEPA determine whether it meets the following three criteria:

- The contaminant may have an adverse effect on human health;
- The contaminant is known to occur or there is substantial likelihood the contaminant will occur in public water systems with a frequency and at levels of public health concern;
- In the sole judgment of the USEPA's administrator, regulation of the contaminant presents a meaningful opportunity for health risk reductions for persons served by public water systems.

Recently, the USEPA and the regulatory risk analysis programs have trended toward establishing health advisory levels (HALs), as opposed to adopting federal maximum contaminant levels (MCLs), with the potential recent exception of strontium. In the last regulatory determination, the USEPA made a preliminary determination in October 2014 to regulate strontium and develop a federal MCL given the potential adverse human health effects, known occurrence and likelihood of occurrence in public water systems at levels of concern, and the opportunity to meaningfully reduce health risks for those using public water supplies (Federal Register, 2014; 40 CFR Part 141 EPA-HQ-OW-2012-0155; FRL 9917-87-OW). After receiving comments on the draft ruling, the USEPA decided to delay the strontium determination to evaluate additional data on relative source contribution, that is, the relative exposure from drinking water versus other exposures such as food, ambient air, and soil (Federal Register, 2016; 40 CFR Part 141, EPA-HQ-OW-2012-0155; FRL-9940-64-OW). A final determination on strontium will be made after additional data on relative source contribution are evaluated.

1.2.2 UNITED STATES DEPARTMENT OF DEFENSE

The DoD is an example of another U.S. federal agency that is working proactively on emerging contaminants. The DoD's Chemical and Material Risk Management Program focuses on identifying and managing risks associated with the chemicals and materials that the DoD uses, or plans to use, and implements a "scan, watch, action" approach. Consequently, it is important to note that this type of due diligence process casts a wide net, and as such mere inclusion on DoD's watch list or action list does not automatically make something an emerging contaminant. In fact, many compounds reflected in these lists have been the subject of remediation and regulation for decades—so while such compounds may certainly be contaminants and they may no longer be emerging, there may be a reason for the DoD to look more closely at them and take some action based on how they are being used. The program concludes that something is an emerging contaminant only if it is ultimately demonstrated to have pathways to enter the environment, presents real or potential unacceptable human health or environmental risks, and either (1) does not have peer-reviewed human health standards, or (2) has standards/regulations that are evolving due to new science, detection capabilities, or pathways (Yaroschak, 2016).

In the first step of their process, DoD scans various sources, such as websites, journals, periodicals, newsletters, articles, and regulatory communications to identify compounds and risk-assessment policies likely to have changing guidance or standards on the international, federal, state, and/or local levels. This list of compounds is then screened to remove those not relevant to the DoD and to prioritize those that are. Each chemical is evaluated based on the five DoD functional areas: (1) health and safety; (2) readiness and training; (3) acquisition/research, development, testing, and evaluation; (4) production, operations, maintenance, and disposal of assets; and (5) environmental cleanup.

Once the "watch list" is developed, qualitative Phase I impact assessments are conducted for each chemical. This identifies the compounds moved to the "action list," from which they are further evaluated within the portfolio via quantitative

Introduction to Emerging Contaminants

Phase II impact assessment with risk management options generated to create strategic investment options for enterprise consideration. The creation of the DoD's watch list and action list provides perspective for others in the industry regarding the emerging contaminants of interest to the DoD.

For perspective on the program, as of early 2016 more than 580 compounds were screened via this process (Yaroschak, 2016). The watch list included compounds such as 1,4-dioxane, dioxin, nanomaterials, PFASs, several metals, flame retardants, hydrofluorocarbons, energetics, and others. Phase I assessments were completed for 39 compounds. Compounds that were moved from the watch list to the action list included 1,3,5-trinitroperhydro-1,3,5-triazine, also known as Royal Demolition Explosive (RDX), hexavalent chromium, lead, naphthalene, beryllium, sulfur hexafluoride, phthalates, and others. As mentioned earlier, not all of these would be considered emerging contaminants or CECs by other stakeholders. Phase II evaluations were completed for 11 chemicals on the action list. This effort resulted in implementation of 66 risk management actions to mitigate the identified risk associated with these chemicals. Some of these actions include RDX toxicity studies, funded research into perchlorate and hexavalent chromium, sulfur hexafluoride capture and recycling, and development of a naphthalene dosimeter for fuel handlers.

1.2.3 UNITED STATES GEOLOGIC SURVEY

The USGS is a third U.S. federal agency that plays a role in the evolving understanding of CECs, mainly through its research programs. The USGS has seven mission areas on which it focuses: climate and land use change, core science systems, ecosystems, energy and minerals, environmental health, natural hazards, and water resources. Within the environmental health mission area, the Toxic Substances Hydrology Program includes research on CECs in the environment. Rather than focusing on the listing of CECs, the USGS has focused its studies on understanding "the actual versus the perceived health risks to humans or wildlife due to low-level exposures from understudied chemical contaminants in the environment" (USGS, 2017). The USGS publishes the results of this work, which has ranged from broad nationwide reconnaissance of drinking water resources, to studies focused on specific compounds and exposure scenarios. These publications play a role in informing the broader dialogue around CECs.

1.2.4 STATE AGENCIES IN THE UNITED STATES

In the U.S., state agencies are also engaging in the dialogue over CECs. As an example, California's Water Resources Control Board has sponsored three expert panels related to CECs. The state's challenges with water availability and increasing reliance on recycled water is well known. Because of the potential for CECs to be present in such water, the state sponsored an expert panel in 2009 to focus on recycled water related human health risks (Anderson et al., 2010). In 2017, California convened a second panel to update the results of the first. The goal was to complete an update on the state of current scientific knowledge related to CECs, confirm the compounds that should be monitored in recycled water, and confirm the best methods for doing so. The report identifies approximately 500 CECs (Drewes et al., 2018). In addition, another panel was sponsored by the state of California in 2012 to focus on aquatic ecosystems (Southern California Coastal Water Research Project, 2012). This panel was tasked to review existing scientific literature on CECs, determine the current state of knowledge regarding ecological risks, and provide recommendations for improving the understanding of CEC presence and protecting the environment.

The 2010 and 2012 reports recommended a risk-based screening framework to identify CECs for recycled water monitoring, coupled with an adaptive monitoring approach that includes initial screening and comparison to benchmarks. These recommendations were incorporated into the California Recycled Water Policy and a monitoring pilot study that began in 2015 (Dodder et al., 2015). The allocation of resources and research being completed by states such as California represents a large investment related to CECs.

1.2.5 Stockholm Convention on Persistent Organic Pollutants

The Stockholm Convention on persistent organic pollutants (POPs) is an example of a global-level treaty that contributes to the catalog of emerging contaminants as new compounds are identified and added to the POP list. The Stockholm Convention on POPs is an international environmental treaty that originated from a request by the Governing Council of the United Nations Environmental Program in 1995 and recommendations from the Intergovernmental Forum on Chemical Safety in 1996. The Stockholm Convention was signed in 2001; it was effective as of 2004 and aims to eliminate or restrict production and use of POPs (Stockholm Convention, 2008). POPs, as defined by the Stockholm Convention, include organic compounds that:

- remain intact and/or persist for long periods of time;
- are widely distributed throughout the environment, via soil, water, or air and are present in locations far from their source;
- are bioaccumulative in aquatic and other living organisms to a sufficient extent to be a concern (generally resulting in successively higher concentrations at higher levels of the food chain); and
- have adverse effects or are otherwise toxic to humans or the environment.

Approximately 30 POPs are identified in the Stockholm Convention. Some of these are actually derivatives of other chemicals or classes of compounds, which puts the total number of individual compounds at much more than 30. As with the DoD's watch list, many contaminants listed in the Stockholm Convention have been the subject of remediation and regulation for decades, so while there may be a reason for their inclusion here based on how they have been used, such contaminants may already be phased out or no longer considered emerging. The POPs listed in the Stockholm Convention include pesticides, industrial chemicals such as polychlorinated biphenyls or polychlorinated naphthalenes, or chemical by-products such as dioxins and chlorinated dibenzofurans. These are grouped into three categories for action: Annex A (elimination) where the intent is to eliminate production; Annex B (restriction) where the intent is to limit production and use; and Annex C (unintentional product) where the intent is to reduce unintentional release. Other compounds can also be proposed for listing and, as of the time of this writing, four were on the list for review, including perfluorooctanoic acid (PFOA) and perfluorohexane sulfonic acid (and its salts), which are discussed in detail in this book.

More than 180 countries adopted the Stockholm Convention and several of these, such as Canada, Australia, and countries of the European Union, are using the definition of "POP" to guide their concept of emerging contaminants and associated regulation. This provides a useful starting point that allows for those countries to readily identify new emerging contaminants, such as those on the review list, but it does not cover all compounds for which focus may be warranted.

The Stockholm Convention identifies a number of obligations that all parties should (or are encouraged to) undertake, including designating a national focal point; fostering information exchange; providing technical assistance; promoting and facilitating public awareness and participation; consultation and education; stimulating research and monitoring; and reporting at periodic intervals. These responsibilities include development of a National Implementation Plan that details how the signing party will meet the obligations set forth by the Stockholm Convention. These reports typically start with a profile of the country, including geographical, socioeconomic, and environmental information. They also identify the country's applicable regulatory framework. Then, the majority of the National Implementation Plan includes an assessment of the current POP inventory in the country as well as a strategy and action plan for implementing the directives of the Stockholm Convention.

1.2.6 EUROPEAN UNION

The European Union has contributed to the discussion on the emerging contaminants, notably through the REACH regulation that went into effect on June 1, 2007, and applies to all chemical substances—meaning those used in industrial processes as well as those used in everyday products such as cleaners, paints, clothes, furniture, electrical appliances, etc. To comply with the regulation, companies must identify and manage the risks linked to the substances they manufacture and market in the European Union. They have to demonstrate to the European Chemical Agency how the substance can be safely used, and they must communicate the risk management measures to the users. If the risks cannot be managed, authorities can ban a substance, restrict its use, or make its use subject to special authorization (European Chemicals Agency, 2018a).

As of May 2018, there were more than 500 substances on the REACH restriction list (European Chemicals Agency, 2018b). Additionally, there were over 200 substances on the REACH candidate list (European Chemicals Agency, 2018c). According to the "Roadmap for substances of very high concern identification and implementation of REACH Risk Management measures from now to 2020," the European Union is committed to having all known substances of very high concern identified in the candidate list by 2020, so this list may get much longer and contribute further to identification of new emerging contaminants. This roadmap focuses on certain groups of substances based on the nature of their potential effects including carcinogens, mutagens, reprotoxicants (Categories 1A/1B); sensitizers; persistent, bioaccumulative, and toxic (PBT) or very persistent, very bioaccumulative (vPvBs) compounds; endocrine disruptors; and petroleum/coal stream substances with certain properties (European Chemicals Agency, 2018d).

In addition, some countries within the European Union are looking to expand the criteria used in the assessment procedure to identify substances of very high concern. For example, the UBA issued a position paper (German Environment Agency, 2017) that seeks to establish persistent, mobile, and toxic (PMT) criteria, as well as very persistent and very mobile (vPvM) criteria to the PBT and vPvB criteria already included in order to protect sources of drinking water. The assessment process put forth by the UBA includes two steps: (1) comparison of candidate properties to the PMT and vPvM criteria; and (2) emission characterization. The consideration of toxicity and emissions or occurrence is common to the USEPA process of identifying contaminants for regulation, and the persistence criteria is similar to the Stockholm Convention. The inclusion of specific mobility criteria is a unique addition being considered by UBA. A preliminary assessment of the PMT/vPvM criteria, known occurrence, and relative emissions rank was developed for UBA by the Norwegian Geotechnical Institute, a private foundation (NGI, 2018). This preliminary assessment evaluated 9,741 REACH-registered substances and identified 240 potential PMT/vPvM substances. The preliminary, unofficial list of PMT/vPvM substances from the NGI assessment is provided in Appendix D. Fewer than 10 percent of these substances were detected in drinking water or groundwater in the studies referenced by the assessment and less than half had significant, known emissions, indicating which substances may become emerging contaminants as awareness progresses.

1.2.7 AUSTRALIAN NATIONAL ENVIRONMENT PROTECTION COUNCIL

The Australian NEPC serves two primary functions: to make National Environment Protection Measures (NEPMs), and to assess and report on the implementation and effectiveness of NEPMs in participating jurisdictions (Commonweath of Australia, 2017). Among the NEPMs made, the Assessment of Site Contamination NEPM serves "to provide adequate protection of human health and the environment, where site contamination has occurred, through the development of an efficient and effective national approach to the assessment of site contamination" (National Environment Protection Council, 2013). Subsequent to this, a National Remediation Framework is being developed through the Cooperative Research Centre for Contamination Assessment and Remediation of the Environment (CRC CARE) to provide practical guidance to practitioners and regulators. This includes identification of and development of guidance for CECs (CRC CARE, 2014). The first-tier priority contaminants were identified for CRC CARE at a forum held in February 2012. The first-tier priority contaminants include perfluorooctanesulfonic acid (PFOS), PFOA, MTBE, benzo(a)pyrene, weathered hydrocarbons, and polybrominated diphenyl ethers (PBDEs). The 2014 report (CRC CARE, 2014) sought to review the available guidance related to site assessment and remediation of these CECs, while later work focused on filling in gaps in current knowledge, such as for PFOS/PFOA.

1.3 WHAT IS THE LIFE CYCLE OF AN EMERGING CONTAMINANT?

The amount of time that a compound remains classified as an emerging contaminant varies widely from compound to compound but, as indicated earlier, is tied to our ability to clearly understand environmental occurrence and toxicity, and our ability to effectively treat or remove the compound from environmental media. Initially, an emerging contaminant may rise into the public awareness based on some combination of new toxicological information, the observation of occurrence in the environment and possible risks to receptors, and the often-limited capability for treatment.

Typically, a compound no longer qualifies as an emerging concern once it has achieved some degree of uniform regulation matched by the capability for detection and treatment. If regulatory values for a given emerging contaminant remain widely variable (orders of magnitude), it is likely that the toxicology surrounding it continues to evolve, so the compound may still be an emerging concern to some. The same is true if the capability for treatment has not fully developed. In addition, new scientific information can spur the reconsideration of a "well-known" contaminant. In this way, the term "emerging contaminant" does not necessarily mean only newly introduced compounds. There are also "off-ramps" for emerging contaminants; for example, when it is determined that a compound has no adverse effect on human health or the environment, or when data collection concludes that the compound is not present in the environment to an extent that could adversely affect human or ecological receptors.

The emerging contaminant life cycle can be broken down into four main categories, depicted in Figure 1.2, and summarized here. Each of these categories or areas of focus that follow must be resolved for an emerging contaminant to exit the life cycle.

Awareness

As previously noted, the awareness of any emerging contaminant often begins with a realization regarding its occurrence in the environment combined with the potential toxicity of the compound which together can pose a risk to human and ecological receptors.

The toxicity of an emerging contaminant is often evaluated through surrogate organisms when human exposure data are not available, with the results from the surrogates extended to estimate human toxicity both in the short term (i.e., acute toxicity) and long term (i.e., chronic toxicity and carcinogenicity). Sometimes readily available studies for a chemical are all that is needed to confirm potential risks associated with a compound that has been recently detected in the environment and qualify it as an emerging contaminant.

Alternatively, as new studies become available and our understanding of the human health and ecological effects of a particular compound improves, we can find that

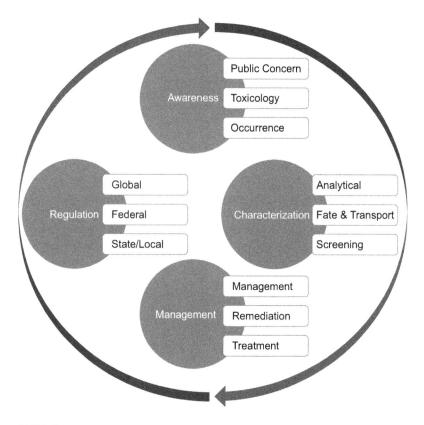

FIGURE 1.2 Emerging contaminants life cycle: A visual depiction of the various stages in the life cycle of an emerging contaminant. Each of the high-level stages (circles) may include several subsets (rectangles). While each of the high-level stages typically progress from one to another, an emerging contaminant can fall "backwards" in the life cycle as additional information becomes available.

environmental concentrations previously thought to be safe may not be. This new understanding may lead to more stringent regulation and control of the compound involved. The opposite, however, is also possible: improved toxicological understanding may indicate a compound poses less risk than previously thought and less stringent standards and criteria may be appropriate. In this way, continuing developments in the toxicological understanding of various compounds determine their inclusion in the emerging contaminant life cycle.

Characterization

As a compound is established as an emerging contaminant, there is understandably a corresponding focus on the ability to characterize its presence in the environment. This area of the life cycle includes (1) the advancement of laboratory analytical methods and, in some cases, field-screening methods to ensure the quality and representative nature of environmental data being collected; (2) a deeper understanding of the fate and transport behavior of the emerging contaminant and linking that back to the development of analytical methods where appropriate; and (3) a better picture of the nature and extent of an emerging contaminant's presence in the environment through the collective effort of many organizations to investigate and screen for it.

Improvements in the ability to detect the occurrence of compounds in the environment, or simply an increased effort to measure the occurrence of compounds in the environment, is a driver for the identification of emerging contaminants. An example of this is the presence of PFASs in class B firefighting foams. Firefighting foams are commonly used at airports, defense installations, petroleum terminals, and other large facilities where they have played a critical role in safety to prevent catastrophic losses associated with large-scale fires. PFASs represent a class of compounds that includes thousands of individual chemicals. When PFASs came to light as a potentially harmful category of compounds, only a handful could be identified and quantified using available analytical techniques (e.g., USEPA Method 537 for drinking water). This analytical method was simply not able to detect or "see" the large number of other compounds in the PFAS family. Advances in analytical methods were needed to create a more accurate picture of PFAS occurrence, which also supported a better understanding of their fate and transport characteristics, and over time a better understanding of their prevalence.

Management

The ability to manage or respond to the risks created by emerging contaminants becomes critical as awareness of them matures. This of course encompasses the treatment of water (drinking water, wastewater, etc.) as well as the treatment of other environmental media. For some emerging contaminants, well-established technologies can be found that are compatible with the compound's unique physical and chemical characteristics. For other emerging contaminants, this is not as straightforward, and while there may be existing technologies that can work to some degree, a much more significant effort must be mounted to find and commercially develop technologies that are both more effective and efficient.

The area of emerging contaminant management also involves product stewardship and the development of best management practices for manufacturing. Product stewardship programs have been developed to mitigate the impacts of commercial chemical products on humans and the environment and work to find replacements for compounds known to adversely affect the environment. In 1971, the President's Council on Environmental Quality said, "We should no longer be limited to repairing damage after it has been done; nor should we continue to allow the entire population or the entire environment to be used as a laboratory" (Council on Environmental Quality, 1971). Although this statement represents a long-standing recognition that product stewardship is essential to ensure that the introduction of new products into society does not adversely affect human health and the environment, there continue to be examples over the last 50 years where various compounds became widely used without all of the potential risks being fully understood. Flame retardants are one example. Polybrominated biphenyls (PBBs) were used as flame retardants beginning about 1970. Production was quickly discontinued in the U.S. in the mid-1970s following an incident through which PBBs were inadvertently introduced directly into the food chain. This incident resulted in a better appreciation of their risk out of necessity (Agency for Toxic Substances and Disease Registry, 2004). PBDEs were then used in place of PBBs as flame retardants. As our understanding of PBDE environmental fate and potential risks has improved, they too are now regarded as emerging contaminants.

Regulation

Finally, and as previously mentioned, a compound typically no longer fits in the category of an emerging concern once it has achieved some degree of uniform regulation matched by the capability for detection and treatment. In this sense, regulation is most often equated to the establishment of safe concentration limits that must be achieved in environmental media. However, regulation can also include banning the use of a particular compound. Although efforts to support regulation often move in parallel with the prior three areas of focus in the life cycle, the degree of regulation is also informed by them.

To illustrate how emerging contaminants come and go through the life cycle, it is helpful to look at some examples. The compounds trichloroethene (TCE) and MTBE are generally no longer considered to be emerging contaminants because they have effectively moved through each step of the emerging contaminant life cycle and are now regulated.

Public awareness of MTBE as an emerging contaminant came to the forefront in the early 1990s (Deeb et al., 2004). As understanding of the human health and ecological effects of MTBE developed, approximately 20 states instituted regulation by 1999 (Deeb et al., 2004). This sparked a flurry of activity around the characterization of MTBE through the 2000s (Interstate Technology & Regulatory Council, MTBE and Other Fuel Oxygenates Team, 2005). Air stripping was predominantly the treatment option of choice in the mid-1990s with granular activated carbon (GAC), advanced oxidation processes, and biological treatment approaches developed and applied by the early 2000s (Interstate Technology & Regulatory Council, MTBE and Other Fuel Oxygenates Team, 2005).

TCE goes back even further, with awareness as an emerging contaminant occurring in the early to mid-1980s (Russell et al., 1992), with the Phase I rule from the USEPA going into effect in 1989 (USEPA, n.d.). Analytical methods were initially developed in the mid-1980s with USEPA Methods 524.2 and 502.2, thus allowing for characterization of TCE. Treatment technologies began with air stripping in the early 1980s and continued with soil vapor extraction in the late 1980s. GAC gained popularity in the early 1990s, and in situ bioremediation approaches became more accepted later that decade (Russell et al., 1992). At that time, the risk management strategies associated with TCE were generally considered well established, and TCE passed out of the emerging contaminant life cycle. However, in 2011, new toxicological data came out related to the inhalation risk associated with TCE (USEPA, 2011). Consequently, vapor intrusion as a pathway for exposure to TCE gained attention. While the new developments do not revert TCE to emerging contaminant status, they show how improvements to our understanding of any contaminant can influence levels of interest, concern, and action at any point in time.

In contrast to TCE and MTBE, 1,4-dioxane and PFASs are examples of emerging contaminants that are still firmly in the emerging contaminant life cycle because they have not yet resolved each of the life cycle focus areas.

General awareness of 1,4-dioxane began as early as the mid-1980s (Environmental Health Division of Washtenaw County, 2017) with more focus in the early 2000s in California (Mohr et al., 2010) and in 2015 with completion of the UCMR sampling program from 2012 to 2015. State regulation of 1,4-dioxane began in 2004 in Colorado and, as of 2016, more than 30 states had drinking water or groundwater standards (Suthersan et al., 2016). USEPA Method 522 for quantification of 1,4-dioxane in drinking water was developed in 2008 and commonly used groundwater analytical methods (e.g., USEPA Methods 8260 and 8270) were augmented with selective ion monitoring to better quantify 1,4-dioxane. The laboratory analytical techniques associated with 1,4-dioxane are still evolving; thus the characterization area of focus in the emerging contaminant life cycle is still ongoing. Likewise, more efficient technologies for treatment are also still evolving, with the potential to engineer biological degradation gaining traction circa 2015.

Relative to 1,4-dioxane, PFASs have only just started their journey in the emerging contaminant life cycle. These compounds entered into the mainstream public awareness through the UCMR program in the U.S. from 2012 to 2015, as well as through public media coverage. While a provisional HAL was issued in 2009, it has been updated since then (USEPA, 2017a). As of the time of this writing, there are a handful of U.S. states with drinking water or groundwater standards, while PFAS regulation is more firmly established in Europe and Australia. USEPA Method 537 was developed in 2009 to quantify a few PFASs (like PFOS and PFOA) in drinking water. However, the toxicology associated with PFASs, the ability to characterize them in the environment, the ability to treat contaminated media, and the replacements for these chemicals are all subjects of intense research.

The list of what qualifies as an emerging contaminant will continue to be dynamic as new compounds are continuously being produced and science continuously improves our understanding of their potential toxicity and occurrence in the environment.

1.4 WHAT ARE THE KEY CHALLENGES ASSOCIATED WITH EMERGING CONTAMINANTS?

While the topic of emerging contaminants has been acknowledged as a highly popular area of research (Sauve and Desrosiers, 2014), the challenge associated with management and treatment of CECs remains a major opportunity for future innovation by the industrial, scientific, and consulting communities.

Given the advancements in understanding how to tackle new contaminants, the life cycle for new emerging contaminants is likely to be shorter than for prior compounds, like TCE. The main difference today is the combination of issues that together can accelerate the process. As awareness of a new emerging contaminant heightens, the demand from the public increases pressure on all stakeholders to take action. On one hand this can be good in the sense that action is being taken where there is a potential risk to human health and the environment. On the other hand, the perceived need to respond can potentially outpace the state of the science. Some of the key challenges involved in the modern-day dynamic around emerging contaminants include the following:

Visibility and Public Sensitivity

In today's digital age, information is instantly available to the general public. This is true regardless of the accuracy of the information or its potential for bias. When news can be shared this easily, it can achieve "viral" status where the rate of spread accelerates exponentially through Internet sharing, including various forms of social media. Such availability of information is a powerful driver, and the impact of emerging contaminants on the public water supply is a common topic of media stories, focused on raising public awareness.

For example, within the broad category of PFASs, the specific compounds PFOS and PFOA have been the subject of increasing media attention with multiple news items every day for the last few years. A big reason for this is the story of Hoosick Falls, New York, which began in August 2014, when a local resident met with the village mayor to request that samples of water from the municipal water system be analyzed for the presence of PFOA (Village of Hoosick Falls, 2016). Water sampling conducted in the fall of 2014 of three water supply wells and the treated water from the municipal plant showed that PFOA was detected at concentrations ranging from 170 to 540 nanograms per liter (ng/L) (Village of Hoosick Falls, 2016), with several samples at concentrations above the (then) short-term exposure HAL of 400 ng/L (USEPA, 2012).

Stories about Hoosick Falls first hit the newsstands in early 2015 when several residents spoke out about their concerns anonymously (Sanzone, 2015). Media coverage continued sporadically through 2015 until residents contacted the USEPA to raise concerns about whether they should drink, bathe in, or cook with their water (Village of Hoosick Falls, 2016). This prompted the governor of New York to issue an emergency regulation to classify PFOA as a hazardous substance, classify a local manufacturing facility as a State Superfund Site, and unlock state resources to address resident concerns. This action landed Hoosick Falls in the national media spotlight in January 2016. According to Google Trends, more than 1,700 news articles related to Hoosick Falls and PFOA were published in 2016. Much of this was early in the year when actions were being taken to provide residents with bottled water, install appropriate treatment at the municipal drinking water plant, and install hundreds of smaller treatment systems at residential water wells. During this time, the health-based standards related to PFOA were lowered (more than once) and the blood of local residents was tested to evaluate the impact to the public. In the case of Hoosick Falls, public visibility and sensitivity clearly played an important role in the response.

Uncertainty

In spite of the powerful driver in visibility, there can be great uncertainty surrounding short-term decision-making due to both the evolving science and toxicology around emerging contaminants and the potential lack of effective treatment approaches. As alluded to earlier in this chapter, health-based treatment standards developed at the state level in the U.S. can vary widely and change over time—which results in a patchwork of regulation that is both difficult to understand and difficult to comply with. A treatment approach that works to meet one standard may no longer work when that standard is lowered. The disparity represents a challenge for industry and municipalities that are trying to do the right thing to protect human health and the environment. The challenges can be compounded by the fact that conventional technologies for water treatment and groundwater remediation may not be suitable for dealing with certain emerging contaminants. The implication is particularly significant as demands for balanced aquifer storage and recharge or replenishment increase.

Vulnerability

Finally, there are the challenges with making longer-term management decisions and "future proofing." The goal of most long-term decisions is to make the right decision now to carry into the future. Despite best efforts, this can be difficult. At the core of this vulnerability is not understanding the full spectrum of what may need to be managed. A great example of this is the earlier story of a flame-retardant replacement chemical that was thought to be safe, but turned out to be a risk to human health and the environment, similar to the chemical it replaced. Another example is the cooccurrence of 1,4-dioxane and chlorinated solvents. Historically, a common groundwater treatment technique employed for chlorinated solvents was groundwater extraction coupled with aboveground air stripping. While air stripping is effective at removing chlorinated solvents from groundwater, it is not effective at removing 1,4-dioxane from the same groundwater stream. When the awareness of the relationship between 1,4-dioxane and certain chlorinated solvents became known, it created a situation where sites that had been remediated for chlorinated solvents and that were close to regulatory closure found themselves needing to continue remedial efforts due to the presence of 1,4-dioxane. In other cases, expensive retrofits had to be made to remediation systems designs so that they would be effective for both chlorinated solvents and 1,4-dioxane. A final example of vulnerability is the ability to analyze and quantify the presence of PFASs, as a category of contaminants. Many of the thousands of individual compounds that comprise this category have the potential to bio-transform to a few end products, including PFOS and PFOA. Therefore, if a current characterization or monitoring strategy only includes quantification of PFOS and PFOA, it will not accurately capture the full mass of PFASs that may be an issue in the future.

Increased visibility/public sensitivity, increased uncertainty, and increased vulnerability can all create high-pressure conditions for regulators and responsible parties alike, underscoring a need for balance.

1.5 THE NEED FOR BALANCE

Developing management strategies and implementing response actions in the face of an immature understanding of emerging contaminants is clearly challenging. In some circumstances, acting quickly to ensure protection of human health and the environment may be necessary. An example of this may be providing bottled water to communities where emerging contaminants are detected in drinking water. However, considering the long-term course of action once initial emergency response is complete is also important. As evidenced by the emerging contaminant life cycle (Figure 1.2), our understanding of a compound changes over time, illustrating the importance of incorporating provisions for change into the public awareness of how emerging contaminants are strategically managed.

Developing an understanding of the range of toxicological effects of a compound takes time, effort, and the resources, both with regard to human health and the environment. While it may not take long to develop an understanding of the short-term acute toxicological effects of a chemical, estimating the long-term chronic effects is complex and requires more time. Although researchers have developed methods to characterize adverse effects, it is important to consider the assumptions inherent in those methods and whether they apply to a given compound, and to include an opportunity to update the methods when new data are available. While it is certainly more protective of human health and the environment to include conservative assumptions early on, it is then important to communicate effectively with the public, regulators, and other stakeholders when new data indicate those conservative assumptions can be modified and relaxed.

As occurrence data for a new emerging contaminant are collected, the results may indicate that the compound is present in many types of environmental media: drinking water, groundwater, surface water, and more. This may raise concern with the public, regulators, and other stakeholders if it seems that the chemical is "everywhere." However, the occurrence of a compound does not necessarily mean it is of human health or environmental concern. It may be that the compound is detected only at very low concentrations, concentrations that do not exhibit an effect on humans or the environment. An example of this is the presence of some pharmaceuticals in domestic wastewater, but at concentrations below the dose that elicits a response in the human body.

When it comes to regulations, there may also be a tendency to adopt a more conservative approach when less is known about a new emerging contaminant. The USEPA issues HALs for new compounds to communicate to the public a potential concern regarding the effect a compound may have on human health. These are commonly issued early in the emerging contaminant life cycle and may change with time as new information becomes available. Many state regulatory agencies use HALs as a starting point for developing local standards or guidance. However, HALs are not promulgated, enforceable values that public water suppliers are held to: MCLs are required for that. The process for developing MCLs takes more time and balances occurrence data (e.g., UCMR results) with toxicological information (i.e., does it cause harm to humans) and other factors to be sure that the value chosen is protective and reasonable. Even after this process, there have been instances of changes in MCLs as new information becomes available (e.g., chromium).

The challenges associated with finding a balanced and measured response to emerging contaminants are felt particularly strongly by the water supply industry. This industry is tasked with providing clean, safe drinking water to the public. Conventional water treatment processes were developed to remove a variety of compounds. However, as the list of chemicals that need to meet MCLs or state-specific standards grows, there is a need to determine if conventional techniques also address these new compounds, or a need to develop and install processes that do. In some cases, new methods for detecting these new chemicals are needed; in other cases, new treatment technologies are needed. These needs provide new opportunities for enhanced collaboration between the water supply industry, academia, and other industries. Now, more than ever, a multidisciplinary approach is needed to ensure clean, safe drinking water for the public.

1.6 THIS BOOK

The topic of emerging contaminants is vast and, as evidenced by the preceding discussion, can be complex and fraught with sensitivity. This book was by no means meant to be comprehensive. Instead, the intent was to create a handbook that could serve as a reference point reflective of the time in which it was written. Within the book, we have compiled up-to-date information related to the sources, human health toxicity, regulatory status, characterization, and treatment options for a subset of the contaminants found in Appendix C. We formatted the book to facilitate finding relevant information in practice. In many instances, it relies on the use of tables and figures to present information as resources for future reference.

While the book aims to incorporate global perspectives associated with emerging contaminants, some topics are more focused on developments within the U.S. Most of the treatment discussions are focused on water, although some information on soil is also provided. It is also important to note that the discussions in this book center on several specifically selected compounds that affect groundwater and water supply (i.e., drinking water and wastewater); and that these discussions do not extend to biological contaminants.

The compounds we specifically selected to expand on, each in their own chapters, are as follows:

- 1,4-dioxane (Chapter 2),
- PFASs (Chapter 3),
- hexavalent chromium (Chapter 4), and
- TCP (Chapter 5).

The chapters dedicated to each of these compounds touch on the toxicological research completed to date, regulatory status, and current status of technology to support characterization and treatment. Table 1.1 serves to provide the context for selecting these compounds with regard to occurrence, understanding of human health effects, and regulatory status. While hexavalent chromium is no longer an emerging contaminant, similar to the story previously given for TCE, it deserves focus given the emergence of new and evolving toxicology data and the evolving state of regulations. Treatment technologies to address hexavalent chromium in water have previously been established, but the development of new treatment techniques may be needed to achieve the lower levels being considered for new regulatory standards. By comparison, 1,4-dioxane is fairly advanced in the emerging contaminant life cycle, but it still exhibits considerable variability in levels of regulation and the availability of effective treatment options. Finally, PFASs and TCP deserve particular attention as they are emerging contaminants attracting more recent and accelerating regulatory focus for which many conventional treatment approaches are ineffective or have limited applicability.

After focusing on these four specific contaminants, to round out the book, in the last chapter (Chapter 6) we touch on broader concepts for consideration relative to categorization and management of the wide range of other emerging contaminants that currently exist.

TABLE 1.1
Contributing factors for selecting featured emerging contaminants

Chapter	Chemical(s)	Occurrence	Understanding of Human Health Effects	Regulatory Status (Primarily in the United States)	Concern for Drinking Water	Concern for Wastewater	Concern for Groundwater Remediation
2	1,4-Dioxane	UCMR3[1]: Detected in 21.9% of public water supplies tested with 6.9% above the reference concentration of 0.35 μg/L.	Human health toxicological profile developed in 2012–2013. Studies in 2014 and 2017 elucidated new information on mode of action for tumor growth and suggested revisions to health-based guidelines.	No federal MCL, but 1,4-dioxane is on the CCL for possible future regulation. Many states have established drinking water and groundwater guidance based on the 2012–2013 toxicity data.	X	X	X
3	PFASs	UCMR3[1]: – PFOS detected in 1.9% of public water supplies tested with 0.9% above the reference concentration of 0.07 μg/L – PFOA detected in 2.4% of public water supplies tested with 0.3% above the reference concentration of 0.07 μg/L.	There is extensive research on the sub chronic effects of PFOS and PFOA on humans. Results are conflicting in terms of PFAS exposure and reproductive/developmental health.	Increased recognition of PFAS exposure routes has led to stricter individual and cumulative PFAS regulations, including shorter-chain PFASs. USEPA has focused regulation on PFOS and PFOA, while many states have considered other PFASs.	X	X	X

Introduction to Emerging Contaminants   21

Chapter	Chemical(s)	Occurrence	Understanding of Human Health Effects	Regulatory Status (Primarily in the United States)	Concern for Drinking Water	Concern for Wastewater	Concern for Groundwater Remediation
4	Hexavalent Chromium	UCMR3[1]: Detected in 89% of public water supplies with 1.8% of public water supplies tested above a concentration of 10 µg/L (2014 California MCL). California UCMR[2]: Detected in 12% of public water supplies above the 2014 California MCL of 10 µg/L.	Updating the 1998 risk assessment, taking into consideration developments in the literature on the carcinogenicity via ingestion.	USEPA is considering establishing an MCL, but is waiting the results of the updated risk assessment. California set the first and only state drinking water standard specific to hexavalent chromium in 2014. In May 2017, the Superior Court of California granted a petition and ordered the MCL to be withdrawn and reestablished due to questions on the economic feasibility for small water systems.	X		X
5	TCP	UCMR3[1]: Detected in 1.4% of public water supplies tested with 1.4% above the reference concentration of 0.04 µg/L California UCMR[3]: Detected in 891 of public water supply wells tested at a concentration greater than 0.005 µg/L (California MCL).	Classified as "reasonably anticipated to be a human carcinogen" by the United States National Toxicology Program in 2016.	There is no federal MCL, but TCP is on the CCL for possible future regulation California adopted an MCL of 0.005 µg/L MCL in 2017.	X		X

Notes:
1 USEPA UCMR3 results, collected from 2012–2015; reference concentrations provided give context for detections above relevant human health risk criteria (USEPA, 2017b).
2 California UCMR results for hexavalent chromium from 2000 to 2012 (California State Water Resources Control Board, 2014).
3 California UCMR results for TCP from 2000 to 2012 (California State Water Resources Control Board, 2016).
µg/L = micrograms per liter

REFERENCES

Agency for Toxic Substances and Disease Registry. 2004. *Toxicological Profile for Polybrominated Biphenyls.* Atlanta, Georgia: U.S. Department of Health and Human Services, Public Health Service, Agency for Toxic Substances and Disease Registry.

Anderson, P., N. Denslow, J. E. Drewes, A. Olivieri, D. Schlenk, and S. Snyder. 2010. "Monitoring Strategies for Chemicals of Emerging Concern in Recycled Water." Final Report to the State Water Resources Control Board, 220.

British Geological Survey. 2011. *Science Briefing: Emerging Contaminants in Groundwater.* British Geological Survey.

California State Water Resources Control Board. 2016. *1,2,3-Trichloropropane Maximum Contaminant Level Development Process.* July 20. Accessed June 11, 2017. www.waterboards.ca.gov/drinking_water/certlic/drinkingwater/documents/123-tcp/tcp_mcl_presentation.pdf.

—. 2014. *Chromium-6 in Drinking Water Sources: Sampling Results.* July 11. Accessed February 26, 2017. www.waterboards.ca.gov/drinking_water/certlic/drinkingwater/Chromium6sampling.shtml.

Commonwealth of Australia. 2017. *National Environment Protection Council.* Accessed May 24, 2018. www.nepc.gov.au/.

Council on Environmental Quality. 1971. "Toxic Substances." *Staff House Committee on Interstate and Foreign Commerce, 94th Congress, 2nd Session, Legislative History of the Toxic Substances Control Act (TSCA Legislative History) at 760 760 (Comm. Print 1976).* 776.

CRC CARE. 2014. *Development of Guidance for Contaminants of Emerging Concern, CRC CARE Technical Report no. 32.* Adelaide, Australia: CRC for Contamination Assessment and Remediation of the Environment.

Deeb, Rula A., Amparo E. Flores, Andrew J. Stocking, Scott E. Thompson, Michael C. Kavanaugh, Daniel N. Creek, and James M. Davidson. 2004. "Evaluation of MTBE Remediation Options." *A Report Written for: The California MTBE Research Partnership.* Fountain Valley, CA: Center for Groundwater Restoration and Protection, National Water Research Institute, April.

DoD. 2009. *Emerging Contaminants.* Instruction, Number 4715.18, United States of America Department of Defense.

Dodder, Nathan G., Alvine C. Mehinto, and Keith A. Maruya. 2015. "Monitoring of Constituents of Emerging Concern (CECs) in California's Aquatic Ecosystems—Pilot Study Design and QA/QC Guidance." Technical Report 854 for Southern California Coastal Water Research Project Authority.

Drewes, J. E., P. Anderson, N. Denslow, W. Jakubowski, A. Olivieri, D. Schlenk, and S. Snyder. 2018. "Monitoring Strategies for Constituents of Concern (CECs) in Recycled Water." Recommendations of a Science Advisory Panel. Convened by the State Water Resources Control Board, 178.

Environmental Health Division of Washtenaw County. 2017. *CARD: Coalition for Action on Remediation of Dioxane.* Accessed June 4, 2017. www.ewashtenaw.org/government/departments/environmental_health/card.

European Chemicals Agency. 2018a. *Understanding REACH.* Accessed May 24, 2018. https://echa.europa.eu/regulations/reach/understanding-reach.

—. 2018b. *Substances Restricted under REACH.* Accessed May 24, 2018. https://echa.europa.eu/substances-restricted-under-reach.

—. 2018c. *Candidate List of Substances of Very High Concern for Authorisation.* Accessed May 24, 2018. https://echa.europa.eu/candidate-list-table.

—. 2018d. *SVHC Roadmap to 2020 Implementation.* Accessed May 24, 2018. https://echa.europa.eu/svhc-roadmap-to-2020-implementation.

Federal Register. 2016. "Announcement of Final Regulatory Determinations for Contaminants on the Third Drinking Water Contaminant Candidate List." *40 CFR Part 141, EPA–HQ–OW–2012–0155; FRL–9940–64–OW*. Environmental Protection Agency, January 4.

—. 2014. "Announcement of Preliminary Regulatory Determinations for Contaminants on the Third Drinking Contaminants on the Third Drinking Water Contaminant Candidate List." *40 CFR Part 141. EPA–HQ–OW–2012–0155; FRL–9917–87–OW*. Vol. 79, no. 202. Environmental Protection Agency, October 20.

German Environment Agency. 2017. *Position: Protecting the Sources of Our Drinking Water*. Dessau-Roblau, Germany: German Environment Agency.

Interstate Technology & Regulatory Council, MTBE and Other Fuel Oxygenates Team. 2005. "Overview of Groundwater Remediation Technologies for MTBE and TBA: MTBE-1." Washington, D.C.: Interstate Technology & Regulatory Council, February.

Mohr, Thomas K.G., Julie A. Stickney, and William H. DiGuiseppi. 2010. *Enviromental Investigation and Remediation: 1,4-Dioxane and Other Solvent Stabilizers*. Boca Raton, FL: CRC Press.

National Environment Protection Council. 2013. *National Environment Protection (Assessment of Site Contaminantion) Measure*. May 15. Accessed May 24, 2018. www.nepc.gov.au/nepms/assessment-site-contamination.

NGI. 2018. *Technical Note: Preliminary Assessment of Substances Registered under REACH that could Fulfil the Proposed PMT/vPvM Criteria*. Document no. 20160526-TN-01, Trondheim, Norway: Norwegian Geotechnical Institute.

NORMAN. 2012. *Emerging Substances*. Accessed June 6, 2018. www.norman-network.net/?q=node/19.

Russell, Hugh H., John E. Matthews, and Guy W. Sewell. 1992. "TCE Removal from Contaminanted Soil and Ground Water." *EPA Ground Water Issue*. Washington, D.C.: Technology Innovation Office, Office of Solid Waste and Emergency Response, United States Environmental Protection Agency, January.

Sanzone, Danielle. 2015. "Hoosick Falls Residents Raise Concerns after Low Levels of Carcinogen Found in Water Supply." *Troy, The Record*. January 24. www.troyrecord.com/article/TR/20150124/NEWS/150129787.

Sauve, Sebastien, and Melanie Desrosiers. 2014. "A Review of What Is an Emerging Contaminant." *Chemistry Central Journal*. Vol. 8, no. 15.

Southern California Coastal Water Research Project. 2012. "Monitoring Strategies for Chemicals of Emerging Concern (CECs) in California's Aquatic Ecosystems." Technical Report 692, 215.

Stockholm Convention. 2008. *What Are POPs?* Accessed June 4, 2017. http://chm.pops.int/TheConvention/ThePOPs/tabid/673/Default.aspx.

Suthersan, Suthan, Joseph Quinnan, John Horst, Ian Ross, Erica Kalve, Caitlin Bell, and Tessa Pancras. 2016. "Making Strides in the Management of 'Emerging Contaminants'." *Groundwater Monitoring and Remediation*, 15–25.

United States Environmental Protection Agency (USEPA). 2017a. *Drinking Water Health Advisories for PFOA and PFOS*. April 14. Accessed June 4, 2017. www.epa.gov/ground-water-and-drinking-water/drinking-water-health-advisories-pfoa-and-pfos.

—. 2017b. "The Third Unregulated Contaminant Monitoring Rule (UCMR3): Data Summary, January 2017." Accessed June 20, 2017. www.epa.gov/sites/production/files/2017-02/documents/ucmr3-data-summary-january-2017.pdf.

—. n.d. *Trichloroethylene Fact Sheet*. Accessed June 4, 2017. https://safewater.zendesk.com/hc/en-us/sections/202346177-Trichloroethylene.

—. 2016. *Technical Fact Sheets about Emerging Contaminants at Federal Facilities*. May 5. Accessed June 6, 2018. https://19january2017snapshot.epa.gov/fedfac/emerging-contaminants-and-federal-facility-contaminants-concern_.html.

—. 2012. "Edition of the Drinking Water Standards and Health Advisories. EPA 822-S-12-001." Washington, D.C.: Office of Water, U.S. Environmental Protection Agency.

—. 2011. "Integrated Risk Information System: Chemical Assessment Summary: Trichloroethylene." September 28.

USGS. 2017. *Contaminants of Emerging Concern in the Environment.* June 16. Accessed March 5, 2018. https://toxics.usgs.gov/investigations/cec/index.php.

—. 2014. *Contaminants of Emerging Concern in Ambient Groundwater in Urbanized Areas of Minnesota, 2009–12.* Scientific Investigations Report 2014-5096, U.S. Department of the Interior.

Village of Hoosick Falls, The. 2016. *Municipal Water.* Accessed June 4, 2017. www.villageofhoosickfalls.com/Water/timeline.html.

Yaroschak, Paul. 2016. "DoD's Emerging Contaminants Program." *Emerging Contaminants Summit.* Winchester, CO.

Acronyms

Australian NEPC	Australian National Environment Protection Council
CCL	Contaminant Candidate List
CECs	chemicals of emerging concern
CRC CARE	Cooperative Research Centre for Contamination Assessment and Remediation of the Environment
DoD	United States Department of Defense
GAC	granular activated carbon
HAL	health advisory level
MCL	Maximum Contaminant Level
MTBE	methyl tertiary butyl ether
NEPMs	National Environment Protection Measures
NGI	Norwegian Geotechnical Institute
ng/L	nanograms per liter
PBBs	polybrominated biphenyls
PBDEs	polybrominated diphenyl ethers
PBT	persistent, bioaccumulative, and toxic
PFASs	per- and polyfluoroalkyl substances
PFOA	perfluorooctanoic acid
PFOS	perfluorooctanesulfonic acid
PMT	persistent, mobile, and toxic
POPs	persistent organic pollutants
REACH	Registration, Evaluation, Authorisation, and Restriction of Chemicals
RDX	1,3,5-trinitroperhydro-1,3,5-triazine, also known as Research Department Explosive
SDWA	Safe Drinking Water Act
TCE	trichloroethene
TCP	1,2,3-trichloropropane
UBA	Umweltbundesamt, German Environment Agency
UCMR	Unregulated Contaminants Monitoring Rule
USEPA	United States Environmental Protection Agency
USGS	United States Geological Survey
vPvB	very persistent, very bioaccumulative
vPvM	very persistent and very mobile

2 1,4-Dioxane

Caitlin H. Bell
(Arcadis U.S., Inc., caitlin.bell@arcadis.com)

Norman D. Forsberg
(Arcadis U.S., Inc., norman.forsberg@arcadis.com)

2.1 INTRODUCTION

1,4-Dioxane is an emerging contaminant that is most commonly known for its use as a stabilizer in some chlorinated solvents. As an environmental contaminant, 1,4-dioxane has been the focus of regulation and treatment since the early 2000s and was immediately recognized as both difficult to quantify via conventional laboratory analyses and difficult to treat using conventional techniques for chlorinated solvents. The associated science has come a long way since that time, but is still evolving on all fronts including quantification, toxicology, regulation, and treatment. These are intertwined, as 1,4-dioxane regulatory levels are low, creating further challenges for treatment.

2.2 BASIC INFORMATION

1,4-Dioxane (Figure 2.1) is most commonly known for its historical use as a stabilizer in the chlorinated solvent 1,1,1-trichloroethane (TCA). For the environmental consultant, this is probably the most common source of 1,4-dioxane to soil and groundwater. However, 1,4-dioxane is also used as a primary solvent, a manufacturing additive, and can be a by-product in the manufacturing of other commonplace chemicals. Table 2.1 provides a concise summary of 1,4-dioxane sources and the environmental considerations associated with each of them. While any of these sources could lead to the occurrence of 1,4-dioxane in the environment (or human/ecological exposure), they can also lead to unintended cross contamination of samples being analyzed for 1,4-dioxane. For example, certain classes of surfactants that may be present in decontamination detergents contain sufficient amounts of 1,4-dioxane to create an issue. Use of these products in decontamination at environmental cleanup sites can lead to false-positive results when quantifying 1,4-dioxane.

FIGURE 2.1 Chemical structure of 1,4-dioxane

TABLE 2.1
Common sources of 1,4-dioxane and environmental relevance

1,4-Dioxane Source	Description/Additional Information	Environmental Relevance
Chlorinated solvents	Used as stabilizer, particularly in TCA; present at 3.2 weight percent	TCA was commonly used interchangeably with TCE, so many chlorinated solvent sites include 1,4-dioxane
Terephthalate esters (polyester), resins, and associated manufacturing	By-product of the esterification process	Releases from facilities can result in environmental impacts
Ethoxylated surfactants/emulsions	By-product of the sulfonation reaction with alcohol ethoxylates	Present in a large number of products that contain surfactants
Polyethylene glycol production	By-product of the glycol production process	Releases from facilities can result in environmental impacts; use of products (e.g., antifreeze and deicing fluids) can result in sources of environmental impacts
Personal care products: detergents, shampoos, cosmetics, deodorants, etc.	Impurity in products that incorporate ethoxylated fatty alcohol sulfates (e.g., sodium laureth sulfate)	Could be introduced to the environment through wastewater discharge
Wetting and dispersing agent	Direct use as solvent	Releases from facilities can result in environmental impacts
Dye baths and stain printing	Direct use as solvent	Releases from facilities can result in environmental impacts
Adhesives	Impurity in solvents used to facilitate spreading/drying	Releases from facilities can result in environmental impacts
Pesticides, herbicides, and fumigants	Identified as inert ingredient; impurity of polyoxyethyleneamine	Releases from facilities can result in environmental impacts; use of products can result in sources of environmental impacts
Lacquers, paints, varnishes, and paint/varnish remover	Additive in resin-related products as well as impurity in the resin	Releases from facilities can result in environmental impacts; use of products can result in sources of environmental impacts
Solvent for fats, oils, waxes, and natural/synthetic resins	Direct use as solvent	Releases from facilities can result in environmental impacts
Pharmaceutical manufacturing	Direct use as solvent in product purification	Releases from facilities can result in environmental impacts
Reaction media for various organic synthesis reactions	Direct use as solvent	Releases from facilities can result in environmental impacts

1,4-Dioxane Source	Description/Additional Information	Environmental Relevance
Cellulose acetate membranes	Used to fluidize manufacturing materials in reverse osmosis and kidney dialysis filters	Releases from facilities can result in environmental impacts; waste disposal considerations
Liquid scintillation cocktails	Used in analyses related to the medical and life sciences	Waste disposal from facilities can result in environmental impacts
Ink and printing operations	Used as a solvent in printing inks since the early 1950s	Releases from facilities can result in environmental impacts
Laboratory solvent	Used in preservation, digestion, purification drying, etc.	Waste disposal from facilities can result in environmental impacts
Flame retardant production	Solvent used in production of brominated fire retardants	Releases from facilities can result in environmental impacts
Food additives and products that include the following ingredients:	Alcohol ethyoxylate, alcohol ethoxysulfate, polyoxyethylene, compounds with "laureth" in the name, sodium laureth sulfate, sodium lauryl ether sulfate (but not sodium lauryl sulfate), ammonium laureth sulfate, triethanolamine laureth sulfate, polyethylene glycol compounds, compounds with "ceteareth" in the name, compounds with "oleth" in the name, compounds with "xynol" in the name, polysorbates, propylene glycol, and compounds with the molecular structure $(C_2H_4O)_n$	Use and disposal of products can result in sources of environmental impacts; could be introduced to the environment through wastewater discharge

Source: Adapted from content in Mohr et al. (2010)

Regulation of 1,4-dioxane is relatively recent, but many states are now establishing action levels for soil, groundwater, and drinking water (see section 2.4 for more information on regulations). The following has become a common scenario:

> A facility that historically used chlorinated solvents, such as trichloroethene (TCE) and TCA, identified groundwater contamination and installed a remedial system to treat it. This remedial system likely included groundwater extraction with aboveground treatment (i.e., pump and treat [P&T]), perhaps coupled with reinjection of the groundwater into the subsurface. This P&T system probably had granular activated carbon (GAC) or an air stripper as the aboveground treatment. Groundwater concentrations decreased with time to below the applicable cleanup criteria. The facility initiates discussions with the regulatory agency regarding case closure. The regulatory agency requests that the facility sample the groundwater for 1,4-dioxane, likely for the first time. Laboratory analysis reveals the presence of 1,4-dioxane in groundwater at concentrations above the regulatory limit. Now the remedial system needs to be retrofitted to include a treatment process for 1,4-dioxane, and the site cannot reach closure at this time.

The reasons for this increasingly common scenario are the chemical and physical properties of 1,4-dioxane. A comparison table, shown in Table 2.2, describes

TABLE 2.2
Comparison of relative physical and chemical properties of 1,4-dioxane with other chemicals

Physical or Chemical Property	Units	Benzene	TCE	TCA	11DCE	1,4-Dioxane
Molecular formula	—	C_6H_6	C_2HCl_3	$C_2H_3Cl_3$	$C_2H_2Cl_2$	$C_4H_8O_2$
Molecular mass	grams per mole	78.11	131.4	133.4	96.95	88.11
Boiling point	°C	80	87	74	32	101*
Water solubility	grams per liter	1.8 (Moderate)	1.1 (Moderate)	0.91 (Moderate)	5.06 (High)	Miscible
Vapor pressure	millimeters of mercury	95.2 (Moderate)	72.6 (Moderate)	124 (High)	234 (High)	38.1 (Low)
Henry's law constant	atmosphere × meter³/mole	5.48×10^{-3} (Moderate)	9.1×10^{-3} (Moderate)	1.6×10^{-2} (High)	5.8×10^{-3} (Moderate)	4.8×10^{-6} (Very Low)
Organic carbon partitioning coefficient	unitless - as log K_{oc}	1.92 (Moderate)	1.81 (Low)	2.18 (Moderate)	1.48 (Low)	0.54 (Very Low)

Source: Developed from data in Suthersan and Payne (2005).
* 1,4-Dioxane forms an azeotrope with water in which 48.5 mole percent of 1,4-dioxane in water will boil at 87.59 °C (Schneider and Lynch, 1943).

the environmentally relevant properties of 1,4-dioxane, benzene, TCE, TCA, and 1,1-dichloroethene (11DCE). Of environmental relevance, 1,4-dioxane is less volatile (i.e., Henry's law constant of 4.8×10^{-6} atmosphere × meter^3/mole [see also the note in Appendix C]; vapor pressure of 38.1 millimeters of mercury), less sorptive (i.e., log K_{oc} of 0.54), and has a higher boiling point (i.e., 101 °C) than the chlorinated compounds with which it is typically colocated. This is why the GAC and air-stripping processes employed in the aforementioned scenario were not effective at removing 1,4-dioxane from groundwater. Specifically, GAC may be ineffective (when originally designed to remove chlorinated solvents), because the 1,4-dioxane will break through the GAC before the chlorinated solvents. Historically, 1,4-dioxane was not sampled as part of routine performance monitoring to know when this occurred.

According to the manufacturing records available, 1,4-dioxane was a stabilizing agent added to TCA at approximately 3.2 weight percent (Mohr et al., 2010). Additionally, 1,4-dioxane is expected to have concentrated during chlorinated solvent usage in vapor degreasers up to 15 to 20 percent (Mohr et al., 2010). The Lewis base properties of 1,4-dioxane (i.e., has an electron pair to donate from the oxygen) allows it to react with, and passivate, metals and their salts, notably aluminum chloride, thus preventing the premature decomposition of TCA in metal tanks (Mohr et al., 2010). Therefore, 1,4-dioxane may be present in groundwater that also contains TCA; its abiotic dehydrohalogenation product, 11DCE; or its biological degradation product, 1,1-dichloroethane. These compounds can be used as chemical indicators for the presence of 1,4-dioxane. Despite a few historical patents that may suggest the use of 1,4-dioxane as a stabilizer in TCE (Morrison and Murphy, 2015), manufacturers do not identify 1,4-dioxane as a stabilizer for TCE (Mohr et al., 2010), largely because TCE is not as susceptible to the same metal reactions as TCA. Meta data evaluations (Adamson et al., 2014) identified 1,4-dioxane present at facilities where TCE was present. While TCE has been used historically as the dominant industrial solvent, TCA replaced it for a period of time until being banned in the 1990s. Therefore, the presence of 1,4-dioxane with TCE is likely attributed to the historical use of TCA and subsequent attenuation of TCA to nondetectable levels.

The presence of 1,4-dioxane at chlorinated solvent sites continues to be a challenge due to the evolving and increasing regulation around 1,4-dioxane concentrations in environmental media, such as groundwater. More than 30 American states now have regulatory standards or guidelines for 1,4-dioxane. Most of these are related to drinking water or groundwater, but some states have provided standards or guidelines related to soil and soil vapor, as well. Regulation of 1,4-dioxane began in the late 1990s, but many of the initial values were put into place in the early 2000s with many of them updated in the 2010s as the toxicological understanding changed (see section 2.3 for more detail). Approximately half of the 1,4-dioxane regulatory values are enforceable (e.g., promulgated standards) while the other half are not (e.g., guidelines or screening levels).

As the number of states with regulatory standards or guidelines for 1,4-dioxane increases, two strategies can be adopted relative to evaluating its presence in a portfolio: reactive or proactive. Under a reactive strategy, a facility may look for 1,4-dioxane only if requested by the regulatory agency, for example. A reactive management strategy may be attractive in situations where the contamination profile does not directly point to the potential for 1,4-dioxane, and there are no site characteristics that would increase the level of risk (e.g., no human health or ecological risk). A proactive management strategy may be attractive where the contamination profile strongly suggests the presence of 1,4-dioxane in the subsurface (i.e., TCA; 11DCE; 1,1-dichloroethane present) and there are site characteristics that could increase the level of risk such as potential receptors. Ultimately, the decision of which approach to take will depend on the risk management practices of the individual site owner.

Another challenge associated with 1,4-dioxane is its already pervasive presence in drinking water supplies (United States Environmental Protection Agency [USEPA], 2017). Of the 4,915 public water supplies tested for 1,4-dioxane as part of the USEPA's Third Unregulated Contaminant Monitoring Rule 3 sampling conducted between 2012 and 2015, 21.9 percent of public water supplies (1,077 of 4,915) had detections of 1,4-dioxane and 6.9 percent (341 of 4,915) were above the reference concentration of 0.35 micrograms per liter (µg/L; USEPA, 2017). Additionally, the presence of 1,4-dioxane in public water supplies with groundwater sources were higher than those with surface water sources (Adamson et al., 2017). The presence of 1,4-dioxane in drinking water supply wells may come from groundwater remediation sites with known, or unknown, levels of 1,4-dioxane. It may also come from less-expected commercial sources, like those associated with car washes that use detergents containing 1,4-dioxane. The drinking water challenge may also be exacerbated by the discharge of 1,4-dioxane from household products to wastewater treatment plants, 1,4-dioxane passing through the treatment processes, and water supply intakes with influence from wastewater discharge.

2.3 TOXICITY AND RISK ASSESSMENT

This section touches on the human health toxicity and risk associated with 1,4-dioxane. It is intended to identify the relevant exposure pathways (as well as the pathways that are not of concern). It also discusses the toxicity information that is the basis for the current 1,4-dioxane regulations. At the time of this writing, the industry's understanding of 1,4-dioxane toxicity was still evolving and influencing the related regulatory standards or guidelines.

Human health toxicity is evaluated from two perspectives: potential cancer effects and potential noncancer effects. The primary routes of human exposure to 1,4-dioxane include oral contact (i.e., ingestion), inhalation, and dermal contact. The absorption of 1,4-dioxane following oral or inhalation exposure is expected to be rapid and nearly complete and to occur via passive diffusion (Agency for Toxic Substances and Disease Registry [ATSDR], 2012). In contrast, absorption following dermal exposure is not expected to contribute significantly to systemic doses of 1,4-dioxane (Marzulli et al., 1981; ATSDR, 2012), and as such is not carried forward as a relevant exposure pathway. After exposure, humans metabolize 1,4-dioxane extensively (ATSDR, 2012; USEPA, 2013). While the exact metabolic pathway has not been fully resolved, substantial evidence indicates that 1,4-dioxane is metabolized primarily to β-hydroxyethoxyacetic acid (HEAA) by cytochrome P450 oxidases (USEPA, 2013; Dourson et al., 2014). At environmentally or occupationally relevant exposure levels, it is expected that most of a 1,4-dioxane dose will be excreted as HEAA in the urine (ATSDR, 2012; USEPA, 2013; Dourson et al., 2014). A summary of the relevant exposure pathways associated with potential cancer effects and potential noncancer effects, as evaluated by USEPA (2013) and ATSDR (2012), follows.

2.3.1 Potential Noncancer Effects

USEPA (2013) and ATSDR (2012) both evaluated the toxicological profile of 1,4-dioxane and reached similar conclusions regarding the spectrum and magnitude of noncancer health effects. In deriving daily human exposure levels expected to be without appreciable risk of noncancer health effects, USEPA (2013) and ATSDR (2012) both identified the hepatotoxic effects reported by Kociba et al. (1974) as the critical effects associated with oral exposure to 1,4-dioxane (Table 2.3). While

TABLE 2.3
Summary of toxicological reference values developed by USEPA and ATSDR for use in evaluating chronic oral and inhalation non-carcinogenic exposures to 1,4-dioxane along with the basis of their derivation

Parameter	USEPA (2013)	ATSDR (2012)
	Chronic Oral Exposure	
Toxicity Reference Value	RfD, Chronic	Minimal Risk Level, Chronic (≥365 days)
Value	0.03 mg/kg/d	0.1 mg/kg/d
POD	9.6 mg/kg/d	9.6 mg/kg/d
POD type	NOAEL	NOAEL
Critical effect	Degenerative effects in the livers of male rats at a LOAEL of 94 mg/kg/d	Liver degeneration, necrosis and signs of regeneration in male rats at a LOAEL of 94 mg/kg/d
Key study	Kociba, et al. (1974)	Kociba, et al. (1974)
Uncertainty factor	300 ($UF_A = 10$, $UF_H = 10$, $UF_D = 3$)	100 ($UF_A = 10$, $UF_H = 10$)
	Chronic Inhalation Exposure	
Toxicity Reference Value	Reference Concentration	Minimal Risk Level, Chronic (≥365d)
	0.009 ppm (eq. to 0.03 mg/m^3)	0.03 ppm
POD	50 ppm	50 ppm
POD type	LOAEL	LOAEL
Critical effect	Atrophy and respiratory metaplasia of the olfactory epithelium in male rats	Atrophy of the olfactory epithelium in male rats
Key study	Kasai, et al. (2009)	Kasai, et al. (2009)
Duration adjustment	Yes, 6/24 hours × 5/7 days	Yes, 6/24 hours × 5/7 days
Human equivalent dosimetric adjustment factor	1 (default)	1 (default)
Uncertainty factor	1,000 ($UF_A = 3$, $UF_H = 10$, $UF_L = 10$, $UF_D = 3$)	300 ($UF_A = 3$, $UF_H = 10$, $UF_L = 10$)

Sources: Adapted from USEPA (2013) and ASTDR (2012).

USEPA (2013) and ATSDR (2012) concluded that limiting exposure to 0.1 milligram per kilogram body weight per day (mg/kg/d) protects against development of hepatotoxic effects following exposure to 1,4-dioxane, USEPA (2013) concluded that an additional uncertainty factor of 3 was warranted to account for the lack of a multigeneration reproduction study, resulting in a reference dose (RfD) of 0.03 mg/kg/d. Thus, it is expected that chronic oral exposure to 1,4-dioxane at a dose between 0.03 and 0.1 mg/kg/d does not pose an appreciable risk of noncancer effects (including potentially sensitive subpopulations) over a lifetime of exposure. USEPA (2013) and ATSDR (2012) identified damage to the olfactory epithelium (Kasai et al., 2009) as the critical noncarcinogenic effect associated with inhalation exposure to 1,4-dioxane (Table 2.3). Limiting chronic inhalation exposures to less than or equal to 0.03 parts per million (ppm) 1,4-dioxane protects against adverse impacts to olfactory epithelium. However, as with the oral RfD, USEPA (2013) concluded that an additional uncertainty factor of 3 should be applied to the point of departure to account for the lack of a multigeneration reproduction study. Thus, limiting chronic inhalation exposures to concentrations less than 0.009 to 0.03 ppm is not expected to pose an appreciable risk of noncancer effects over a lifetime of exposure (including potentially sensitive subpopulations).

2.3.2 POTENTIAL CANCER EFFECTS

After updates to the USEPA assessment in 2010 and 2013, USEPA (2013) concluded that 1,4-dioxane is "likely to be carcinogenic to humans" and justified this conclusion based on inconclusive findings in humans and evidence of carcinogenicity in several two-year bioassays (Table 2.4). USEPA (2013) identified the two-year drinking water study performed by Kano et al. (2009) as the key study for deriving the oral cancer slope factor (CSF) for 1,4-dioxane (Table 2.4). USEPA (2013) identified increased incidence of liver tumors as the critical effect in rats and mice exposed to 1,4-dioxane in the Kano et al. (2009) study and derived a CSF of 0.1 $(mg/kg/d)^{-1}$ using standard USEPA guidance and a default linear low-dose extrapolation approach. In regard to potential carcinogenic risks associated with inhalation exposures, USEPA (2013) identified the two-year inhalation study performed by Kasai et al. (2009) as the key study for deriving the inhalation unit risk factor (IUR) for 1,4-dioxane. Because chronic inhalation exposure to 1,4-dioxane was associated with development of multiple types of tumors at multiple sites, USEPA (2013) identified total tumors combined across all sites as the critical effect in rats (most sensitive species) and derived a IUR of 5×10^{-6} per microgram per cubic meter $(\mu g/m^3)^{-1}$ using standard USEPA guidance and a default linear low-dose extrapolation approach.

Although it is generally understood that 1,4-dioxane's carcinogenicity is not mediated by a mutagenic mode of action (MOA), USEPA (2013) concluded that the MOA by which 1,4-dioxane produces liver tumors is "unknown, and available evidence in support of any hypothetical mode of carcinogenic action for 1,4-dioxane is inconclusive." Thus, the USEPA (2013) proceeded to develop the conservative toxicity values described earlier using default linear low-dose extrapolation procedure. It is important to note that this conservative approach has resulted in state drinking water criteria in the subpart billion concentration range (see section 2.4). However, scientific

TABLE 2.4
Summary of USEPA's oral cancer slope factor and inhalation unit risks for 1,4-dioxane and their derivation process

Parameter	Oral Exposures	Inhalation Exposures
Toxicity Reference Value	CSF	IUR
Value	0.1 (mg/kg/d)$^{-1}$	5×10^{-6} (μg/m^3)$^{-1}$
POD	4.95 mg/kg/d	19,500 μg/m^3
POD type	Benchmark dose, lower 95 percent confidence limit at 50 percent extra risk, human equivalent dose	Benchmark concentration, lower 95 percent confidence limit at 10 percent extra risk
Critical effect	Hepatocellular adenoma or carcinoma in female mice	Tumors combined across all sites in male rats
Key study	Kano et al. (2009)	Kasai et al. (2009)

Source: Adapted from USEPA (2013).

investigations by Dourson et al. (2014; 2017) identified an MOA for the generation of tumors that was previously unknown. Importantly, the 2017 study by Dourson and others performed in collaboration with the Alliance for Risk Assessment and several state agencies, specifically sought to determine if data gaps identified by USEPA (2013) relating to 1,4-dioxane's cancer MOA could be resolved. The work provides evidence supporting the conclusion that 1,4-dioxane's cancer MOA has a threshold below which 1,4-dioxane does not accumulate in the liver and tumors do not form, thus supporting the use of a nonlinear low-dose extrapolation procedure rather than the default low-dose extrapolation procedure previously used. A threshold approach is similar to that taken by the European Union (European Communities, 2002) and Canada (Environment Canada/Health Canada, 2010). Application of the more relevant nonlinear low-dose extrapolation approach results in a risk-based drinking water criterion in the hundreds of parts-per-billion compared to the subparts-per-billion levels adopted to date. This is an area of ongoing discussion among toxicologists.

2.4 REGULATORY STATUS

As of the time of this writing, USEPA had not issued a federal drinking water maximum contaminant level (MCL) for 1,4-dioxane. Traditionally, the USEPA's MCL has been the basis of various regulatory standards or guidelines. Most drinking water standards or guidelines are set to the MCL. Likewise, for groundwater sources considered actual, or potential, drinking water sources, the MCL is also commonly the groundwater cleanup standard or guideline. Soil vapor and indoor air standards and guidelines may be developed via fate and transport models based on the groundwater standards or guidelines. Soil cleanup standards or guidelines may be calculated based on contaminant leaching into groundwater. Surface water standards or guidelines may be based on whether they feed drinking water supplies. Therefore, in the absence of a federal MCL, other approaches must be employed to develop relevant standards or guidelines.

TABLE 2.5
Federal guidelines and health standards for 1,4-dioxane

Guideline/Health Standard	Value (with units)	Source
Drinking water concentration representing a 10^{-6} cancer risk level[1]	0.35 µg/L	USEPA (2013)
Screening level in residential tap water, based on 10^{-6} cancer risk[2]	0.46 µg/L	USEPA (2018)
Lifetime health advisory	200 µg/L	USEPA (2012)
Soil screening levels: residential; industrial; and soil-to-groundwater	5.3 mg/kg; 24 mg/kg; 9.4×10^{-5} mg/kg	USEPA (2018)
Indoor air screening levels: residential; industrial	0.56 µg/m^3; 2.5 µg/m^3	USEPA (2018)

Notes:
[1] The risk-based drinking water concentration considers the ingestion exposure route, adult exposure, and does not account for annual expected exposure duration.
[2] The residential tap water screening value accounts for dermal, inhalation, and ingestion exposure routes, adjusts to consider both children and adults, and adjusts for annual expected exposure duration.

As described in section 2.3, a risk-based approach is a common method for developing standards or guidelines for 1,4-dioxane. These commonly consider the health-based advisories that are issued by various federal agencies (Table 2.5). As of the time of this writing, more than 30 states issued state-specific standards or guidelines for 1,4-dioxane. These values include several environmental media (i.e., soil, groundwater, etc.) and include both enforceable (e.g., promulgated standards) and unenforceable values (e.g., guidelines, screening levels). Due to the nature of chemical releases of 1,4-dioxane into the environment, as well as the fate and transport characteristics, many states focused on developing drinking water and/or groundwater values.

As seen in Figure 2.2, drinking water/groundwater values vary widely from state to state and span three orders of magnitude. This is largely due to state-specific application of the toxicity values discussed in section 2.3. States are also revising these values as new data become available. For example, Michigan implemented an emergency rule in 2016 (with promulgated new rule in 2017) to lower the allowable level of 1,4-dioxane in residential drinking water from 85 µg/L to 7.2 µg/L, three years after USEPA's issuance of the revised toxicity values. The value of 7.2 µg/L was calculated based on the updated CSF of 0.1 (mg/kg/d)$^{-1}$, a target risk level of 10^{-5}, and the calculation used by the Michigan Department of Environmental Quality to calculate the risk associated with drinking water ingestion (MCL 324.20120a[5]; R 299.10). Similarly, New Jersey's groundwater quality criterion of 10 µg/L decreased to 0.4 µg/L (rounded up from 0.35 µg/L) in 2015. The NJ Department of Environmental Protection also used the updated CSF of 0.1 (mg/kg/d)$^{-1}$, but calculated the updated groundwater quality criterion using a 10^{-6} risk level and a

1,4-Dioxane 37

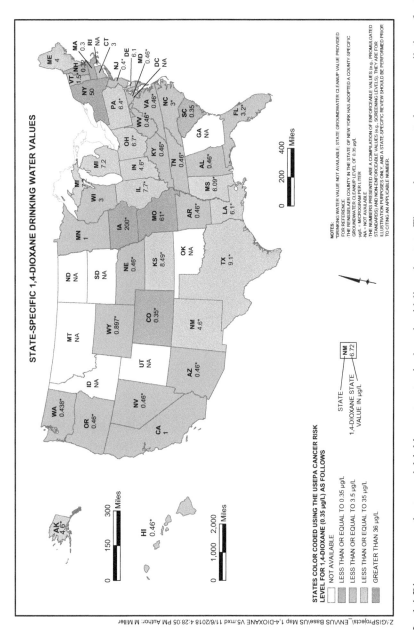

FIGURE 2.2 1,4-Dioxane groundwater and drinking water standards and guidelines, by state: The numbers presented are a compilation of enforceable values (e.g., promulgated standards) and non-enforceable values (e.g., screening levels). They are for illustration purposes only, and a state-specific review should be performed prior to citing an applicable number. Adapted from Suthersan et al. (2016).

different set of default assumptions in the calculation (New Jersey Department of Environmental Protection, Water Monitoring and Standards, 2015).

In addition to individual states, the USEPA developed a risk-based drinking water concentration of 0.35 µg/L representing a 10^{-6} incremental cancer risk. Similarly, USEPA developed a 1,4-dioxane screening value of 0.46 µg/L for residential tap water (USEPA, 2018) as part of their regional screening levels (RSLs). These screening levels are a merger of USEPA Region 3's Risk-Based Concentrations Table, Region 6's Human Health Medium-Specific Screening Levels Table, and Region 9's Preliminary Remediation Goals Table and is now used for all USEPA regions' Superfund site screening, and is adopted by many of the states. The residential tap water screening value of 0.46 µg/L is greater than the risk-based drinking water concentration of 0.35 µg/L because it accounts for dermal, inhalation, and ingestion exposure routes, adjusts to consider both children and adults, and adjusts for annual expected exposure duration. The risk-based drinking water concentration of 0.35 µg/L only considers the ingestion exposure route, adult exposure, and does not account for annual expected exposure duration.

Both standards and guidelines are expected to continue to change over time. This may be especially true if states adopt Dourson et al.'s (2014; 2017) proposed cancer MOA and updated RfD. Therefore, it is imperative to have an up-to-date understanding of the current regulatory guidance on 1,4-dioxane for a given site or state, as well as potential changes and current regulatory climate. Knowing the regulatory agencies involved, the state-specific guidance, applicable USEPA regional guidelines, and other regulatory factors up front can avoid project upsets and challenges down the road. However, despite the best up-front planning, these types of upsets and challenges may be unavoidable as the industry works toward a more unified view on regulating 1,4-dioxane.

Outside of the United States, there are some guidelines associated with 1,4-dioxane at the time of this writing. As summarized at the 2016 Emerging Contaminants Summit (Westminster, Colorado, in March 2016), Germany has a precautionary guideline limit of 0.1 µg/L in groundwater. France has a potable water limit of 6.6 µg/L and a risk threshold of 37.5 µg/L. Japan has adopted the World Health Organization's suggested threshold of 50 µg/L, and Australia has a groundwater cleanup level of 77 µg/L. The values being adopted outside of the United States are similar to those being adopted within the United States.

2.5 SITE CHARACTERIZATION

Environmental sampling and laboratory analysis associated with site characterization of 1,4-dioxane can be challenging. This is largely due to its chemical and physical characteristics. Specifically, 1,4-dioxane is not compatible with some sampling techniques commonly employed (e.g., passive diffusion bags), there is potential for cross contamination from field decontamination detergents, and methods for accurate quantification are still evolving.

For the purposes of this discussion, we will focus on two elements related to site characterization: investigation approach—including use of high-resolution site characterization—and analytical methods. Both of these are critical elements to ensuring an accurate and representative understanding of the nature and extent of 1,4-dioxane presence in various environmental media.

First, however, it is helpful to consider how the chemical and physical characteristics of 1,4-dioxane impact planning an investigation to confirm the presence of, or delineate the extent of, 1,4-dioxane. As described in Table 2.2 and Appendix C, 1,4-dioxane does not readily sorb to soil, nor does it volatilize significantly from water. This means it is expected to travel quickly along the groundwater flow path and remain in groundwater, rather than transfer to the vapor phase to cause vapor intrusion concerns. These tendencies become particularly important when 1,4-dioxane is present with chlorinated solvents. The chlorinated solvents will tend to sorb to soil and readily volatilize from groundwater. Based on these differences, it is reasonable to expect chlorinated solvents to produce smaller groundwater plumes with residual impacted soil mass while 1,4-dioxane would produce a larger, more dilute, plume without a significant soil mass source. This dichotomy is reminiscent of the differences between petroleum products and their fuel oxygenates, such as methyl tertiary butyl ether.

Despite the above, as with all generalities there are exceptions to the rule and factors that can influence what is actually observed. The work by Adamson et al. (2014) evaluated >2,000 sites with both chlorinated solvents and/or 1,4-dioxane and revealed that only 21 percent of 1,4-dioxane plumes were larger than the colocated chlorinated solvent plume. This may be attributable to both physical and biological natural attenuation processes that are discussed in more detail in section 2.7.3, or it may be an artifact of incomplete delineation of the 1,4-dioxane plume extent.

Consequently, it may be important to take a multifaceted approach to site characterization. This includes first sampling for 1,4-dioxane within the source area to identify if it is present and what the upper bounds of the groundwater concentrations are. If the groundwater plume is relatively old, the highest concentrations of 1,4-dioxane may be found downgradient of the source, as the center of mass will travel with time. To identify the downgradient extent of the 1,4-dioxane plume, it may be most efficient to first sample near the downgradient edge of the existing chlorinated solvent plume. If 1,4-dioxane is not detected, step upgradient toward the source area until the length of the 1,4-dioxane plume is delineated. If 1,4-dioxane is detected at the downgradient edge of the existing plume, it may be necessary to employ additional infrastructure to delineate the length of the 1,4-dioxane plume. A similar approach can be taken to define the lateral extents of the 1,4-dioxane in relation to the existing groundwater plume.

2.5.1 Investigation Approaches

For facilities that already have monitoring infrastructure in place, and the presence of 1,4-dioxane in groundwater is being explored after the delineation of other chemicals, it is often most advantageous to begin the investigation by simply sampling that same infrastructure. The extent to which the existing monitoring well network can be used may quickly become limited because monitoring wells may not already be present to define the downgradient extent of 1,4-dioxane if the 1,4-dioxane plume has traveled farther downgradient than the chlorinated solvents it is associated with.

1,4-Dioxane groundwater samples can be collected using most traditional methods (e.g., low-flow purge, HydraSleeve™, Snap Sampler®, SUMMA canisters, etc.); however, passive diffusion bags are not suitable for 1,4-dioxane groundwater sampling because the molecule will not readily diffuse into the bag. An important consideration for field sampling is that of cross contamination. As mentioned in Table 2.1, 1,4-dioxane is present in many detergents, including those commonly used for decontamination of sampling equipment (e.g., Liquinox®). Particular care should be taken to use a decontamination method that will not produce false-positive results in analytical samples.

The use of high-resolution site characterization can be highly effective at identifying the mass transport zones within the subsurface as well as the extent of contamination. This approach also applies to characterization of 1,4-dioxane and associated chlorinated solvents. As described in the case study to follow, cone penetrometer testing (CPT) was coupled with whole core soil sampling and vertical aquifer profiling (VAP). Soil and groundwater samples were analyzed in real time using an on-site laboratory equipped with direct sampling ion trap mass spectrometry (DSITMS). This investigation led to a three-dimensional understanding of the dominant groundwater flow paths, the chlorinated solvent source area, and the (larger) 1,4-dioxane impacts.

Case Study: High-Resolution Site Characterization of 1,4-Dioxane and Chlorinated Solvents

A 30-acre facility was historically operated as a chemical manufacturing facility from the 1960s until 2012, and historical investigations led to the identification of chlorinated solvents and 1,4-dioxane in environmental media. High-resolution site characterization methods were used to efficiently map soil, groundwater, and hydrostratigraphy at the site. The approach combined dynamic, high-density soil and groundwater sampling with hydrostratigraphic interpretations and permeability mapping in three dimensions. By correlating high-resolution concentration data with hydrostratigraphy and permeability data, it was possible to map, and distinguish, mass transport zones from mass storage zones and classify the scales of variability that controlled transport in the source and distal segments of plumes in real time.

The investigation was designed to evaluate the presence of chlorinated solvents in soil beneath former facility structures, evaluate the magnitude and extent of 1,4-dioxane and chlorinated solvents in the source area in both soil and groundwater at a high resolution, conduct detailed mapping of the site permeability to identify potential preferential pathways, and correlate chlorinated solvent and 1,4-dioxane distribution to geological features. This level of understanding was needed to develop a remedial approach that would target the area of highest chlorinated solvent and 1,4-dioxane concentrations. The investigation included the following: 23 CPT borings were installed to map the site hydrostratigraphy and identify groundwater sample intervals based on high relative permeability, 21 borings were advanced and samples were collected on approximately 0.6-meter

(2-foot) intervals for analysis of chlorinated solvents and 1,4-dioxane in soil and groundwater, 18 VAP borings were installed with 39 sample intervals based on review of the CPT data.

Soil and groundwater samples were analyzed using an onsite laboratory with DSITMS, capable of analyzing approximately 80 samples per day, in the first commercial application of DSITMS for both chlorinated solvents and 1,4-dioxane, completed in 2014. Onsite laboratory results were verified by a subset of analytical laboratory analyses. During the site investigation, field and laboratory data were entered in real time into the three-dimensional modeling software Environmental Visualization System, developed by C Tech Development Corporation. The rendered three-dimensional models and related cross-sectional views generated in the software allowed for dynamic adjustment of the network of proposed sampling locations. For example, once the approximate extent of the source area was identified to the east, remaining borings further east were not advanced.

The investigation resulted in a refined conceptual site model, with the following key conclusions. Previous data collection methods (soil borings) indicated primarily low-permeability intervals within the source area, previously described as primarily low-permeability silts and clays. Data generated from this work indicated that former investigations missed higher-permeability soils present within this zone that could act as potential transport pathways.

Concentration data indicated transport of 1,4-dioxane occurred along the preferential flowpaths identified during the CPT investigation. The extent of chlorinated solvent and 1,4-dioxane source mass was identified (Figure 2.3). This information allowed for development of viable remedial alternatives that met the time frame of the pending sale of the property and focused on the areas of greatest mass flux.

2.5.2 ANALYTICAL METHODS

The ability to quantify 1,4-dioxane accurately is evolving, and as of the time of this writing several challenges existed, particularly with groundwater analysis. The groundwater analytical methods typically used include USEPA Method 8260 for volatile organic compounds (VOCs; but specifically with selected ion monitoring [SIM]), USEPA Method 8270 for semi-volatile organic compounds (SVOCs; also specifically with SIM), and USEPA Method 522 for 1,4-dioxane in drinking water. The challenges associated with these analytical methods are largely related to the matrix extraction method. For example, Chiang et al. (2016) summarized findings of an analytical comparability study that found significantly lower concentrations (up to a factor of three) when USEPA Method 8270 was used compared to USEPA Method 8260. However, USEPA Method 8270 typically gives lower laboratory reporting limits than USEPA Method 8260. Likewise, USEPA Method 522 is specifically designed for drinking water and may not always be appropriate for groundwater analysis. Table 2.6 summarizes various extraction methods, analytical methods, and the benefits and challenges associated with each of these; while, Table 2.7 goes into more detail on the topic. In summary, it may be necessary

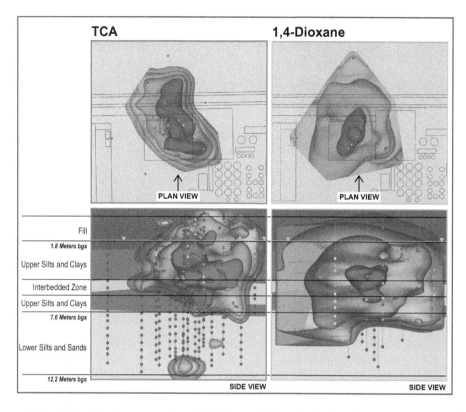

FIGURE 2.3 Visual representation of 1,4-dioxane high resolution investigation: Distribution of 1,4-dioxane (right) and TCA (left) identified during high-resolution site characterization. Generalized geology is shown on the left based on the depth below ground surface (e.g., 1.8 meters bgs). Concentration shells are results from whole soil samples (i.e., soil and porewater) and represent greater than 10,000 μg/kg (pink); 1,000 to 10,000 μg/kg (red); 100 to 1,000 μg/kg (orange); 10 to 100 μg/kg (yellow); 1 to 10 μg/kg (green); and less than 1 μg/kg (blue).

to use different analytical methods depending on the data quality objectives. For example, a lower detection limit (e.g., USEPA Method 8270 with SIM) may be preferred when first looking for the presence of 1,4-dioxane in groundwater, when trying to define the extents of a groundwater plume, or when confirming concentrations have met regulatory criteria. Higher replicability (e.g., USEPA Method 8260 with SIM) may be preferred when 1,4-dioxane concentrations are elevated and active remediation is ongoing. Adjustment for matrix recovery via use of isotopic dilution by the analytical laboratory is an approach that mitigates many of the extraction- and concentration-related challenges, but is not widely available. Until a unified quantification method is developed for 1,4-dioxane, it is important to communicate with the analytical laboratory prior to sampling to understand their capabilities and reporting structure. The analytical challenges described in Tables 2.6 and 2.7 should also be considered when comparing data analyzed at different laboratories, or with different methods.

1,4-Dioxane

TABLE 2.6
Summary of 1,4-dioxane groundwater analytical methods

Typical Extraction/Analytical Combinations	Advantages	Challenges
Common VOC analysis: USEPA Method 5030B (purge and trap extraction) USEPA Method 8260C (VOC analysis)	– Commonly used analytical method that can also identify 1,4-dioxane; could be used as a screening tool	– Typically has high detection limits that do not meet regulatory requirements – The presence of chlorinated VOCs causes interference with the 1,4-dioxane quantification – The extraction method is not efficient at removing 1,4-dioxane from groundwater, so the results may be biased low
VOC analysis with SIM: USEPA Method 5030B (purge and trap extraction) USEPA Method 8260C-SIM (VOC analysis with SIM)	– Lower detection limits than 8260C full scan – SIM mode is specific for quantifying 1,4-dioxane, so there is no interference from chlorinated VOCs – May be more accurate than other methods at concentrations of 1,4-dioxane on the order of 100s or 1,000s µg/L – Request laboratory to utilize heated purge and trap extraction and isotopic dilution for improved extraction and quantification of losses	– May have detection limits that do not meet regulatory requirements; confirm with laboratory – The extraction method is not efficient at removing 1,4-dioxane from groundwater, so the results may be biased low – May need to run separate analysis for quantification of other compounds, if needed – Modifications to analytical method (e.g., isotopic dilution or heated purge and trap) may require regulatory approval for use
Common SVOC analysis: USEPA Method 3510C/3520C (liquid-liquid extraction) USEPA Method 8270D (SVOC analysis)	– Commonly used analytical method that can also identify 1,4-dioxane; could be used as a screening tool	– May have detection limits that do not meet regulatory requirements; confirm with laboratory – The extraction and/or concentration step of the preparation method may result in loss of 1,4-dioxane, so the results may be biased low

(Continued)

TABLE 2.6 (Continued)

Typical Extraction/Analytical Combinations	Advantages	Challenges
SVOC analysis with SIM: USEPA Method 3510C/3520C (liquid-liquid extraction) USEPA Method 8270D-SIM (SVOC analysis with SIM)	– Typically has reporting limits that meet regulatory requirements, making it an attractive option for determining whether 1,4-dioxane is present at a site (i.e., initial investigation) – Request laboratory to utilize isotopic dilution for improved quantification	– The extraction and/or concentration step of the preparation method may result in loss of 1,4-dioxane, so the results may be biased low – May be less accurate than other methods at concentrations of 1,4-dioxane on the order of 100s or 1,000s µg/L – May need to run separate analysis for other compounds, if needed
Drinking water method: USEPA Method 522	– Typically has lowest reporting limits of the methods discussed here, making it an attractive option for determining whether 1,4-dioxane is present at a site (i.e., initial investigation) or for routine analysis of 1,4-dioxane – Extraction method does not incur significant losses of 1,4-dioxane from the sample (i.e., less bias) – The method was developed specifically to quantify 1,4-dioxane	– May be subject to interferences when applied to groundwater (e.g., high levels of suspended solids, total organic carbon, and/or mineral content) that result in elevated reporting limits – Use of a drinking water method may require regulatory approval for groundwater quantification – Does not also quantify other compounds

TABLE 2.7
Detailed review of 1,4-dioxane groundwater analytical methods

Method Description	Method Numbers	Advantages	Challenges
Extraction Methods—Used to remove/concentrate the target analyte(s) from the environmental matrix			
Purge and trap—conventional (bubbling an inert gas through water to transfer target analytes)	USEPA Method 5030B	– Commonly employed method	– Because of 1,4-dioxane's low volatility, it will not readily transfer from water to the purge gas, resulting in low recoveries

1,4-Dioxane

Method Description	Method Numbers	Advantages	Challenges
Purge and trap—heated (heating and bubbling an inert gas through water to transfer target analytes)	USEPA Method 5030B	– Enhanced extraction of 1,4-dioxane over conventional purge and trap (i.e., added heat helps transfer 1,4-dioxane from water to the purge gas)	– Inconsistent purge efficiency – May not be available at all laboratories
Distillation (increasing temperature to separate target analytes from water sample)	USEPA Method 5031 (azeotropic distillation) USEPA Method 5032 (vacuum distillation)	– Azeotropic distillation is specifically identified as a useful extraction method for 1,4-dioxane	– Not commonly employed by analytical laboratories – Vacuum distillation is not effective due to 1,4-dioxane's miscibility in water
Liquid-liquid extraction (transferring target analytes from the water sample into an organic solvent, like methylene chloride)	USEPA Method 3510C or 3520C	– Commonly employed method – 1,4-Dioxane is transferred into extraction solvent	– Low recoveries if the extraction solvent is concentrated above 101°C; 1,4-dioxane will boil off
Solid phase extraction via cartridge (capture of target analytes on cartridge filter from water sample)	Part of USEPA Method 522	– Loss of 1,4-dioxane during the extraction is minimal	– May be subject to interferences when applied to groundwater (e.g., high levels of suspended solids, total organic carbon, and/or mineral content)
Analytical Methods—Quantifies the target analyte(s)			
VOCs via GC/MS (analyte separation on GC with quantification using MS)	USEPA Method 8260C	– Commonly employed method – Simultaneously quantifies other VOCs	– High levels of chlorinated solvents interfere with 1,4-dioxane quantification – Reporting limits may not be low enough to meet regulatory criteria (typically 25 to 200 µg/L) – Poor recovery during extraction step may bias results

(*Continued*)

46   Emerging Contaminants Handbook

TABLE 2.7 (Continued)

Method Description	Method Numbers	Advantages	Challenges
VOCs via GC/MS with SIM (analyte separation on GC with quantification using MS; looks specifically for 1,4-dioxane)	USEPA Method 8260C-SIM	– Use of SIM increases ability to quantify 1,4-dioxane accurately – Reporting limits typically 0.5 to 2 µg/L	– Less commonly used method that may also cost more per sample – Poor recovery during extraction step may bias results
SVOCs via GC/MS (analyte separation on GC with quantification using MS)	USEPA Method 8270D	– Commonly employed method – Reporting limits typically 2 to 5 µg/L – Not affected by high levels of chlorinated compounds	– Reporting limits may not be low enough to meet regulatory criteria
SVOCs via GC/MS with SIM (analyte separation on GC with quantification using MS; looks specifically for 1,4-dioxane)	USEPA Method 8270D-SIM	– Use of SIM increases ability to quantify 1,4-dioxane accurately – Reporting limits typically 0.2 to 1 µg/L – Not affected by high levels of chlorinated compounds	– Less commonly used method that may also cost more per sample
GC/MS/MS with SIM (analyte separation on GC with quantification using MS/MS; looks specifically for 1,4-dioxane)	USEPA Method 8270E	– Very low reporting limits (e.g., 0.02 µg/L)	– New technology with limited availability
1,4-Dioxane in drinking water via GC/MS with SIM (analyte separation on GC with quantification using MS; looks specifically for 1,4-dioxane)	USEPA Method 522	– Specifically designed to analyze for 1,4-dioxane with optimized quantification – Reporting limits typically <0.4 µg/L for groundwater and <0.2 µg/L for drinking water	– Drinking water method and may not be applicable to groundwater (i.e., interferences and regulatory acceptance) – Higher cost than other analytical methods
Analytical Surrogates—Added to the environmental sample; helps quantify losses during extraction			
Toluene-d8, 4-bromofluorobenzene, and 1,2-dichloroethane-d	USEPA Method 8260C (recommended list)	– Commonly employed surrogates; recommended in the method guidance	– More volatile than 1,4-dioxane and do not behave similarly during extraction (i.e., purge and trap)

1,4-Dioxane

Method Description	Method Numbers	Advantages	Challenges
Phenol-d6, 2-fluorophenol, 2,4,6-tribromophenol, nitrobenzene-d5, 2-fluorobiphenyl, and p-terphenyl-d14	USEPA Method 8270D (recommended list)	– Commonly employed surrogates; recommended in the method guidance	– Less soluble in water than 1,4-dioxane and do not behave similarly during extraction (i.e., liquid-liquid)
1,4-Dioxane-d8 (also referred to as isotopic dilution)	USEPA Method 522 May be used with other analytical methods	– Compound exhibits similar properties during the extraction and analysis; yields better quality assurance/quality control and evaluation of usability of the data	– Not widely applied to other analytical methods; discuss with laboratory before analysis – May be considered a method modification and may not be accepted by all agencies
Analytical Internal Standards—Added to the sample after extraction and before analysis; helps verify correct instrument operation			
Fluorobenzene, chlorobenzene-d5, and 1,4-dichlorobenzene-d4	USEPA Method 8260C (recommended list)	– Commonly employed internal standards; recommended in the method guidance	– Do not emulate the analytical response of 1,4-dioxane, thus do not allow for useful corrections
1,4-Dichlorobenzene-d4, naphthalene-d8, acenaphthene-d10, phenanthrene-d10, chrysene-d12, and perylene-d12	USEPA Method 8270D (recommended list)	– Commonly employed internal standards; recommended in the method guidance	– Do not emulate the analytical response of 1,4-dioxane, thus do not allow for useful corrections
Tetrahydrofuran-d8	USEPA Method 522	– Internal standard exhibits properties similar to 1,4-dioxane during the analysis; yields better quality assurance/quality control and evaluation of usability of the data	– Not widely applied to other analytical methods; discuss with laboratory before analysis – May be considered a method modification and may not be accepted by all agencies

2.5.3 Advanced Investigation Techniques

In addition to direct analysis of 1,4-dioxane in environmental media, there are advanced analyses specifically related to 1,4-dioxane that can shed more light on its behavior in the environment. These include molecular biology tools (MBTs), directed at microorganisms implicated in biodegradation of 1,4-dioxane, and compound specific isotope analysis (CSIA) that can provide other lines of evidence related to intrinsic biodegradation. The Interstate Technology & Regulatory Council (ITRC) produced a guidance document on environmental molecular diagnostics (2013); the reader is referred to this guidance document for background information on both MBTs and CSIA. The application of these tools to gain a better understanding of the fate of 1,4-dioxane in the subsurface is relatively new and is still developing.

MBTs include use of genetic targets that can be applied to deoxyribonucleic acid (DNA) or messenger ribonucleic acid (mRNA). Evaluation of DNA provides practitioners with insight into what microbial populations are present and what their genetic capabilities are. However, the presence of a microorganism does not guarantee that the desired microbial process is being carried out. Evaluation of mRNA provides additional information about whether the microorganisms that are present are actively utilizing their genetic capabilities. While mRNA results also do not guarantee that the desired reactions are occurring (i.e., degradation of 1,4-dioxane), they do get closer to answering that question. Quantitative polymerase chain reaction (qPCR) is technique that quantifies the number of copies of either DNA or mRNA genetic targets in an environmental sample. The genetic targets described here focus on functional steps of the different biodegradation mechanisms that have been identified to date (see section 2.7.1.2 for more detail). These genetic targets have only recently been developed for, or applied to, evaluation of biodegradation of 1,4-dioxane, and it is anticipated that other relevant targets will be developed in the future. Currently, these analyses can be performed by academic laboratories, or a handful of commercial laboratories, such as Microbial Insights (Knoxville, Tennessee). Genetic targets commonly employed, and commercially available, include:

- 1,4-Dioxane/tetrahydrofuran monooxygenase (DXMO/THFMO) and aldehyde dehydrogenase (ALDH): enzymes involved in aerobic metabolism of 1,4-dioxane. These targets have been shown to be correlated to metabolic biodegradation (Gedalanga et al., 2014).
- Soluble methane monooxygenase (sMMO), propane monooxygenase (PPO), ring-hydroxylating toluene monooxygenases (RMO and RDEG), phenol hydroxylase (PHE): enzymes involved in aerobic cometabolism of 1,4-dioxane in the presence of a primary substrate, such as methane, propane, toluene, or phenol, respectively. Some studies have concluded that cometabolic genetic targets, specifically sMMO, are ubiquitous in the environment and do not make for meaningful targets when evaluating biodegradation of 1,4-dioxane (Gedalanga et al., 2016; Sadeghi et al., 2016).

- Other soluble di-iron monooxygenases, such as the group 6 soluble di-iron monooxygenase described in He et al. (2017); this genetic target is commonly referred to as a propane monooxygenase, but differs from the PPO target described above and can be induced in the presence of 1,4-dioxane alone (i.e., without propane).

Results from these sorts of analyses are commonly used as part of a multiple-lines-of-evidence approach in evaluating whether biodegradation of 1,4-dioxane is occurring, or has the potential to occur (see section 2.7.3 for examples in natural attenuation). Alternatively, they can be used to determine whether the desired microbial population is present prior to implementing an active bioremediation treatment strategy or as part of troubleshooting an existing remedy (section 2.7.1.2). It should be noted that benchmarks for these types of data are still being developed, and the applicability of them to understanding field conditions is still being evaluated.

Stable isotope probing (SIP) is another MBT that can be used to better understand the biodegradation potential for 1,4-dioxane in the subsurface. SIP includes a family of techniques that use target constituents enriched with stable isotopes (such as carbon-13 in 1,4-dioxane) to characterize contaminant-specific biodegradation processes. In its simplest form, SIP includes the addition of an isotopically enriched compound into the test system to trace the fate of the target isotope within the microbial system. For example, the carbon-13 in 1,4-dioxane acts as a carbon-atom tracer to track the biotransformation of 1,4-dioxane into mineralized carbon dioxide and/or cellular biomass to identify the fate of 1,4-dioxane under test conditions.

CSIA for 1,4-dioxane is a tool that can identify shifts in the ratio between naturally occurring stable isotopes, such as carbon-12/13 and hydrogen-1/2. 1,4-Dioxane CSIA data are typically used for two purposes: identification of multiple 1,4-dioxane sources based on a difference in isotopic signature; and confirmation of biodegradation of 1,4-dioxane based on an isotopic shift over time or along the plume centerline. The bond strength between lighter isotopes (e.g., carbon-12 bonded to carbon-12) is slightly less than that created by heavier isotopes (e.g., carbon-12 bonded to carbon-13). Because of this difference, microorganisms tend to break the bonds between lighter isotopes before those of heavier isotopes. For example, carbon-12 bonds will be broken before carbon-13 bonds and 1,4-dioxane will become enriched in carbon-13. Physical transformations (e.g., dilution) do not create this same enrichment. That is why CSIA can be a powerful tool for identifying when biodegradation is occurring (ITRC, 2013). CSIA differs from SIP in that it evaluates the naturally occurring presence of target isotopes (e.g., carbon-13), rather than using a compound that has been anthropogenically enriched with a specific isotope (e.g., isotopically enriched 1,4-dioxane).

1,4-Dioxane CSIA analytical capabilities are evolving, as well as the ability to apply these types of data in a meaningful way. 1,4-Dioxane CSIA is available from academic laboratories, as well as from a few commercial laboratories, such as Pace Analytical Services (specifically the Pittsburgh, Pennsylvania, location). Some laboratories may only have the capability to analyze carbon isotopes, while others have the capability to analyze both carbon and hydrogen isotopes. Likewise, different analytical laboratories have different detection limits associated with CSIA, some of which may be too high for application at sites with low groundwater concentrations. The capabilities of the laboratory should be fully vetted before selection.

Research related to the isotopic signature of 1,4-dioxane is ongoing. Wang (2016) quantified the carbon and hydrogen isotopic ratios in seven stock solutions. Carbon isotope ratios ranged between −29‰ (per mil) and −34‰, and hydrogen isotope ratios ranged from −90‰ to −150‰. Pornwongthong et al. (2011) quantified a site-specific enrichment factor for carbon (−1.73±0.14‰) during metabolic biodegradation of 1,4-dioxane at the bench-scale. Likewise, Bennett et al. (2018) quantified enrichment factors for carbon and hydrogen for pure cultures grown on propane (carbon = −2.7±0.3‰; hydrogen = −21±2‰), isobutane (carbon = −2.5±0.3‰; hydrogen = −28±6‰), and tetrahydrofuran (THF; carbon = −4.7±0.9‰; hydrogen = −147±22‰). Field applications of CSIA have begun to provide benchmarks for evaluating biodegradation, and they similarly suggest that the carbon ratio does not change much under conditions where biodegradation is inferred by other lines of evidence: 1.6‰ by Sadeghi et al. (2016); 1.5 to 2‰ by Gedalanga et al. (2016); and 3‰ by Bell and Heintz (2016). The last example of this is detailed in the case study in section 2.7.3. While carbon isotope CSIA does not appear to provide a decisive line of evidence for 1,4-dioxane biodegradation, the hydrogen isotope ratio may be a more useful tool for evaluating biodegradation of 1,4-dioxane in the field. Sadeghi et al. (2016) reported a hydrogen CSIA value of −76.34‰ in an area of active 1,4-dioxane bioremediation and a value of −103.68‰ in an area that did not exhibit biodegradation of 1,4-dioxane, a difference of approximately 30‰ compared to the approximately 2‰ difference seen in carbon isotopes. While these investigations help provide benchmarks by which to compare future field data, additional research is needed to better identify what isotopes change meaningfully, what isotopic shifts are indicative of biodegradation (e.g., 2‰ for carbon or 30‰ for hydrogen), and if different biodegradation mechanisms (see section 2.7.1.2 for more details) result in different isotopic signatures.

2.6 SOIL TREATMENT

Based on the chemical and physical properties of 1,4-dioxane (Table 2.2 and Appendix C), it most commonly requires treatment in groundwater, rather than in soil. However, 1,4-dioxane may be present at elevated concentrations in soil and soil vapor in some cases, perhaps in the vicinity of the initial release. Likewise, soil/soil vapor treatment may be required for colocated contaminants, like chlorinated solvents, and the effectiveness of treating 1,4-dioxane needs to be considered. Soil vapor extraction is a commonly employed vadose zone treatment technology for chlorinated solvents and other VOCs. Conventional soil vapor extraction can remove some 1,4-dioxane, but substantial residual is usually left behind, because it preferentially partitions into the soil moisture rather than the vapor phase. As an aside, the tendency for 1,4-dioxane to partition into soil moisture in the vadose zone is also why it has not emerged as a significant vapor intrusion concern, except in situations where soil moisture content is low or source mass is large. To address 1,4-dioxane in the vadose zone, the concept of extreme soil vapor extraction (XSVE) was developed through Strategic Environmental Research and Development Program (SERDP) project ER-201326. XSVE employs a combination of increased air flow and temperature increases to facilitate removal of 1,4-dioxane from the vadose zone by decreasing soil moisture content and increasing the pore volume exchange rate.

As described in a presentation on SERDP project ER-201326 at the March 2016 Emerging Contaminants Summit held in Westminster, Colorado (Hinchee, 2016), hot air (120 °C) was injected at approximately 100 standard cubic feet per minute into each of four injection wells and extracted at approximately 100 standard cubic feet per minute from one well—estimated as 1,000 pore volume exchanges per day. The subsurface vadose zone temperatures were raised 30 to 90 °C. The results of this project suggested that both the increase in temperature and resultant decrease in soil moisture and the elevated pore volume exchange rate facilitated 1,4-dioxane removal. The relative contribution of each of these mechanisms is an ongoing area of study.

2.7 GROUNDWATER TREATMENT

Historically, groundwater remediation activities did not include investigation and treatment of 1,4-dioxane. However, as more regulatory agencies are requiring these actions, practitioners are exploring application of existing treatment technologies to also address 1,4-dioxane. Early on in this process, few treatment technologies were considered appropriate. As more research has been conducted and more field-scale demonstrations completed, there are a variety of treatment technologies identified as viable options to treat 1,4-dioxane in groundwater. The purpose of this section is to provide information on the most relevant, applied, and/or developing treatment technologies that are specific to 1,4-dioxane treatment in groundwater. The considerations and remedial design elements associated with many existing treatment technologies that are specific to 1,4-dioxane are highlighted.

2.7.1 IN SITU TREATMENT

In situ treatment technologies are attractive, as they have evolved to be solutions that are relatively faster and cheaper than their predecessors. They are less-invasive technologies capable of achieving stringent cleanup objectives. They can also be sustainable technologies that complement the natural environment and leverage nature's capacity to degrade contaminants. In situ treatment technologies are being applied more and more for 1,4-dioxane and associated chlorinated solvents. In particular, in situ chemical oxidation (ISCO) and in situ bioremediation are attractive options for 1,4-dioxane.

2.7.1.1 In Situ Chemical Oxidation

ISCO includes injection of strong chemical oxidants into the subsurface to facilitate direct oxidation of environmental contaminants, or generate reactive radicals that destroy those contaminants. Common ISCO oxidants include permanganate, hydrogen peroxide, ozone, persulfate, percarbonate, and calcium peroxide. There are other references that go into great detail on the various ISCO chemistries (Suthersan et al., 2017), so this discussion will focus on high-level observations related to 1,4-dioxane. In general, ISCO strategies that generate reactive radicals (i.e., catalyzed hydrogen peroxide or activated persulfate) are more effective at treating 1,4-dioxane than those that rely on direct oxidation (i.e., permanganate). In fact, some oxidants are not theoretically able to degrade 1,4-dioxane, because their oxidation potential is below the threshold needed (2 electron volts). These include hydrogen peroxide alone, permanganate, chlorine, and oxygen (Mohr et al., 2010). However, other sources have reported various levels of success with these oxidants (e.g., hydrogen peroxide with mineral catalysis from Siegrist et al. [2011] and permanganate from Waldemer and Tratnyek [2006]).

ISCO is an attractive in situ treatment technology for 1,4-dioxane because it will destroy 1,4-dioxane and can also address comingled compounds, such as chlorinated solvents. Selection of the appropriate chemical oxidant is crucial to the success of an ISCO strategy. Table 2.8 provides a comparison of the effectiveness of different chemical oxidants and activation strategies for 1,4-dioxane and commonly colocated compounds, such as chlorinated ethenes, chlorinated ethanes, and chlorinated methanes. Selection of an oxidant and activation method may depend on the chemical types commingled with the 1,4-dioxane. While many oxidants are effective at destroying 1,4-dioxane, some have limited effectiveness for other compounds (e.g., chlorinated ethanes), so all groundwater constituents should be considered when selecting an oxidant. As shown in Table 2.8, hydrogen peroxide (including Fenton's chemistry), persulfate, and ozone/peroxide rise to the top of chemical oxidants that effectively treat 1,4-dioxane, chlorinated ethenes, and hlorinated ethanes simultaneously.

While detailed ISCO design criteria are beyond the scope of this work, there are a few general items that are important to keep in mind when considering ISCO (more detailed information can be found in Suthersan et al. [2017]).

- ISCO is a contact sport, meaning that success is reliant on delivery of the chemical oxidant so that it achieves complete contact with the targeted contaminants in the subsurface. Effective contact will limit the potential for contaminant concentration rebounding and the need for additional injections.
- Many naturally occurring and humanmade compounds are present in the subsurface that will scavenge the chemical oxidant (e.g., soil matrix components, chloride, other contaminants). This ambient oxidant demand needs to be considered in the design.
- ISCO can cause temporary and spatially limited changes in the redox and pH conditions of the subsurface that can create secondary water quality concerns within the treatment area. This can be anticipated and managed as part of the application.
- Bench-scale testing is highly recommended for evaluating the efficacy of chemical oxidation for a given site, as well as for identifying optimized oxidant dosing.

TABLE 2.8
Summary of chemical oxidant efficacy on 1,4-dioxane and common chlorinated compounds

Oxidant and Activation Technique	Permanganate	Hydrogen Peroxide				Persulfate					Ozone		
		Chelated Iron	Ambient Mineral Catalysis	Iron/Acid	Alkaline pH	Thermal	Iron	Chelated Iron	None	Peroxide	Ozone Only	Ambient Iron	With Peroxide
1,4-Dioxane	F	E	E	E	E	E	E	E	E	E	F	G	E
Chloroethenes													
Tetrachloroethene	E	E	E	E	E	E	E	E	E	E	E	E	E
Trichloroethene	E	E	E	E	E	E	E	E	E	E	E	E	E
Dichloroethenes	E	E	E	E	E	E	E	E	E	E	E	E	E
Vinyl chloride	E	E	E	E	E	E	E	E	E	E	E	E	E
Chloroethanes													
Tetrachloroethanes	NR	F	F	P	F	F	F	F	F	F	F	F	F
Trichloroethanes	NR	G	G	P	G	G	F	F	P	G	G	G	G
Dichloroethanes	NR	G	G	F	G	G	F	F	P	G	G	G	G
Chloroethane	NR	G	G	F	G	G	F	F	P	G	G	G	G
Chloromethanes													
Carbon tetrachloride	NR	E	G	P	E	F	NR	NR	NR	E	NR	NR	NR
Chloroform	NR	G	G	P	G	NR	NR	NR	NR	G	P	P	P
Dichloromethane	NR	G	F	P	F	NR	NR	NR	NR	E	P	P	P
Methylene chloride	NR	G	F	P	G	F	P	P	P	G	P	P	P

E = Excellent; G = Good; F = Fair; P = Poor; NR = Not recommended
Source: Adapted from Siegrist et al. (2011) and USEPA (2006).

- A rigorous health and safety plan rooted in practical experience and a behavior-based health and safety system are cornerstones to mitigating the risk of injuries associated with ISCO implementation.
- Some states have Homeland Security limitations on the storage mass of particular oxidants. This is not federally mandated and should be confirmed with the governing fire department in the area where ISCO is to be implemented.

2.7.1.2 Bioremediation

1,4-Dioxane was historically believed to be recalcitrant and not amenable to biodegradation (Alexander, 1973; Fincher and Payne, 1962); now it has garnered enough attention to warrant review articles on the topic (Zhang et al., 2017). In the 1990s and 2000s, studies focused on the use of pure cultures to understand the degradation rates and the degradation pathways (Parales et al., 1994; Mahendra and Alvarez-Cohen, 2005; Burback and Perry, 1993; Bernhardt and Diekmann, 1991; Kim et al., 2009). Since that time, microcosm studies have also been used to assess the biodegradation rates using field groundwater and soil samples collected from a wide range of subsurface environments with and without biostimulation or bioaugmentation (Zenker et al., 2000; Li et al., 2010; Li et al., 2014; Li et al., 2015). It has become apparent that 1,4-dioxane can be biodegraded under a variety of conditions and via different mechanisms (i.e., metabolic and cometabolic). Generally, 1,4-dioxane biodegradation has not been observed under anaerobic conditions (Arve, 2015; Steffan et al., 2007), with only limited evidence that 1,4-dioxane may be biodegraded anaerobically via an iron-mediated pathway (Shen et al., 2008). However, there is evidence that 1,4-dioxane is biodegraded under aerobic conditions by a range of bacteria who produce monooxygenase enzymes (Mahendra and Alvarez-Cohen, 2006). In the case of metabolic biodegradation, 1,4-dioxane is the primary substrate and specific monooxygenase enzymes target this substrate for energy and growth. These monooxygenase enzymes couple the oxidation of 1,4-dioxane with the reduction of oxygen; therefore, effective aerobic 1,4-dioxane biodegradation is dependent on oxygen availability. In the case of cometabolic 1,4-dioxane biodegradation, the microorganisms support their growth and energy demands using a different primary substrate, and the monooxygenase enzymes degrade 1,4-dioxane in a side reaction that does not provide additional energy. Cometabolic biodegradation occurs as a result of the low substrate specificity of many monooxygenase enzymes and is an important process governing biodegradation of many common groundwater constituents (Hazen, 2010). Examples of primary substrates linked to 1,4-dioxane cometabolism include methane, propane, toluene, phenol, THF (Mahendra and Alvarez-Cohen, 2006), isobutane (Rolston et al., 2017), and ethane (Hatzinger et al., 2017). The degradation pathway is expected to be similar in metabolic or cometabolic biodegradation, with HEAA as a notable intermediate (pathway included in Gedalanga et al., 2016).

Methane is the most abundant alkane in the environment and is a commonly collected groundwater parameter. Therefore, methane monooxygenase enzymes (e.g., sMMO) are particularly relevant to cometabolism because the primary substrates (methane and oxygen) are present in many environments, the enzymes are strong oxidizers, and the enzymes have broad cometabolic substrate specificity (Hazen, 2010; Semrau, 2011). As a result of relatively low apparent half-saturation constants of sMMO for methane (Dunfield et al., 1999), these enzymes have the potential to be active even at low methane concentrations. As of the date of this publication, the importance of the methane-mediated cometabolic biodegradation pathway is not well understood. Some laboratories have reported no biodegradation of 1,4-dioxane under a variety of rigorously tested methane biostimulation conditions (Hatzinger et al., 2017). However, observations of field data (Sadeghi et al., 2016) and anecdotal evidence from bench-scale and pilot-scale testing suggest otherwise.

This is of particular interest because of the potential synergy between the reducing conditions typically created for bioremediation of chlorinated solvents and the methane-mediated cometabolic biodegradation pathway for 1,4-dioxane. In situ bioremediation of chlorinated solvents typically includes addition of a carbon substrate into the subsurface. This carbon substrate is then fermented, and hydrogen is produced for reductive dechlorination. At the same time, various terminal electron acceptors (e.g., oxygen, nitrate, sulfate) are depleted in the subsurface, creating conditions conducive to both reductive dechlorination and methanogenesis. Methane is commonly produced coincident with chlorinated solvent reduction. This methane may be transported to the downgradient edge of the plume where aquifer geochemical conditions are transitioning to be more oxic, or air could be injected for active remediation of the 1,4-dioxane. Under this scenario, methane, oxygen, and 1,4-dioxane are all present and conditions are conducive to cometabolic biodegradation of 1,4-dioxane.

Application of a bioremediation approach for treatment of 1,4-dioxane at the field scale is gaining acceptance and is being considered more frequently as a viable remedial alternative. Metabolic bioremediation approaches may be considered appropriate when 1,4-dioxane concentrations are high enough to sustain growth. Cometabolic approaches may be more fitting when 1,4-dioxane concentrations are relatively lower or need to meet low cleanup criteria. As with bioremediation of other constituents, there must be an adequate understanding of the subsurface biogeochemistry, the limiting parameters (e.g., oxygen), potential inhibitors [e.g., chlorinated compounds (Zhang et al., 2016)], high concentrations of metals (Pornwongthong et al., 2014), and the most appropriate biodegradation pathway to pursue (e.g., metabolic or cometabolic). Table 2.9 provides examples of what these approaches may include. Similar to other in situ bioremediation approaches, 1,4-dioxane bioremediation may be implemented as biostimulation or may include bioaugmentation. The necessary substrates are delivered to the subsurface via means that will result in distribution in the site-specific lithology. The following case study (also described in Bell et al., 2016) is an example of application of in situ bioremediation for 1,4-dioxane. As of the time of this writing, the propanotrophic cometabolic bioaugmentation culture of *Rhodococcus ruber* ENV425 was available commercially, with the commercial availability of the metabolically degrading *Pseudonocardia dioxanivorans* CB1190 pending.

TABLE 2.9
Considerations for bioremediation of 1,4-dioxane

	Biostimulation Approach	Bioaugmentation Approach
Metabolic Biodegradation	*Consider when:* 1,4-Dioxane concentrations are high (i.e., >1,000 µg/L) Inhibitors (i.e., 11DCE or metals) are not present, or can be controlled Lines of evidence indicate the presence of microorganisms that can metabolically degrade 1,4-dioxane (e.g., DXMO/ALDH) *Design elements:* Deliver adequate oxygen (e.g., >2 mg/L) Prepare for degradation stalling at lower 1,4-dioxane concentrations (e.g., 200 µg/L)	*Consider when:* 1,4-Dioxane concentrations are high (i.e., >1,000 µg/L) Inhibitors (i.e., 11DCE or metals) are not present, or can be controlled Lines of evidence do not indicate the presence of microorganisms that can metabolically degrade 1,4-dioxane Addition of bioaugmentation culture is feasible and economical *Design elements:* Deliver adequate oxygen (e.g., >2 mg/L) Prepare for degradation stalling at lower 1,4-dioxane concentrations (e.g., 200 µg/L) Distribution of bioaugmentation culture
Cometabolic Biodegradation	*Consider when:* 1,4-Dioxane concentrations are in the 100s or 10s µg/L Lines of evidence indicate the presence of microorganisms that can cometabolically degrade 1,4-dioxane (e.g., propanotrophs) *Design elements:* Deliver adequate oxygen (e.g., >2 mg/L) Deliver adequate primary substrate (e.g., >100 µg/L propane) Provide additional nutrients that might be limiting (e.g., nitrogen or phosphorous)	*Consider when:* 1,4-Dioxane concentrations are in the 100s or 10s µg/L Addition of bioaugmentation culture is feasible and economical *Design elements:* Deliver adequate oxygen (e.g., >2 mg/L) Deliver adequate primary substrate (e.g., >100 µg/L propane) Provide additional nutrients that might be limiting (e.g., nitrogen or phosphorous) Distribution of bioaugmentation culture

Note: The operational parameters suggested in this table are based on the author's personal experience and may differ from one application to another.

Case Study: In Situ Biodegradation of 1,4-Dioxane via Propane Biosparging

Chlorinated solvent use at a site on Vandenberg Air Force Base resulted in a comingled plume of chlorinated solvents and 1,4-dioxane. Starting circa 2013, several bioremediation demonstration projects were conducted, all focused on using an in situ propane biosparge injection system to facilitate cometabolic biodegradation of 1,4-dioxane. While the biodegradation mechanism was proven in bench-scale testing conducted prior to the first demonstration project, early field results were less straightforward in revealing the dominant mechanism for decreases in 1,4-dioxane concentrations (Lippencott et al., 2015). Therefore, SIP with carbon-13-labeled 1,4-dioxane was employed to better articulate the transformation mechanism.

The pilot-scale propane and air injection system included an air compressor, compressed propane storage tank, an explosion-proof mixing system, and a lower explosive limit (LEL) meter to measure the concentration of propane mixed into the air line (Figure 2.4). Several safety measures were included to relay key operational information (e.g., LEL and pressures) to a

FIGURE 2.4 Propane biosparge mixing system: Photograph of the propane biosparge mixing system at Vandenberg Air Force Base in California.

control panel that would shut down operation before hazardous conditions were reached. The mixed gas line fed a single sparge point in one of the pilot test areas. A propanotrophic bioaugmentation culture was injected into the sparge point, and dissolved diammonium phosphate was added as a nutrient source to facilitate growth of the propanotrophic bacteria. The propane biosparging system operated for more than two months, and concentrations of 1,4-dioxane in groundwater decreased approximately 45 to 85 percent in that time (Figure 2.5).

BioTrap™ samplers amended with carbon-13-labeled 1,4-dioxane were deployed in monitoring wells near the sparge point. The carbon-13 acted as a tracer to identify how the 1,4-dioxane was transformed within the subsurface microbial system (see section 2.5.3 for more detail). The SIP testing quantified transformation of 1,4-dioxane into both carbon dioxide and microbial biomass (Figure 2.6). This confirmed the link between 1,4-dioxane concentration reductions and biodegradation. This in situ bioremediation approach will go full scale in 2019.

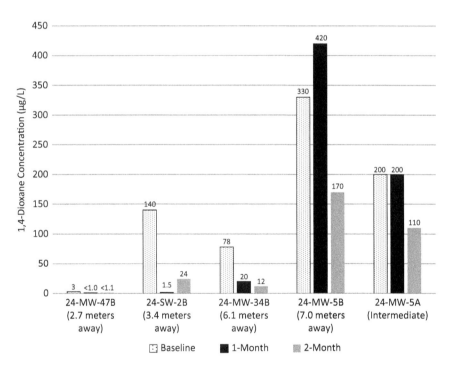

FIGURE 2.5 1,4-Dioxane results from propane biosparge treatment at Vandenberg Air Force Base in California: Monitoring well 24-MW-5A is located approximately 4.9 meters from the sparge point, but is screened within a different portion of the aquifer (marked as "Intermediate"). Used with permission from Bell et al. (2016).

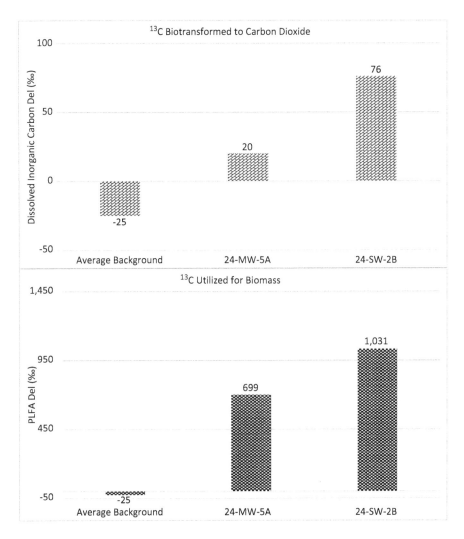

FIGURE 2.6 Stable isotope probing results from propane biosparge treatment at Vandenberg Air Force Base in California: The results indicate transformation of 1,4-dioxane into carbon dioxide and microbial biomass and confirmed the link between 1,4-dioxane concentration reductions and biodegradation. Used with permission from Bell et al. (2016).

2.7.1.3 Phytoremediation

Phytoremediation relies on the processes of plants to naturally remove environmental contaminants from the subsurface. The mechanisms by which plants do this include biodegradation in the rhizosphere (rhizodegradation); immobilization of the contaminants via plant-produced compounds (phytostabilization); uptake and accumulation (phytoextraction); adsorption or precipitation onto plant roots (rhizofiltration);

transpiration through plant leaves (photovolatilization); metabolism and destruction within the plant (phytodegradation); and/or hydraulic control of groundwater movement (USEPA, 1999a). Based on research and examples of field application, phytovolatilization and hydraulic control are thought to be the most relevant mechanisms for phytoremediation of 1,4-dioxane. Once transferred to the atmosphere, the half-life of 1,4-dioxane is estimated to be on the order of 8.8 to 22.4 hours (Mohr et al., 2010). Several studies have demonstrated that phytoremediation may be a viable remedial option, under the appropriate site conditions (e.g., shallow depth to groundwater; growing climate). Gestler et al. (2017) described several phytoremediation applications for treatment of 1,4-dioxane using the TreeWell® System. This patented system targets specific groundwater by directing root growth downward, uses capillary action to draw deeper groundwater up to the treatment depth, and allows for amendment addition from the ground surface. The case studies shared included installation of approximately 150 to 230 TreeWell® Systems to treatment depths of approximately 5 to 17 meters below ground surface (bgs)—approximately 15 to 50 feet bgs. The types of trees planted included pine, sycamore, willow, cypress, and popular. Results suggested that both hydraulic control was attained, as well as 1,4-dioxane concentration decreases in groundwater, after two growing seasons. Bench-scale tests with concentrations of 1,4-dioxane of greater than 1,000 milligrams per liter (mg/L) did not appear to have an adverse effect on tree growth (Gestler et al., 2017).

2.7.1.4 Thermal Treatment

In situ thermal treatment is a remedial technology that can be considered appropriate for both soil and groundwater. In general, thermal remediation may be considered for hard-to-treat areas: source zones, areas with non-aqueous phase liquids (both light and dense), low permeability lithologies, and/or fractured bedrock. As the subsurface temperature increases during thermal treatment, contaminant removal mechanisms include evaporation (i.e., expediting transfer from the water to the gas phase at the water surface) and boiling (i.e., complete transfer from the water to gas phase throughout the liquid). In situ thermal remediation methods increase the subsurface temperature in different ways: electrical resistance heating delivers an electrical current to the subsurface that generates heat; steam-enhanced extraction includes the injection of steam into the subsurface to raise the temperature; and thermal conduction heating includes installation of heaters into the subsurface.

The boiling point of pure 1,4-dioxane is 101 °C (Appendix C) which is close to, but slightly higher, than that of water. This would suggest that for thermal remediation to be effective at removing 1,4-dioxane from the subsurface, the temperature would need to be raised above the boiling point of water, and the water would need to be volatilized prior to removal of 1,4-dioxane. However, 1,4-dioxane in water has an azeotropic decrease in boiling point, down to a minimum value of 87.59 °C when a maximum of 48.5 mole percent (82 percent by mass) of 1,4-dioxane is present (Schneider and Lynch, 1943). Additionally, there is a 50-fold increase in the Henry's law constant between the standard testing temperature (i.e., 25 °C) and the boiling point of water (i.e., 100 °C [Stantec Corporation—Treatability Testing Services, 2013]). Both of these attributes may facilitate removal of 1,4-dioxane from water at temperatures below the boiling point of water.

1,4-Dioxane

Oberle et al. (2015) conducted a bench-scale test to evaluate the removal of 1,4-dioxane (5.4 milligrams per kilogram [mg/kg]) during heating from soil containing 21.9 percent moisture. Results indicated removal of 20 percent of the soil moisture corresponded to 87.6 percent removal of 1,4-dioxane (reduced to 0.67 mg/kg), and that after removal of 45 percent of the soil moisture, 1,4-dioxane could not be detected above the laboratory reporting limit of 0.58 mg/kg. At the field scale (Oberle et al., 2015), electrical resistance heating was used to treat a site with TCE, TCA, and 1,4-dioxane present. Prior to treatment, groundwater concentrations of 1,4-dioxane ranged from 1 to 90 mg/L. After 186 days of operation, 1,4-dioxane concentrations were reduced to less than 50 µg/L, a 99.8 percent removal. At another site, 1,4-dioxane concentrations we reduced from 140 µg/L to 1.4 µg/L, thus indicating that thermal treatment may be applicable at both high and low concentrations of 1,4-dioxane. One of the common considerations associated with thermal treatment is the cost. Oberle et al. (2015) indicated that use of electrical resistance heating to treat 1,4-dioxane costs from $150 to $300 per cubic yard, compared to an estimated $100 to $250 per cubic yard for a typical electrical resistance heating site, thus reflecting the added challenge with removal of 1,4-dioxane by this technology.

2.7.2 Ex Situ Treatment and Dynamic Groundwater Recirculation

While ex situ treatment is typically the term used to refer to the aboveground treatment process(es) used to reduce concentrations of contaminants in extracted groundwater, it is important to consider the subsurface portion of the remedial approach as well. In the classic P&T approach, groundwater is extracted from the subsurface from any number of extraction wells and combined prior to treatment and/or discharge. Many sites have operated P&T strategies for years—or decades—with lingering concentrations of compounds still present in the subsurface due to back-diffusion of contaminants from lower-permeability aquifer materials, as described in Payne et al. (2008).

The concept of dynamic groundwater recirculation (DGR) builds on the concepts associated with P&T. DGR includes extraction of groundwater from areas with the highest chemical concentrations/mass flux. Extracted groundwater is treated aboveground to remove chemicals; it is then strategically reinjected at both the periphery of the plume, as well as the plume interior. This approach manipulates groundwater elevations to create gradients across less permeable zones and establish flow through zones where contaminant mass is stored. This approach maximizes contaminant mass removal and shortens remedial time frames when compares to conventional P&T systems. System performance is further enhanced through dynamically varying pumping rates and locations in response to groundwater concentrations in extraction wells to maximize contaminant mass removal rates, while maintaining hydraulic control of the plume. Many historical P&T systems can be converted to DGR systems, making it an attractive option for historical chlorinated solvent plumes that are now also 1,4-dioxane plumes (see section 2.2 for more detail on this scenario). Additionally, DGR can be combined with in situ treatment approaches by amending the reinjection stream with chemical oxidants or biological substrates. Specific design considerations are discussed in more detail in Suthersan et al. (2017).

While the subsurface hydraulics associated with either a conventional P&T system or an enhanced DGR system directly influence the ability to achieve groundwater cleanup criteria, the ex situ treatment approach directly influences the ability to meet cleanup requirements and project economics. Depending on the discharge requirements for a P&T system, it may not be necessary to perform any 1,4-dioxane treatment. However, if a DGR approach is implemented, it is critical to remove 1,4-dioxane (and other constituents) from the water stream prior to reinjection. Ex situ treatment technologies relevant to 1,4-dioxane are detailed in section 2.9.

2.7.3 Natural Attenuation

While not specifically listed as an in situ treatment technology, natural attenuation is an attractive management strategy for many of the same reasons. Natural attenuation is the reduction in mass or concentration of constituents in groundwater over time and/or distance from source areas due to naturally occurring processes. Attenuation processes include nondestructive physical processes (e.g., dilution, volatilization) and destructive chemical and biological processes. Biodegradation is often a favored natural attenuation process, particularly when it results in the transformation of potentially harmful constituents to innocuous end products. For many common groundwater contaminants, the success of this process can be predicted and evaluated based on analysis of geochemical, microbiological, and/or isotopic parameters. Natural attenuation is an attractive remediation alternative because it is often a low-cost and highly effective technology for mitigating groundwater impacts. Based on its chemical and physical properties, as well as the historical understanding that 1,4-dioxane was not readily biodegraded, natural attenuation was not typically considered for 1,4-dioxane. Concurrent with the relatively recent advances in the understanding of 1,4-dioxane biodegradation (see section 2.7.1.2), so too has come a better understanding of 1,4-dioxane natural attenuation.

1,4-Dioxane natural attenuation potential in the field was first documented by Chiang et al. (2008) based on an industrial 400-acre site using numerical modeling to simulate and predict 1,4-dioxane plume behavior under natural conditions. Further field-based evidence of 1,4-dioxane natural attenuation has been documented, and environmental molecular diagnostic tools (including MBTs and CSIA) have been applied (Chiang et al., 2012; Gedalanga et al., 2014; Wang, 2016). Additionally, 1,4-dioxane attenuation rates based on large datasets have been published (Adamson et al., 2015).

Demonstration of natural attenuation of organic compounds, with biodegradation, is commonly presented as a tiered, lines-of-evidence approach (see Chapter 4 for natural attenuation of inorganic compounds). As described in the USEPA document (1999b), this approach includes:

- Tier 1: Historical groundwater data that demonstrate a clear and meaningful trend of decreasing contaminant mass and/or concentration over time. This is typically referred to as a plume stability evaluation to demonstrate that the groundwater plume is stable or decreasing. Methods for evaluating plume stability may include statistical analyses (e.g., linear regression, Mann-Kendall/Sen's slope), or plume contour maps over time. In all cases, data sets over an adequate period of time are needed. This is an example of when a proactive management strategy (see section 2.2) can readily provide the data needed.

- Tier 2: Hydrogeological and geochemical data that can be used to demonstrate, indirectly, the type of natural attenuation processes occurring. Many times, this focuses on a geochemical evaluation of soil and/or groundwater to determine if conditions are conducive to biodegradation. This typically includes quantification of parameters such as dissolved oxygen, pH, nitrate, iron, manganese, and sulfate. For 1,4-dioxane, this evaluation should also include primary substrates that could contribute to cometabolic biodegradation, specifically dissolved gases like methane, ethane, and propane.
- Tier 3: Data from field-scale or bench-scale studies that directly demonstrate the occurrence of a particular attenuation mechanism. This may be confirmation of biodegradation through laboratory microcosms, MBTs, or CSIA. In some cases, the Tier 3 evaluation may not be needed. However, it can provide powerful information related to the pathway associated with biodegradation of 1,4-dioxane. Microcosms using site soil and groundwater can be monitored in a controlled environment to verify that biodegradation is occurring and better understand what parameters influence it. MBTs (see section 2.5.3) can provide information related to the identification and activity of the indigenous microbial population and compare that information to known 1,4-dioxane biodegradation pathways. CSIA (see section 2.5.3) can identify if there is an isotopic shift in 1,4-dioxane composition, confirming microbial influence on the compound. The Tier 3 lines of evidence are still evolving with better evaluation tools and understanding of results on the horizon.

The following case study describes the application of these lines of evidence as part of evaluation of natural attenuation of 1,4-dioxane.

Case Study: 1,4-Dioxane Natural Attenuation Evaluation

Due to historical operations, chlorinated solvents and 1,4-dioxane were partially colocated in groundwater over an approximately 158,000-square-meter (39-acre) plume, in both shallow and lower saturated units. Groundwater concentration trends and geochemical data supported natural attenuation of the chlorinated solvents. Likewise, initial data suggested natural attenuation of 1,4-dioxane may also be occurring. Advanced analytical techniques, including qPCR and CSIA, were used to provide additional lines of evidence in support of biodegradation of 1,4-dioxane (see section 2.5.3).

A stepwise approach was employed to evaluate the potential for, and activity related to, biodegradation of 1,4-dioxane. First, chemical concentrations were quantified alongside various geochemical parameters, such as dissolved oxygen, nitrate, iron, sulfate, and pH. Second, an expanded list of dissolved gases was quantified, including methane and propane. The results of these analyses provided insight into monitoring locations that had appropriate conditions to support cometabolic biodegradation of 1,4-dioxane (i.e., presence of oxygen with methane or propane as primary substrates). Next, qPCR targets were analyzed to characterize the microbial population under existing site conditions. These included DNA and mRNA analyses for two cometabolic biodegradation targets: sMMO and PPO. Lastly, carbon-13 CSIA of 1,4-dioxane was employed at a subset of monitoring locations to corroborate the qPCR results.

The dissolved gas analyses yielded elevated concentrations of methane (>1,000 µg/L) in the shallower plume, and moderate concentrations (100 µg/L) in the lower plume. Dissolved oxygen was lower in the shallower plume (0.2 mg/L) and higher in the lower plume (>1 mg/L). Propane was not detected above the laboratory reporting limit at the sample locations, shallow or lower. DNA-based analyses indicated potential for both sMMO and PPO production by indigenous microorganisms, with greater potential in the shallower plume. mRNA-based analyses identified sMMO production at most locations, but not PPO production (as expected based on the dissolved gas analyses). These results may provide an additional line of evidence in support of a natural attenuation approach, particularly within the shallow unit, and suggest feasibility in implementing a bioremediation approach in both shallow and lower units, using either methane or propane as the cometabolic substrate. When considering these lines of evidence, it is important to note that some studies have concluded cometabolic genetic targets, specifically sMMO, are ubiquitous in the environment and do not make for meaningful targets when evaluating biodegradation of 1,4-dioxane (Gedalanga et al., 2016; Sadeghi et al., 2016).

CSIA can be a powerful tool to understand the influence of biodegradation on groundwater concentrations (see section 2.5.3). As of the time of this writing, the use of CSIA for 1,4-dioxane in evaluating environmental samples was still maturing. Analytical advancements and additional research are needed to better understand the isotopic signature of 1,4-dioxane at the time it is released to the environment (i.e., stock signature) as well as the influence of chemical and biological degradation on the isotopic signature. With those considerations in mind, it is notable that the CSIA results fell into two general categories: approximately −33‰ and approximately −30‰. CSIA results compare the carbon-13 content in the sample to an international standard. As biodegradation occurs, a sample will have a less negative (i.e., closer to zero) carbon-13 CSIA result. Therefore, the more negative group of values (i.e., approximately −33‰) are less enriched in carbon-13 and more likely to be similar to the stock signature; while, the more positive group of values (i.e., approximately −30‰) are more enriched in carbon-13 and more likely to be influenced by biodegradation mechanisms. Two locations included in the dissolved gas and qPCR analyses were also subject to CSIA. Results from these locations were enriched in carbon-13 (i.e., −31.30‰ and −30.90‰), further supporting the contribution of biodegradation as part of a natural attenuation approach.

2.8 DRINKING WATER AND WASTEWATER TREATMENT

The primary objective of drinking water treatment is to remove substances that could result in acute human sickness or create unpalatable water. Therefore, drinking water treatment unit processes typically focus on removal of suspended solids, bacteria, algae, viruses, fungi, and minerals such as iron and manganese. A review of the conventional water treatment process (i.e., coagulation, flocculation, settling, sand filtration, and disinfection) reveals a limited number of unit processes expected to remove 1,4-dioxane. Because of 1,4-dioxane's low sorption tendency, it is not expected to be readily removed with suspended solids. Likewise, the disinfection process would not be expected to chemically destroy 1,4-dioxane. That said, the use of advanced unit processes in drinking water treatment, some of which are described in section 2.9, could provide beneficial treatment of 1,4-dioxane.

The same thing holds true to conventional wastewater treatment. Conventional wastewater treatment plants that include separation, settling, aerobic/anaerobic biological treatment, clarification, and disinfection may not have many opportunities for destruction or removal of 1,4-dioxane. Again, due to its chemical characteristics, 1,4-dioxane is not expected to be removed during the physical treatment processes. The potential for destruction of 1,4-dioxane in the aerobic/anaerobic biological treatment processes is not well studied. Only about a 50 percent reduction in 1,4-dioxane has been observed from conventional biological wastewater treatment processes in the lab (Lee et al., 2014) and the field (Zenker et al., 2003). Further consideration of aerobic biological treatment in wastewater treatment plants suggests that unless there are indigenous 1,4-dioxane-degraders, or the presence of a substrate to facilitate cometabolic biodegradation of 1,4-dioxane (section 2.7.1.2), notable removal of 1,4-dioxane is not expected. Likewise, there aren't many studies that support significant removal of 1,4-dioxane via anaerobic biological treatment processes. After investigating a variety of anaerobic 1,4-dioxane biodegradation pathways, Arve (2015) did not observe anaerobic biodegradation of 1,4-dioxane when it was provided as the sole electron donor. However, there was evidence of potential anaerobic biodegradation in the presence of other substrates. Conversely, Shen et al. (2008) observed biotransformation of 1,4-dioxane to carbon dioxide under iron-reducing conditions with a microbial culture originating from a wastewater treatment plant, thought to be the result of the presence of extracellular compounds that facilitated the reaction. Similar to the disinfection step in drinking water treatment, there is not expected to be an opportunity for chemical 1,4-dioxane destruction in wastewater treatment. Therefore, the most likely options for achieving significant decreases in 1,4-dioxane concentrations during wastewater treatment come from more advanced treatment processes, some of which are described in section 2.9.

A quick note is needed here on the use of septic systems for removal of 1,4-dioxane. Similar to the preceding discussion regarding aerobic/anaerobic biological wastewater treatment, there is not expected to be significant removal of 1,4-dioxane via septic systems. While some biologically mediated degradation may occur, either via direct metabolism or cometabolism, a septic system is not specifically engineered for 1,4-dioxane removal. This could present a challenge due to the presence of 1,4-dioxane in many household soaps, detergents, and personal care products.

2.8.1 POINT-OF-USE AND POINT-OF-ENTRY TREATMENT

With the ever-growing number of instances of 1,4-dioxane present in private drinking water wells, the need for point-of-use (POU) and point-of-entry (POE) treatment technologies is becoming more relevant. Common household POU/POE treatment processes include particulate filters, GAC or other sorptive media, aeration, ion exchange, reverse osmosis, and ultraviolet (UV) lamps or ozonation for disinfection. Each of these treatment techniques is discussed in section 2.9, as is its application to 1,4-dioxane.

A complicating factor to POU/POE treatment of 1,4-dioxane is the lack of commercially available POU/POE treatment devices that have consistently demonstrated, or been certified to, remove 1,4-dioxane in a variety of water quality scenarios. The ex situ technologies used to readily treat 1,4-dioxane are not yet robust enough, or compact enough, to be sold as POU/POE devices. Many times, POU/POE treatment processes are designed to be easily mixed and matched, and treat reasonably expected contaminants for an acceptable period of time before replacement. Use of GAC for POU/POE treatment could become particularly challenging for residents to manage. As discussed previously, 1,4-dioxane does not readily sorb to GAC; therefore, longer contact times and more frequent changeouts (i.e., higher carbon usage rates) are needed to adequately remove 1,4-dioxane and ensure treatment goals are met.

2.9 1,4-DIOXANE TREATMENT TECHNOLOGIES FOR DRINKING WATER TREATMENT AND EX SITU GROUNDWATER REMEDIATION

Compared to the development of in situ treatment technologies for 1,4-dioxane, technologies for drinking water and ex situ groundwater remediation could be considered better established. For example, the use of advanced oxidation processes (AOPs) for water treatment is considered the most established treatment technology for 1,4-dioxane. However, a better understanding of how other water treatment technologies can be applied to 1,4-dioxane (e.g., GAC), as well as development of new options (e.g., synthetic media), has come about more recently. The technologies described in this section can be applied to different types of water that need to be treated: municipal drinking water, household drinking water, domestic wastewater, industrial wastewater, and/or extracted groundwater at a remediation site. The application of technologies to these treatment scenarios may differ, with certain technologies being more appropriate than others under the design conditions (e.g., flow rates, 1,4-dioxane concentrations, cleanup criteria). Table 2.10 includes a list of common water treatment technologies, provides a description of each, and identifies the ones that are, or are not, applicable to 1,4-dioxane. A relevant subset of these technologies is described in more detail in this section.

As with any treatment strategy, it is important to consider removal of 1,4-dioxane in the context of other unwanted constituents that might be present, or restrictions on generation of by-products. As discussed throughout this chapter, 1,4-dioxane is commonly colocated with chlorinated solvents. Conventional ex situ water treatment technologies for chlorinated solvents include use of air-stripping towers or GAC.

TABLE 2.10
Water treatment technologies and their application to 1,4-dioxane

Treatment Process	Description	Application to 1,4-Dioxane
GAC and other sorbent media	Passing the water stream through a highly porous media that allows for both absorption and/or adsorption of chemical to the media.	1,4-Dioxane does not readily sorb (i.e., low log K_{oc} value). GAC can be effective at removing 1,4-dioxane; however, treatment units may be larger and more frequent carbon replacement may be needed than for other chemicals. Some other sorbent media are specially designed to enhance sorption of 1,4-dioxane; therefore, they can be effective options for treatment.
AOPs	A reaction—typically between various oxidants and sometimes paired with UV light—that produces highly reactive radicals (e.g., hydroxyl radicals). These reactive radicals aggressively react with other compounds, destroying the other compound in the process.	1,4-Dioxane is readily destroyed by reactive radicals to form carbon dioxide and water. Therefore, AOPs are very effective treatment options.
Electrochemical oxidation	Application of an electrical current to the water. Complete destruction of chemicals occurs through both direct oxidation of the chemical at the anode, as well as indirect oxidation through the reactive species that are generated (e.g., hydroxyl radicals).	1,4-Dioxane can effectively be destroyed by hydroxyl radicals; therefore, electrochemical oxidation is a relevant treatment technology.
Ion exchange	Reversible chemical reactions that remove ions (either positive or negative) from water and replaces them with other, similarly charged, ions. A common example of ion exchange in water treatment is the exchange of calcium or magnesium ions for sodium ions (i.e., water softening).	Because 1,4-dioxane is not commonly present as an ion, use of traditional ion exchange is not applicable.
Air stripping	The contact between the water stream and a gaseous stream (e.g., air) to facilitate transfer of chemicals from the water phase into the gaseous phase, for removal.	Due to the low Henry's law constant for 1,4-dioxane (i.e., low volatility), it is not readily transferred from the water phase to the gaseous phase; thus, air stripping is not an effective treatment technology.

(Continued)

TABLE 2.10 (*Continued*)

Treatment Process	Description	Application to 1,4-Dioxane
Reverse osmosis and Nanofiltration	A process that separates dissolved salts (monovalent and divalent ions), particles, colloids, organics, bacteria, and pyrogens from water by filtering through a semi-permeable membrane at a pressure greater than the osmotic pressure caused by the dissolved salts in the water stream. Reverse osmosis is effective at removing molecules with a molecular mass greater than 200 grams per mole, and ionic compounds. Nanofiltration is similar to reverse osmosis, but generally removes only larger divalent ions.	1,4-Dioxane's molecular mass is 88.11 grams per mole, which is less than the threshold of 200 grams per mole; also, it is not ionically charged. Therefore, reverse osmosis and nanofiltration are not effective treatment technologies for 1,4-dioxane.
Microfiltration and Ultrafiltration	A pressure-driven process that forces water through a micro-porous separating layer (the filter) which captures the particles present in the water stream, but allows dissolved components to pass through.	Because 1,4-dioxane is miscible in water, it will not be removed via physical filtration techniques.

Unfortunately, these technologies are less effective at removing 1,4-dioxane, particularly when they are designed to treat the chlorinated solvents present. However, design adjustments could be made to make them more effective at removing 1,4-dioxane. For example, adjusting the empty bed contact time (EBCT) or the breakthrough monitoring and carbon changeout schedule may be a way to manage 1,4-dioxane using GAC.

Another treatment technology that is not typically effective in treating 1,4-dioxane is air stripping, and by extension in situ air sparging. Air stripping typically refers to ex situ treatment systems that force air through/over water that is distributed onto media designed to maximize the surface area of air-to-water contact. The degree to which a compound can be effectively stripped from water is predominantly described by the Henry's law constant. An optimistic estimate of the air-to-water ratio required to achieve the desired removal of a given compound can be described by Equation 2.1.

$$A/W = (C_o - C_f)/(H \times C_f) \qquad (2.1)$$

where:

A/W = volumetric air-to-water ratio
C_o = initial contaminant concentration
C_f = final concentration
H = Henry's law constant (unitless)

1,4-Dioxane

The air-to-water ratio needed to achieve 90 percent removal of 1,4-dioxane (using a unitless Henry's law constant of 2×10^{-4}) is 45,000 compared to 14 for TCA (using a unitless Henry's law constant of 0.663). Based on this analysis, even at very high air-to-water ratios, the required residence times for 1,4-dioxane removal would be impracticable. This is further exacerbated when a 1,4-dioxane reduction of several orders of magnitude is required to meet treatment criteria.

2.9.1 Advanced Oxidation Processes

AOPs have been employed for water treatment since the 1980s. They were initially applied as tertiary drinking water treatment but expanded to treatment of municipal and industrial wastewater as well as part of groundwater remediation treatment strategies. The fundamental goal of AOPs is to produce highly reactive radicals, which are molecules with an unpaired electron typically denoted by a dot next to the chemical formula for the molecule (e.g., ·OH for the hydroxyl radical). As a rule, a radical needs to pair its unpaired electron with another. This is the source of its oxidative power, and it will aggressively react with other molecules to obtain this missing electron. The chemistry of AOPs can be applied both in situ and ex situ, but the term "AOP" most commonly refers to an ex situ unit process. Over time, AOPs have evolved from Fenton's reagent chemistry (i.e., hydrogen peroxide with a ferrous iron catalyst to generate hydroxyl radicals) to include oxidant combinations that maximize destructive capability while minimizing formation of unwanted by-products.

Some radicals are less reactive than others, but most radicals are highly reactive and thus very short lived. Hydroxyl radicals can destroy most organic pollutants in water and are particularly effective at destroying 1,4-dioxane, making AOPs among the most used treatment technologies for destruction of 1,4-dioxane. There are a variety of commercially available AOPs based on the formation of hydroxyl radical. These include different combinations of UV light, hydrogen peroxide, ozone, catalysts, and hypochlorite. Table 2.11 describes several AOPs that have applicability to the treatment of 1,4-dioxane, the basis of their oxidation process, advantages, and challenges associated with each. Technology selection is dependent on site-specific water quality, target reductions, and flow rates.

To date, UV light/hydrogen peroxide and ozone/hydrogen peroxide are the most common AOP technologies used given their cost-effectiveness for large-scale applications. The advantages of a UV light/hydrogen peroxide approach over an ozone-based approach include: (1) no bromate formation, and (2) lower costs for waters with low UV absorbance. Likewise, the challenges include: (1) the need to incorporate quenching of excess hydrogen peroxide, and (2) potential increases in disinfection by-products. The advantages of an ozone/hydrogen peroxide approach over a UV-based approach include: (1) lower power requirements, especially with waters with high UV absorbance, and (2) the ability to minimize hydrogen peroxide residual because the ozone is dosed stoichiometrically. However, the challenges include: (1) formation of compounds that are then regulated, specifically bromate; (2) considerations for storage/handing of liquid oxygen; and (3) potential increases in disinfection by-products.

More generally, key challenges associated with implementation of AOPs are typically associated with formation of unwanted by-products, energy consumption, and a

TABLE 2.11
Comparison between commercially available ex situ AOPs

AOP System and Manufacturer	Method to Create Hydroxyl Radicals	Advantages	Challenges
HiPOx® by APTwater	Ozone and hydrogen peroxide	Less sensitive to poor water quality No iron pretreatment required (<10 mg/L) Lower power consumption compared to UV-based technologies	Potential bromate production Narrower range of treatment (does not handle wide fluctuations in concentrations very well) Higher chemical usage than UV-based technologies
RayOX® by CalgonCarbon®	UV light and hydrogen peroxide	No potential bromate formation Simple design of system (low number of lamps)	Requires higher water quality (pretreatment may be needed) Low number of lamps increases failure consequence High electrical usage
TrojanUVPhox™ by TrojanUV®	UV light and hydrogen peroxide	No potential bromate formation Available systems have wide range of flow rates Simple system design (lower number of lamps)	Requires higher water quality (pretreatment may be needed) High electrical usage, but use of low-pressure lamps reduces electrical usage compared to other UV-based systems
PhotoCat® by Purifics	UV light, titanium dioxide catalyst, and oxidant	No potential bromate formation Multiple lamps reduce consequence of lamp failure	Requires higher water quality (pretreatment may be needed) Catalyst is sensitive to fouling, pretreatment is typically required (especially for iron) High electrical usage Mechanically more complex than other AOP systems
Wedeco® MiPro™ eco$_3$ and Pro mix® by Xylem	Ozone and peroxide	Less sensitive to poor water quality Lower power consumption compared to UV-based technologies	Potential bromate production Higher chemical usage than UV-based technologies
Wedeco® MiPro™ photo by Xylem	UV light and hydrogen peroxide or UV light and hypochlorite	No potential bromate formation Hypochlorite can be used in waters with low pH	Requires higher water quality (pretreatment may be needed) High electrical usage
Wedeco® MiPro™ eco$_3$ plus by Xylem	UV light, ozone, and hydrogen peroxide	No potential bromate formation Efficient for treatment of multiple contaminants, based on use of multiple oxidants	Requires higher water quality (pretreatment may be needed) High electrical usage

high level of maintenance and balancing needed during operation. Potential by-product formation is a major consideration, especially in drinking water applications. However, the production of unwanted by-products can be managed through use of pretreatment technologies (e.g., aeration and sand filtration for iron removal) or adjustments to the AOP operational parameters (e.g., ozone dosing to limit bromate formation).

When considering AOPs as a treatment technology for 1,4-dioxane, cost is a contributing factor in the decision-making process. Absolute costs will vary depending on the technology selected, the size of the treatment system, the need for pretreatment of the water, among other factors. Likewise, the cost for chemicals and power usage may lead to high annual costs associated with this type of treatment system. In many cases, AOPs are likely to remain the most appropriate treatment technology for 1,4-dioxane due to their effectiveness and acceptance, particularly when retrofitting existing treatment systems that may already include processes that focus on treatment of chlorinated solvents.

2.9.2 Bioreactors

The term "bioreactor" includes a large number of treatment processes, particularly when one considers their application to domestic and industrial wastewater treatment, as well as groundwater remediation. Bioreactors include unit treatment processes that capitalize on intrinsic biodegradation processes. Bioreactors can be designed to remove a wide variety of targets, including biological oxygen demand; inorganic compounds (e.g., nitrate, phosphate); metals (e.g., selenium, chromium); organic compounds (e.g., petroleum hydrocarbons, perchlorate, methyl tertiary butyl ether); and emerging contaminants (e.g., bis[2-chloroethyl] ether or 1,4-dioxane). They can be set up to operate as aerobic, anoxic, or anaerobic systems, depending on the removal target(s).

Bioreactors generally fall into one of two categories depending on how the biological culture is maintained: suspended growth or attached growth (also called fixed film). In a suspended growth reactor, the biomass is suspended in the water being treated. Examples of suspended growth reactors include activated sludge tanks and sequencing batch reactors. In a fixed film reactor, the biomass attaches itself to a solid media in the reactor and the water being treated flows over it. Examples of fixed film reactors include trickling filters, rotating biological contactors, fluidized bed reactors, and moving bed bioreactors. The advantages of using an ex situ bioreactor to treat impacted water are the relatively low initial capital costs and long-term operating costs, compared to AOPs. These cost savings are driven by the lack of UV lamps, chemical oxidants, and high power demands. Because bioreactors are living systems, many of the challenges with using them are related to establishing, and maintaining, the desired microbial community. Careful testing of the effectiveness of a bioreactor to treat water to an acceptable level is needed to ensure project success. Where drinking water purity is needed, additional treatment may be required to address residual organics, suspended solids, or biomass in the effluent.

Bioreactors are commercially available for use in full-scale treatment applications for a variety of contaminants but are still in development for treatment of 1,4-dioxane. Several bench-scale bioreactor configurations have been evaluated. Sock (1993) enriched a mixed culture capable of utilizing 1,4-dioxane as its sole carbon and energy source and employed it in an aerobic fluidized bed reactor where 1,4-dioxane was reduced from 100 mg/L to 1 mg/L. Work performed by Sandy et al. (2001) included testing various combinations of anaerobic bioreactors, aerobic bioreactors, distillation, and AOP technologies for treatment of industrial wastewater with high concentrations of 1,4-dioxane (e.g., 35 mg/L). Treatment of 1,4-dioxane to less than 10 µg/L was consistently achieved when biological treatment was followed by AOP polishing. Only one of the anaerobic/aerobic bioreactor combinations achieved effluent concentrations of 40 µg/L, while the others still had 1 to 25 mg/L 1,4-dioxane in the effluent. Zenker et al. (2004) evaluated 1,4-dioxane degradation in the presence of THF in a trickling bioreactor at the bench-scale and found 1,4-dioxane removal efficiencies of 93 percent (influent concentrations ranging from 20 µg/L to 1.25 mg/L) with the lowest effluent concentration of 10 µg/L, under limited conditions. The most relevant example of full-scale treatment of 1,4-dioxane using a bioreactor is the Lowry Landfill Superfund Site near Denver, Colorado. Here, a full-scale, fixed film moving bed biological treatment system was deployed to remove 1,4-dioxane from recovered groundwater/landfill leachate. Fortuitously, the THF also present provided a primary substrate for cometabolic biodegradation of 1,4-dioxane (Cordone et al., 2016). As of 2015/2016, the bioreactor consistently decreased 1,4-dioxane from an average of 1,326 µg/L to an average of 93 µg/L, the required discharge limit, at a flow rate of 19 gallons per minute.

Some of these bioreactor examples relied on the biodegradation of 1,4-dioxane in the presence of THF. However, it may be challenging to attain regulatory approval to add a primary substrate like THF, due to its own environmental concerns. Therefore, other bioreactor configurations were pilot tested in Lansing, Michigan, as described in the case study to follow. It is noteworthy to contrast the 1,4-dioxane concentrations treated, and the water sources, as part of the studies described above (i.e., parts per million levels from industrial wastewater/landfill leachate) with those in the following case study (i.e., approximately 200 µg/L, from groundwater).

Case Study: Ex situ Bioreactor Pilot Study for Treatment of 1,4-Dioxane

Historical use of chlorinated solvents resulted in up to approximately 3,000 µg/L of 1,4-dioxane in groundwater. An ex situ biological treatment approach was piloted, using fixed film bioreactors. Three bioreactors were piloted to treat extracted groundwater prior to reinjection: one designed for metabolic biodegradation of 1,4-dioxane and two designed for propane-mediated cometabolic biodegradation of 1,4-dioxane (at two different hydraulic residence times). Each bioreactor pilot system consisted of propane and/or oxygen cylinders, gas injection points, a bioreactor filled with approximately 40 percent media, an LEL meter, and associated controls. The bioreactors also received micro- and macronutrients to promote healthy bacterial colonization. The metabolic bioreactor was seeded with a known 1,4-dioxane degrading organism, *Pseudonocardia dioxanivorans* CB1190. The cometabolic bioreactors were seeded with a propanotrophic culture, *Rhodococcus ruber* ENV425. This culture uses propane as its primary substrate, which has been used elsewhere to facilitate cometabolic biodegradation of 1,4-dioxane. Over the course of the three-month pilot test, the 1,4-dioxane concentration in one of the propane bioreactors was reduced from 240 µg/L to 40 µg/L. However, the other bioreactor configurations were not able to attain meaningful 1,4-dioxane removal. Additional testing would be needed to understand how to tailor the bioreactor conditions to achieve maximum 1,4-dioxane removal at these environmentally relevant concentrations.

These pilot test bioreactors were set up as constant feed systems, or near-constant feed systems. The highest 1,4-dioxane removal was observed in the propane bioreactor where the propane was only fed into the bioreactor half of the time. This allowed propane concentrations to decrease and the microbially generated enzymes to degrade the 1,4-dioxane. A negative growth system configuration may be another attractive option for cometabolic bioreactors. In a negative growth system, the consortium would be grown up on the primary substrate (e.g., propane), then switched to a 1,4-dioxane-bearing water stream. 1,4-Dioxane would be degraded, although the microbial population would decline due to lack of an effective food source. Negative growth systems work in parallel: while one bioreactor is treating 1,4-dioxane-bearing water, another is being grown up on the primary substrate.

2.9.3 GRANULAR ACTIVATED CARBON AND OTHER SORBENT MEDIA

GAC is a commonly employed treatment technology for both drinking water and groundwater. It is also a popular treatment option for POU/POE systems. As described in section 2.2, the chemical characteristics of 1,4-dioxane suggest that it will not readily sorb to organic material, including carbon-based sorbents. This has led to the increasingly common scenario where chlorinated solvents may be effectively treated by existing systems, but 1,4-dioxane rapidly breaks through the carbon. While vapor-phase GAC is not discussed in this chapter, it is important to note that removal of 1,4-dioxane from the vapor phase using GAC does not suffer from the same challenges as removal from water.

There are examples of both bench-scale and field-scale studies to understand the effectiveness of GAC for removal of 1,4-dioxane, in addition to chlorinated solvents, present in the water. Bench-scale work presented by Fotta (2012) sought to compare breakthrough of 1,4-dioxane and various chlorinated solvents for different types of GAC (Figure 2.7). After approximately 200 to 850 bed volumes (depending on GAC type), 30 percent breakthrough of 1,4-dioxane was detected (influent concentration of 2.32 µg/L). After initial breakthrough, 1,4-dioxane concentrations in the effluent continued to increase to those above the influent, suggesting that sorbed 1,4-dioxane was displaced by more strongly sorbing compounds (e.g., the other chlorinated solvents). After approximately 3,500 bed volumes, the influent and effluent concentrations were similar. Contrast this with no detectable breakthrough of TCE after 30,000 bed volumes. Johns et al. (1998) evaluated the effectiveness of agricultural by-products for GAC composition. The results indicated that GAC made from pecan or walnut shells sorbed more 1,4-dioxane than commercial GACs. This included a

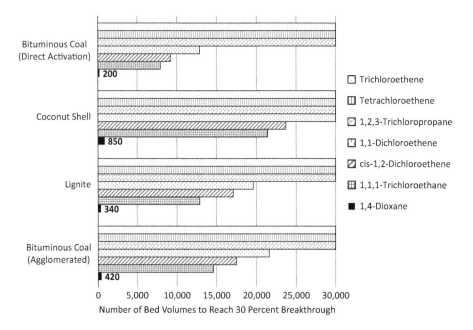

FIGURE 2.7 1,4-Dioxane breakthrough in various types of GAC: Comparison of the number of bed volumes to reach 30 percent breakthough for 1,4-dioxane and various chlorinated solvents for different types of GAC. Adapted from Fotta (2012).

more than 50 percent decrease in 1,4-dioxane from an aqueous solution, with 80 mg/L initially present. This corresponded to 4.5 grams of 1,4-dioxane sorbed per gram of GAC. Eigenbrodt and Rooney (2014) investigated the equilibrium sorptive capacity of seven commercially available GAC products and found that a coconut-shell-based GAC outperformed the other products. They also tested the efficacy of UV regeneration on titanium-dioxide-embedded GAC and found a minimal increase in 1,4-dioxane sorption.

While GAC is certainly not the most effective treatment approach for 1,4-dioxane, it may be applicable under certain conditions (i.e., smaller-scale systems). GAC systems expected to treat both chlorinated solvents and 1,4-dioxane must be designed to the limiting factor: the 1,4-dioxane. This may result in larger GAC vessels, additional vessel redundancy, more frequent monitoring for breakthrough, and/or more frequent carbon changeouts as compared to a system targeting only chlorinated solvents. This design approach is expected to be successful on smaller scales, specifically for POE treatment of drinking water for residential use. Larger applications of GAC treatment for 1,4-dioxane treatment (e.g., groundwater remediation systems or municipal drinking water systems) will likely become cost prohibitive.

As described in Mohr et al. (2010), zeolites have been explored as sorptive media for 1,4-dioxane. Specifically, a surfactant-modified zeolite-only product was tested alongside a surfactant-modified zeolite/zero-valent iron product and a GAC product. The results of partition testing suggest that measurable amounts of 1,4-dioxane sorbed to the surfactant-modified zeolite/zero-valent iron product at low starting concentrations and low media to water ratios. No measurable sorption of 1,4-dioxane to the surfactant-modified zeolite-only product was observed. While the surfactant-modified zeolite/zero-valent iron product showed potential application as a sorptive media for 1,4-dioxane, the GAC product was chosen for further testing.

Another commercially available sorptive media is AMBERSORB™ 560 (Ambersorb). Ambersorb is a carbon-based sorbent produced by the partial pyrolysis of a sulfonated styrene-divinylbenzene copolymer. Similar to other carbon-based sorbent materials, Ambersorb has high surface area particles. Unlike GAC, Ambersorb is designed with more specificity and selectivity for 1,4-dioxane. Also, because the starting material for Ambersorb is a manufactured product rather than a natural resource, it has more predictable sorption and regeneration properties because pore size distribution and other properties are controlled (Woodard et al., 2014). Compared to GAC, Ambersorb has a higher ratio of meso- and macropores as compared to micropores; it also has more hydrophobicity allowing for greater sorption potential for organic contaminants. 1,4-Dioxane has molecular functional groups that give it both polar and nonpolar characteristics (Figure 2.1). While the exact mechanism of 1,4-dioxane removal has not been established, it is believed to be an adsorption process that relies on weak van der Waals forces and/or electrostatic attraction (Woodard et al., 2014). The characteristics of Ambersorb allow for shorter EBCTs than GAC to give similar times for chemical breakthrough. Regeneration of Ambersorb is performed via low-pressure steam, microwave radiation, solvents, or hot gases that can be performed on-site (Woodard et al., 2014), rather than off-site reactivation as is common with GAC. The highly concentrated condensate from the steam regeneration is often treated through a small GAC vessel to minimize the volume of waste to be disposed. This approach takes advantage of the greater 1,4-dioxane sorptive capacity of GAC at elevated concentrations. Theoretically, destructive approaches could also be used to treat the highly concentrated condensate.

While Ambersorb may seem like an attractive option for retrofitting GAC systems associated with existing P&T systems, the cost of the synthetic media can make it prohibitive. At the time of this writing, field applications of Ambersorb for treatment of 1,4-dioxane were still limited. Some applications of Ambersorb for treatment of 1,4-dioxane have shown flow limitations (Chiang et al., 2016) or decreased sorptive capacity over time (i.e., after six months of regeneration). The latter appears to be associated with the naturally occurring presence of competitive compounds, such as natural organic matter. Long-term application of Ambersorb for ex situ treatment of 1,4-dioxane is still evolving and up-front water quality testing can provide valuable predictions of operational success.

2.9.4 Electrochemical Oxidation

Electrochemical oxidation includes the application of an electrical current across an anode and cathode placed within the water that needs to be treated. This results in two destruction mechanisms for chemicals in the water: direct oxidation at the anode, and indirect oxidation through the generation of reactive radicals (hydroxyl radicals, in particular). Electrochemical oxidation could be considered another AOP; however, due to its significantly different nature, it is considered separately here.

Successful application of electrochemical oxidation relies highly on the nature of the electrode material. Research has been conducted into the effectiveness of various electrode materials (Martinez-Huitle and Ferro, 2006). Each material has direct oxidation benefits, indirect oxidation benefits, corrosivity considerations, cost considerations, etc. Boron-doped diamond electrodes, in particular, have been shown to effectively oxidize 1,4-dioxane at the bench scale (Jasmann et al., 2016; De Clercq et al., 2010). The addition of other catalysts, such as titanium dioxide, may enhance oxidation efficiency of this approach (Jasmann et al., 2016). Full-scale application of electrochemical oxidation for treatment of water, and measurement of the efficacy of destruction of 1,4-dioxane, are less documented. The use of expensive electrodes and associated catalysts may make electrochemical oxidation a cost-prohibitive option at large scales.

2.10 CONCLUSION

The evolving understanding of the toxicology of 1,4-dioxane, uncertainty in the regulatory landscape, and the increased attention from regulatory and community stakeholders contribute to 1,4-dioxane's status as an emerging contaminant. Its historical presence as a stabilizer in chlorinated solvents has led to a high frequency of occurrence in the environment. Testing performed under the Unregulated Contaminant Monitoring Rule 3 indicated that 1,4-dioxane was present in 21 percent of public water supplies, was the second-most detected compound on the analyte list, and occurred more frequently when the source of the drinking water supply was groundwater (Adamson et al., 2017). While 1,4-dioxane is no longer expected to be released to the environment through chlorinated solvent use and disposal, addressing that which is already present is a challenge. Additionally, other sources of release (e.g., consumer product use and direct use/generation in common manufacturing processes) may continue to add to the challenge.

Many advances related to 1,4-dioxane have been made in recent years, including understanding the toxicology; developing routine and advanced analytical methods; and identifying viable treatment technologies. Further developments in these areas are expected. Some of these may include:

- Gaining consensus on the toxicology of 1,4-dioxane and the use of a threshold concentration, below which carcinogenic effects are not observed.
- Revision of regulatory standards or guidelines to reflect this updated understanding in toxicology. Many of the standards and guidelines in place in 2018 (i.e., <1 µg/L) may increase.
- Advances in analytical methods to allow for more accurate quantification, and reproducible results, of 1,4-dioxane in water. Specifically, the use of isotopic dilution is expected to become commonplace with existing USEPA Methods 8260 and 8270 (both with SIM).
- Continued research and field application of advanced analytical techniques (i.e., MBTs and CSIA) to develop benchmarks and heuristics for comparison.

- Moving away from ex situ AOP technologies as the standard treatment method for 1,4-dioxane to approaches that are more economical and sustainable.
- Continued research and field testing of bioremediation and natural attenuation approaches. Areas of development are expected to include demonstrations of metabolic and/or cometabolic in situ bioremediation, ex situ bioreactors, and development of predictive guidelines for evaluating natural attenuation.

The presence of 1,4-dioxane in the environment is a concern to regulatory bodies, responsible parties, environmental consultants, drinking water suppliers, and the general public. While the efforts made to date to understand its impact to society and develop ways to eliminate it have provided the industry with invaluable insight and guidance, future efforts will reveal the path to a solution.

REFERENCES

Adamson, David T., Elizabeth A. Pina, Abigail E. Cartwright, Sharon R. Rauch, R. Hunter Anderson, Thomas Mohr, and John A. Connor. 2017. "1,4-Dioxane Drinking Water Occurrence Data from the Third Unregulated Contaminant Monitoring Rule." *Science of the Total Environment*, 236–245.

Adamson, David T., Hunter R. Anderson, Shaily Mahendra, and Charles J. Newell. 2015. "Evidence of 1,4-Dioxane Attenuation at Groundwater Sites Contaminated with Chlorinated Solvents and 1,4-Dioxane." *Environmental Science and Technology*, 6510–6518.

Adamson, David T., Shaily Mahendra, Jr., Kenneth L. Walker, Sharon R. Rauch, Shayak Sengupta, and Charles J. Newell. 2014. "A Multisite Survey to Identify the Scale of the 1,4-Dioxane Problem at Contaminated Groundwater Sites." *Environmental Science and Technology Letters*, 254–258.

Agency for Toxic Substances and Disease Registry. 2012. *Toxicological Profile for 1,4-Dioxane.* April. www.atsdr.cdc.gov/toxprofiles/tp187.pdf.

Alexander, M. 1973. "Nonbiodegradable and Other Recalcitrant Molecules." *Biotechnology and Bioengineering* (15): 611–647.

Arve, Philip. 2015. *Microcosm Study of 1,4-Dioxane Biotransformation: Theses: Paper 2193.* Clemson, SC: Clemson University.

Bell, Caitlin, and Monica Heintz. 2016. "The Search for 1,4-Dioxane Biodegradation in the Field." *Battelle's Tenth International Conference on Remediation of Chlorinated and Recalcitrant Compounds.* Palm Springs, May.

Bell, Caitlin, Jeff McDonough, Kelly S. Houston, and Kathleen Gerber. 2016. "Stable Isotope Probing to Confirm Field-Scale Co-Metabolic Biodegradation of 1,4-Dioxane." *Remediation Journal*, 47–59.

Bennett, Peter, Michael Hyman, Christy Smith, Humam El Mugammar, Min-Ying Chu, Michael Nickelsen, and Ramon Aravena. 2018. "Enrichment with Carbon-13 and Deuterium during Monooxygenase-Mediated Biodegradation of 1,4-Dioxane." *Environmental Science and Technology Letters*, 148–153.

Bernhardt, Dirk, and Hans Diekmann. 1991. "Degradation of Dioxane, Tetrahydrofuran, and Other Cyclic Ethers by an Environmental Rhodococcus strain." *Applied Microbiology and Biotechnology*, 120–123.

Burback, B. L., and J. J. Perry. 1993. "Biodegradation and Biotransformation of Groundwater Pollutant Mixtures by Mycobacterium Vaccae." *Applied and Environmental Microbiology*, 1025–1029.

Chiang, Sheau-Yun (Dora), Richard (Hunter) Anderson, Michael Wilken, and Claudia Walecka-Hutchison. 2016. "Practical Perspectives of 1,4-Dioxane Investigation and Remediation." *Remediation Journal*, 7–27.

Chiang, Dora Sheau-Yun, Rebecca Mora, William H. DiGuiseppi, Greg Davis, Kerry Sublette, Phillip Gedalang, and Shaily Mahendra. 2012. "Characterizing the Intrinsic Bioremediation Potential of 1,4-Dioxane and Trichloroethene Using Innovative Environmental Diagnostic Tools." *Journal of Environmental Monitoring*, 2317–2326.

Chiang, Dora Sheau-Yun, Everett W. Glover Jr., Jeff Peterman, Joseph Harrigan, Bill DiGuiseppi, and David S Woodward. 2008. "Evaluation of Natural Attenuation at a 1,4-Dioxane-Contaminated Site." *Remediation Journal*, 19–37.

Cordone, Leslie, Chris Carlson, William Plaehn, Timothy Shangraw, and David Wilmoth. 2016. "Case Study and Retrospective: Aerobic Fixed Film Biological Treatment Process for 1,4-Dioxane at the Lowry Landfill Superfund Site." *Remediation Journal*, 159–172.

De Clercq, Jeriffa, Evelien Van de Steene, Kim Verbeken, and Marc Verhaege. 2010. "Electrochemical Oxidation of 1,4-Dioxane at Boron-Doped Diamond Electrode." *Journal of Chemical Technology and Biotechnology*, 1162–1167.

Dourson, Michael, Jeri Higginbotham, Jeff Crum, Heather Burleigh-Flayer, Patricia Nance, Norman Forsberg, and Mark Lafranconi. 2017. "Update: Mode of Action (MOA) for Liver Tumors Induced by Oral Exposure to 1,4-Dioxane." *Regulatory Toxicology and Pharmacology* (88): 45–55. http://allianceforrisk.org/14-dioxane-analysis/.

Dourson, Michael, John Reichard, Patricia Nance, Heather Burleigh-Flayer, Anna Parker, Melissa Vincent, and Ernest E. McConnell. 2014. "Mode of Action Analysis for Liver Tumors from Oral 1,4-Dioxane Exposures and Evidence-Based Dose Response Assessment." *Regulatory Toxicology and Pharmacology*, 387–401.

Dunfield, Peter F., Werner Liesack, Thilo Henckel, Roger Knowles, and Ralf Conrad. 1999. "High-Affinity Methane Oxidation by a Soil Enrichment Culture Containing a Type II Methanotroph." *Applied and Environmental Microbiology*, 1009–1014.

Eigenbrodt, Caroline, and Eric Rooney. 2014. "Adsorption of 1,4-Dioxane on Activated Carbon with Regeneration by Titanium Dioxide/Ultraviolet Light." *MQP at Worcester Polytechnic Institute*.

Environment Canada/Health Canada. 2010. "Screening Assessment for the Challenge: 1,4-Dioxane." Ottawa, Ontario.

European Communities. 2002. *1,4-Dioxane: Risk Assessment*. Luxembourg: Office for Official Publications of the European Communities.

Fincher, Edward L., and W. J. Payne. 1962. "Bacterial Utilization of Ether Glycols." *Applied and Environmental Microbiology* (10): 542–547.

Fotta, Meredith Elyse. 2012. "Effect of Granular Activated Carbon Type on Adsorber Performance and Scale-Up Approaches for Volatile Organic Compound Removal." *Thesis*. Raleigh, NC: North Carolina State University.

Gedalanga, Phillip, Andrew Madison, Yu (Rain) Miao, Timothy Richards, James Hatton, William H. DiGuiseppi, John Wilson, and Shaily Mahendra. 2016. "A Multiple Lines of Evidence Framework to Evaluate Intrinsic Biodegradation of 1,4-Dioxane." *Remediation Journal*, 93–114.

Gedalanga, Phillip B, Peerapong Pornwongthong, Rebecca Mora, Dora Sheu-Yun Chiang, Brett Baldwin, Dora Ogles, and Shaily Mahendra. 2014. "Identification of Biomarker Genes to Predict Biodegradation of 1,4-Dioxane." *Applied and Environmental Microbiology*, 3209–3218.

Gestler, R., E. Gatliff, P. Thomas, and P. J. Linton. 2017. "Phytoremediation of 1,4-Dioxane Contaminated Aquifers: Case Studies and Lessons Learned." *Fourth International Symposium on Bioremediation and Sustainable Environmental Technologies.* Miami, FL.

Hatzinger, Paul B., Rahul Banerjee, Rachael Rezes, Sheryl H. Streger, Kevin McClay, and Charles E. Schaefer. 2017. "Potential for Cometabolic Biodegradation of 1,4-Dioxane in Aquifers with Methane or Ethane as Primary Substrates." *Biodegradation*, 453–468.

Hazen, Terry C. 2010. "Cometabolic Bioremediation in Handbook of Hydrocarbon and Lipid Microbiology." 2505–2514.

He, Ya, Jacques Mathieu, Yu Yang, Pingfeng Yu, Marcio L.B. da Silva, and Pedro J.J. Alvarez. 2017. "1,4-Dioxane Biodegradation by Mycobacterium dioxanotrophicus PH-06 is Associated with a Group-Soluble Di-Iron Monooxygenase." *Environmental Science and Technology Letters*, 494–499.

Hinchee, Rob. 2016. "1,4-Dioxane Remediation by Extreme Soil Vapor Extraction (XSVE)." *Emerging Contaminants Summit.* Westminster, CO.

Interstate Technology & Regulatory Council. 2013. *Environmental Molecular Diagnostics, New Site Characterization and Remediation Enhancement Tools, EMD-2.* Washington D.C. www.itrcweb.org/emd-2/.

Jasmann, Jeramy R., Thomas Borch, Tom C. Sale, and Jens Blotevogel. 2016. "Advanced Electrochemical Oxidation of 1,4-Dioxane via Dark Catalysis by Novel Titanium Dioxide (TiO2) Pellets." *Environmental Science and Technology*, 8817–8826.

Johns, Mitchell M., Wayne E. Marshall, and Christopher A. Toles. 1998. "Agricultural By-products as Granular Activated Carbons for Adsorbing Dissolved Metals and Organics." *Journal of Chemical Technology and Biotechnology*, 131–140.

Kano, Hirokazu, Yumi Umeda, Tatsuya Kasai, Toshiaki Sasaki, Michiharu Matsumoto, Mazunori Yamazaki, Kasuke Nagano, Heihachiro Arito, and Shoji Fukushima. 2009. "Carcinogenicity Studies of 1,4-Dioxane Administered in Drinking Water to Rats and Mice for 2 Years." *Food and Chemical Toxicology*, 2776–2784.

Kasai, Tatsuya, Hirokazu Kano, Yumi Umeda, Toshiaki Sasaki, Naoki Ikawa, Nishizawa Tomoshi, Nagano Kasuke, Heihachiro Arito, Hiroshi Nagashima, and Shoji Fukushima. 2009. "Two-Year Inhalation Study of Carcinogenicity and Chronic Toxicity of 1,4-Dioxane in Male Rats." *Inhalation Toxicology*, 889–897.

Kim, Young-Mo, Jong-Rok Jeon, Kumarasamy Murugesan, Eun-Ju Kim, and Yoon-Seok Chang. 2009. "Biodegradation of 1,4-Dioxane and Transformation of Related Cyclic Compounds by a Newly Isolated Mycobacterium sp. PH-06." *Biodegradation*, 511–519.

Kociba, R. J., S. B. McCollister, C. Park, and T. R. Tork. 1974. "1,4-Dioxane. I. Results of a 2-Year Ingestion Study in Rats." *Toxicology and Applied Pharmacology*, 275–286.

Lee, Jae-Ho, Jeung-Jin Park, Im-Gyu Byun, Tae-Joo Park, and Tae-Ho Lee. 2014. "Anaerobic Digestion of Organic Wastewater from Chemical Fiber Manufacturing Plant: Lab and Pilot-Scale Experiments." *Journal of Industrial and Engineering Chemistry*, 1732–1736.

Li, Mengyan, E. Tess Van Orden, David J. DeVries, Zhong Xiong, Rob Hinchee, and Pedro J. Alvarez. 2015. "Bench-Scale Biodegradation Tests to Assess Natural Attenuation Potential of 1,4-Dioxane at Three Sites in California." *Biodegradation*, 39–50.

Li, Mengyan, Jacques Mathieu, Yuanyuan Lui, E. Tess Van Orden, Yu Yang, Stephanie Fiorenza, and Pedro J. J. Alvarez. 2014. "The Abundance of Tetrahydrofuran/Dioxane Monooxygenase Genes (thmA/dxmA) and 1,4-Dioxane Degradation Activity Are Significantly Correlated at Various Impacted Aquifers." *Environmental Science and Technology Letters*, 122–127.

Li, Mengyan, Stephanie Fiorenza, James R. Chatham, Shaily Mahendra, and Pedro J. J. Alvarez. 2010. "1,4-Dioxane Biodegradation at Low Temperatures in Arctic Groundwater Samples." *Water Research*, 2894–2900.

Lippincott, D., Streger, S. H., Schaefer, C. E., Hinkle, J., Stormo, J., & Steffan, R. J. (2015). Bioaugmentation and propane biosparging for in situ biodegradation of 1,4-dioxane. *Groundwater Monitoring and Remediation*, 1–12.

Mahendra, Shaily, and Lisa Alvarez-Cohen. 2006. "Kinetics of 1,4-Dioxane Biodegradation by Monooxygenase-Expressing Bacteria." *Enviromental Science and Technology*, 5435–5442.

Mahendra, Shaily, and Lisa Alvarez-Cohen. 2005. "Pseudonocardia dioxanivorans sp. nov., a Novel Actinomycete That Grows on 1,4-Dioxane." *International Journal of Systematic and Evolutionary Microbiology*, 593–598.

Martinez-Huitle, Carlos A., and Sergio Ferro. 2006. "Electrochemical Oxidation of Organic Pollutants for the Wastewater Treatment: Direct and Indirect Processes." *Chemical Society Reviews*, 1324–1340.

Marzulli, Francis Nicholas, D. M. Anjo, and Howard I. Maibach. 1981. "In Vivo Skin Penetration Studies of 2,4-Toluenediamine, 2,4-Diaminoanisole, 2-Nitro-p-Phenylenediamine, p-Dioxane, and N-Nitrosodiethanolamine in Cosmetics." *Food and Cosmetics Toxicology*, 743–747.

Mohr, Thomas K. G., Julie A. Stickney, and William H. DiGuiseppi. 2010. *Enviromental Investigation and Remediation: 1,4-Dioxane and Other Solvent Stabilizers*. Boca Raton, FL: CRC Press.

Morrison, Robert, and Brian Murphy. 2015. "Source Identification and Age Dating of Chlorinated Solvents." In *Introduction to Environmental Forensics*, Third Edition, by Robert Morrison and Brian Murphy, New York, NY: Elsevier, 311–345.

New Jersey Department of Environmental Protection, Water Monitoring and Standards. 2015. "Ground Water Quality Standard: 1,4-Dioxane." October. www.nj.gov/dep/wms/bears/docs/1,4%20dioxane%20final%20draft%20for%20posting2.pdf.

Oberle, Daniel, Emily Crownover, and Mark Kluger. 2015. "In Situ Remediation of 1,4-Dioxane Using Electrical Resistance Heating." *Remediation Journal*, 35–42.

Parales, R. E., J. E. Adamus, N. White, and H. D. May. 1994. "Degradation of 1,4-Dioxane by an Actinomycete in Pure Culture." *Applied and Environmental Microbiology*, 4527–4530.

Payne, Fred C., Joseph A. Quinnan, and Scott T. Potter. 2008. *Remediation Hydraulics*. Boca Raton, FL: CRC Press.

Pornwongthong, Peerapong, Anjali Mulchandani, Phillip B. Gedalanga, and Shaily Mahendra. 2014. "Transition Metals and Organic Ligands Influence Biodegradation of 1,4-Dioxane." *Applied Biochemistry and Biotechnology*, 291–306.

Pornwongthong, P., G. L. Paradis, and S. Mahendra. 2011. "Stable Carbon Isotope Fractionation During 1,4-Dioxane Biodegradation." *Proceedings of the Water Environment Federation, WEFTEC 2011: Session 1 through Session 10*, 111–116.

Rolston, H., M. Azizian, L. Semprini, and M. Hyman. 2017. "Modeling Aerobic Cometabolism of 1,4-Dioxane and Chlorinated Solvents by Isobutane-Utilizing Bacteria." *Fourth International Symposium on Bioremediation and Sustainable Environmental Technologies*. Miami, FL.

Sadeghi, Venus, Rebecca Mora, Priya Jacob, and Sheau-Yun (Dora) Chiang. 2016. "Characterizing in Situ Methane-Enhanced Biostimulation Potential for 1,4-Dioxane Biodegradation in Groundwater." *Remediation Journal*, 115–132.

Sandy, Tom, C. P. Leslie Grady Jr., Steve Meininger, and Randy Boe. 2001. "Biological Treatment of 1,4-Dioxane in Wastewater from an Integrated Polyethylene Terephthalate (PET) Manufacturing Facility." *7th Annual Industrial Wastes Technical and Regulatory Conference*. Water Environment Federation, 1–30.

Schneider, Charles H., and Cecil C. Lynch. 1943. "The Ternary System: Dioxane-Ethanol-Water." *Journal of the American Chemical Society*, 1063–1066.

Semrau, Jeremy D. 2011. "Bioremediation via Methanotrophy: Overview of Recent Findings and Suggestions for Future Research." *Frontiers in Microbiology*, 1–7.

Shen, WeiRong, Hong Chen, and Shanshan Pan. 2008. "Anaerobic Biodegradation of 1,4-Dioxane by Sludge Enriched with Iron-Reducing Microorganisms." *Bioresource Technology*, 2483–2487.

Siegrist, Robert L, Michelle Crimi, and Thomas J Simpkin. 2011. *In Situ Chemical Oxidation for Groundwater Remediation*. New York, NY: Springer-Verlag.

Sock, S. M. 1993. "A Comprehensive Evaluation of Biodegradation as a Treatment Alternative for the Removal of 1,4-Dioxane." *Master of Science Thesis*. Clemson, SC: Clemson University.

Stantec Corporation—Treatability Testing Services. 2013. "Research Performed in 2007 by Dr. Angus McGrath and David Schroder." Sylvania, OH.

Steffan, Robert J., Kevin R. McClay, Hisako Masuda, and Gerben J. Zylstra. 2007. "Biodegradation of 1,4-Dioxane: SERDP Project ER-1422." Arlington, VA.

Suthersan, Suthan S., John Horst, Matthew Schnobrich, Nicklaus Welty, and Jeff McDonough. 2017. *Remediation Engineering: Design Concepts*, Second Edition. Boca Raton, FL: CRC Press.

Suthersan, Suthan, Joseph Quinnan, John Horst, Ian Ross, Erica Kalve, Caitlin Bell, and Tessa Pancras. 2016. "Making Strides in the Management of 'Emerging Contaminants'." *Groundwater Monitoring and Remediation* (National Groundwater Association), 15–25.

Suthersan, Suthan S., and Fred C. Payne. 2005. *In Situ Remediation Engineering*. Boca Raton, FL: CRC Press.

United States Environmental Protection Agency. 2017. "The Third Unregulated Contaminant Monitoring Rule (UCMR3): Data Summary, January 2017." Accessed June 20, 2017. www.epa.gov/sites/production/files/2017-02/documents/ucmr3-data-summary-january-2017.pdf.

—. 2018. "Regional Screening Level (RSL) Summary Table." May. https://www.epa.gov/risk/regional-screening-levels-rsls-generic-tables.

—. 2012. 2012. "Edition of the Drinking Water Standards and Health Advisories. EPA 822-S-12-001. Office of Water. Washington, D.C. April.

—. 2013. *Toxicological Review of 1,4-Dioxane (with Inhalation Update) (CAS No. 123-91-1) in Support of Summary Information on the Integrated Risk Information System (IRIS)*. Washington, DC.: EPA-635/R-11/003-F.

—. 2006. *Engineering Issue: In-Situ Chemical Oxidation: EPA/600/R-06/072*. Cincinnati, OH: Office of Research and Development National Risk Management Research Laboratory.

—.1999a. "Phytoremediation Resource Guide." *EPA 542-B-99-003*. Office of Solid Waste and Emergency Response., June.

—. 1999b. *Use of Monitored Natural Attenuation at Superfund, RCRA Corrective Action, and Underground Storage Tank Sites, OSWER Directive Number: 9200.4-17P*. Washington, DC: Office of Solid Waste and Emergency Response.

Waldemer, Rachel H., and Paul G. Tratnyek. 2006. "Kinetics of Contaminant Degradation by Permanganate." *Environmental Science and Technology*, 1055–1061.

Wang, Yi. 2016. "Breakthrough in 2D-CSIA Technology for 1,4-Dioxane." *Remediation Journal*, 61–70.

Woodard, Steven, Thomas Mohr, and Michael G. Nickelsen. 2014. "Synthetic Media: A Promising New Treatment Technology for 1,4-Dioxane." *Remediation Journal*, 27–40.

Zenker, Matthew J., Robert C. Borden, and Morton A. Barlaz. 2004. "Biodegradation of 1,4-Dioxane Using a Trickling Filter." *Journal of Environmental Engineering*, 926–931.

Zenker, Matthew J., Robert C. Borden, and Morton A. Barlaz. 2003. "Occurrence and Treatment of 1,4-Dioxane in Aqueous Environments." *Environmental Engineering Science*, 423–432.

Zenker, Matthew J., Robert C. Borden, and Morton A. Barlaz. 2000. "Mineralization of 1,4-Dioxane in the Presence of a Structural Analog." *Biodegradation*, 239–246.

Zhang, Shu, Phillip B. Gedalanga, and Shaily Mahendra. 2017. "Research Article: Advances in Bioremediation of 1,4-Dioxane-Contaminated Waters." *Journal of Environmental Management*, 765–774.

Zhang, Shu, Phillip B. Gedalanga, and Shaily Mahendra. 2016. "Biodegradation Kinetics of 1,4-Dioxane in Chlorinated Solvent Mixtures." *Environmental Science & Technology*, 9599–9607.

Acronyms

‰	per mil
11DCE	1,1-dichloroethene
ALDH	aldehyde dehydrogenase
AOP	advanced oxidation process
ASTDR	Agency for Toxic Substances and Disease Registry
bgs	below ground surface
CPT	cone penetrometer testing
CSF	cancer slope factor
CSIA	compound specific isotope analysis
DGR	dynamic groundwater recirculation
DNA	deoxyribonucleic acid
DSITMS	direct sampling ion trap mass spectrometry
DXMO	1,4-dioxane monooxygenase
EBCT	empty bed contact time
g/L	gram per liter
GAC	granular activated carbon
GC	gas chromatography
HEAA	β-hydroxyethoxyacetic acid
ISCO	in situ chemical oxidation
ITRC	Interstate Technology & Regulatory Council
IUR	inhalation unit risk
K_{oc}	organic carbon partition coefficient
LEL	lower explosive limit
LOAEL	lowest-observable-adverse-effect level
MBT	molecular biology tools
MCL	maximum contaminant level
mg/kg	milligram per kilogram
mg/kg/d	milligram per kilogram body weight per day
mg/L	milligrams per liter
mg/m³	milligrams per cubic meter
MOA	mutagenic mode of action
mRNA	messenger ribonucleic acid
MS	mass spectrometry
NOAEL	no-observable-adverse-effect level
P&T	pump and treat
PHE	phenol hydroxylase
POD	point of departure
POE	point-of-entry
POU	point-of-use
ppm	parts per million
PPO	propane monooxygenase
qPCR	quantitative polymerase chain reaction

RfD	reference dose
RMO/RDEG	ring-hydroxylating toluene monooxygenases
RSL	regional screening level
SERDP	Strategic Environmental Research and Development Program
SIM	selected ion monitoring
SIP	stable isotope probing
sMMO	soluble methane monooxygenase
SVOC	semi-volatile organic compound
TCA	1,1,1-trichloroethane
TCE	trichloroethene
THF	tetrahydrofuran
THFMO	tetrahydrofuran monooxygenase
UF$_A$	interspecies extrapolation factor
UF$_D$	toxicological database uncertainty factor
UF$_H$	intraspecies variability factor
UF$_L$	LOAEL-to-NOAEL extrapolation factor
µg/kg	micrograms per kilogram
µg/L	micrograms per liter
µg/m^3	micrograms per cubic meter
USEPA	United States Environmental Protection Agency
UV	ultraviolet
VAP	vertical aquifer profiling
VOC	volatile organic compound
XSVE	extreme soil vapor extraction

3 Per- and Polyfluoroalkyl Substances

Ian Ross, PhD
(Arcadis U.K., Ian.Ross@arcadis.com)

Erica Kalve
(Arcadis U.S., Inc., Erica.Kalve@arcadis.com)

Jeff McDonough
(Arcadis U.S., Inc., Jeff.McDonough@arcadis.com)

Jake Hurst
(Arcadis U.K., Jake.Hurst@arcadis.com)

Jonathan A L Miles
(Arcadis U.K., Jonathan.Miles@arcadis.com)

Tessa Pancras
(Arcadis NL., Tessa.Pancras@arcadis.com)

3.1 INTRODUCTION

Per- and polyfluoroalkyl substances (PFASs) are a broad group of several thousand xenobiotic chemicals (Wang et al. 2017; Organization for Economic Cooperation and Development [OECD] 2018a) that have been widely used in industrial and consumer applications since the 1940s and are classified as emerging contaminants.

The almost ubiquitous presence of PFASs in water, air, sediment, wildlife, and humans at low levels, their extreme persistence, long-range transport potential, and toxic effects has triggered an increasing concern over the potential harm PFASs pose to human health and the environment. PFASs are currently the subject of increasing environmental regulatory concern in many countries. The prior regulatory focus has been on a selection of PFASs termed long-chain perfluoroalkyl acids (PFAAs) such as perfluorooctane sulfonic acid (PFOS), perfluorooctanoic acid (PFOA), and perfluorohexane sulfonic acid (PFHxS), but an expanding range of differing PFASs are now regulated in many locations.

PFASs comprise two main groups, (1) perfluoroalkyl and (2) polyfluoroalkyl substances, both of which contain a perfluoroalkyl group. The perfluoroalkyl substances comprise a fully fluorinated molecule, while the polyfluoroalkyl substances are not fully fluorinated. PFASs may also be divided into short- and long-chain compounds based on the length of the perfluoroalkyl chain, which can result in different physicochemical properties that influence the substances' behavior in the environment and potential for bioaccumulation in organisms, which can influence their (eco)toxicity. Long-chain PFASs are generally being replaced in commercial products by short-chain PFASs varieties and a new class of perfluoroalkyl substances, termed perfluoroalkyl ethers.

PFASs are thermally stable and repel oils and water with impressive surface tension levelling properties. For example, they have been used in some firefighting foams, for coating fabrics and textiles, in non-stick surfaces, paints, polishes and waxes. Some PFASs have been the key ingredient in "film forming" Class B firefighting foams used to extinguish liquid hydrocarbon fires. Since the mid-1960s these foams have been used for repeated fire training events at fire training areas and in building and fuel storage fire suppression systems. Major sources of PFASs can also include manufacturing facilities such as fluoroplastics, dust/mist suppressants, metal plating, leather tanneries, paper and fabric coatings producers, landfills and biosolids derived from sewage sludges.

The detections of PFOS and PFOA in drinking water in the United States, combined with the U.S. Environmental Protection Agency (U.S. EPA) issuing a long-term health advisory level of 70 parts per trillion (ppt) (for a combination of PFOS and/or PFOA) (U.S. EPA 2016), led to 6.5 million people's drinking water being out of compliance in 2016 (Hu et al. 2016a). Acceptable guidance concentrations for PFOS and PFOA in drinking water are in the ppt range, which is significantly lower than the parts per billion (ppb) criteria established for many other contaminants. PFASs are known to threaten drinking water supplied in many countries with increased awareness and regulatory scrutiny being most evident in Scandinavia, Germany, Australia, and Canada.

Globally, environmental regulations considering PFASs are rapidly being promulgated to conservative (i.e., low) levels. The evolving regulatory attention has generally focused on perfluorinated compounds and is related to the restricted production and use of PFOS under the 2009 international Stockholm Convention on persistent organic pollutants (POPs).

None of the several thousand PFASs can biodegrade and cannot be referred to as being biodegradable (Colosi, Pinto et al. 2009; Liu and Mejia Avendano 2013; Luo, Lu et al. 2015; Ochoa-Herrera, Field et al. 2016). Biodegradation by definition, refers to conversion of the organic compound to carbon dioxide, water, and ammonia etc. (OECD 2002). Polyfluoroalkyl substances can biotransform in the environment and in higher organisms to form the perfluorinated substances, as "dead end" daughter products which appear to persist indefinitely, and PFASs have been termed "forever chemicals" (Allen 2018).

PFASs differ from many common contaminants such as fuel hydrocarbons and chlorinated solvents because they are extremely persistent. Fuel hydrocarbons and chlorinated solvents are subject to microbial attack and can be metabolized and naturally detoxified in the environment. PFASs are also distinct from other POPs such as many highly chlorinated compounds such as polychlorinated biphenyl (PCBs). These POPs have been subject to international regulations, as a result of a combination of their persistence (P), bioaccumulation (B) potential, and toxicity (T). The long-chain PFASs generally exhibit all the PBT properties, however in addition they are water soluble and hence significantly more mobile in the environment. Most

POPs, such as PCBs, are not highly water soluble as they are hydrophobic and bioaccumulate through interaction with lipids. Long-chain PFASs can bioaccumulate and biomagnify through interaction with proteins, while short-chain PFASs concentrate in the edible portion of crops. So PFASs differ from these other contaminant classes as they exhibit extreme persistence, in addition to being mobile and accumulating or concentrating in the biosphere through a distinctly differing mode of protein interaction, not observed with other contaminant classes.

As mentioned earlier, PFASs are generally soluble and hence very mobile in the environment, as opposed to the organochlorine POPs, so PFASs have the potential to impact a greater number of receptors. Depending on the site setting, they can be transported with groundwater well beyond the original source area, and form large plumes.

The long-chain PFASs accumulate in humans through consumption of impacted drinking water. Replacement PFASs are short-chain (often termed C6 or C4) and while the understanding of their toxicology and bioaccumulation potential is evolving, there is some evidence that short-chained PFASs accumulate in the edible portion of crops and are more mobile in the environment than long chain compounds, making them a potentially larger threat.

Given growing evidence of human health risks and potential ecological harm, more countries are now regulating an increasing number of PFASs, including both long- and short-chain compounds, while the latter are still commonly used as commercial replacements (e.g., C6 in firefighting foams).

As PFASs emerge as contaminants of concern, regulatory focus is expanding from an initial focus on long-chain PFAAs to other shorter-chain PFAAs, perfluoroalkyl ethers and further to a limited number of the thousands of polyfluorinated PFASs.

There are many more proprietary PFASs present in commercial products than are regulated. These polyfluoroalkyl substances can evade detection by common analytical methods but will ultimately be transformed in the environment, or in higher organisms, to perfluoroalkyl substances, which persist indefinitely. Firefighting foams, for example, comprise hundreds of individual polyfluoroalkyl PFASs, which have not been accounted for until recent analytical advances have enabled the total amount of PFASs to be measured using a novel technology termed the total oxidizable precursor (TOP) assay. Regulators in Australia have recently adopted this advanced analytical tool for sampling environmental matrices and compliance.

The increased regulatory focus of PFASs has created the requirement for solutions for their environmental management. It is very difficult to remove and destroy PFASs using conventional water treatment technologies, particularly the more mobile the short-chain PFASs, so they can recirculate within the water cycle indefinitely.

Current attention is focused on ensuring protection of drinking water. However, it is important to recognize that low levels of PFAS have been detected in our environment for many years. The global transport of PFASs via atmospheric evaporation and condensation, leading to aerial deposition, in addition to transit via groundwater, surface water and ocean currents, means that many locations that are remote from point sources of PFASs will have low-level detections of PFASs in multiple matrices.

This is in part due to the fact that we live in advanced industrialized societies and a wide array of consumer products have incorporated PFASs in their formulation for decades and continue to do so.

This chapter is split into several sections that describe the significance of many aspects of the complex chemistry of PFASs, their environmental effects, and their relevance to practitioners managing PFAS-impacted sites.

The chemistry of this large group of xenobiotics influences their fate and transport, transformation in environmental matrices and higher organisms, and potential to bioaccumulate in both humans and differing biota. The aim is to describe the PFAS's chemistry, to enable some generalizations to be made, where possible, about the differing members and subclasses of the group, which enable a better understanding of how to characterize them, evaluate their fate and transport in the environment, and pragmatically understand how to manage the risks they may pose to human health and environmental receptors. The remedial technologies that are evolving to manage PFASs are also described in relation to their ability to treat different PFASs in the class, be that to remove and concentrate, stabilize, or mineralize PFASs.

3.2 PFASs CHEMISTRY

PFASs are a very diverse class of chemicals united by the common structural element of a fully fluorinated carbon (alkyl) chain, known as the perfluoroalkyl group. The carbon-fluorine bond is the strongest in organic chemistry, so when multiple fluorine atoms saturate an alkyl chain to create the perfluoroalkyl moiety, this imparts biological, thermal, and chemical stability to the molecule. The extreme stability of the perfluoroalkyl group is responsible for its ultra-persistence in the environment. The perfluoroalkyl group can also impart both hydrophobic and lipophobic properties and decreased surface tension, which are sought-after commercial properties.

PFASs comprise both polymeric and nonpolymeric forms with the nonpolymer PFASs of more current concern to environmental regulators. Polymers may use nonpolymeric PFASs in their manufacture, such as the use of PFOA and perfluoroalkyl ethers such as GenX in the production of polytetrafluorethylene (PTFE). PFAS polymers may also present long-term sources of nonpolymeric PFASs, as they biotransform in landfills over time and evolve nonpolymeric PFASs such as the PFAAs and intermediate compounds into leachates, described further in section 3.8.3.

The nonpolymeric PFASs comprise two major classes: perfluoroalkyl substances and polyfluoroalkyl substances. A summary of the different subgroups of the nonpolymeric PFASs is provided in Figure 3.1.

The perfluoroalkyl group chain lengths reported can range from a one-carbon backbone in length to over 20 carbons long (OECD 2018a).

The perfluoroalkyl group is attached to a functional group (or groups) that contains a range of other atoms and functionality. The bonds between the perfluoroalkyl group and functional groups can be direct or they can be via an ethyl (or propyl group), which has been described as a "spacer" in molecules termed fluorotelomers.

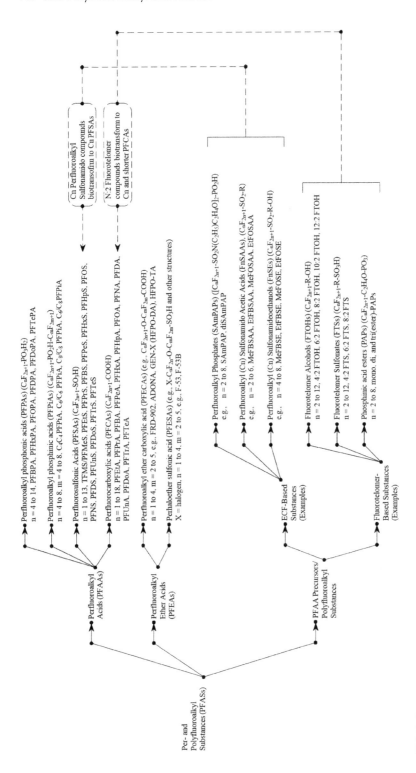

FIGURE 3.1 PFAS families and subgroups.

Buck et al. (2011) defined PFASs as compounds containing the perfluoroalkyl moiety $C_nF_{2n+1}^-$, or more specifically:

- Perfluoroalkyl substances: aliphatic substances for which all of the H atoms attached to C atoms in the nonfluorinated substance from which they are notionally derived have been replaced by F atoms, except those H atoms whose substitution would modify the nature of any functional groups present (e.g., hydroxyl —OH).
- Polyfluoroalkyl substances: aliphatic substances for which all H atoms attached to at least one (but not all) C atoms have been replaced by F atoms in such a manner that they contain the perfluoroalkyl moiety $C_nF_{2n+1}^-$.

The nonpolymeric compounds are generally based on two structural components: (1) the perfluoroalkyl group, which is generally hydrophobic (water repellent) and lipophobic (oil repellant) and is sometimes referred to as a "tail," and (2) a hydrophilic (water-soluble) polar functional group, such as a sulfonate or carboxylate group, sometimes described as the "head" group. For fluorotelomers, an ethyl (or propyl) group joins the perfluoroalkyl group to the functional group (Figure 3.2). There are exceptions to this generalization, such as where neutral, alcohol functional groups are present, as for fluorotelomer alcohols.

A perfluoroalkyl group with an elongated chain (approximately greater than C6) confers increasing hydrophobic/lipophobic properties, which in turn, when combined with the hydrophilic functional group, creates a surfactant. Surfactants may be described as being amphiphilic, meaning they contain both hydrophobic and hydrophilic groups. Surfactants lower the surface tension (or interfacial tension) between two liquids, between a gas and a liquid, or between a liquid and a solid, and may act as detergents, wetting agents, emulsifiers, foaming agents, and dispersants. The perfluoroalkyl group is both lipophobic and hydrophobic, and tends to drive migration of surfactant PFASs to the gas water interface. The water-soluble polar functional group can be made up of a wide range of different moieties such as: (a) anionic, such as carboxylates, sulfonates, and phosphonates; (b) cations, such as quaternary ammonium; (c) nonionic, such as acrylamide oligomers and polyethylene glycols; and (d) amphoteric, such as betaines and sulfobetaines (Buck et al., 2011).

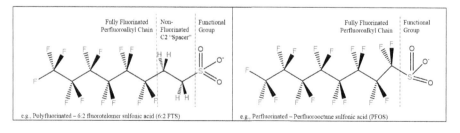

FIGURE 3.2 Structure of 6:2 FTS and PFOS.

3.2.1 Ionic State

Many PFASs possess terminal ionic functional groups in their structure, which can exist as charged anions, cations, or protonated acids as they can be in differing ionic states. These terms can be used interchangeably, as, for example, PFOS can be termed perfluorooctane sulfonic acid (a salt) or perfluorooctane sulfonate (a dissociated acid). When a PFAA salt or acid exists in water or other liquids, it will dissociate to form the anionic PFAA. As such, most PFASs exist in their charged ionic state (i.e., perfluorooctane sulfonate) under environmental conditions. Although laboratories may be reporting the carboxylic or sulfonic acids, the species actually measured will usually be the anionic form (e.g., octanoate or sulfonate).

3.2.2 Linear and Branched Isomers

Many PFASs may comprise isomers, which are mixtures of linear and branched forms. Isomers are compounds with the same empirical chemical formula but different molecular structures, and they contain the same number of atoms of each element but have different arrangements of their atoms. PFASs may contain linear perfluoroalkyl groups, meaning that the carbon atoms in the perfluoroalkyl chain are connected together in a single straight line. There can be only one linear isomer in the perfluoroalkyl moiety for PFASs of a specific perfluoroalkyl chain length. Alternatively, branched isomers can contain multiple branches in the carbon backbone of the perfluoroalkyl moiety, meaning at least one carbon atom is bonded to more than two carbon atoms. There can be multiple branched chain isomers of each chain length, which increases with the length of the carbon chain. PFOS is reported to be routinely present in many environmental samples as a mixture of the linear isomer and 10 branched isomers (Riddell et al., 2009; Rayne and Friesen 2008), whereas 89 isomers are theoretically possible (Rayne and Friesen 2008). Examples of the structural formula of linear and branched PFOS are presented in Figure 3.3.

The presence of branched or straight chain isomers in environmental samples is relevant as it may be indicative of the manufacturing process used to synthesize specific PFASs. The electrochemical fluorination (ECF) process creates branched chain isomers, whereas the fluorotelomerization (FT) process generates straight chain isomers, as described in more detail in section 3.4.2.

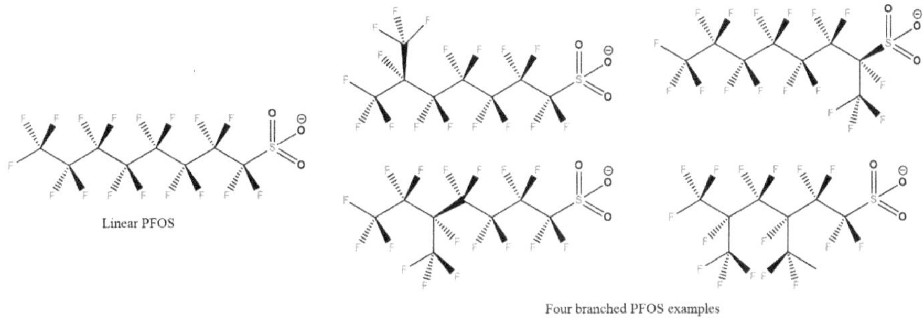

FIGURE 3.3 Examples of branched and linear PFOS isomers.

3.2.3 Perfluoroalkyl Substances

Perfluoroalkyl substances have previously been referred to as perfluorinated compounds (PFCs), but are now more commonly termed PFAAs. PFAAs comprise some of the simplest molecules in the class of PFASs though there are examples of di-substituted PFAAs, where two perfluoroalkyl chains are linked to the same functional group, such as di-substituted perfluoroalkyl phosphinic acids (PFPiAs) (Lang et al. 2017). A summary of the naming conventions for perfluoroalkyl carboxylic acids (PFCAs; also referred to as perfluoroalkyl carboxylates) and perfluoroalkyl sulfonic acids (PFSAs; also referred to as perfluoroalkyl sulfonates) is provided in Figure 3.4, with examples of PFAAs presented in Table 3.1.

PFAAs generally include a perfluoroalkyl or perfluoroalkyl ether chain with a terminal polar, acidic, terminal functional group and comprise six major groups (Wang et al. 2017). The more commonly measured PFAAs in the environment, and thus the subject of research, include the PFCAs, such as PFOA and PFSAs, such as PFOS. Further examples of PFCAs and PFSAs are given in Figure 3.4 for PFAAs containing between 1 and 12 carbons.

Perfluoroalkyl ether sulfonic acids (PFESAs) such as F-53B are also included as perfluorinated compounds by Wang et al. (2017), which also contains a terminal chlorine atom.

Perfluoroalkyl acid naming references the number of fluorinated (n) carbons.

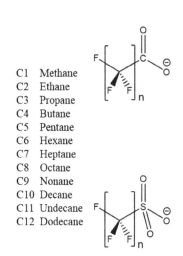

C1 Methane
C2 Ethane
C3 Propane
C4 Butane
C5 Pentane
C6 Hexane
C7 Heptane
C8 Octane
C9 Nonane
C10 Decane
C11 Undecane
C12 Dodecane

Perfluoroalkyl Carboxylates (PFCA)*
n=1 Perfluroethanoic acid (PFEtA)
n=2 Perfluoropropanoic acid (PFPrA)
n=3 Perfluorobutanoic acid (PFBA)
n=4 Perfluoropentanoic acid (PFPeA)
n=5 Perfluorohexanoic acid (PFHxA)
n=6 Perfluoroheptanoic acid (PFHpA)
n=7 Perfluorooctanoic acid (PFOA)
n=8 Perfluorononanoic acid (PFNA)
n=10 Perfluorodecanoic acid (PFDA)
n=11 Perfluoroundecanoic acid (PFUnDA)
n=12 Perfluorododecanoic acid (PFDoA)

Perfluoroalkyl Sulfonates (PFSA)
n=1 Perfluoromethanesulfonic acid (PFMeS)
n=2 Perfluoroethanesulfonic acid (PFEtS)
n=3 Perfluoropropanesulfonic acid (PFPrS)
n=4 Perfluorobutanesulfonic acid (PFBS)
n=5 Perfluoropentanesulfonic acid (PFPeS)
n=6 Perfluorohexanesulfonic acid (PFHxS)
n=7 Perfluoroheptanesulfonic acid (PFHpS)
n=8 Perfluorooctanesulfonic acid (PFOS)
n=10 Perfluorodecanesulfonic acid (PFDS)
n=11 Perfluoroundecansulfonic acid (PFUnDS)
n=12 Perfluorododecansulfonic acid (PFDoS)

* The final carbon atom in a PFCA compound is part of the carboxyl functional group and is therefore, not part of the perfluoroalkyl chain.

FIGURE 3.4 Naming convention for PFCAs and PFSAs.

TABLE 3.1
Common perfluoroalkyl acids

Perfluoroalkyl Acid	Acronym	Terminal functional group where $R = C_nF_{2n+1}^-$	Example PFASs	Example PFASs Chemical Drawing	Documented Uses*
Perfluoroalkyl carboxylic acids / Perfluoroalkyl carboxylates	PFCAs	$R-COO^-/H^-$	Perfluorooctanoic acid (PFOA) / Perfluorooctanoate (PFOA)	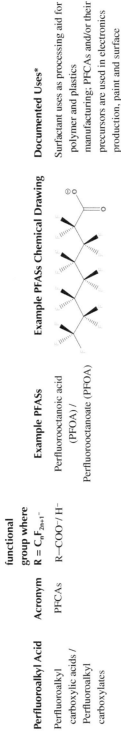	Surfactant uses as processing aid for polymer and plastics manufacturing; PFCAs and/or their precursors are used in electronics production, paint and surface coatings, agro-chemical, paper coatings, textiles, personal care products, etc.
Perfluoroalkyl sulfonic acids / Perfluoroalkyl sulfonates	PFSAs	$R-SO_3^-/-SO_3H$	Perfluorooctane sulfonic acid (PFOS) / Perfluorooctane sulfonate (PFOS)		Surfactant uses in Class B firefighting foams, metal plating, carpets, furniture, outdoor clothing, leather impregnation, food packaging, copy paper, paints/varnish, pesticide formulations, hydraulic fluids for the aviation industry, electronic equipment and components, etching solutions for semiconductor manufacture, photographic film coating and processing, acid catalyst and raw materials for ionic liquids, personal care products, etc.

(*Continued*)

TABLE 3.1 (Continued)

Perfluoroalkyl Acid	Acronym	Terminal functional group where $R = C_nF_{2n+1}^-$	Example PFASs	Example PFASs Chemical Drawing	Documented Uses*
Perfluoroalkyl phosphonic acids	PFPAs	$R-P(=O)(OH)_2$	Perfluorooctyl phosphonic acid (C8-PFPA)	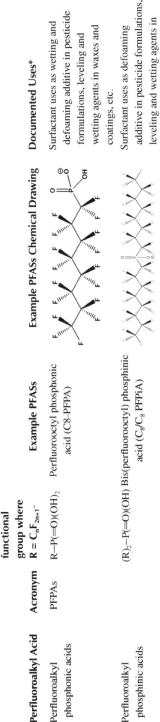	Surfactant uses as wetting and defoaming additive in pesticide formulations, leveling and wetting agents in waxes and coatings, etc.
Perfluoroalkyl phosphinic acids		$(R)_2-P(=O)(OH)$	Bis(perfluorooctyl) phosphinic acid (C_8/C_8 PFPiA)		Surfactant uses as defoaming additive in pesticide formulations, leveling and wetting agents in waxes and coatings, etc.
Perfluoroalkyl ether acids / Perfluoroether carboxylic acids	PFECAs	Example: $C_nF_{2n+1}-O-C_m$ $F_{2m+1}-COO^-/H$	Hexafluoropropylene oxide-dimer acid (HFPO-DA) / GenX	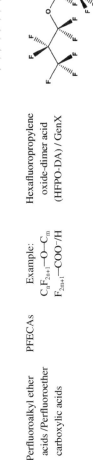	Surfactant uses as processing aid for plastics manufacturing; non-critical applications in the semiconductor manufacture process, other specialty agrochemical and pharmaceutical applications, etc.
Perfluoroalkyl ether sulfonic acids / Perfluoroether sulfonic acids	PFESAs	Example: $X-C_nF_{2n+1}-O-C_m$ $F_{2m+1}-SO_3^-/H$	6:2 Chlorinated perfluoroalkyl ether sulfonate (6:2 Cl-PFESA)/ F-53B (or 6:2 Cl-PFESA)		Surfactant uses in metal plating, etc.

Note:
* The reader is directed to Figure 3.14 for more information related to documented uses, including references.

It is noted that perfluoroalkyl ether carboxylic acids (PFECAs) and PFESAs have been included in the class of perfluoroalkyl substances by Wang et al. (2017). PFECAs and PFESAs have not been reported to be biodegradable; therefore, in terms of their environmental properties, it seems appropriate to group them with PFAAs. There are many emerging PFECAs and PFESAs, some of which are described as potential precursors (OECD 2018a).

When a polar terminal functional group is combined with a longer perfluoroalkyl chain (above approximately six carbons), this can provide unique surface-active properties to the molecule as a whole. The shorter chains are less effective as surfactants because they are less amphiphilic more water soluble and thus less likely to adhere to surfaces and more mobile in the environment (Kjølholt et al., 2015).

3.2.3.1 Perfluoroalkyl Sulfonic Acids

PFSAs have the generic structure $C_nF_{2n+1}SO_3H$, with perfluoroalkyl chain lengths described as low as one carbon such as trifluoromethane sulfonate) (TFMS) (De Voogt 2010), with two and three carbon compounds, such as perfluoroethane sulfonate (PFEtS) and perfluoropropane sulfonate (PFPrS) (Barzen-Hanson and Field 2015; Dreyer et al. 2010) also reported. Chain lengths up to 18 carbons have also been reported (Brice 1956) depending on the hydrocarbon starting material used for production.

PFOS is a PFSA and comprises a chain of eight fully fluorinated carbon atoms with a sulfonate group as the functional group bonded to the terminal carbon. It is the most well-studied and commonly assessed PFSA in the environment (Giesy et al. 2010; Kjølholt 2015). The structure of PFOS is shown in Figure 3.5. PFOS is generally manufactured as a salt of ammonium, diethanolamine, potassium, or lithium, but is found as an anionic species in water. PFOS comprises a mixture of linear and branched isomers, as a result of being manufactured using the ECF process, with approximately 70 percent linear and approximately 30 percent branched (Pancras et al. 2016; Benskin et al. 2010).

PFSAs of differing chain lengths can also be production by-products of manufacture. Production of the longer perfluoroalkyl chain sulfonates has largely been phased out, except in China and India, and PFSAs bearing a shorter C4 perfluoroalkyl chain have been introduced as alternatives (Hori et al. 2006; Ritter 2010). For example, perfluorobutane sulfonate (PFBS) is one important replacement substance for PFOS (Herzke et al. 2007). The C6 PFSA (PFHxS) is classed as long chain, as it shows significant bioaccumulation potential.

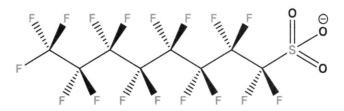

FIGURE 3.5 Chemical structure of the PFOS linear isomer.

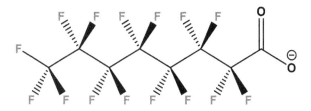

FIGURE 3.6 Chemical structure of linear PFOA.

3.2.3.2 Perfluoroalkyl Carboxylic Acids

PFCAs have the generic structure $C_nF_{2n+1}CO_2H$, with perfluoroalkyl chain lengths described with two carbons such as perfluoroethanoic acid (PFEtA) (Barzen-Hanson and Field 2015; Dreyer et al. 2010) and up to 18 carbons (Moller et al. 2010) and longer have been reported (OECD 2018a).

PFOA is a PFCA and comprises a fully fluorinated chain of seven carbon atoms (perfluoroalkyl moiety) with a nonfluorinated carbon atom within the carboxylate group bonded to the terminal carbon on the perfluoroalkyl moiety; it is also commonly termed "C8" (Lyons 2007). PFOA is the most commonly encountered PFCA (Lindstrom, Strynar, and Libelo 2011). Its structure is presented in Figure 3.6.

PFOA is used as either the undissociated carboxylic acid or one of its carboxylate salts such as the ammonium perfluorooctanoate (APFO) as an example PFOA salt. APFO has several synonyms, including C8 acid, FC 143, and perfluoroammonium octanoate (Kissa 2001).

PFCAs may have been manufactured using the ECF process or the FT process, such that they may be found as a mixture of linear and branched isomers if produced via synthesis and manufacturing using the ECF process, or be solely linear isomers via manufacture using the FT process (Pancras et al. 2016) (as described in section 3.4).

Products containing PFCAs often comprise a mixture of perfluoroalkyl chain lengths, not just C8 (Guo et al. 2009); therefore, a range of differing PFCA chain lengths may be expected to be present in environmental samples. Additionally, as production of the longer perfluoroalkyl chain carboxylates is being phased out, shorter chains such as C6 compounds are increasingly being used (Ritter 2010).

3.2.3.3 Perfluoroalkyl Phosphonic and Phosphinic Acids

Perfluoroalkyl phosphonic acids (PFPAs) have been described as one of the three major groups of anionic fluorinated surfactants, along with PFCAs and PFSAs (Xiao et al. 2017). They have a generic structure of $C_nF_{2n+1}PO(OH)_2$. Phosphinic acids (PFPiAs) are dimers (i.e., a molecule consisting of two identical molecules linked together) with the generic structure $(C_nF_{2n+1})(C_mF_{2m+1})PO(OH)$ where n,m equals 6, 8, 10, or 12, though some PFPiA dimers are reported as comprising C_2/C_2 and C_4/C_4 perfluoroalkyl groups (Wang, Cousins et al. 2016). The C_2/C_2 and C_4/C_4 PFPiAs are reported to be manufactured via the ECF process, while the remaining PFPAs and PFPiAs with C_6, C_8, C_{10}, or C_{12} perfluoroalkyl groups are reported to be synthesized via

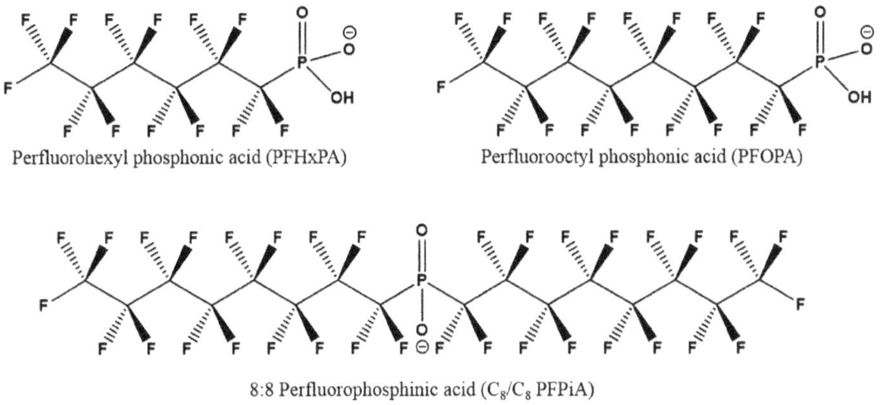

Perfluorohexyl phosphonic acid (PFHxPA)

Perfluorooctyl phosphonic acid (PFOPA)

8:8 Perfluorophosphinic acid (C_8/C_8 PFPiA)

FIGURE 3.7 Chemical structure of PFHxPA, PFOPA, and C_8/C_8 PFPiA.

the FT process (Wang, Cousins et al. 2016). More recently, C_4/C_4 PFPiA and its derivatives have been presented as alternatives to long-chain PFASs in certain applications.

The structure of perfluorooctano phosphonic acid (PFOPA) and C_8/C_8 perfluorophosphonic acid (C_8/C_8 PFPiA) is shown in Figure 3.7. Given their fully fluorinated structure, it is considered that PFPiAs will be equally as resistant to biodegradation as the other PFAAs.

3.2.3.4 Perfluoroalkyl Ether Carboxylates and Perfluoroalkyl Ether Sulfonates

PFECAs have oxygen atoms inserted in the perfluoroalkly chain with an example structure $C_nF_{2n+1}-O(C_mF_{2m})OCHF(C_pF_{2p})COOH$ and PFESAs have the example structure $X-C_nF_{2n+1}-O-C_mF_{2m+1}-SO_3^-/H$. There are several examples of PFECAs such as hexaflropropylene hexafluoropropylene oxide-dimer acid (HFPO-DA) commonly known as GenX, hexafluoropropylene oxide-trimer acid (HFPO-TA), and ammonium 3H-perfluoro-3-[(3-methoxy-propoxy)propanoic acid] (ADONA), shown in Figure 3.8. Multiple structurally similar PFECAs have been developed to replace PFOA in fluoropolymer production by different companies (Wang, Cousins et al. 2013). Many novel PFECAs are listed in the OECD global database of PFASs (OECD 2018a). Consequently, there is a proliferation of new PFASs on the market, which will likely continue in the foreseeable future (Wang et al. 2017).

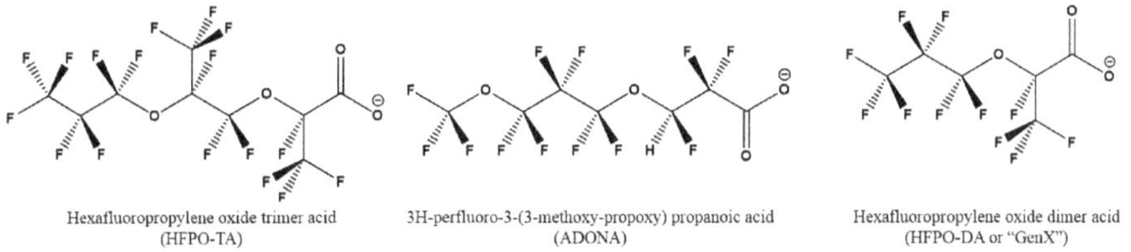

Hexafluoropropylene oxide trimer acid (HFPO-TA)

3H-perfluoro-3-(3-methoxy-propoxy) propanoic acid (ADONA)

Hexafluoropropylene oxide dimer acid (HFPO-DA or "GenX")

FIGURE 3.8 Chemical structure of ADONA, HFPO-DA, and HFPO-TA.

3.2.4 POLYFLUOROALKYL SUBSTANCES

Polyfluoroalkyl substances contain the perfluoroalkyl moiety ($C_nF_{2n+1}^-$) but also contain carbon to hydrogen bonds within their molecular structure, such as fluorotelomers and sulfonamide compounds. They may also have more complex structures and terminal functional groups which can be neutral, anionic, cationic, or zwitterionic. Polyfluoroalkyl substances comprise by far the more diverse group of compounds when compared to perfluoroalkyl substances, with a wider range of molecular weights and potential functional groups, as thousands of polyfluoroalkyl compounds have been synthesized for a broad array of commercial uses.

Polyfluorinated compounds can be broadly separated into two main classes: (1) those manufactured via the ECF process, which for PFSA production, involves perfluoroalkane sulfonyl fluorides (PASFs) as intermediates; and (2) those manufactured via the fluorotelomerization process, which involves perfluoroalkyl iodides (PFAIs) as process intermediates (OECD 2013; Buck et al. 2011; Wang et al. 2017).

3.2.4.1 ECF-Derived Polyfluoroalkyl Substances

Electrochemical fluorination (ECF) has been used to manufacture a variety of polyfluoroalkane sulfonamide (FASA) based substances. This process attached additional functional groups and alkyl chains to a perfluoroalkyl sulfonamide moiety to create hundreds of differing polyfluorinated compounds (Wang et al. 2017). A selection of the FASA-based compounds characterized to date include the following (with some molecular structures presented in Figure 3.9):

- FASAs, such as perfluorooctane sulfonamide (FOSA) and methyl/ethyl derivatives, such as N-methyl perfluorooctane sulfonamide (N-MeFOSA) and N-ethyl perfluorooctane perfluorooctane sulfonamide (N-EtFOSA).
- Perfluoroalkane sulfonamidoethanols (FASEs), such as perfluorooctane sulfonamidoethanol (FOSE) and their methyl and ethyl derivatives, such as N-methyl perfluorooctane sulfonamidoethanol (N-MeFOSE) and N-ethyl perfluororooctane sulfonamidoethanol (N-EtFOSE).
- Perfluoroalkane sulfonamidoacetates, such as perfluorooctane sulfonamidoacetate (FOSAA) and their methyl and ethyl derivatives, such as N-methyl perfluorooctane sulfonamidoacetate (N-MeFOSAA) and N-ethyl perfluororooctane sulfonamidoacetate (N-EtFOSAA).
- EtFOSE-based phosphate esters such as sulfonamidoethanol-based phosphate esters (SAmPAP) including mono, di and triesters, such as 2-(N-ethylperfluorooctane-1-sulfonamideo)ethyl phosphate (SAmPAP) and bis-[2-(N-ethylperfluorooctane-1-sulfonamido)ethyl] phosphate (diSAmPAP).
- Perfluoroalkyl sulfonamide amines such as N,N-dimethyl-3-{[(trideca-perfluorohexyl)sulfonyl]amino}propan-1-aminium (PFHxSaAm), N,N-dimethyl-3-{[(trideca-perfluorooctyl)sulfonyl]amino}propan-1-aminium

(PFOSaAm), 3-(*N*-(2-carboxyethyl)-trideca-perfluorohexylsulfonamido)-*N,N*-dimethylpropan-1-aminium (PFHxSaAmA) and 3-(*N*-(2-carboxyethyl)-trideca-perfluorohoctylsulfonamido)-*N,N*-dimethylpropan-1-aminium (PFOSaAmA).

These compounds have been classed as potential precursors to PFSAs or PFOS (assuming a C8 perfluoroalkyl chain) and are also termed pre-FOS and PFOS-related substances (Martin et al. 2010; Benskin et al. 2013). In the environment and within biota they all will ultimately transform to PFSAs as a terminal product given adequate time for biotic or abiotic reactions to occur. There are numerous potential PFOS-precursors that have not been previously detected in the environment and for which little is known about their stability (Benskin, Ikonomou, Gobas et al., 2012), though some PFSA precursors discovered in fire-fighting foams have been characterized (Place and Field 2012; Backe et al. 2013), as described in section 3.2.9.

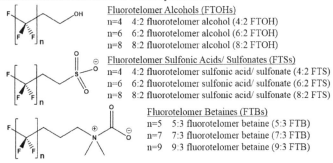

FIGURE 3.9 Naming conventions for FASA-based and fluorotelomer-based PFASs.

3.2.4.2 Fluorotelomerization-Derived Polyfluoroalkyl Substances

Fluorotelomers are created using PFAIs as feedstock chemicals that are converted into fluorotelomer iodides, which adds an alkyl group (usually ethyl) to the perfluoroalkyl chain to create a fluorotelomer iodide (OECD 2018a), which is the feedstock to further modify and create hundreds (Wang et al. 2017), if not potentially thousands, of fluorotelomer molecules which are PFCA or PFPA/PFPiA precursors (Wang et al. 2014a; Wang et al. 2014b; OECD 2018a). There is a huge diversity of fluorotelomers in many products with many recently characterized in firefighting foams (Barzen-Hanson, Roberts et al., 2017).

Some examples of fluorotelomer compounds include:

- Fluorotelomer alcohols (FTOHs) such as 4:2 fluorotelomer alcohol (4:2 FTOH) and 6:2 fluorotelomer alcohol (6:2 FTOH).
- Fluorotelomer sulfonic acids (FTSs) such as 6:2 fluorotelomer sulfonate/sulfonic acid (6:2 FTS) and 8:2 fluorotelomer sulfonate/sulfonic acid (8:2 FTS).
- Fluorotelomer carboxylic acids (FTCAs) such as 5:3 fluorotelomer carboxylate/carboxylic acid (5:3 FTCA) commonly known as the "5:3 acid."
- Fluorotelomer betaines (FTBs) such as 5:3 fluorotelomer betaine (5:3 FTB) and 9:3 fluorotelomer betaine.
- Polyfluoroalkyl phosphates (PAPs) such as the diPAPs including 6:2 di-substituted polyfluoroalkyl phosphate (6:2 diPAP) and 8:2 di-substituted polyfluoroalkyl phosphate (8:2 diPAP).

The FTOHs tend to be volatile, whereas the FTSs and FTCAs are ionic and so nonvolatile and water soluble. The FTSs and FTCAs are also stable or meta-stable partial transformation intermediates of multiple other fluorotelomer compounds with more complex structures and functional groups. All of these polyfluorinated compounds will eventually transform to create extremely persistent PFAAs, with PFCAs generally being the transformation products formed from fluorotelomers. The term "PFOA-related substances" has also been used to describe precursors that contain a perfluoroalkyl group that will transform in the environment to form PFOA (OECD 2005; OECD 2007a; European Chemicals Agency [ECHA] 2014; ECHA 2015).

A small number of polyfluoroalkyl substances are shown in Table 3.2. Note that fluorotelomer compounds are named using a differing system to PFAAs and polyfluoroalkyl sulfonamide compounds. The naming system identifies both the number of perfluorinated carbons (n) in the perfluoroalkyl group and the number of nonfluorinated carbons (m) within the carbon chain; thus an eight-carbon chain comprising six perfluorinated carbons and two nonfluorinated carbons is referred to as a 6:2 fluorotelomer (such as 6:2 FTS and 6:2 FTOH), a similar eight-carbon alkyl chain with five perfluorinated carbons and three nonfluorinated carbons (such as 5:3 FTCA). The naming systems for the two main precursor classes are shown in Figure 3.9.

TABLE 3.2
Selection of known polyfluoroalkyl substances

Class	Group	Acronym(s)	Terminal structure and functional group $C_nF_{2n+1}R$, where R=	Example(s)	Example PFASs Chemical Drawing	Documented Use*
Fluorotelomers	n:2 Fluorotelomer alcohols	n:2 FTOH	$-CH_2CH_2OH$	6:2 Fluorotelomer alcohol (6:2 FTOH)		Associated with manufacture of textile and leather manufacture, paper- and food-packaging metal coatings and paints, cookware, cleaning agents, photographic film, etc.; also associated with biotransformation of FTSs in firefighting foams.
	n:2 Fluorotelomer sulfonic acids	n:2 FTSs / n:2 FTSA	$-CH_2CH_2SO_3H / -SO_3^-$	6:2 Fluorotelomer sulfonic acids (6:2 FTS / 6:2 FTSA)		Found as a component of firefighting foams, used in coated fabrics, copy paper, printed circuit boards, and as an alternative to PFOS in the metal plating industry, etc.
	Fluorotelomer carboxylic acids	FTCAs	$-CH_2COOH$	6:2 Fluorotelomer carboxylic acid (6:2 FTCA)		Used in textiles and paper coatings; also found as a transformation intermediate associated with firefighting foams.
			$-CH_2CH_2COOH$	5:3 Fluorotelomer carboxylic acid (5:3 Acid)		Found as a stable transformation intermediate associated with firefighting foams.

(Continued)

TABLE 3.2 (Continued)

Class	Group	Acronym(s)	Terminal structure and functional group $C_nF_{2n+1}R$, where R=	Example(s)	Example PFASs Chemical Drawing	Documented Use*
	n:2 Fluorotelomer Betaines	n:2 FTBs	$-CH_2CH_2N^+(CH_3)_2CH_2COOH$	7:2 fluorotelomer betaine		Found as a component of firefighting foams
		n:2 FTBs	$-CHFCH_2CH_2N^+(CH_3)_2CH_2COOH$	7:1:2 fluorotelomer betaine		Found as a component of firefighting foams
	n:2 Fluorotelomer thioether amido sulfonic acids/ n:2 Fluorotelomer captoalkylamido sulfonate	n:2 FTTAoSs/ n:2 FTSASs	$-CH_2CH_2SCH_2CH_2$ $CONHC(CH_3)_2CH_2SO_3H$	6:2 fluorotelomer thioether amido sulfonic acid		Found as a component of firefighting foams
	n:2 Fluorotelomer sulfonamide betaine	n:2 FTSBs	$-CH_2CH_2SO_2NHCH_2$ $CH_2CH_2N^+(CH_3)_2CH_2COOH$	6:2 fluorotelomer sulfonamide betaine		Found as a component of firefighting foams
	mono-, di-, and tri-substituted polyfluoroalkyl phosphates	n:2 PAPs	$(-CH_2)_2-PO_3H_2$	6:2 mono-substituted polyfluoroalkyl phosphate		Found in food paper packaging, other surface treatments, etc.
		n:2 diPAPs	$(-CH_2CH_2O)_2-PO_2H$	6:2 di-substituted polyfluoroalkyl phosphate		

Class	Group	Acronym(s)	Terminal structure and functional group $C_nF_{2n+1}R$, where R=	Example(s)	Example PFASs Chemical Drawing	Documented Use*
Perfluoroalkyl sulfonamido substances	N-Alkyl perfluoroalkane sulfonamides	N-MeFASAs, N-EtFASAs, N-BuFASAs	$-SO_2NH(R')$ where $R' = C_mH_{2m+1}$ ($m = 1,2,4$)	N-Methyl perfluorooctane sulphonamide (MeFOSA)		Major raw material for surface protection products; also found as breakdown intermediates.
				N-Ethyl perfluorobutane sulfonamide (EtFBSA)		
				N-Butyl perfluorooctane sulphonamide (BuFOSA)		
	N-alkyl perfluoroalkane sulfonamidoacetic acids	N-MeFASAAs, N-EtFASAAs, N-BuFASAAs	$-SO_2NH(R')CH_2COOH$ where $R' = C_mH_{2m+1}$ ($m = 0,1,2,4$)	N-Ethyl perfluorooctane sulfonamidoacetic acid (EtFOSAA)		Found as an intermediate environmental transformation product.
	Perfluoroalkyl sulfonamido amines	PFBSaAm PFPeSaAm PFHxSaAm PFHpSaAm PFOSaAm	$-SO_2NH(C_3H_6)NH_4^+$	Perfluorohexane sulfonamide amine (PFHxSaAm)		Found as cations in ECF-derived firefighting foams, etc.

(Continued)

TABLE 3.2 (Continued)

Class	Group	Acronym(s)	Terminal structure and functional group $C_nF_{2n+1}R$, where R=	Example(s)	Example PFASs Chemical Drawing	Documented Use*
	Perfluoroalkyl sulfonamide amino carboxylates	PFBSaAmA PFPeSaAmA PFHxSaAmA PFHpSaAm APFOSaAmA	$-SO_2N(C_2H_4COO^-)C_3H_6N^+(CH_3)_2$	Perfluorohexane sulfonamide amino carboxylate (PFHxSaAmA)	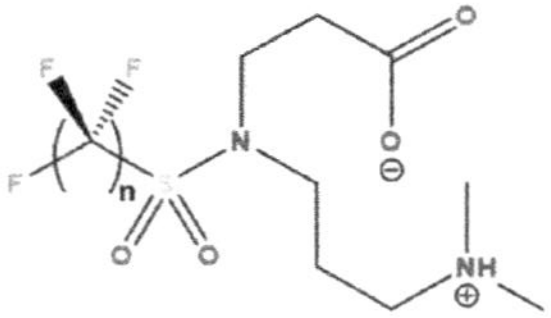	Found as zwitterions in ECF derived firefighting foams, etc.
	N-ethyl perfluoroalkyl sulfonamido ethanol-based phosphate esters (SAmPAP mono-, di-, and triester)	SAmPAP monoester SAmPAP diester SAmPAP triester	Monoester $-SO_2N(CH_2CH_3)CH_2CH_2OPO(OH)_2$ Diester $[-SO_2N(CH_2CH_3)CH_2CH_2O]_2PO_2H$ Triester $[-SO_2N(CH_2CH_3)CH_2CH_2O]_3PO$	N-ethyl perfluorooctane sulfonamido ethanol-based phosphate triester (SAmPAP triester)	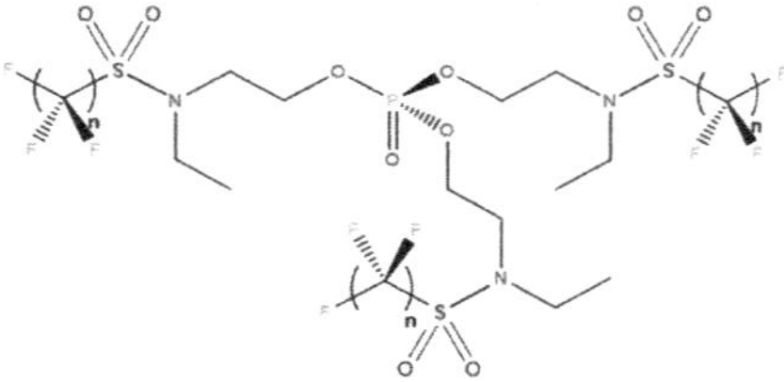	Found in food paper packaging, other surface treatments/protectants, etc.

Note:

* The reader is directed to Figure 3.14 for more information related to documented uses, including references.

3.2.5 LONG- AND SHORT-CHAIN PFASs

PFSAs and PFCAs and their potential precursors may be subdivided into two broad classes, short-chain and long-chain PFASs (OECD 2013). As shown in Figure 3.10, long chain refers to:

- Perfluoroalkyl carboxylic acids, with seven or more perfluoroalkyl carbon atoms e.g. PFOA (with eight carbons including the carbon on the carboxyl functional group) and perfluorononanoic acid PFNA (with nine carbons including the carbon on the carboxyl functional group).
- Perfluoroalkyl sulfonates, with six or more perfluoroalkyl carbon atoms (e.g., PFHxS, PFHpS and PFOS).

Long-chain PFAS terminology is also applied to substances that have the potential to degrade to long-chain PFCAs or PFSAs (i.e., precursors such as PASF- and fluorotelomer-based compounds).

As shown in Figure 3.10, short chain refers to:

- Perfluoroalkyl carboxylic acids with six or fewer perfluoroalkyl carbon atoms (e.g., perfluoroheptanoic acid [PFHpA] and perfluorohexanoic acid [PFHxA]).
- Perfluoroalkyl sulfonates with five or fewer perfluoroalkyl carbon atoms e.g., perfluoropentanoic sulfonate (PFPeS) and PFBS.

There are also polyfluorinated PFASs that transform to produce PFCAs or PFSAs which are short-chain PFASs (i.e., PFCAs with two to six fluorinated alkyl carbon atoms or PFSAs with one to five fluorinated alkyl carbon atoms).

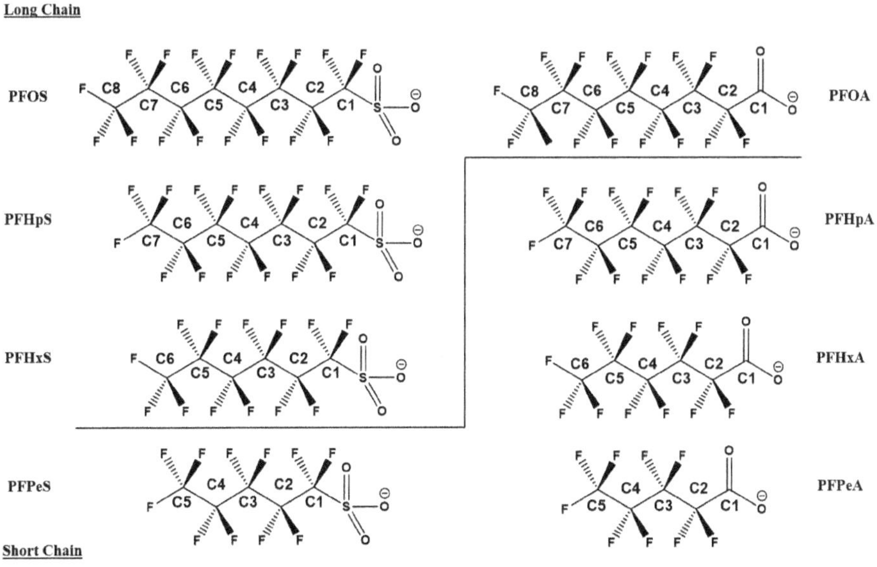

FIGURE 3.10 Examples of long- and short-chain PFAAs.

Studies have been conducted that indicate long-chain PFASs have a higher potential to bioconcentrate and bioaccumulate through trophic levels as compared to shorter-chain PFASs (Stahl et al. 2012). The long-chain PFASs are described as exhibiting biopersistence, which describes compounds that tend to remain within an organism, rather than being expelled or broken down, and demonstrate slow clearance from organisms. In 2013, the OECD/United Nations Environment Programme (UNEP) Global PFC Group reported that it distinguished between long-chain and short-chain PFASs based on toxicity and bioaccumulation differences between them (OECD 2013). However, it should be noted that polyfluorinated precursors to PFAAs have been described as more toxic than the PFAAs they transform into (Phillips et al. 2007).

PFAS typically bioaccumulate via interaction with proteins, such as albumin, rather than (the more usual) fats and lipids where highly chlorinated organic compounds tend to partition to and have been described as proteinophillic (Department of Toxic Substances Control [DTSC] 2018), as described further in section 3.6.3. Short-chain PFASs are found to bioaccumulate in the edible portion of crops such as fruits (Blaine et al. 2013; Blaine et al. 2014a; Blaine et al. 2014b; Sepulvado et al. 2011). Human serum elimination data has documented that some short-chain PFASs are eliminated by humans on the order of days whereas long-chain PFASs are eliminated on the order of several years or longer (OECD 2013). As such, it is important not to make generalizations about PFAA behavior within organisms based only on chain length. Other factors surrounding complex interactions with proteins may affect bioaccumulation potential of PFASs (Ng and Hungerbuhler 2013). Additionally, short-chain PFASs typically have higher water solubility compared to long-chain PFASs which makes them more mobile in aquifers and less effectively treated by granular activated carbon (GAC) (Appleman et al. 2013; Appleman et al. 2014).

Note that other classes of PFASs such as PFPAs, PFPiAs, PFECAs, and PFESAs are not specifically defined in terms of long-chain versus short-chain PFASs (OECD 2015). Some PFECAs, such as hexafluoropropylene oxide trimer acid (HFPO-TA), have bioconcentration factors that are significantly higher than that of legacy PFOA, suggesting greater bioaccumulation potential in aquatic organisms (Pan et al. 2017).

3.2.6 Polymeric PFASs

Polymeric PFASs are high molecular weight structures that contain small repetitive perfluoroalkyl moieties polymerized together to form the fluoropolymer. OECD defines the polymeric PFASs group into three distinct groupings (OECD 2013):

- Fluoropolymers—In which the fluorine atoms are directly bound to the carbon backbone. Examples include polyvinylidene fluoride (PVDF), polytetrafluoroethylene (PTFE), fluorinated ethylene propylene (FEP), perfluoroalkoxy polymer (PFA) etc. PFCAs, such as PFOA (for PTFE) and PFNA (for PVDF), have been used extensively as processing aids in the polymerization process of certain types of fluoropolymers, with perfluoroalkyl ethers often now being used as the replacement chemistry (Prevedouros et al., 2006, National Institute for Public Health and the Environment [RIVM] 2017a).

- Side-chain fluoropolymers—In which a perfluoroalkyl group is contained on the side chains and bound to a backbone of carbon with hydrogen substituents, so there is potential that some are PFAA precursors. Examples include fluorinated (meth)acrylate polymers and fluorinated urethanes. The rates associated with biotransformation of the polymers to create PFAAs has been studied, demonstrating PFAA formation (Russell et al. 2008; Washington et al. 2014; Washington et al. 2015).
- Perfluoropolyethers (PFPEs)—Comprise fluorinated polymers with a backbone of ether-linked carbon atoms, with the fluorine atoms directly attached to the carbon atoms. They cannot degrade to PFAAs, but PFAAs may be used in manufacturing or present as potential impurities (DTSC 2018).

The majority of PFOS-related substances (i.e., PFOS precursors) have been described as high molecular-weight polymers (Stahl et al. 2012).

3.2.7 Replacement PFASs

Manufacturers of PFASs have developed replacements for the long-chain PFAAs, which has generally involved shorter-chain perfluoroalkyl or polyfluoroalkyl substances. The main manufacturing processes now make compounds such as PFBS (as a replacement for PFOS), shorter perfluoroalkyl chain fluorotelomers (i.e., 6:2 FTOH) and perfluoroalklyethers (e.g., GenX and ADONA) and multiple other novel PFASs (Hori et al. 2006; OECD 2007b; OECD 2007a; Herzke et al. 2012; Wang et al. 2015; Wang, Cousins et al., 2013; Holmquist et al. 2016).

The replacement products are structurally similar to their predecessors (Pancras et al. 2016; Wang et al. 2015) with suggestion that there is uncertainty regarding the hazards associated with some of the replacements compared to the long-chain PFASs, as independently published information on many replacements is limited (Wang et al. 2015; Beekman et al. 2016).

3.2.8 Chemistry of PFASs in Class B Firefighting Foams

Fluorosurfactant PFASs contained in firefighting foams historically have been produced by both ECF and FT. Class B firefighting foams produced by both methods have been found to contain highly diverse mixtures of PFASs (Barzen-Hanson, Roberts et al., 2017), many of which are considered proprietary, so their molecular formula is unknown. Foams containing PFASs from the ECF process are dominated by a mixture of PFAAs, primarily PFSAs, but also some PFCAs. FT produces Class B foam formulations dominated by polyfluorinated fluorotelomer compounds with lesser amounts of PFAAs. Testing of archived Class B foam formulations suggests precursor compounds account for 41 to 100 percent of the total molar concentration of PFASs (Houtz et al. 2013; Backe et al. 2013) in these concentrates. The PFAS composition data for a range of ECF and fluorotelomer-derived concentrates is presented as Tables 3.3a and 3.3b, respectively (Backe et al. 2013).

The types of PFASs in Class B foam formulation vary by year of production and manufacturer; however, all aqueous film forming foam (AFFF) formulations that have

TABLE 3.3A
Concentrations of PFASs in electrochemical fluorination foams

	Year Manufactured				
Compound	1989 (mg/L)	1993 (mg/L)	1993 (mg/L)	1998 (mg/L)	2001 (mg/L)
PFBSaAm	9	120 ± 2.0	180	140	110
PFPeSaAm	8	140 ± 1.8	180	140	110
PFHxSaAm	189	660 ± 8.1	850	743	690
PFHpSaAm	ND	12 ± 0.40	15	30	24
PFOSaAm	9.9	62 ± 1.1	75	67	37
PFBSaAmA	ND	140 ± 3.1	120	110	150
PFPeSaAmA	4	200 ± 6.3	170	140	130
PFHxSaAmA	ND	930 ± 13	850	850	960
PFHpSaAmA	ND	17 ± 0.16	17	34	44
PFOSaAmA	ND	72 ± 0.81	58	53	65
PFBS	380	220 ± 2.0	160	210	250
PFPeS	210	120 ± 1.5	80	90	120
PFHxS	1,700	910 ± 14	760	850	900
PFHpS	410	120 ± 2.0	120	93	140
PFOS	15,000	8,000	9,300	6,700	7,900
PFNS	160	53 ± 0.97	56	9	27
PFDS	102	51 ± 0.34	52	11	27
PFBA	37	24 ± 0.48	35	31	38
PFPeA	47	36 ± 0.14	52	43	48
PFHxA	170	99 ± 1.1	110	99	170
PFHpA	54	25 ± 0.28	22	26	37
PFOA	150	83 ± 1.3	93	86	170
PFNA	ND	ND	ND	ND	ND
PFDA	ND	ND	ND	ND	ND
PFUdA	ND	ND	ND	ND	ND
PFDoA	ND	ND	ND	ND	ND
PFTrA	ND	ND	ND	ND	ND
PFTeA	ND	ND	ND	ND	ND
% precursors	1.2	19.5	18.8	21.9	19.1
% PFOS	80.5	66.1	69.6	63.5	65.0
% cationic	1.2	8.2	9.7	10.6	8.0
% anionic	98.8	80.5	81.2	78.1	80.9
% zwitterionic	0	11.2	9.1	11.2	11.1

Source: Adapted with permission from Backe, W. J., T. C. Day, and J. A. Field. Zwitterionic, Cationic, and Anionic Fluorinated Chemicals in Aqueous Film Forming Foam Formulations and Groundwater from U.S. Military Bases by Nonaqueous Large-Volume Injection HPLC-MS/MS. Environmental Science & Technology. 2013, 47, 5226-5234. Copyright 2013 American Chemical Society.

Note: Bold text used to indicate anionic compounds, italicized text used to indicate cationic compounds, and underlined text to indicate zwitterionic compounds.

TABLE 3.3B
Concentrations of PFASs in fluorotelomer foams

	Year Manufactured					
Compound	2005 (mg/L)	2010 (mg/L)	2002 (mg/L)	2003 (mg/L)	2009 (mg/L)	Not Recorded (mg/L)
4:2 FtTAoS	26	ND	25	ND	ND	ND
6:2 FtTAoS	6,100	11,000	4,900	ND	ND	ND
8:2 FtTAoS	1,100	24	170	ND	ND	ND
4:2 FTS	ND	ND	ND	ND	ND	ND
6:2 FTS	ND	ND	ND	42	ND	53
8:2 FTS	ND	ND	ND	19	ND	56
6:2 FTTHN	ND	ND	2,200	ND	ND	ND
6:2 FTSaB	ND	ND	ND	4,600	ND	4,800
8:2 FTSaB	ND	ND	ND	540	ND	1,800
10:2 FTSaB	ND	ND	ND	450	ND	830
12:2 FTSaB	ND	ND	ND	210	ND	430
6:2 FTSaAm	ND	ND	ND	2,100	ND	3,400
8:2 FTSaAm	ND	ND	ND	450	ND	720
5:1:2 FTB	ND	ND	ND	ND	2,000	ND
7:1:2 FTB	ND	ND	ND	ND	4,700	ND
9:1:2 FTB	ND	ND	ND	ND	1,900	ND
5:3 FTB	ND	ND	ND	ND	530	ND
7:3 FTB	ND	ND	ND	ND	610	ND
9:3 FTB	ND	ND	ND	ND	430	ND
% precursors	100	100	100	100	100	100
% PFOS	0	0	0	0	0	0
% cationic	0	0	30.2	30.3	0	34.1
% anionic	100	100	69.8	0.7	0	0.9
% zwitterionic	0	0	0	69.0	100	65.0

Source: Adapted with permission from Backe, W. J., T. C. Day, and J. A. Field. Zwitterionic, Cationic, and Anionic Fluorinated Chemicals in Aqueous Film Forming Foam Formulations and Groundwater from U.S. Military Bases by Nonaqueous Large-Volume Injection HPLC-MS/MS. Environmental Science & Technology. Copyright 2013 American Chemical Society.

Note: Bold text used to indicate anionic compounds, italicized text used to indicate cationic compounds, and underlined text to indicate zwitterionic compounds.

been previously characterized as containing a significant percentage (41 to 100 percent) of PFAA precursors (Place and Field, 2012; Houtz et al. 2013). PFAS constituents present in Class B foams have roughly followed the following timeline:

- Class B foams manufactured by the ECF process reportedly contained PFCAs during the 1960s and early 1970s and PFSAs from the 1970s to 2001, with the main U.S. manufacturer ceasing production of long-chain PFSAs by 2003 (Prevedouros et al. 2006; Santoro 2008). These were all largely PFOS-based foams; however, they also contained multiple chain-length PFAAs and polyfluorinated PFAA precursors (Place and Field 2012; Backe et al. 2013), as described in Table 3.3a.

- Foams containing fluorotelomers were introduced in the 1970s (Cheremisinoff 2017), and this production method now forms the basis of modern foams, with the potential exception of some produced in China and India (Cheremisinoff 2017), with a chemistry based mainly on 6:2 fluorotelomer derivatives commonly referred to as "C6" formulations) with some older concentrates employing 8:2, 10:2, and 12:2 homologues, as shown in Table 3.3b.
- From the 1970s to 2001, a number of manufacturers made fluorotelomer-based foams. Of the foams that have been characterized, they were always primarily C6-based, but they also contained up to 20 percent C8-based PFASs (Houtz et al. 2013).
- From 2001 to 2015, most fluorotelomer-based AFFF formulations were predominantly C6-based or shorter, containing 95 percent or more C6- and shorter PFASs (Houtz et al. 2013).
- Since 2015, most suppliers of PFASs to foam blenders are expected to have adhered to the U.S. EPA PFOA Stewardship Program and to have completely phased out production of long-chain PFASs. Residuals of long-chain PFASs in post-2015 foams are expected to be minimal. The C6 content of the PFAS-based foams currently in the marketplace have been reported to be precursors to PFHxA, perfluoropentanoic acid (PFPeA), perfluorobutanoic acid (PFBA), 5:3 FTCA, 4:3 FTCA, and 3:3 FTCA as stable and semi-stable daughter products (Liu and Mejia Avendano 2013).

It is notable that 40 new classes of polyfluorinated PFAA precursors were identified in AFFF-impacted groundwater in 2017 (Barzen-Hanson, Roberts et al., 2017), with each class containing precursors of variable perfluoroalkyl chain length. Several hundred proprietary PFASs have been identified either as components of Class B foams or as their environmental breakdown products in soil and water which vary by manufacturer (Place and Field 2012; Barzen-Hanson, Roberts et al., 2017; Field et al. 2003). These are often proprietary foam formulations, meaning their exact chemical content is not public knowledge due to their commercial value. The presence of ultra-short-chain PFASs in firefighting foams, including perfluoropropane sulfonate (PFPrS) and perfluoroethane sulfonate (PFEtS) in five AFFFs, has recently been reported (Barzen-Hanson and Field 2015).

3.3 PHYSICAL, CHEMICAL, AND BIOLOGICAL PROPERTIES

The diverse and unique properties of PFASs contribute substantially to their environmental distribution and recalcitrance. The carbon fluorine bond is the strongest in organic chemistry so when multiple fluorine atoms replace hydrogens to form the saturated alkyl chain of the perfluoroalkyl group, this imparts very significant biological, thermal, and chemical stability and hence persistence to the PFAAs and perfluoroalkyl moiety of the polyfluoroalkyl substances. This stability results from the fluorine atoms shielding the carbons in the alkyl chain from attack by the chemical species present in the environment that break down more conventional contaminants. Furthermore, although the carbon-fluorine bond is strongly polarized, the electronegativity of fluorine makes changes in this polarization difficult, making attack on the bond energetically unfavorable. PFAAs are not degraded by strong acids or alkalis, and resist both reduction and oxidation even at high temperatures (Kissa 2001). They cannot be degraded by UV light or photolysis (Stedingk and Bergman 2004).

Longer-chain PFAAs typically consist of a hydrophobic, perfluoroalkyl chain and a hydrophilic functional group such as a sulfonate or carboxylate. This amphiphilic characteristic of the longer-chain PFASs makes them ideal for use as surfactants and can also cause them to accumulate at soil-groundwater-air interfaces. The longer-chain PFAS, which exhibit surfactant properties, can preferentially form film at the air water interface, with the lipophobic and hydrophobic perfluoroalkyl chains oriented into the air and the hydrophilic functional groups remaining in the aqueous phase (Krafft and Riess 2015). The surface activity of PFAS surfactants is higher than analogous hydrocarbon surfactants and they exhibit extremely low surface tension (Lau et al. 2004). However, not all PFASs exhibit surfactant properties. For example, the hydroxyl group found on fluorotelomer alcohols is not charged so these compounds do not act as a surfactant and, in general, the shorter the perfluoroalkyl chain is, the lower its hydrophobicity and the molecules' surfactant properties. Hydrophobicity is related to perfluorinated chain length (Higgins and Luthy 2006b), but this trend is not followed by all homologues (Guelfo and Higgins 2013) as discussed in section 3.8.

The longer-chain PFASs with polar functional groups can act as surfactants, so micelles can form in a solution above a critical micelle concentration (CMC) (Bhhatarai and Gramatica 2011). It was considered that the CMC of PFAAs was mostly dependent on chain length and not counterion with a reported CMC between 3.2 g/L and 4.45 g/L for PFOS (Shinoda 1972; Yu et al. 2009) and 15.7 g/L for PFOA (Yu et al. 2009). It should be noted, however, that hemi-micelles can form at liquid-solid surfaces at concentrations in the range of 0.001 to 0.01 of the CMC (Schwarzenbach 2003) and that ionic strength is expected to influence the CMC. There has also been a description of a "molecular brush" effect where perfluoroalkyl chains, which are stiffened by fluorination, can self-assemble at interfaces and pack closely together. This is reported to create a dense layer of PFAAs, which is described to be more pronounced as the perfluoroalkyl chains lengthen and allows closer packing of PFAAs. Large molecular macro-aggregates of PFAAs were reported to form in the intraparticle pores of ion exchange resins indicating alternative sorption mechanisms (Zaggia et al. 2016). This "molecular brush" mechanism of surface interaction may be a process in which PFASs can bind to surfaces. For PFAAs, sorption typically increases with decreasing pH and with increasing ionic strength (Higgins and Luthy 2006b; McKenzie et al. 2015). At increasing ionic strengths, the authors have observed a diminished ratio of straight chains to branched chain PFOS, which may be accounted for by the selective aggregation of straight-chain PFAAs via the aforementioned "molecular brush" sorption mechanism.

PFAAs are typically relatively soluble in water (e.g., solubilities of PFOS and PFOA are 520 mg/L and 3,400 mg/L at 20°C), although the solubility can reduce significantly in brackish or saline water. Aqueous solubility increases as the length of the perfluoroalkyl chain diminishes. The octanol and water partition coefficient, K_{ow}, is difficult to determine for the PFAAs (Kissa 2001). Most PFASs do not readily partition from groundwater into air due to low vapor pressures and high solubilities. FTOHs are examples of volatile PFASs. Additionally, the FTOHs are very hydrophobic and are of relatively low solubility in water.

The acid dissociation constants (pKa values) for common PFAAs are generally extremely low or negative (Pancras et al. 2016), meaning PFAAs are likely to be fully dissociated and anionic under common environmental conditions. However, the range of reported pKa values for PFOA is between negative 0.16 and 3.8, which indicates that PFOA could become protonated and thus potentially volatile at low pHs, such as around or below a pH of 4.

All PFASs contain a perfluoroalkyl group, but beyond that, especially for the polyfluoroalkyl substances, their chemical structures and resulting physical and chemical properties can vary significantly. Due to the wide range of specific chemicals within the PFAS class, many do not have published values for common properties such as pKa, solubility, Henry's law constant, and environmental partitioning equilibrium constants. A description of the common physiochemical properties of some PFASs, where data is available, is provided in section 3.8.

3.3.1 Biological Activity Towards PFASs

Biodegradation is defined as the process by which organic compounds are decomposed by microorganisms (mainly aerobic bacteria) into substances such as carbon dioxide, water, and ammonia (OECD 2002a). None of the several thousand PFASs can biodegrade as there has not been a credible report of mineralization, which involves evolution of stoichiometric quantities of fluoride, carbon dioxide, and, potentially, sulfate for degradation of PFSA and their PFSA precursors. Some PFASs have been reported to exhibit biodegradation or to be biodegradable (D'Agostino and Mabury 2017a; Dasu et al. 2012), but this refers to the PFAAs precursors that biotransform to create PFAAs, which persist. So the parent PFAA precursors may not be detected via chemical analysis as a result of microbial action, but they have not biodegraded, as they are biotransformed through multiple intermediates to form the recalcitrant PFAAs. The replacement chemistries for the PFCAs, the perfluoroalkyl ethers, also do not appear to have increased biodegradation potential. The introduction of an oxygen to the perfluoroalkyl chain, via an ether linkage, seems to have no effect on its potential for biological attack as perfluoroalkyl ethers have not been reported to be biodegradable (Gordon 2011; ECHA 2013; Wang, Cousins et al., 2013).

The perfluoroalkyl chains in PFASs are halogenated and thus fully oxidized; therefore from a thermodynamic perspective, PFASs will serve as the terminal electron acceptor (TEA) in metabolic processes or could be metabolized via cometabolic processes. The perfluoroalkyl group, being fully oxidized, cannot act as an electron donor and carbon source to support metabolism, as for many hydrocarbons. However, the nonfluorinated functional groups present in the polyfluorinated compounds can potentially act as an electron donor and source of carbon or nitrogen.

Microbial metabolism of PFAAs could potentially be analogous to the biological reductive dechlorination mechanisms now well understood for organo-chlorine compounds, such as perchloroethene (PCE). From a thermodynamic perspective, reductive dehalogenation of PFASs, used as a TEA, may appear favorable when an electron donor is supplied. It has been reported that the thermodynamics for biodegradation of some perfluorinated alkanes (such as hexafluoroethane or octafluoropropane) do not present a barrier for mineralization (Parsons et al. 2008). The C-F bond strength (as reflected by its dissociation energy of 536 kJ/mol) is reported to be the strongest organic bond in chemistry, but this has not prevented enzymatic attack on C-F bonds as with fluoroacetate (Goldman 1965); however, biological attack on trifluoroacetate has not been demonstrated (Matheson et al., 1995), indicating that multiple fluorine atoms on a PFAA molecule pose significant challenges for microbial attack. The C-F bond strength has been estimated to increase as additional fluorine atoms bond to the same carbon, with the heat for formation of the C-F bond reported to increase from 448 kJ/mol for CH_3F to 486 kJ/mol for CF_4 (Kissa 2001). The atomic radius of a C-F bond is very low, and has been described to shield a perfluorinated carbon-to-carbon bond in the perfluoroalkyl group without steric stress. This shielding has been described as a coating of fluorine atoms which provide kinetic stability, as the three tightly bound lone pairs of electrons per fluorine atom and partial negative charge are an effective electrostatic and steric shield against nucleophilic attack targeted at the central C-C bonds in the perfluoroalkyl group (Kirsch 2004). The extreme stability of fluoroorganic compounds was also described to increase as additional fluorine atoms bond to the same carbon atom, as reflected by the length of the C-F bond which has been calculated to decrease from 140 pm (pico meters) for CH_3F to 133 pm for CF_4 (Kirsch 2004). The nearly optimal overlap between the fluorine 2s and 2p orbitals and corresponding orbitals of carbon was reported, which enables the formation of a dipolar resonance structure for multiple fluorine atoms bonded to the same carbon, which promotes "self-stabilization" of multiple fluorine atoms bonded to the same carbon atom.

The C-F bond strength and increased stability of multiple C-F bonds in PFAAs is an important factor when considering their extreme persistence in the environment. However, there are other significant factors to consider between organofluorine compounds and other halogenated compounds such as organochlorine compounds, namely, their relative natural abundance. This links to the length of time microbial communities have been exposed to biogenic compounds versus xenobiotics, with longer exposure allowing evolution of the metabolic capacity to recognize a molecule as a TEA or an electron donor and develop genetic and enzymic systems to allow biological attack and metabolism. Many organochlorine compounds have been present in the environment over geological time, both from biogenic sources and via volcanic activity (van Pée and Unversucht 2003). There are over 2,000 biogenic organochlorine compounds (Gribble 2002), and the millions of years of exposure time for microbial communities to evolve to metabolize them explains the ability of many micro-organisms to have developed a reductive dechlorination mechanism. A limited number of organofluorine compounds have been reported to have been exuded from volcanoes, but in comparison to organochlorine compounds, the diversity of naturally occurring organofluorine compounds is very limited (Gribble 2002; Harper 2005), with only around 30 examples of natural organofluorine compounds described, which all have a single fluorine substituent. The elevated heat of hydration of the fluoride ion, compared to other halide ions, has been described as the main reason for the high redox potential necessary to form the halonium ion (F^+) from F^- compared to that required to form other halonium ions from their respective halides. This difference has been reported to preclude the incorporation of fluorine into natural products by the haloperoxidase reaction, which is considered to be the major route by which organohalogens are formed naturally (Harper and Murphy 2005). Therefore, as a very limited number of organofluorine compounds have been created naturally, their biodegradation potential will be vastly diminished as compared to other halogenated organics. Then as PFASs are xenobiotics, microorganisms have not had the exposure time to evolve enzymic systems to attack and degrade them. This will be confounded as a result of a lack of naturally occurring organofluorine compounds to stimulate evolution of a catabolic operon and associated enzymes to attack organofluorine compounds. With no exposure of the perfluoroalkyl group to microbial communities until it was synthesized by humans, the strength of the C-F bond and the added stability that multiple fluorine atoms bring to the perfluoroalkyl group and time required to evolve a reductive defluorination process, biological attack on PFAAs seems very unlikely. As PFAAs will serve as a TEA in strongly reducing aquifer systems, such as those that would support reductive defluorination, it's considered that methanogenesis would likely occur in preference to reductive defluorination. So even in situations where reductive dehalogenation is viable, the potential for reductive defluorination to progress would likely be limited as carbon dioxide would be a preferable TEA, to support methanogenesis.

If a reductive defluorination process could be force-evolved by increasing mutation rates and using exposed anaerobic microbial communities via application of a selective pressure (such as providing a PFAA as a sole TEA), there will be numerous intermediate sequential defluorination metabolites. Microorganisms would need to adapt to defluorinate not just the C8 PFAAs but the sequential 17 fluorine atom that needs removal to mineralize PFOS. Evolution of a metabolic pathway to mineralize PFAAs via anaerobic conditions appears highly unlikely, with the potential to form intermediates that may be more toxic, in a manner analogous to the formation of vinyl chloride from PCE.

3.3.2 Transformation of Polyfluoroalkyl Substances

Polyfluorinated substances transform in the environment through abiotic and biologically mediated reactions and, in higher organisms, often via attack by cytochrome P450 enzymes (Vestergren et al. 2008) to create PFAAs. Because of their transformation potential, they are often referred to as PFAA Precursors compounds, and PFAAs can be referred to as "terminal transformation products" or "dead-end daughter products" as they persist indefinitely under environmental conditions. The PFAAs formed from precursors, including PFOS and PFOA, are all extremely persistent and can be more (or sometimes less) mobile in groundwater than the precursors from which they originated. None of the PFASs biodegrade; the polyfluoroalkyl substances are attacked at their nonfluorinated functional groups to create PFAAs, which persist indefinitely. These reactions cannot be termed *biodegradation* as PFAAs are formed from polyfluoroalkyl substances as dead-end daughter products which persist. These transformation reactions may occur through biotic or abiotic mechanisms (such as hydrolysis, photolysis, and oxidation). The biotransformation processes converting polyfluoroalkyl substances to PFAAs have been reported to occur more quickly under aerobic conditions compared to anaerobic conditions (Zhang et al. 2013b; Zhang et al. 2016; Wang et al. 2005; Lui, Wang, Buck et al., 2010; Lee et al. 2010; Dasu et al. 2012; Dinglasan et al. 2004).

Precursors themselves and their biotransformation intermediates can go undetected in the environment as most are not identified by conventional commercial analytical techniques. The application of advanced analytical approaches (described in section 3.5) has indicated that a significant proportion of PFASs present in some environment matrices comprise PFAA precursors (D'Agostino and Mabury 2014), and the corollary of this is that precursors can represent a considerable latent mass of PFAAs.

3.3.2.1 Abiotic Transformation

The formation of PFAAs from PFAA precursors by abiotic mechanisms includes indirect photolysis of fluorotelomer alcohols in the atmosphere, which can be a significant contributor to atmospheric deposition of PFCAs (Armitage 2009; Yarwood 2007). Both polyfluoroalkyl sulfonamide- and fluorotelomer-based PFAA precursors are susceptible to indirect photolytic reactions (Gauthier and Mabury 2005; Plumlee et al. 2008). For example, fluorotelomers have been found to undergo atmospheric conversion to their corresponding PFAAs, with 8:2 FTOH converted to PFOA via oxidative reactions in the atmosphere. Similar processes have been reported for 6:2 FTOH and 4:2 FTOH (Ellis et al., 2004; Wallington 2006). The indirect photolysis of 8:2 FTOH to form PFOA as a major by-product was reported in surface water samples (Gauthier and Mabury 2005). Polyfluoroalkyl sulfonamides can also be transformed via abiotic mechanisms in the atmosphere to form PFCAs (Martin et al. 2006). Gaseous transformation of polyfluoroalkyl sulfonamides to form a combination of PFCAs and PFSAs has also been observed (D'Eon et al. 2006). The pathways of transformation of polyfluoroalkyl substances, for example, the polyfluoroalkyl sulfonamides, will create multiple transient intermediates, so PFAAs may not initially form as intermediates could dominate, before ultimate transformation to create PFAAs (Martin et al. 2010).

The use of wastewater disinfection processes such as the application of chlorination or ozone has been reported to generate PFOS and PFOA from PFAA precursors during disinfection and water treatment (Xiao et al. 2018). An understanding of the PFAA-precursor loading may be an important consideration when developing water treatment solutions.

3.3.2.2 Biotic Transformation

All PFAA precursors have the potential for biotransformation to ultimately create PFAAs as persistent daughter products. The rates of transformation of PFAA precursors to intermediates and terminal PFAAs vary by a number of factors. Highly aerobic microbially active media such as activated sludge (Rhoads et al. 2008; Weiner et al., 2013), tend to result in more rapid transformation of some PFAA precursors to terminal PFAA products than soils (Harding-Marjanovic et al. 2015; Wang et al. 2005).

Some precursors transform more rapidly than others under the same conditions (i.e., rate-limiting transformation steps are apparent). In general, sulfonamide-based PFAA precursors tend to transform more slowly than fluorotelomer-based PFAA precursors (Mejia-Avendano et al. 2016; Harding-Marjanovic et al. 2015; Ritter 2010) and transformation under aerobic conditions is more rapid than transformation under anaerobic conditions (Zhang et al. 2016; Yi et al. 2018). Biotransformation occurs through multiple intermediates, some of which may have appreciable lifetimes in the environment depending on the geochemical environment (i.e., aerobic or anaerobic) and conditions which can sustain microbial communities. For example, generation of 6:2 FTS as a "meta-stable" transformation intermediate generated from fluorotelomer

PFAA precursors has been described. 6:2 FTS is more stable under anoxic conditions compared to aerobic, but ultimately will transform to create PFAAs. However, recent research suggests that 5:3 FTCA and its analogues may be relatively stable and therefore potentially persistent within the environment (Wang et al. 2012) as well.

There have been relatively few studies examining transformation rates of PFAA precursors. The majority of those studies were carried out under aerobic conditions with only limited work done using anaerobic conditions; a summary of transformation rates provided in the literature is provided in Table 3.4. Fluorotelomer-based PFAA precursors that contain an alkyl linkage (e.g., the spacer next to the perfluoroalkyl group), tend to form PFCAs following biotransformation, and this breakdown pathway can occur via biotransformation mechanisms (Wang et al. 2005; Lee et al. 2010; Wang, Liu et al., 2011; Wang et al. 2009; Liu, Wang, Szostek et al., 2010; Liu, Wang, Buck et al., 2010; Dasu et al. 2012), or via attack with oxidative radicals (Gauthier and Mabury 2005; Ellis et al. 2004). Biotransformation of many fluorotelomers can occur via formation of FTOHs as intermediates, and further transformation of FTOHs has been shown to proceed via a complex series of intermediates and multiple pathways, as shown in Figure 3.11.

TABLE 3.4
Degradation rates of PFAA precursors

PFAA Precursor	Degradation Half Life (days)	PFASs Generated	Reference
EtFOSE	44 at 25°C 160 at 4*°C	PFOS	Benskin et al. 2013
EtFOSE	0.71 at 30°C	PFOS	Rhoads et al. 2008
SAmPAP Diester	3,400 at 4°C >379 at 25°C	PFOS	Benskin et al. 2013
6:2 FTOH	1.3 with 6% mineralization in 84 days	PFBA, PFPeA, PFHxA	Liu et al. 2010
8:2 FTOH	0.2 days/mg to 0.8 days/mg*	PFOA	Dinglasan et al. 2004
8:2 FTOH	145	PFOA	Zhang et al. 2013
4:2 PAP	100% reduction in <5 days	–	Lee et al. 2010
6:2 PAP	100% reduction in <5 days	PFPeA, PFHxA, PFHpA	Lee et al. 2010
6:2 diPAP	100% reduction in 80 days	PFPeA, PFHxA, PFHpA	Lee et al. 2010
6:2 diPAP	85% reduction in 165 days	PFBA, PFPeA, PFHxA	Lee et al. 2010
6:2 FTS	63.7% reduction in 90 days	PFBA, PFPeA, PFHxA	Wang et al. 2011
8:2 FTAC	≤5	PFHxA, PFHpA, PFOA	Royer et al. 2015
8:2 FTMA	≤15	PFHxA, PFHpA, PFOA	Royer et al. 2015
8:2 FTS$_{ME}$	10.3 (initial half life)	PFHxA, PFHpA, PFOA	Dasu et al. 2012
8:2 FTS$_{ME}$	28	PFHxA, PFHpA, PFOA	Dasu et al. 2013

Source: Held and Reinhard (2016).

* Based on results of a microorganism inoculation study that indicate an initial rate of 0.2 days/mg followed by slower degradation

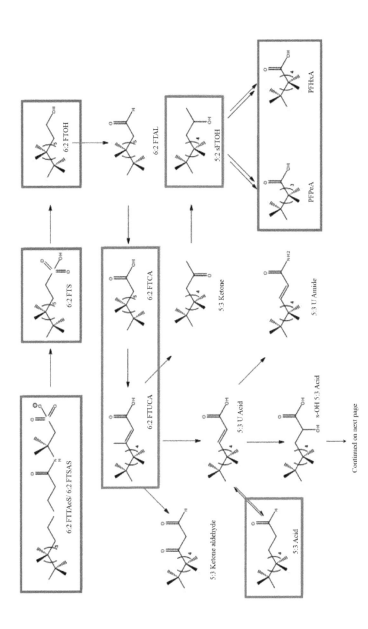

FIGURE 3.11 Published biotransformation pathways for fluorotelomers (Liu and Avendaño 2013; Wang, Buck et al. 2012; Weiner et al. 2013).

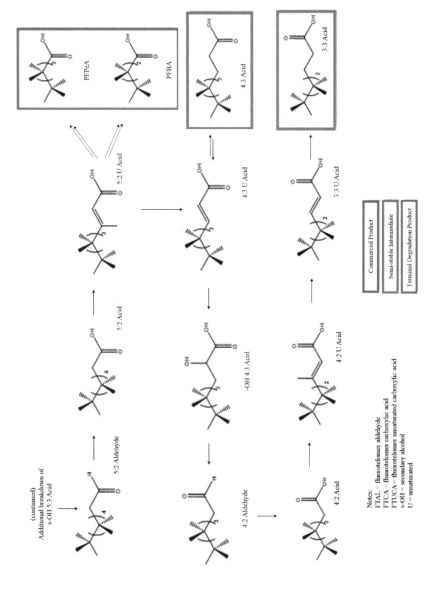

FIGURE 3.11 (Continued)

TABLE 3.5
Examples of biotransformation pathways

Substrate	Hydrolysis Products	Reference
8:2 Fluorotelomer stearate monoester	8:2 FTOH + stearin acid	Dasu et al. 2012; Dasu et al. 2013
8:2 Fluorotelomer citrate triester	8:2 FTOH + citrate	Benskin et al. 2013
6:2 diPAP	6:2 FTOH + 6:2 monoPAP	Lee et al. 2014
6:2 monoPAP	6:2 FTOH + phosphate	Lee et al. 2014
8:2 Fluorotelomer acrylate	8:2 FTOH + carboxylic acid	Royer et al. 2015
8:2 Fluorotelomer methacrylate	8:2 FTOH + carboxylic acid	Royer et al. 2015

This illustrates that biotransformation of 6:2 FTOH produces of PFHxA but also produces shorter-chain PFAAs including PFPeA, and PFBA. The presence of a nonfluorinated alkyl group connected to the perfluoroalkyl chain enables shortening of the perfluoroalkyl chain from the adjacent perfluorinated carbon atom resulting in the sequential loss of CF_2 groups, but a PFAA appears to be the terminal daughter product as biodegradation has not been demonstrated (Wang et al. 2012; Wang et al. 2005). A selection of the reported biotransformation reactions are presented in Table 3.5.

The sulfonamide-based PFAA precursors, that contain a sulfonamide linkage adjacent to the perfluoroalkyl group, tend to form PFSAs following biotransformation in the presence of aerobic bacteria (Mejia-Avendano et al. 2016; Rhoads et al. 2008). The perfluoroalkyl chain lengths are not typically shortened during biotransformation reactions involving sulfonamide-based PFSA precursors. For example, the biotransformation of N-MeFOSE and N-EtFOSE to PFOS within the environment is documented (Rhoads et al. 2008). An example biotransformation pathway for a polyfluoroalkyl sulfonamide-based PFSA precursors is presented in Figure 3.12.

Biotransformation of PFAA precursors to create PFAAs is commonly encountered in industrial or domestic wastewater treatment plants (WWTPs), with higher concentrations of PFAAs encountered in the effluent following biological treatment of the influent (Oliaei et al. 2006; Frömel et al., 2016).

The transformation of PFAA precursors has also been documented *in vitro* (Nabb et al. 2007; Bull et al. 2014). Blood levels in workers that apply ski wax to winter sports equipment suggest that 8:2 FTOHs form PFOA *in vivo* in humans (Nilsson et al. 2013). The *in vivo* transformation of a fluorotelomer to a PFAA was first reported in rat plasma in 1981 (Hagen et al., 1981).

3.4 PFASs PRODUCTION AND USE

PTFE was discovered in 1938 by Roy Plunkett while working for E.I. du Pont de Nemours and Company at their Jackson Laboratory in Deepwater, New Jersey. Since then, PFASs have been synthesized for use in a large range of industrial and commercial applications (Banks 2000; Banks et al. 1994).

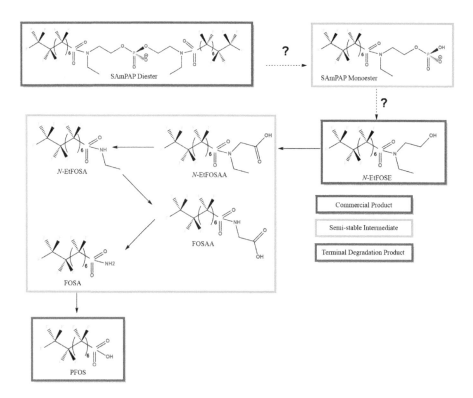

FIGURE 3.12 Published biotransformation pathways for polyfluoroalkyl sulfonamide-based PFSA precursors (Rhoads et al., 2008; Benskin et al. 2013; Liu and Avendaño 2013).

3.4.1 Manufacturing Processes and Uses

The production of PFASs has been via three main processes: (1) electrochemical fluorination, (2) fluorotelomerization, and (3) oligomerization. A process flow diagram of the differing manufacturing processes is shown in Figure 3.13. ECF was used to manufacture both PFSAs, such as PFOS, and PFCAs, such as PFOA (Wang et al. 2014b), whereas FT was used to make fluorotelomers, which are precursors to PFCAs.

3.4.2 Electrochemical Fluorination

The Simons Process or electrochemical fluorination process was devised during the 1940s with the process license first acquired by 3M in 1946 (Cheremisinoff 2017) and was used to manufacture long-chain PFASs until 2003 (Santoro 2008). PFCAs using ECF was reported during the 1960s and early 1970s (Prevedouros et al. 2006). To varying extents, commercial exploitation of ECF by some 15 other companies has been described (Santoro 2008). 3M was the major global producer of PFOS with manufacturing bases in the U.S. and Belgium. ECF was used to synthesize both PFSAs and PFCAs. The ECF process is still understood to be used to manufacture shorter-chain PFASs based around butyl (C4) chemistry, such as PFBS, in the U.S. and other countries (Herzke et al. 2007).

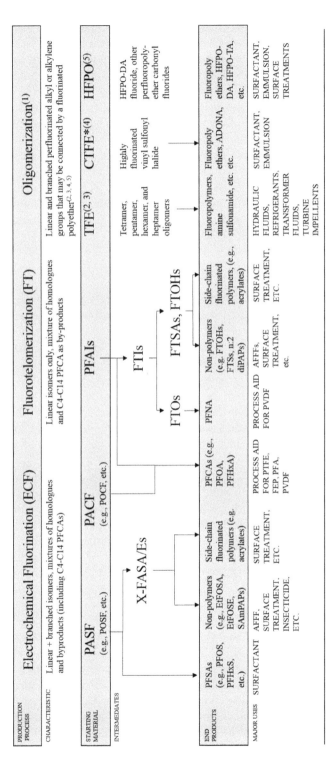

FIGURE 3.13 Three major PFAS manufacturing processes, starting material, intermediates, end products, and major uses. Modified from Wang et al. (2014a); additional references include (1) Banks et al., 1994; (2) Lagow and Inoue 1978; (3) Kissa 2001; (4) Fukushi et al. 2014; and (5) Flynne et al. 2003; Hintzer et al. 2006.

Acronyms not previously defined: PASF = perfluoroalkane sulfonyl fluoride; PACF = perfluoroalkyl carbonyl fluoride; PFAI = perfluoroalkyl Iodide; FTOs = fluorotelomer olefin; TFE = tetrafluoroetheylene; CTFE* = chlorotrifluoroethylene and other unsaturated fluoromonomers; X-FASA/Es = (N-methyl/ethyl) perfluoroalkyl sulfonamide/ sulfonamidoethanol

In the most commonly used ECF process, an electric current is passed through a solution of anhydrous hydrogen fluoride (HF) and an organic feedstock (e.g., octane sulfonyl fluoride), resulting in the substitution of the hydrogen atoms of the feedstock with fluorine (Kissa 2001). The reaction yielded PASFs as the main products (e.g., perfluorooctane sulfonyl fluoride [POSF]) (Banks et al. 1994). The resulting PASFs can be hydrolyzed to form PFSAs or reacted to produce other related compounds. PFPiA such as the C_2/C_2 and C_4/C_4 PFPiAs have been reported to have been manufactured using ECF (Wang, Cousins et al., 2016).

ECF produces both linear (n-) and branched (br-) isomers even if the feedstock material used in the reaction contained just the linear isomer. PFOA produced by ECF typically had an isomer composition of 78 percent linear (n-PFOA), with the remaining 22 percent of the product made up of a mixture of various branched (br-PFOA) isomers. For PFOS, the distribution was approximately 70 percent linear (n-PFOS) and 30 percent branched (br-PFOS) (Benskin et al. 2010).

3.4.3 FLUOROTELOMERIZATION

The fluorotelomerization process was developed in the late 1960s and has been widely used by manufacturers (Wang et al. 2014b; Wang et al. 2014a). FT is used to synthesize PFCAs and their fluorotelomer PFAA precursors; it has not been reported to synthesize PFSAs. FT is a chemical polymerization process that involves the sequential addition of a taxogen, such as perfluoroethylene ($CF_2 = CF_2$), to a telogen such as PFAI (e.g., perfluoroethyl iodide (CF_3-CF_2I) to selectively increase the perfluoroalkyl chain length of the resulting perfluoroiodide adduct by two CF_2 units (e.g., $CF_3CF_2CF_2CF_2$I) per perfluoroethylene addition. The PFAIs are then used as a feedstock to produce linear isomers of PFCAs, or can be coupled with ethene (ethylene) to form fluorotelomer compounds (Lindstrom, Strynar, and Libelo 2011) (see Figure 3.13).

FT produces PFASs containing only linear perfluoroalkyl chains (Lindstrom, Strynar, and Libelo 2011) and an even number of carbon atoms, although odd numbers of carbon may be formed through biotransformation processes in the environment (Liu, Wang, Buck et al., 2010; Liu, Wang, Szostek et al., 2010; Wang, Liu et al., 2011; Wang et al. 2009). Additionally, the longer chain (>C4) PFPA and/or PFPiA homologs have been manufactured through use of FT (Wang et al. 2014b; Wang et al. 2014a; Wang, Cousins, Berger et al. 2016).

3.4.4 OLIGOMERIZATION

Oligomerization has been reported to be one of the three manufacturing processes typically employed for the manufacture of PFASs (OECD 2018b). Oligomerziation is reported to produce highly branched oligomers through oligomerization of tetrafluoroethylene and can be controlled to yield short-chain compounds (Jarnberg et al. 2006). The process results in branched products—tetramer C_8F_{16}; pentamer $C_{10}F_{20}$; and hexamer $C_{12}F_{24}$—and these intermediates can be converted to other PFASs (Deem 1973; Jarnberg et al. 2006). Oligomerization using direct fluorination is also used in industrial processes to generate PFECAs and PFESAs through a polymerization reaction

(Lagow and Inoue 1978; Costello and Moore 1988). Oligomerization is also used to process hexafluoropropylene oxide (HFPO) into a perfluoropolyether carbonyl fluoride that can be further converted into an acid, acid salt, ester, amide or alcohol such as the ammonium salt of HFPO-DA (Flynn et al. 2003).

3.4.5 Uses

The perfluoroalkyl group confers a range of potential properties to PFASs including thermal stability and the ability to repel both oil and water, with impressive surface tension leveling properties. This has led to the widespread use of PFASs in many industrial and consumer product applications. There is a broad array of major uses of PFASs in multiple products and applications, including as surfactants, coatings and polymers, plus other potential applications as shown in Figure 3.14. A brief description of major PFAS uses is provided below.

3.4.5.1 Use as Surfactants

Surfactants are used to lower the interfacial tension of various liquids or surfaces, with fluorosurfactants providing an unparalleled surface tension leveling, which is combined with chemical and thermal stability and the unique property of being both hydrophobic and oleophobic. As a result of these properties, PFAS use as surfactants has a wide variety of applications including emulsifiers in the polymerization of fluorinated monomers such as used in the production of PTFE. PFOA was previously used in the manufacturing of PTFE but has been replaced by perfluoroalkyl ethers such as GenX (RIVM 2017a). PFNA has been used in PVDF production (Scheirs 1997; Pancras et al. 2016; Pan et al. 2018). PFASs are also used as a surfactant in firefighting foams (described in section 3.2.9), metal plating as wetting agents and mist-suppressing agents, photographic industry for the manufacture of film, film paper, and photographic plates, and as spreading agents for paints, inks, and waxes; cleaning agents and shampoos, and more.

3.4.5.2 Use as Surface Coatings

PFASs are used as surface treatment agents in a range of products to impart both water and stain resistance and oil repellency, including clothing carpets, upholstery, outdoor equipment, shoes, and leather items, household cleaning products and impregnation sprays for upholstery, fabric, and carpets. PFASs are also commonly used as a greaseproof surface for food wrappers and coated papers (Ritter et al. 2017). Food-contact packaging is treated with PAPs, diPAPs, and FTOHs to provide the material with water and grease resistance (Schaider et al. 2017; Yuan et al. 2016; Lee et al. 2010). Historically SAmPAPs have also been used in surface coatings; however, these have been phased out in favor of diPAPs (Benskin, Ikonomou, Gobas et al., 2012; Gebbink et al. 2013).

3.4.5.3 Other Uses

PFASs can be used as inert fluids that are typically used for their favorable thermal and electrical properties. These include multiple applications including chlorofluorocarbon (CFC) replacement in electronics, such as high-voltage transformers

		ECF Process	FT Process	Oligomer Process
Plastics	Non-polymeric PFASs used as a processing aid for manufacturing several types of fluoropolymers such as PTFE (e.g., PFOA, C_4 and C_6 PFPAs, C_4/C_3 and C_6/C_5 PFPiAs, HFPO-DA, and HFPO-TA), PVDF (e.g., PFNA), FEP, PFA, or PVDF. FT-based PFASs such as FTOH also used as a building block to synthesize FT-based polymeric products.	C4 to C13 PFCAs; C_4/C_4 PFPiAs	C4 to C13 PFCAs; 6:2, 8:2, 10:2, 12:2 FTOHs; 6:2 FTCA, C_4 and C_6 PFPAs; C_6/C_6 PFPiAs; other FT-based compounds	HFPO-DA, HFPO-TA, ADONA, other PFECAs
Textiles	Used for textile treatment products to achieve water, oil and dirt repellency for carpet, furniture, and personal apparel; leather impregnation; personal protective garments and surgical gowns. Noted polymeric PFASs include long- and short-chain polyfluorinated polymers with acrylates and PFPEs.	C4 to C10 PFSAs, C4 to C8 alkyl-FASAs, -FASAAs, -FASEs, -FASEA based; C4 to C18 PFCAs	C4 to C18 PFCAs; 3:1, 5:1, 4:2, 6:2, 8:2, 10:2, 12:2 FTOHs; 4:2, 6:2, and 8:2 FTSs; DiPAPs (n=4, 6, 8, 10); FTCAs	Not identified in the literature *possibly PFPEs
Paper Coatings	Used as a surface treatment for food contact materials such as microwave popcorn bags, pizza boxes, fast food packaging, paper bowls, and other paper products. Also used in copy paper and packaging including cardboard packaging. Noted polymeric PFASs include side-chain fluorinated polymers with acrylates, PFPEs, and fluorocarbon resins.	C4 to C10 PFSAs (and/or their precursors), PFSAA, PFASE, alkyl-FASAAs, FASE, FASEA, mono-, di-, and tri-SAmPAPs, C4 to C18 PFCAs (and/or their precursors)	C4 to C18 PFCAs; 6:2, 8:2, 10:2, 12:2, 14:2, 16:2, and 18:2 FTOHs; FTCAs, FT-based mono- and di-PAPs, C_6/C_6 PFPiAs; possibly perfluoroalkyl phosphinyl amides (PFPiAMs)	Not identified in the literature *possibly PFPEs
Other Coatings	Used for a variety of solvent and aqueous coating systems for coating plastics, metals, etc.; surface treatment for glasses, natural stones, wood, and ceramics. Prevents paint defects. Also used for polish, varnish, ink, dyes, adhesives, cookware, ski wax, etc.	C4, C6 to C8 PFSAs; PFASEs, alkyl-FASACs and FASMACs; C4 and C8 PFCAs	C4 and C8 PFCAs; FTOHs; fluorotelomer acrylates and methacrylates	Not identified in the literature *possibly aromatic PFASs, and/or ultrashort PFASs, and/or PFPEs
Tech Industry	Used on printed circuit boards; in semiconductor manufacturing; as a photoacid generator for top coatings in photolithography; and chemical-mechanical polishing slurries, etc. Also useful in electrochemical devices such as lithium batteries and fuel cells. Polymeric PFASs include PTFE, methacrylate polymers, PFPEs, etc.	C4 to C10 PFSAs; PFOSA, N-alkyl perfluoroalkyl sulfonamides; C4 and C8 PFCAs	C4 and C8 PFCAs; 6:2 FTS; other fluorotelomer-based compounds	Not identified in the literature *possibly aromatic PFASs, and/or ultrashort PFASs, and/or PFPEs
Metal Plating	Used as wetting agents and mist suppressants in decorative and non-decorative hard plating; as a surfactant in electroplating and etching baths; and as metal cleaners and metal surface treatment products.	C4, C6, and C8 PFSAs; FOSA, FOSE, N-alkyl sulfonamides, C_2/C_2 and C_4/C_4 PFPiAs; PFNA, possibly other PFCAs historically	PFNA, FTOHs, 6:2 FTS *possibly 4:2, 6:2, 8:2, and 10:2 Cl-PFESA (e.g., F-53B) and other PFESAs such as F-53	Not identified in the literature *possibly 4:2, 8:2, and 10:2 Cl-PFESA or other PFESAs
Chemical	Used in various chemical manufacturing products as a surfactant, acid catalyst, and raw material for ionic liquids, detergents, and alkaline cleaning agents. Also used in insecticides and herbicides as well as in the medical industry in certain medications and as a resin for certain types of implants (e.g., PTFE).	C4 and C8 PFCAs; FOSA, PFOSA, FOSE, N-alkyl sulfonamides, -FOSA, PFOA and possibly other PFCAs	PFOA and possibly other PFCAs, FTOHs	Not identified in the literature *possibly aromatic PFASs, and/or ultrashort PFASs such as Cl PFSA precursors
Fire Fighting	Most commonly associated with Class B firefighting foams such as AFFF; also used in FFFP, FP, and alcohol-resistant variants. Fluoropolycarbonate resins also marketed as a fire retardant.	C1 to C10 PFSAs; FOSA, PFOSA, C4 to C8 SaAms, C4 to C8 SaAmAs, early documented use of PFCAs	6:2, 8:2, and 10:2 FTOH, 6:2 FTAB, 4:2, 6:2, 8:2, and 10:2 FTSs, FTBs; FTSaAm, FTTHN, FTTAoS, other FT-based compounds	Not identified in the literature
Other	Used in hydraulic fluids and lubricants, mining and oil surfactants, photographic and xray films; personal care products such as cosmetics, lotions, shampoos, and other cleansing products. Also used to improve fuel delivery systems and prevent seepage (e.g., fluoropolymers and fluoroelastomers used in gaskets and seals). Historical documented use in gasoline and oil.	PFOS, FOSE, low molecular weight sulfonamide derivatives, N-alkyl sulfonamides, PFOA	PFOA, 8:2 FTOHs	PFECAs; PFPEs *possibly aromatic PFASs, and/or ultrashort PFASs

FIGURE 3.14 PFAS uses and associated manufacturing processes as identified in literature and patent information. References cited as follows: 3M Company 1999; 3M 2018; Arkema 2018; Backe et al., 2013; Banks et al., 1994; Barzen-Hanson and Field 2015; Beekman et al., 2016; Begley et al., 2005; Begley et al., 2008; Benskin, Ikonomou et al., 2012; Björnsdotter et al., 2018; Buck et al., 2011; Caporiccio 1994; Caporiccio 1996; D'eon et al., 2009; DEPA 2015c; Daikin Global 2018; ECHA 2015; Gebbink et al., 2013; Hamrock and Pham 2000; Herke et al., 2012; Kannan et al., 2004; Lang et al., 2016; Lee et al. 2010; Lu et al. 2017; Pan et al., 2018; Pancras et al., 2016; Place and Field 2012; Prevedouros et al., 2006; Ritter 2010; Schellenberger et al., 2018; Sheng et al., 2018; Solvay Specialty Polymers 2016; Swedish Chemicals Agency 2015; Trier et al., 2018; UNEP 2014; Wang et al., 2006; Wang, Cousins et al., 2013; Wang, Cousins, et al., 2016; Yuan et al., 2016; Yueng et al., 2016; OECD 2013.

(Banks et al. 1994). PFASs are also used in agrochemical products, such as PFOS in sulfuramid to control leaf cutting ants (OECD 2013). PFPAs and PFPiAs have also been described as wetting agents and defoaming additives in pesticide formulation products (Wang, Cousins et al., 2016; DAS 2011; Chen et al. 2018).

3.5 SAMPLING AND ANALYSIS

In addition to typical best practices for field sampling of environmental media, specific precautions need to be taken to avoid contamination of samples with PFASs from consumer products, technical field gear, as well as sampling containers and equipment. Table 3.6 provides a summary of prohibited and acceptable materials to use during PFAS sample collection. Supplemental sampling guidelines are summarized in the section to follow.

3.5.1 General Sampling Guidelines

- All sampling equipment should be free of PFAS-containing materials such as Teflon and Viton. High-density polyethylene (HDPE) is the preferred material for sample collection, and HDPE and silicon are preferred materials to be used for tubing, plastic sheeting, and pump bladder needs.
- No glass containers.
- Sample in the suspected order of least PFAS contamination (e.g., upgradient wells) to most PFAS contamination.
- Do not reuse disposable materials (e.g., disposable bailers) between samples.
- Avoid potential food packaging-related PFAS contact.
- Avoid potential clothing-related PFAS contact; do not wear outdoor performance wear that is water or stain resistant.
- Avoid use of personal care products such as lotions, makeup, and perfumes. The use of sunscreen and insect repellent may be necessary for health and safety reasons; these products may contain PFASs and products that are PFASs free should be employed.
- Equipment blanks, when using reusable equipment, are a useful indicator of inadvertent PFAS contamination and should be included in addition to regular samples and corresponding QCs.
- Detergents suitable for decontamination are specified in Table 3.6.

3.5.1.1 Soil and Sediment Sampling
- Between samples, equipment should be scrubbed with a plastic brush or steam cleaned with potable water; triple rinsing of equipment in PFAS-free water.
- Soil and sediment core samples must be collected directly from single-use PVC liners that must not be decontaminated or reused at different locations.

3.5.1.2 Surface Water and Groundwater Sampling
- Filtration of aqueous samples is prohibited (PFASs adsorb to filters).
- Surface water must be collected by inserting a capped HDPE sampling container with the opening pointing down to avoid the collection of surface

TABLE 3.6A
Summary of acceptable sampling equipment and materials items for PFAS sampling

Sampling Materials	Additional Considerations	References
Water Sampling Materials		
High density polyethylene (HDPE) or silicone tubing materials	–	DER 2016; U.S. ACE 2016; NHDES 2016; MassDEP 2017
HDPE HydraSleeves™	Low density polyethylene (LDPE) HydraSleeves™ are not recommended	USACE 2016; MassDEP 2017
Drilling and Soil Sampling Materials		
PFAS-free drilling fluids	–	DER 2016
PFAS-free makeup water	Confirm PFAS-free water source via laboratory analysis prior to investigation	–
Acetate liners	For use in soil sampling	U.S. ACE 2016
Sample Containers and Storage		
HDPE sample containers with HDPE lined lids for soil and water samples	Laboratory should provide; whole bottle analysis of aqueous samples combined with a solvent rinse of bottle is recommended	DER 2016, MassDEP 2017
Ice contained in plastic (polyethylene) bags (double bagged)	–	DER 2016; U.S. ACE 2016; NHDES 2016; MassDEP 2017
Field Documentation		
Ball point pens	–	MassDEP 2017
Standard paper and paper labels	–	DER 2016; U.S. ACE 2016; NHDES 2016; MassDEP 2017
Decontamination		
Water-only decontamination	Confirm PFAS-free water source via laboratory analysis prior to investigation	DER 2016
Alconox®, Liquinox® or Citranox® followed by deionized water or PFAS-free water rinse	Alconox® known to contain trace levels of 1,4-dioxane	NHDES 2016; U.S. ACE 2016; MassDEP 2017
Methanol, isopropanol, or acetone	Special health and safety precautions are necessary	UNEP 2015; U.S. ACE 2016

Note: Due to the potential for PFASs to be present in a wide array of sampling products, it is important to follow a methodical evaluation process of materials to be used and confirm acceptance prior to implementation of field activities. Acronyms not previously defined include: DER = Government of Western Australia, Department of Environment Regulation; MassDEP = Massachusetts Department of Environmental Protection; NHDES = New Hampshire Department of Environmental Services; U.S. ACE = U.S. Army Corps of Engineer.

TABLE 3.6B
Summary of sampling equipment and materials not recommended for PFAS site investigations

Sampling Materials	Known PFAS-Containing Materials	Suspected PFAS-Containing Materials	Materials with Potential to Retain PFASs	References
Water Sampling Materials				
Teflon® or PTFE-containing or coated field equipment (e.g., tubing, bailers, tape, plumbing paste)	x			DER 2016; U.S. ACE 2016; NHDES 2016; MassDEP 2017
Passive diffusion bags			x	MassDEP 2017
LDPE HydraSleeves ™			x	USACE 2016; MassDEP 2017
Water particle filters			x	MassDEP 2017
Drilling and Soil Sampling Materials				
Aluminum foil			x	DER 2016; U.S. ACE 2016; NHDES 2016; MassDEP 2017
Drilling fluid containing PFASs	x	x		DER 2016
Sample Containers and Storage				
Glass sample containers with lined lids			x	DER 2016; U.S. ACE 2016; NHDES 2016; MassDEP 2017
LDPE containers and lined lids			x	U.S. ACE 2016
Teflon® or PTFE- lined lids on containers (e.g., sample containers, rinsate water storage containers)	x			DER 2016; U.S. ACE 2016; NHDES 2016; MassDEP 2017
Reusable chemical or gel ice packs (e.g., BlueIce®)		x		DER 2016; U.S. ACE 2016; NHDES 2016; MassDEP 2017
Field Documentation				
Self-sticking notes and similar office products (e.g., 3M Post-it-notes)		x		DER 2016; U.S. ACE 2016; NHDES 2016; MassDEP 2017
Waterproof paper, notebooks, and labels	x			DER 2016, MassDEP 2017
Non-Sharpie® markers		x		NHDES 2016
Decontamination				
[Some] detergents and decontamination solutions (e.g., Decon 90® Decontamination Solution)	x	x		DER 2016; NHDES 2016; MassDEP 2017

Note: For materials that are suspected of containing PFASs, or have the potential to retain PFASs, project specific considerations may provide adequate justification for use during the field event. For example, further evaluation may be conducted in the form of pre-field equipment blank sample analysis.

films. At the time of container opening, the container must be more than 10 centimeters (cm) from the sediment bed and more than 10 cm below the surface water level and as close to the center of the channel as possible, where practicable. Point the container up to fill so that gloved hands, sample container, and sampler are downstream of where the sample is being collected.
- Drilling equipment should be scrubbed with a plastic brush or steam cleaned with potable water; triple rinsing of equipment in Grade 3 or better distilled water is also required.
- Low flow groundwater sampling is done with peristaltic pumps and disposable silicone or HDPE tubing. If the depth to groundwater prevents the use of peristaltic pumps, then bladder pumps may be considered; however, bladders and other internal parts (check balls, O-rings, compression fittings) must not be made of PTFE or other fluoropolymer materials. Bladders must be changed between sample locations, and it is recommended that O-rings also be changed between sample locations. HDPE Hydrasleeves and Snap Samplers are also available as another PFAS sampling technique.

3.5.1.3 Storage and Hold Times
- Samples can be transported (double bagged) in a cooler with ice.
- If samples cannot be shipped the same day as collected, an appropriate means of refrigerating the samples overnight must be arranged to maintain the temperature between 0 to 10°C for the first 48 hours after collection and 0 to 6°C, thereafter.
- The sample holding time between collection and sample preparation is 14 days (U.S. EPA 2009a). Laboratories have an additional 28 days to complete the analysis after sample preparation.

3.5.2 Chemical Analysis Methods

Standard analytical methods have been developed to evaluate a large number of PFCAs and PFSAs, and a growing number of PFAA precursors are being added to the list of analytes quantified by those methods via modifications. However, many polyfluorinated compounds are present in commercial products that cannot be detected by conventional analytical methods. Unlike the PFAAs for which analytical measurement methods are well established, few standards are available for the PFAA precursors. Several thousand PFAA precursors have been manufactured (OECD 2018a; Wang et al. 2017), but methods for quantitative direct measurement are available for fewer than 30 of these compounds.

Because so many PFAA precursors have been manufactured and because they transform in the environment via a variety of intermediate precursors (such as 6:2 fluorotelomer sulfonate [6:2FTS]), it is impractical to individually measure all precursors that may be relevant to an environmental sample, especially as many of the intermediates in the biotransformation pathways, like many parent precursor molecules, are unknown. For this reason, advanced analytical methods have been developed. The following two subsections provide an overview of standard methods and advanced analytical methods available for PFASs analysis.

3.5.2.1 Overview of Standard Methods

A variety of standard methods available are applicable for the analysis of PFAAs, including the International Standards Organization (ISO) 25101:2009(E) for the analysis of PFOS and PFOA (ISO 2009) and U.S. EPA Method 537 (U.S. EPA 2009a) for the analysis of 12 PFAAs and 2 PFAA precursors (N-EtFOSAA and N-MeFOSAA). These standard analytical methods apply liquid chromatography with tandem mass spectrometry (MS) detectors (LC-MS/MS).

Many laboratories have adapted U.S. EPA Method 537, which was developed for drinking water analysis to other matrices and added additional analytes. Commonly up to 14 to 31 compounds (Casson and Chiang 2018) are included, primarily PFAAs, with a handful of known polyfluorinated precursors for which standards are available. These methods are referred to as "modified 537" methods with laboratory-specific standard operating procedures.

The U.S. Department of Defense Quality Systems Manual (QSM) 5.2 provides an approach to standardize the procedures for measurement of PFASs in groundwater, surface water, soil, and sediments. The guideline stipulates whole bottle analysis of aqueous samples, quantification using isotope dilution, sample cleanup with a product called ENVI-CARB, and specific ion transitions to be used for quantification.

The analysis of branched PFOS (br-PFOS) isomers is also now considered best practice and typically undertaken using mass-labeled standards used to quantify and report concentrations of both linear PFOS (n-PFOS) and br-PFOS isomers. Analysis of branches isomers of PFOA, PFOSA, and PFHxS is also becoming more commonly available.

Laboratories have been working toward achieving lower reporting limits in response to ever more stringent regulatory standards with reporting limits typically around 1 nanogram per liter (ng/L) in water and 5 to 10 microgram per kilogram (μg/kg) in soils. Some analytical laboratories are currently capable of reporting to 0.09 ng/L to achieve detection to one seventh of the 0.65 ng/L for PFOS (which is the European Union [EU] Annual Average Inland Surface Water Environmental Quality Standards [EQS] for PFOS) as quantification to 0.09 ng/L is required for surface water testing in the EU (see Appendix D for additional detailed information on standards established for PFASs in various matrices (current as of July 2018)). In many commercial laboratories it is currently common to achieve limits of detection of 1 ng/L and 0.1 μg/kg.

3.5.2.2 Advanced Analytical Techniques

While standard methods are available for the analysis of PFAAs, the quantitative analysis of most other PFASs is often difficult due to the large number of compounds and lack of appropriate reference standards. To address this difficulty, analytical techniques have been developed which can make a more complete estimate of the total mass of PFASs present in environmental matrices. There are four advanced analytical methods currently available to estimate the total concentration of PFAS in multiple matrices, as described in detail below.

3.5.2.2.1 Total Oxidizable Precursor Assay

To estimate the range of PFAA-precursor compounds in samples, a method was developed to indirectly measure the total precursor concentrations by converting polyfluorinated precursors to PFAAs and measuring the PFAA concentrations using standard LC-MS/MS analyses, known as TOP Assay (Houtz 2013; Houtz and Sedlak 2012). Detection limits for TOP Assay are generally the same as the detection limits in the LC-MS/MS method (i.e. ng/L water and µg/kg soil). The following description of TOP Assay is adapted from "Guidance on Use and Interpretation of the TOP Assay," a document written by Arcadis and currently in press through the Queensland Department of Environment and Heritage Protection.

As mentioned in section 3.2.2, the rates of transformation of PFAA precursors to intermediates and terminal PFAAs vary by a number of factors. The TOP Assay converts PFAA precursors to PFAAs via a hydroxyl radical-based oxidative mechanism while also exploiting the non-reactivity of PFAAs to hydroxyl radicals. The concepts used to develop the TOP Assay were apparent in previous hydroxyl-radical based experiments: C8 sulfonamide compounds and 8:2 fluorotelomer alcohol formed PFOA via hydroxyl radical-mediated reactions in aqueous solutions (Plumlee et al. 2008; Gauthier and Mabury 2005) and C4 sulfonamide precursors formed PFBA from exposure to hydroxyl radical and other oxidative radicals in smog chamber experiments (Martin et al. 2006). In each of these experiments, the PFCAs were not oxidized significantly by hydroxyl radical (•OH).

An example of the use of TOP Assay is presented in Figure 3.15, which shows the indirect detection of fluorotelomer compounds in a surface water using this method (Holmes 2018).

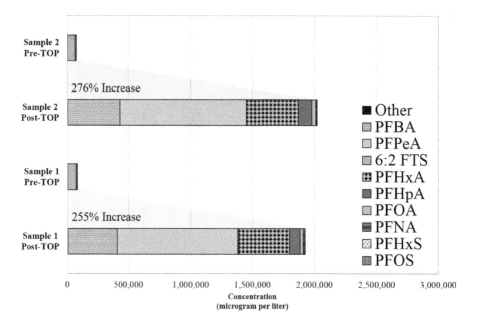

FIGURE 3.15 Example water samples from a recent C6 fluorotelomer foam spill to surface water and analytical results using the pre-TOP and post-TOP Assay digest. Data courtesy of Nigel Holmes, Principal Advisor Incident Management, Queensland Department of Environment and Science.

The TOP Assay digest involves reacting a sample at an elevated temperature with persulfate under highly basic conditions. Thermolysis of persulfate ($S_2O_8^{2-}$) results in the formation of sulfate radical ($SO_4^-\bullet$), which is rapidly converted to hydroxyl radical at the high pH values (pH > 12) maintained during TOP Assay (Tsao 1959) The excess of hydroxyl radicals generated can fully convert PFAA precursors to PFAAs (Houtz 2013; Houtz and Sedlak 2012).

$$S_2O_8^{2-} + \text{heat} \rightarrow 2\ SO_4^- \bullet \qquad \text{Reaction 1}$$

$$SO_4^- \bullet + OH^- \rightarrow SO_4^{2-} + \bullet OH \qquad \text{Reaction 2}$$

By measuring a sample for PFAAs using LC-MS/MS before and after the TOP Assay digest, the increase in the PFAA concentration can be approximated as the total concentration of precursors present in a sample. The molar increase of PFAAs is a conservative estimate of the molar concentration of PFAA precursors in the sample. The post-TOP Assay PFAA concentrations are not equivalent to the free concentration of PFAAs in the sample initially.

The maximum chain length of the PFAAs generated is representative of the perfluoroalkyl chain lengths in the precursors, however fluorotelomer compounds are converted to a series of PFCAs with the same or fewer number of perfluoroalkyl carbons during normal operation of the method. The sulfonamide precursors usually form perfluoroalkyl carboxylates with one less perfluorinated carbon than the precursor, as the terminal carbon is converted to a carboxylate (i.e., PFOA will form from C8 sulfonamide precursors) (see Figure 3.16).

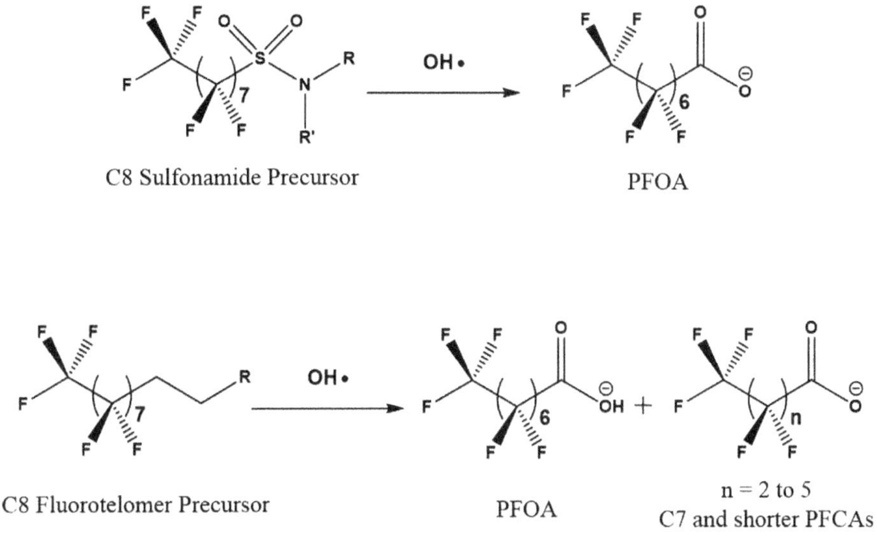

FIGURE 3.16 PFAAs formed by TOP Assay of common precursors.

Partial defluorination of fluorotelomer compounds is also routinely observed in biological systems, where a range of PFAA products with the same or fewer number of perfluoroalkyl carbons are observed (Lee et al. 2010; Liu, Wang, Buck et al. 2010; Liu, Wang, Szostek et al. 2010; Dasu et al. 2012; Zhang et al. 2013b; Dinglasan et al. 2004). Partial defluorination by chain shortening occurs only with fluorotelomers as there is an alkyl group adjacent to the perfluoroalkyl chain and this presents a weak point in the molecule. Only partial perfluoroalkyl chain shortening occurs (i.e., one to three carbons), so complete mineralization of fluorotelomer compounds is not observed, which is analogous to biotransformation reactions. The PFCA yields under TOP Assay digest conditions for specific precursors is presented in Table 3.7 (adapted from Houtz & Sedlak 2012).

3.5.2.2.1.1 Optimizing TOP Assay Conditions Laboratories implementing TOP Assay are basing the method on two peer-reviewed literature publications (Houtz and Sedlak 2012; Houtz et al. 2013), which applied the technique to stormwater runoff, diluted AFFF formulations, and AFFF-contaminated groundwater and soil from a firefighter training site. Pre-treatment steps and TOP Assay reagent concentrations were optimized in these specific samples to achieve no detectable PFAA precursors post-oxidation and to achieve measurement on a calibration curve that ranged from 100 ng/L to 1 microgram per liter (µg/L). Aqueous samples were not filtered to prior to oxidation, and were sometimes diluted prior to oxidation if concentrations were likely to be out of calibration range. Soil samples (500 milligrams (mg)) were extracted in triplicate in basic methanol, centrifuged with ENVI-CARB, decanted, and evaporated.

After these pretreatment steps, all samples or evaporated extracts were amended with 60 millimolar (mM) potassium persulfate and 125 mM sodium hydroxide, thermolyzed in sealed HDPE bottles at 85 degrees Celcius for six or more hours, neutralized with concentrated HCl to a pH between 4 and 10, and amended with isotopically labeled surrogate standards prior to analysis. If a sample was further concentrated with solid phase extraction, the surrogate standards were amended prior to extraction.

Because these methodological steps were sample specific, they may be insufficient to achieve complete oxidation in samples with higher concentrations of organic carbon or other competitive species that may consume the radical concentrations generated. Additional dilution of the samples or sample extracts or adding greater amounts of persulfate may be required to generate sufficient oxidant to fully convert PFAA precursors to PFAAs. Maintaining a highly basic pH as samples become acidic as a result of increased amounts of persulfate may require additional sodium hydroxide or adding a buffer with high pH buffering capacity. The pH needs to be maintained above pH12 because sulfate radical is capable of breaking down PFCAs (Hori et al. 2005), particularly at acidic pH (Bruton et al. 2017). So maintaining a high pH where sulfate radical is converted to hydroxyl radical is important to avoid losses of PFCAs. In general, sample specific adjustments to TOP Assay are anticipated to be required.

3.5.2.2.1.2 TOP Assay Data Interpretation To differentiate between precursors and free PFAAs present in initially in the sample, a sample must be analyzed both pre- and post-TOP Assay. It may be sufficient to measure a sample only after the TOP

TABLE 3.7
Molar yields of perfluoroalkyl carboxylates from precursors subjected to the TOP assay

Starting Precursor Compound	[persulfate]$_0$	[precursor]$_0$	Δ[PFBA]/ [precursor]$_0$	Δ[PFPeA]/ [precursor]$_0$	Δ[PFHxA]/ [precursor]$_0$	Δ[PFHpA]/ [precursor]$_0$	Δ[PFOA]/ [precursor]$_0$	Δ[PFNA]/ [precursor]$_0$
N-EtFOSAA (n=7)	20 mM	25 ng/L, 250 ng/L					92% ± 4%	
N-MeFOSAA (n=8)	20 mM	25 ng/L, 250 ng/L					110% ± 8%	
	5 mM	10 μg/L						
FOSA (n=8)	20 mM	25 ng/L					97% ± 3%	
	5 mM	10 μg/L						
6:2 FtS (n=8)	5 to 60 mM	5 to 10 μg/L	22% ± 5%	27% ± 2%	22% ± 2%	2% ± 1%		
8:2 FtS (n=9)	1 to 60 mM	5 to 10 μg/L	11% ± 4%	12% ± 4%	19% ± 3%	27% ± 3%	21% ± 2%	3% ± 0.1%
6:2 diPAP* (n=6)	20 to 60 mM	25 μg/L	27% ± 3%	47% ± 3%	33% ± 2%	15% ± 3%		
8:2 diPAP* (n=6)	20 to 60 mM	25 μg/L	10% ± 2%	17% ± 1%	24% ± 1%	43% ± 2%	38% ± 2%	13% ± 1%

Source: Reprinted with permission from Houtz, E. F. and D. L. Sedlak. Oxidative Conversion as a Means of Detecting Precursors to Perfluoroalkyl Acids in Urban Runoff. Environmental Science & Technology, 2012, 46, 17, 9342-9349. Copyright 2012, American Chemical Society.

Notes: Initial conditions ranged from 1 to 60 mM persulfate and 25 ng/L to 25 μg/L of the precursor.
Complete disappearance of precursors was observed under all conditions reported.
[precursor]$_0$ – Denotes the initial molar concentration of the precursor.
* Each mole of N:2 diPAP contains two perfluorinated alkyl chains; the maximum yield of PFCAs is 200%.

Assay digest if the total concentration of PFASs is of primary interest. Analyzing samples after TOP Assay will also identify if long-chain PFASs are present in the sample. An example of how TOP Assay data may be reported is presented in Table 3.8. The percent increase in the sample suggests that more than half of the sample is comprised of PFAA precursors, while less than half is comprised of PFAAs. Because PFOA was generated in the sample after TOP Assay, C8 or longer PFAA precursors are present. While the generation of PFSAs is not expected from the TOP Assay, their production is occasionally observed and are presumed to be derived from PFAA precursors in a sample. Figure 3.17 demonstrates conclusions that may be drawn from TOP Assay results, as observed in five individual AFFF-impacted groundwater samples (Battelle book at https://www.arcadis.com/media/4/6/C/%7B46C129DB-9CCE-4BBF-93AA-FCA1E04BD18E%7DAdvances%20in%20Remediation-eBOOK.pdf).

3.5.2.2.1.3 Data Quality Objectives The following metrics should be used to verify the quality of TOP Assay data:

- PFAA precursors such as 6:2 FTS and FOSA that can be individually measured (as analytical standards are available) should not be detectable or are only detected at very low concentrations in the post-TOP Assay sample, compared to concentrations detected in the same samples prior to the oxidative digest; this ensures that oxidation was complete or nearly complete. After the oxidative digest the following practical data quality objectives are advised:
  - For aqueous samples, the sum of measured PFAA precursors divided by the sum of the total PFASs measured is less than 5 percent.
  - For soil samples, the sum of measured PFAA precursors divided by the sum of the total PFASs measured is less than 10 percent.
  - For samples where PFASs were detected near the limits of detection, these metrics may be relaxed.

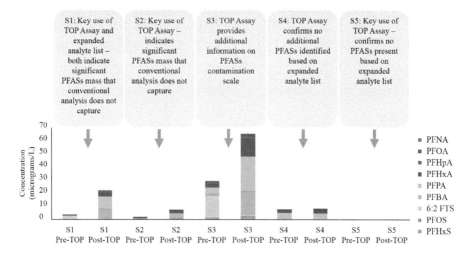

FIGURE 3.17 Interpretation of TOP Assay results in five groundwater samples.

TABLE 3.8
Sample TOP Assay data output

Sample ID	6:2 FTS	PFBA	PFBS	PFPeA	PFHxA	PFHxS	PFHpA	PFHpS	PFOA	Branched PFOS	Linear PFOS	PFOSA	PFNA	PFDA	PFUnA	PFDoA	Sum of PFASs
Pre-TOP Assay	16.7	1.80	<0.1	4.41	2.72	0.17	1.41	<0.1	0.56	0.63	0.76	<0.1	0.19	<0.1	<1	<1	29.35
Post-TOP Assay	0.10	17.65	<0.1	26.75	13.6	0.19	2.67	0.04	1.17	1.44	1.26	<0.1	0.37	<0.1	<1	<1	65.59
Increase in Concentration	−16.59	15.85	0	22.34	10.34	0.03	1.26	0.03	0.61	0.82	0.50	0	0.18	0	0	0	36.25 (122% increase)

Note: All concentrations shown as ng/L.

- Any reaction efficiency surrogates should also be non-detectable after TOP Assay.
- The total PFAS concentration measured after TOP Assay should be at least 90% of the concentration of PFASs measured before TOP Assay; this ensures that PFASs are not lost due to TOP Assay preparation steps.
  - Decreases of up to 10 percent of PFASs might be expected due to normal analytical variability.
  - Typically the total concentration of PFASs measured after TOP Assay should be greater than the concentration measured pre-TOP Assay, because most samples contain measurable concentrations of PFAA precursors.
- PFCAs should remain the same or increase in concentration after TOP Assay; this suggests that PFAA precursors are converted to perfluoroalkyl carboxylates, which is the intent of TOP Assay. If the PFCA concentrations decreases after TOP Assay, this may suggest a sample preparation problem or that the digest conditions became acidic and carboxylates were destroyed.
- The perfluoroalkyl sulfonate concentrations should remain unchanged after the application of TOP Assay; in most matrices, PFAA precursors are not expected to convert to perfluoroalkyl sulfonates under the conditions of TOP Assay. Some slight increases may be observed.

3.5.2.2.1.4 Conversion of TOP Assay Results to Prediction of Likely Terminal Products TOP Assay is not intended to predict the exact eventual end products of precursor transformations in the environment. It does, however, convert precursors to PFAAs, which can then be measured. The rates at which precursors transform to PFAAs in the environment are highly dependent on the site-specific conditions of where the sample is located.

One complicating factor in applying TOP Assay results to eventual precursor end products is that both sulfonamide and fluorotelomer precursors form PFCAs under the conditions of TOP Assay, while sulfonamide precursors are more likely to form PFSAs upon terminal transformation via biological processes in the environment or *in vivo*.

Those caveats stated, TOP Assay results may be used to estimate potential terminal PFAA products. The simplest transfer of TOP Assay results is to assume that the concentration of a Cn perfluoroalkyl carboxylate generated is equivalent to a sample's potential to form either additional Cn perfluoroalkyl carboxylates or Cn perfluoroalkyl sulfonates. Because fluorotelomer compounds form a range of perfluoroalkyl carboxylates under TOP Assay, this approach would allocate, for example, C4 acids generated from a 6:2 fluorotelomer compound to a C4 precursor rather than a C6 precursor.

The use of branched and linear isomeric analytical standards in the quantification of PFCAs could assist in differentiating potential perfluoroalkyl sulfonate precursors from PFCA precursors. The formation of branched perfluoroalkyl carboxylates following TOP Assay is indicative of the presence of sulfonamide

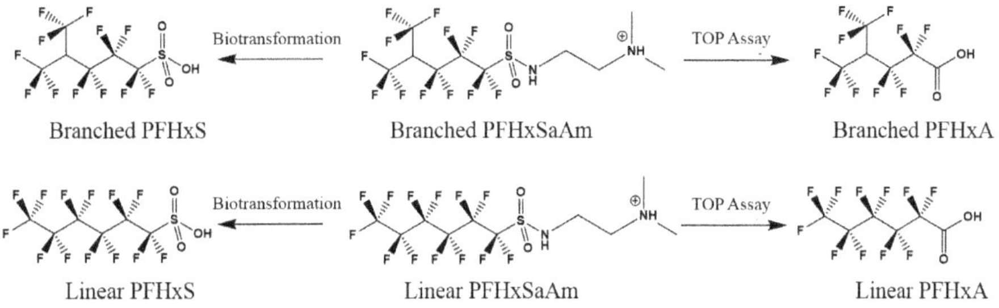

FIGURE 3.18 Branching in PFHxSaAm is retained in PFHxA after TOP Assay digest.

precursors (as shown in Figure 3.18) because sulfonamide compounds contain approximately 30 percent branched perfluoroalkyl groups (Martin et al. 2010) and fluorotelomer compounds contain no branched perfluoroalkyl groups. The concentration of branched carboxylate products could be divided by 30 percent to capture the likely total concentration of potential perfluoroalkyl sulfonate precursors. Differential transport of branched versus linear compounds is known to occur in many systems (Beesoon et al. 2011; Benskin et al. 2010; Benskin, Ikonomou, Woudneh et al., 2012), so the branched versus linear composition is not typically preserved at exactly 30 percent. For example PFHxSaAm which is cationic (Backe et al., 2013), which would be expected to be converted to predominantly PFHxA via the TOP Assay digest (Houtz and Sedlak, 2012), but biologically will transform to PFHxS (Mejia-Avendano et al., 2016;, Rhoads et al., 2008). There have been some reports of poor recovery for some cationic and zwitterionic precursors, such as FtSaBs and 6:2 FtSaAm using published PFAS soil extraction methods (Barzen-Hanson, Davis et al. 2017) with ongoing developments to optimize commercial soil extractions which may be beneficial for PFASs analysis of soils.

Revealing the distribution of the PFAS chain lengths present is a key benefit of TOP Assay, as the potential of a sample to form additional long- or short-chain PFASs is of interest to many practitioners. The use of TOP Assay in addition to discrete quantification of PFAAs and known precursors also provides significantly more confidence in locating and estimating the total PFAS mass at a contaminated site.

Analysis of PFASs in AFFF-impacted groundwater using TOP Assay and Adsorbable Organo Fluorine (AOF) data (described as follows) has shown excellent agreement between the two techniques, as shown in Figure 3.19. This indicates that TOP Assay represented a reliable quantitative method to measure the total fluorinated compounds for the PFASs present at this site.

3.5.2.2.2 Adsorbable Organofluorine Analysis

The Adsorbable Organofluorine (AOF) analysis uses combustion ion chromatography, which involves sorption of PFASs onto a synthetic activated carbon (thus excluding inorganic fluoride) followed by combustion and measurement of evolved fluoride by ion chromatography (Yeung et al. 2008; Dauchy et al.

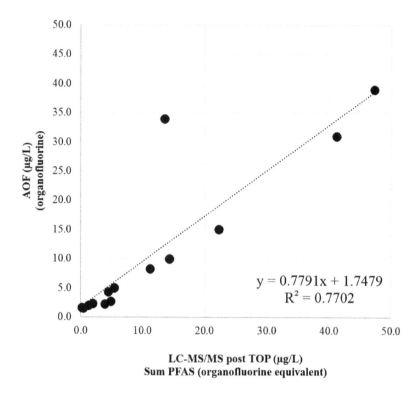

FIGURE 3.19 Comparison of AOF and TOP Assay using AFFF-impacted groundwater.

2017a). AOF analysis provides a single measurement of organofluorine content with a limit of detection of ~1 µg/L. Extractable Organofluorine (EOF) analysis, based on a similar process, has also recently been edited for soils with a level of detection of 50 µg/kg. AOF is still undergoing standardization and is offered in a limited number of commercial laboratories (primarily in Europe and Australia), although no extremely specialized equipment is required to conduct AOF. Due to its relatively high (µg/L) detection limit and the inability to separate the individual PFAS, AOF can only be used as a guideline value. Analysis of individual PFASs would still need to be conducted for comparison to current screening criteria.

3.5.2.2.3 Particle-Induced Gamma Emission

Particle-Induced Gamma Emission (PIGE) spectroscopy involves the initial separation of aqueous samples of PFAS via solid extraction cartridges followed by proton bombardment and the measurement of the unique gamma ray signature emitted from any fluorine present (Ritter et al. 2017). This analysis also provides a single measure of total organofluorine with a limit of detection of ~2 µg/L. PIGE is only provided in a limited number of locations in the U.S., which possess the appropriate equipment and is still undergoing standardization.

3.5.2.2.4 Quadrupole Time of Flight—Mass Spectroscopy

Quadrupole Time of Flight—Mass Spectroscopy (QTOF-MS) can identify multiple precursors via mass ion capture and accurate mass estimation to give likely empirical formulas. However analytical standards are required for unequivocal structural identification (Barzen-Hanson, Roberts et al. 2017; Newton et al. 2017). The identification of PFASs via QTOF-MS can be assisted by library searching, a preliminary identification of potential PFAS sources, and information gathered from patents. Full-scan high-resolution data can be used with a mass defect filter to produce a semi-quantitative estimate of the total concentration fluorine containing compounds in a sample both for known and unknown compounds (Casson and Chiang 2018). This tool is currently used mainly in academic settings and is not likely to be relevant for PFAS site characterization on a broad scale, though it may be useful for specific sites or samples where a detailed delineation of many individual PFASs is desired, for example, with environmental forensics.

3.6 HEALTH CONSIDERATIONS

3.6.1 Exposure Routes

A general conceptual diagram of PFAS exposure to humans and the environment is presented in Figure 3.20. Human exposure to PFASs is mainly by ingestion of contaminated food or water (dietary uptake/oral route) (Fromme et al. 2009; Agency for Toxic Substances and Disease Registry [ATSDR] 2009), for example, where water supplies used for drinking or crop irrigation become contaminated or where PFASs bioaccumulate within animal food sources such as fish. Exposure via food may also occur through eating food that was packaged in material that contains PFASs (ATSDR 2017).

As PFAS have also been found in both air and dust, exposure by breathing air, ingestion of dust, or dermal contact with dusts or aerosols of PFAS may also be a source of exposure (ATSDR 2009a; 2009b).

Research has suggested that exposure to PFOA and PFOS from consumer products containing PFASs, such as nonstick cookware, stain-resistant carpeting, and water-repellant clothing, is usually low, especially when compared to exposures to contaminated drinking water (ATSDR 2017) and only a small amount of PFAS can get into your body through skin contact.

Both in animals and humans, PFOS and PFOA can cross the placenta barrier (Stahl et al. 2011) and so cause exposure to the developing fetus. Breast milk and household dust are potentially important exposure pathways for infants and young children (ATSDR 2017).

3.6.2 Distribution in Tissue

PFAAs such as PFOS and PFOA have, contrary to most other persistent organic pollutants (POPs), a low affinity to lipids and preferentially bind to proteins. PFOS is associated with cell membrane surfaces and accumulates in various, mainly high perfused, body tissues of exposed organisms (Danish Ministry of the Environment, Environmental Protection Agency [DEPA], 2013).

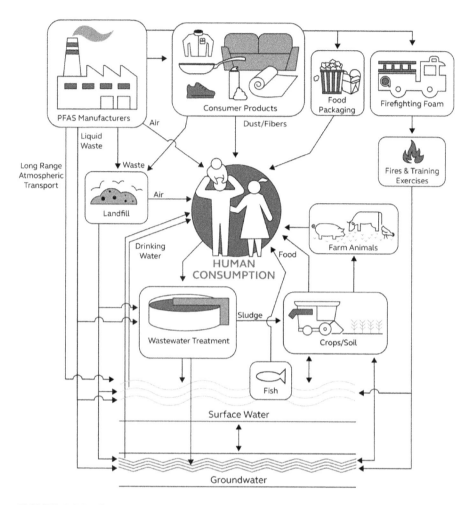

FIGURE 3.20 Conceptual diagram of PFAS exposure to humans and the environment (modified after Oliaei et al. 2013).

The highest concentrations are usually detected in blood, liver, kidneys, lung, spleen, and bone marrow. PFOS concentrations in the blood range between sub-parts per billion (ppb) level up to hundreds of ppb level. PFOA levels in blood are generally lower (sub-ppt level up to tens of ppb level) (Loganathan and Kwan-Sing Lam 2011).

3.6.3 Bioaccumulation

Conder et al. (2008) concluded that: "(1) bioconcentration and bioaccumulation of perfluorinated acids is directly related to the length of each compound's fluorinated carbon chain; (2) PFSAs are more bioaccumulative than PFCAs of the same fluorinated carbon chain length." Other studies have been conducted that indicate long-chain PFASs have a higher potential to bioconcentrate and bioaccumulate through trophic levels as compared to shorter-chain PFAS (Stahl et al. 2012; Asher et al. 2012; Awad et al. 2011; De Silva et al. 2011; OECD 2013; Liu et al. 2011). Short-chain PFAAs have been found to concentrate in the edible portion of some crops (Blaine et al. 2014a; Blaine et al. 2014b).

Information about the bioconcentration, bioaccumulation, and biomagnification for PFOS and PFOA are presented in Tables 3.9 and 3.10, respectively. The numerical criterion under Registration, Evaluation, Authorization and Restriction of Chemicals (REACH) defining that a substance is bioaccumulative if a bioconcentration factor (BCF) in aquatic species is higher than 2,000 liter/kilogram (L/kg). (EU 2011). Bioconcentration factors greater than 1 L/kg indicate bioaccumulative potential only from a scientific standpoint.

The data relating to PFOS in Table 3.9 show BCF greater than 2,000 L/kg, demonstrating the bioaccumulation properties of PFOS. The data relating to PFOA in Table 3.10 show BCF is closer to 10 L/kg. The biomagnification factors in Tables 3.9 and 3.10 highlight that predatory animals are recorded with greater concentrations in their bodies compared to the concentrations in their diets, demonstrating biomagnification of PFOS. As a result, concentrations of PFOS are likely to be elevated within

TABLE 3.9
Bioconcentration, bioaccumulation, and biomagnification information for PFOS

Organism	Bioconcentration Factor (L/kg)	Bioaccumulation Factor (L/kg)
Benthic algae	690; 1,000	
Zooplankton	220	
Amphipods		380; 830
Crayfish		920; 1,200
Zebra mussel		380; 590
Fish species*	450 to 2,800	710 to 9,800
Turtles		11,000

Predator	Prey	Biomagnification Factor (normalized for the trophic level)
Beluga	Arctic cod	0.91
Narwhal	Arctic cod	0.76
Dolphin	Various fish species	0.76 to 14
Various seal species	Arctic cod	1.9 to 6.4

Notes:
Source: Adapted from Conder et al. 2008 and references therein.
* Some values are based on laboratory studies; other reported values are from field exposure studies
Bioconcentration factor/bioaccumulation factor: defined mathematically as the concentration of chemical in an organism (whole body, wet weight basis) divided by the concentration in an abiotic environmental medium, usually water.
Biomagnification factor: expressed mathematically as the concentration of chemical in an organism divided by the concentration of chemical in its food; values greater than 1 may suggest biomagnification potential.

TABLE 3.10
Bioconcentration, bioaccumulation, and biomagnification information for PFOA

Organism	Bioconcentration Factor (L/kg)	Bioaccumulation Factor (L/kg)
Benthic algae	6.8; 23	
Zooplankton		83
Amphipods		170; 570
Crayfish		6.8; 23
Zebra mussel		170; 570
Fish species*	1.8 to 9.4	23 to 230

Predator	Prey	Biomagnification Factor (normalized for the trophic level)
Predator: Dolphin Prey: Red drum	Various fish species	1.8 to 10

Notes:
Source: Adapted from Conder et al. 2008 and references therein.
* Some values are based on laboratory studies; other reported values are from field exposure studies.
Bioconcentration factor/bioaccumulation factor: defined mathematically as the concentration of chemical in an organism (whole body, wet weight basis) divided by the concentration in an abiotic environmental medium, usually water.
Biomagnification factor: expressed mathematically as the concentration of chemical in an organism divided by the concentration of chemical in its food; values greater than 1 may suggest biomagnification potential.

organisms at higher trophic levels. Additionally, the bioaccumulation potential in the terrestrial environment has been shown to be significantly lower than in the aquatic or marine environment (DEPA 2013).

3.6.4 ELIMINATION

A summary of available information on half-lives for PFOA and PFOS are summarized in Table 3.11. As shown, both PFOS and PFOA are very slowly eliminated from the human body. The Toxicological Overview for PFOS and PFOA, published

TABLE 3.11
Elimination half-lives of PFASs

Half-Life (days)	PFBA	PFBS	PFHxA	PFHxS	PFOA	PFOS
Rat	0.3	0.2	0.05–0.2	7	5	25
Monkey	2	4	1	100	21	45
Human	3–4	26	<28	3,000	1,000	1,500

Source: Data sources: Olsen et al. 2009; Butenhoff et al. 2004; Chang et al. 2008; Chang et al. 2012

by the Public Health England (PHE 2009), documents a half-life from the human body of approximately nine years for PFOS and four years for PFOA. In fluorochemical workers, PFHxS had the longest observed elimination half-life (8.5 years), followed by PFOS (5.4 years), and PFOA (2.3–3.8 years) (Olsen et al. 2007). The excretion of PFASs varies with the type of PFAA and also with the animal species and gender (Olsen et al. 2007; Chang et al. 2008; Wong et al. 2014; Wong et al. 2015). The reason for the species and gender differences in elimination are not well understood (U.S. EPA 2009b).

In general, the blood half-lives of perfluorochemicals:

- are longer for sulfonates than for carboxylates;
- are shorter for branched isomers than straight chain;
- are often shorter in females than males. This may be due to the difference in renal clearance (and hormones) (DEPA 2013). Sex differences documented for rats and monkeys are not always found in humans (DEPA 2013);
- increase with chain length for carboxylates;
- vary significantly between species.

While the elimination of PFOS, PFHxS, and PFOA from the human body takes some years (Olsen et al. 2007), elimination of short-chain PFAAs is in the range of days, but repeat exposure to short chain PFAAs can potentially lead to generation of a steady state concentration. In addition, the half-life of PFOS and PFOA in rodents is in the range of months (Chang et al. 2012; Ohmori 2003), which differs significantly from humans and can cause extrapolation issues in tests. The primary clearance route for PFOS and PFOA is urine, rather than fecal elimination (Bull et al. 2014).

The general human population contains measurable PFASs, in particular PFOS and PFOA, in their blood, while significantly higher levels are observed in people who live near primary or secondary PFAS manufacturing facilities or have been occupationally exposed (ATSDR 2016) (e.g., firefighters equipped to handle Class B fires and employees working in coating facilities). The Center for Disease Control and Prevention (CDC) National Health and Nutrition Examination Survey (NHANES) has documented declining concentrations of PFOS and PFOA in the U.S. general population, which is correlated with recent efforts to stop or reduce production and use of PFOS and PFOA in consumer products (ATSDR 2016). Similar trends have been observed in Germany (Yeung et al. 2013a; Yeung et al. 2013b) and are also observed elsewhere, as documented in several publications. Some long-chain PFASs continue to rise in human serum, despite these phase-outs (Yeung et al. 2013a; Yeung et al. 2013b), and short-chain PFASs such as PFHxA have increased in human serum (Glynn et al. 2012). Exposure to increasing amounts and proportions of unknown PFASs has been reported from analysis of human plasma from Germany using AOF. This indicates that the reported diminishing concentrations of PFOS, PFOA, and PFHxS in plasma do not reflect PFAS concentrations in human blood (Yeung and Mabury 2016). It seems likely that detections of the replacement chemistries will start to dominate over PFASs that are widely regulated.

3.6.5 Toxicologic and Epidemiological Studies

While the toxicology of PFAS is not completely understood, the following sections provide an overview of the available human and animal toxicity data and epidemiological studies.

3.6.5.1 Acute Toxicity

Currently, data to assess the acute toxicity following high exposure by means of inhalation, ingestion, dermal, or ocular contact in humans is not available (PHE 2009). Additionally, Bull et al. (2014) provided an extensive literature search which does not identify data on the acute toxicity of PFOS and PFOA.

3.6.5.2 (Sub)Chronic Toxicity

As discussed further in the epidemiological studies section, there is a significant amount of data on the impact of (sub)chronic PFOS and PFOA exposure on reproductive and/or developmental and other types of effects in humans.

Conflicting results appear when one looks at the relationship between PFAS exposure and reproductive parameters as well as with maternal PFAS exposures and indices of fetal growth and development (e.g., birth weight, birth length, gestational age, and pre-term birth). While some studies have found increased PFAS levels to result in decreased indices of fetal growth and development (Fei et al. 2007; Maisonet et al. 2012; Wang, Hsieh et al., 2011), other studies have reported increased PFAS levels resulting in no effect on fetal growth and development (Fei et al. 2008; Hamm et al. 2010; Olsen et al. 2009).

3.6.5.3 Epidemiological Studies

During the past few years, several epidemiological studies were conducted to investigate relations between PFOS/PFOA exposure and various health effects including fertility, growth, and developmental biomarkers. One of the largest studies included community members and workers near the plant in Parkersburg, West Virginia. This large human health screening study determined probable association between PFOA exposure and adult chronic diseases (C8 Science Panel 2017). The C8 Science Panel concluded that probable links existed between PFOA exposure and high cholesterol, thyroid disease, pregnancy-induced hypertension, ulcerative colitis, and kidney and testicular cancer (Winquist et al. 2013; Barry et al. 2013).

Similarly, in the Veneto Region of Italy, contamination of drinking water by PFAS resulted in elevated PFAS serum levels in impacted areas (Ingelido et al. 2018). Based on available serum concentrations and health data, statistically significant higher risk ratios were detected for mortality due to several causes including diabetes, cerebrovascular diseases, myocardial infarction, and Alzheimer's disease (Mastrantonio et al. 2017). Additionally, for females, significantly higher risk ratios were observed for kidney and breast cancer and Parkinson's disease (Mastrantonio et al. 2017). Similar studies are ongoing in Sweden for PFOS.

Rappazzo et al. (2017) performed a systematic review for studies on measured PFAS exposure through serum, blood, or breast milk and child health outcomes, including immunity/infection/asthma, cardio-metabolic, neurodevelopmental/attention, thyroid, renal, and puberty onset. They observed consistent evidence between PFAS concentrations and association with dyslipidemia, immunity (i.e., vaccine response and asthma), renal function, and age at menarche (Rappazzo et al. 2017). A recent review by the National Toxicology Program Office of Health Assessment and Translation on the exposure to PFOA or PFOS and immunotoxicology concluded that PFOA or PFOS is "presumed to be an immune hazard to humans" (National Toxicology Program 2016).

In a recent diet-induced weight-loss trial, higher baseline plasma PFAS concentrations were associated with a greater weight regain, especially in women, considered possibly explained by a slower regression of Resting Metabolic Rate (RMR) levels (Liu et al. 2018). The authors suggest the data illustrates a potential novel pathway through which PFASs interfere with human body weight regulation and metabolism.

While several of the human epidemiological studies have recently reported associations with PFOS and cholesterol, birth weight changes and various thyroid parameters, these studies were reported to show inconsistent conclusions resulting by the U.S. EPA Science Advisory Board noting: "The results of existing epidemiology studies are not adequate for use in quantitative risk assessment" (U.S. EPA 2014a). Following review of more than 200 epidemiological studies, European Food Safety Agency (EFSA) concluded in 2018 that three endpoints associated with PFOS or PFOA exposure are likely to be causal and adverse as a result of consistent data (EFSA 2018). These comprise (1) increased serum cholesterol, (2) impaired vaccination responses in children, and (3) high serum alanine transferase (ALT).

3.6.5.4 Polyfluoroalkyl Substance Toxicity

Initial toxicity studies determined that PFCA precursors are more toxic than PFCA compounds (Phillips et al. 2007; Phillips et al. 2010). Studies looking at chronic toxicity (i.e., growth, reproduction, and survival) in *Chironomus dilutus* (C. Dilutus), *Daphnia magna* (D. magna), and *Lemna gibba* (L. gibba) found that the saturated forms of the telomer acids were more toxic than the unsaturated forms of the telomer acids (Phillips et al. 2007), with toxicity thresholds for fire training areas (FTAs) to D. magna in the low to mid in the µg/L and C. dilutus in the mid µg/L to low mg/L range. Additionally, in all three species, increasing chain length (particularly from C8 or greater fluorinated carbons) lead to an increase in toxicity (Phillips et al. 2007). The increased toxicity observed for PFCAs and FTAs of longer chain lengths could be the result of greater bioaccumulation and slower depuration rates of these compounds (Phillips et al. 2007).

Precursors may have an important role in PFCA exposure for humans as well. The data on uptake and biotransformation factors for PFCAs and their precursors results in uncertainties for total human exposure (Gebbink et al. 2015). These precursors may lead to accumulation of PFCAs, contributing to the overall PFAS burden in humans (Rand and Mabury 2017).

3.6.5.5 Derivation of Reference Doses/Tolerable Daily Intakes

Different cleanup targets are being developed for the same PFASs detected in different parts of the world and they are often developed from the same underlying toxicological research studies (U.S. EPA 2014a; U.S. EPA 2014b; RIVM 2017b; DEPA 2015; ATSDR 2018). Criteria for drinking water vary depending on which critical end point is selected for the point of departure (POD), scaling factors from no observed adverse effect level (NOAEL) to lowest observed adverse effect level (LOAEL), pharmacokinetic modeling to scale from animal to human, uncertainty factors, selection of most appropriate human receptor, and the exposure percentage assumed from drinking water. A summary of NOAEL and LOAEL information is provided in Table 3.12.

While PFOS and PFOA do not behave identically, they do share common sensitive endpoints and may be considered for additive toxicity (U.S. EPA 2016). The U.S. EPA selected developmental outcomes for both PFOA and PFOS as the most sensitive endpoint when calculating the human health advisory; therefore, the lifetime health advisory for drinking water was set at 70 ppt for their combined concentrations (U.S. EPA 2014a; U.S. EPA 2014b; U.S. EPA 2016).

However, several research institutes and authorities have total daily intake values for PFOS and PFOA, which are used as points of departure for the drinking water target values (and other environmental values) in these countries. In the last couple of years these standards have significantly diminished (see Table 3.13). Considering the short-chain PFASs, a recent review described a lack of toxicological information available (DEPA 2015a).

The EFSA is currently discussing revising the TDIs for PFOS and PFOA with EU member states, as part of its "scientific opinion of the risk to human health related to the presence of PFOS and PFOA in food" (EFSA 2015), adopted March 22, 2018. A recent draft of an EFSA report described revised tolerable weekly intake (TWI) levels for PFOS and PFOA (EFSA 2018). The draft indicated TWIs proposed by EFSA were a similar order of magnitude to those produced by a draft report from ATSDR in 2018 (ATSDR 2018); these are presented in Table 3.13. It is also noted that the EFSA is also reviewing the risks to human health related to PFAS apart from PFOS and PFOA (EFSA, 2017) which has a scheduled deadline of December 31, 2018.

Rather than developing human health–based standards for each PFAS individually, there have been efforts to categorize compounds by chain-length or to regulate certain PFAS based on toxicology information regarding other PFAS. For example, Australia, the U.S., Denmark, Sweden, and Italy have developed drinking water criteria based on a sum of multiple PFASs (as described in Appendix D), which helps account for potential cumulative risk of certain PFCAs and PFSAs.

3.6.5.6 Carcinogenic Effects

The U.S. EPA concludes that evidence of carcinogenicity of PFOS is "suggestive," but not definitive, because the tumor incidence does not indicate a dose response (U.S. EPA, 2014a). Based on the risk-assessment study performed in 2005 (U.S. EPA, 2005), PFOA's carcinogenicity was also categorized as "suggestive." In June

TABLE 3.12
Reported NOAELs and LOAELs

Method	NOAEL or LOAEL (mg/kg bw/d)	Effects	Reference
		PFOS	
Mice			
Oral Gavage, 60 days	NOAEL: 0.008 LOAEL: 0.083	Increased liver mass, increased splenic natural killer cell activity, decreased plaque forming cell response to sheep red blood cells	Dong et al. 2009
Oral, 90 days	NOAEL: 0.43 LOAEL: 2.15	Impaired spatial learning and memory	Long et al. 2013
Rats			
Oral Diet, 14 weeks	NOAEL: 0.40 (female) 0.34 (M) LOAEL: 1.56 (female) 1.33 (male)	Liver effects	Seacat et al. 2003
Oral Gavage, 20 days	NOAEL: 1 LOAEL: 2	Developmental effects	Lau et al. 2003
Oral Diet, 90 days	LOAEL: 2	Liver effects	Goldenthal et al. 1978
Oral Gavage, 28 days	LOAEL: 5	Decrease in body weight	Cui et al. 2009
Oral Gavage, 20 days	NOAEL: 0.3 LOAEL: 1.0	Maternal toxicity	Butenhoff et al. 2009
Oral Gavage, 12 weeks 2 generation	NOAEL: 0.1 LOAEL: 0.4	Decreased adult body weight gain, decreased pup body weight	Luebker, Case et al. 2005
1 generation – 63 days female only 42 day dosing prior to GD 21	NOAEL: 0.4 (maternal) LOAEL: 0.8 (maternal) LOAEL: 0.4 (fetal)	Decreased maternal weight gain, decreased gestational length, deceased pup survival; decreased pup body weight	Luebker, York et al. 2005
Oral, Diet, 2 years	NOAEL: 0.120 (female) 0.024 (male) LOAEL: 0.299 (female) 0.098 (male)	(male): Cystic degeneration, centrilobular vacuolation (female): Entrilobular eosinophilic granules Both: increased hepatic necrosis centrilobular vacuolation at higher doses	Thomford 2002/ Buttenhoff, Cheng et al. 2012
Rabbits			
Oral Gavage, 14 days	NOAEL: 0.1 (maternal) NOAEL: 1 (fetal) LOAEL: 1 (maternal) LOAEL: 2.5 (fetal)	Decreased maternal body weight and food consumption; decreased fetal body weight	Case et al. 2001
Cynomologous Monkey			
Oral Gavage, 90 days	LOAEL: 0.5	Diarrhea, anorexia	Goldenthal et al. 1979
Oral Gavage, 6 months	NOAEL: 0.15 LOAEL: 0.75	Increased liver weight and hepatocellular hypertrophy, decrease in total T3, decreased survival	Seacat et al. 2002

(Continued)

TABLE 3.12 (*Continued*)

Method	NOAEL or LOAEL (mg/kg bw/d)	Effects	Reference
		PFOA	
Mice			
Oral Gavage, 14 days	LOAEL: 0.3	Liver weight	Loveless et al. 2006
Oral Gavage, 17 days	LOAEL: 1 (maternal) LOAEL: 3 (fetal) NOAEL: 1 (fetal)	Increased maternal liver weight; decreased maternal weight gain; fetal developmental effects	Lau et al. 2006
Rats			
Oral Gavage, 14 days	LOAEL: 1 NOAEL: 0.3	Effect on hormone values	Loveless et al. 2006
Oral Diet, 14 days	NOAEL: 0.6 (male) NOAEL: 22 (female) LOAEL: 1.7 (male) LOAEL: 76 (female)	Liver effects	Goldenthal et al. 1978
Oral Diet, 13 weeks, Male only	LOAEL: 0.64 NOAEL: 0.06	Liver effects	Perkins et al. 2004
Oral Gavage; 2-generation F0 generation (F): 127 days (M): 84 days	NOAEL: 30 (female) LOAEL: 1 (male)	(female): No effects observed (male): Increased absolute and relative liver and kidney weight, decreased body weight	Butenhoff et al. 2004; York et al. 2010
Oral Gavage; 2-generation F1 generation (F): 10 weeks (M): 16 weeks	NOAEL: 10 (female) LOAEL: 30 (female) 1 (male)	(female): Delay in sexual maturity, decreased body weight and weight gain (male): Decreased body weight and weight gain, increased absolute and relative liver weights, liver hypertrophy, increased absolute and relative kidney weights	Butenhoff et al. 2004; York et al. 2010
Oral Diet, 2 years	NOAEL: 1.6 (female) 1.3 (male) LOAEL: 16.1 (female) 14.2 (male)	(female): Decreased body weight gain (male): Decreased body weight gain, lesions in liver, testes, and lungs	Butenhoff, Kennedy et al. 2012
Cynomologous Monkey			
Oral Gavage, 90 days	NOAEL: 3 (female) LOAEL: 10 (female) 3 (male)	(female): Decreased absolute heart and brain and relative liver weight (male): Increased relative pituitary weight	Goldenthal 1978
Oral Capsule, 26 weeks Male only	LOAEL: 3	Increased absolute liver weight and mean liver to body weight percentages	Butenhoff et al. 2002

Updated from Pancras et al. 2016

TABLE 3.13
The evolution of TDIs over time

Agency, Year	Health-Based Criteria Type	Point of Departure	Uncertainty Factor	Calculated Health-Based Value
PFOS				
UKCOT, 2006	Tolerable daily intake	0.03 mg/kg bw/day	100	300 ng/kg bw/day
EFSA, 2008	Tolerable daily intake	0.03 mg/kg bw/day	200	150 ng/kg bw/day
Swedish EPA, 2012	Derived no effect level (immunotoxicity)[a]	17.8 ng/mL serum	150[d]	0.12 ng/mL serum
DEPA, 2015	Tolerable daily intake	0.033 mg/kg bw/day	1,230	30 ng/kg bw/day
ATSDR, 2015	Minimal risk level	2.52×10^{-3} mg/kg bw/day[b]	90	30 ng/kg bw/day
US EPA, 2016	Reference dose	0.00051 mg/kg bw/day	30	20 ng/kg bw/day
ATSDR, 2018	Minimal risk level	0.000515 mg/kg bw/day	30[c]	2 ng/kg bw/day
EFSA, 2018	Tolerable weekly intake	21 – 26 ng/mL serum	-	13 ng/kg bw/week
PFOA				
UKCOT, 2009	Tolerable daily intake	0.3 mg/kg bw/day	200	1.5 µg/kg bw/day
EFSA, 2008	Tolerable daily intake	150 mg/kg bw/day	200	1.5 µg/kg bw/day
Swedish EPA, 2012	Derived no effect level	150 ng/mL serum	75[d]	2.0 ng/mL serum
DEPA, 2015	Tolerable daily intake	0.003 mg/kg bw/day	30	100 ng/kg bw/day
ATSDR, 2015	Minimal risk level	1.54×10^{-3} mg/kg bw/day	90	20 ng/kg bw/day
US EPA, 2016	Reference dose	0.0053 mg/kg bw/day	300	20 ng/kg bw/day
ATSDR, 2018	Minimal risk level	0.000821 mg/kg bw/day	300	3 ng/kg bw/day
EFSA, 2018*	Tolerable weekly intake	9.2 – 9.4 ng/mL serum	-	6 ng/kg bw/week

Notes:

[a] Higher Derived-No-Effect-Levels were also calculated by the Swedish EPA for hepatoxicity and reproductive toxicity.

[b] Human Equivalent Dose using physiologically based pharmacokinetic modeling.

[c] Another modifying factor of 10 was used for concern that immunotoxicity may be a more sensitive endpoint than developmental.

[d] Assessment factor used in place of the uncertainty factor.

* Indicates that the values are draft and subject to change.

"Tolerable daily intake" is an estimate of the amount of a chemical in food or drinking water, expressed on a body weight basis, that can be ingested daily over a lifetime without appreciable health risk to the consumer.

"Derived no effect level" is the level of exposure to the substance above which humans should not be exposed.

"Minimal risk level" is an estimate of the daily human exposure to a hazardous substance that is likely to be without appreciable risk of adverse noncancer health effects over a specified duration of exposure.

"Reference dose" is an estimate, with uncertainty spanning perhaps an order of magnitude, of a daily oral exposure to the human population (including sensitive subgroups) that is likely to be without an appreciable risk of deleterious effects during a lifetime.

Modified after FSANZ (2016). Updated values from ASTDR (2018) and Schwerdtle et al., (2018).

2014, the International Agency for Research on Cancer (IARC), as part of the World Health Organization, assessed the carcinogenicity of PFOA. PFOA was classified as follows: "possibly carcinogenic to humans (Group B), based on limited evidence in humans that exposure to PFOA is associated with testes and kidney cancer and limited evidence in experimental animals" (World Health Organisation 2014). Currently, PFOS is not yet classified by IARC.

3.7 REGULATION

Recognition that a major route of public exposure to PFASs has been via impacted drinking water has led to regulators in multiple jurisdictions reviewing national and state standards. This has resulted in overall trends toward regulation of a greater number of PFASs, including short-chain species and some polyfluoroalkyl substances, to much lower allowable concentrations, and with some use of cumulative targets across a range of PFASs (Kjølholt 2015; Plassmann and Berger 2010; Plassmann and Berger 2013; Banzhaf et al. 2017; Environment Protection Authority, Victoria 2018, O'Brien et al. 2011; Ding and Peijnenburg 2013; RIVM 2017a, Wilson 2016a; Environment Agency 2007; Post et al. 2017, 2017b; Hoke et al. 2016; Gannon et al. 2016; Gomis et al. 2015; Integral Consulting Inc. [Integral] 2016; Vierke 2017; Environment Agency 2017; DEPA 2015a, 2015b; Gobelius et al. 2018). The underlying drivers of increasingly lower drinking water standards are the diminishing acceptable daily exposure levels considered acceptable for the general population; these values, known as tolerable daily intakes (TDIs) or reference doses (RfDs), have fallen significantly in many jurisdictions as shown in Table 3.13. The evolution of regulatory standards is shown diagrammatically in Figure 3.21. Appendix E provides a summary of current international standards as of July 2018 that are established for PFASs in various matrices.

3.7.1 REGULATION OF PFASs

A brief summary of regulatory considerations for various regions is provided below.

3.7.1.1 Global Treaties and Conventions

In 2009, PFOS was added to Annex B of the Stockholm Convention on POPs, meaning that measures must be taken to restrict its production and use. In June 2015, the EU submitted a proposal to list PFOA, its salts such as APFO, and PFOA-related precursor compounds such as 8:2 FTOH to Annexes A, B, and/or C of the Stockholm Convention. Most recently, during 2017 both PFOA and PFHxS were added to compounds proposed to be listed under the Stockholm Convention (UNEP 2017).

The Stockholm Convention on Persistent Organic Pollutants aims to eliminate or restrict the production and use of persistent organic pollutants (POPs) based on its persistence, bioaccumulation, and toxicity; termed PBT. However, there are additional emerging methods to evaluate environmental hazards posed by many emerging contaminants, including PFASs, such as the assessment of persistence, mobility,

Per- and Polyfluoroalkyl Substances

FIGURE 3.21 Evolution of regulatory standards and awareness of PFASs over time. See Appendix D for additional information regarding international regulatory screening levels for PFASs. Other references cited as follows: Cheremisinoff 2017; UNEP 2009; UNEP 2017; U.S. EPA 2002; U.S. EPA 2009; U.S. EPA 2017.

and toxicity (PMT), or very bioaccumulative and very persistent (vPvB) compounds, or very mobile and very persistent (vPvM) compounds (German Environment Agency [UBA] and Nowegian Geotechnical Institute [NGI] 2018). The use of these additional criteria to assess environmental effects of PFASs may potentially lead to the identification of a far wider range of PFASs to be restricted under future environmental regulations.

The regulatory focus in many locations have been on long-chain PFAAs such as PFOS, PFOA, and PFHxS, but an expanding number of countries are now regulating additional PFASs such as the short-chain PFAAs, select polyfluoralkly substances, and perfluoroalkyl ethers (Environment Protection Authority, Victoria 2018; RIVM 2017a; DEPA 2015a; DEPA 2015b; Kjølholt 2015; Integral 2016; Banzhaf, Filipovic et al. 2017; Environment Agency 2017; Post, Gleason et al. 2017; Vierke 2017; Fire 2018; Gobelius, Hedlund et al. 2018; Land, de Wit et al. 2018).

Several other international conventions and agreements have covered PFASs in some way; of note are the OECD and UNEP Global PFC Group. These organizations aim to develop and promote international stewardship programs and regulatory approaches for PFASs of concern. Work completed includes a process to establish a global emission inventory for PFASs, and promotion of information on alternatives to PFAS in key industries; they have also provided a snapshot of current activities regarding risk reduction approaches for PFASs in a number of countries (OECD 2018b) with 4,730 individual PFASs recently identified and published. The advances in analytical techniques may allow for regulations to expand to include a more holistic and comprehensive measure of the total PFAS loading in environmental matrices (Kotthoff and Bucking 2018), as is being observed in Australia.

3.7.1.2 United States of America

In 2016, the U.S. EPA finalized lifetime health advisory for PFOS and PFOA in drinking water, which are not to exceed 70 ppt (or ng/L) separately or in combination. While this is not a federally enforceable maximum contaminant limit (MCL), most public water systems have taken actions to reduce PFOS and PFOA below these levels. The results of the U.S. EPA Third Unregulated Contaminant Monitoring Rule (UCMR3), which measured six PFASs in drinking water around the U.S. from 2013 to 2015, detected PFOS and PFOA in excess of a 70 ppt combined concentration in 66 water supplies (Hu et al. 2016a).

There are currently no federal advisory levels or enforceable regulations for any PFAS in groundwater. Individual U.S. states have set screening values or limits in drinking water, groundwater, and soil (see Appendix E). Most notably, the following states have proposed or issued screening values and regulations for PFAS as follows:

- New Jersey MCL for drinking water: 13 ppt PFNA, 14 ppt PFOA, 13 ppt PFOS (recommended).
- Vermont drinking water health advisory: 20 ppt combined PFOA, PFOS, PFHxS, PFHpA and PFNA.

- California lists PFOS and PFOA as "known to the state to cause reproductive toxicity," triggering Proposition 65 labeling requirements for products that contain PFOS and PFOA with recommended interim notification levels in drinking water at 13 ppt for PFOS and 14 ppt for PFOA (California State Water Qualty Control Board 2018).
- Texas has set protective concentrations levels (PCLs) for 16 individual PFASs in soil and groundwater.

3.7.1.3 Europe

Pan-European regulation is led by the EU and in 2013, PFOS and its derivatives were included in the Directive on EQS. These standards working under the Water Framework Directive aim to allow member states to work toward achieving good surface water chemical status for priority substances and certain other pollutants.

The EU annual average EQS for PFOS in surface freshwater is set at a very low criterion of 0.65 ng/L, based on the potential for secondary poisoning in humans due to fish consumption and is derived from starting points that have been described as potentially overly conservative (Wilson 2016b). However, the TDIs applied here are those developed by EFSA in 2008, which were revised in 2018 and are significantly lower. It should be noted that 0.65 ng/L is lower than background levels typically recorded in many European surface waters (Loos et al. 2009).

Across Europe, the most stringent drinking water guidelines are found across Scandinavia, where the approach of using a total concentration for a set of PFAS to define a target is also common. In Denmark the guideline is set at 100 ppt for the sum total of 12 individual PFASs, while in Sweden the guideline is set at 90 ppt, again for a sum of 12 PFASs. Certain German Federal States have also followed the Scandinavian approach, with Bavaria, for example, regulating 13 individual PFASs in drinking water to varying acceptance criteria between 100 ppt (for long-chain PFAAs and some precursors) and between 300 ppt and 10 ppb (for short-chain PFAAs) (Bavarian State Office for the Environment 2017). See Appendix E for additional information on regulatory standards in Europe.

3.7.1.4 Australia

In April 2017, following the publication of the Food Standards Australia New Zealand (FSANZ) Hazard Assessment Report (FSANZ 2017), health-based drinking water quality values of 0.07 µg/L for a sum of PFOS and PFHxS, and 0.56 µg/L for PFOA were released for use in site investigations in Australia.

The Queensland Department of Environment and Heritage Protection released its Foam Policy in August 2016, which provides guidelines recommending the use of TOP Assay for the assessment of soils, waters, and foams. The guidance also recommends foams where TOP Assay C7 to C14 results are greater than 50 mg/kg are withdrawn and prohibits the release of wastewater with TOP C4 to C14 results greater than 1.0 µg/L (Queensland Department of Environment and Heritage Protection 2016). TOP Assay is increasingly being used as an assessment tool to manage environmental impacts from PFASs in Queensland and across Australia, and is now widely available at commercial laboratories.

3.7.2 REGULATION OF PERFLUOROALKYL ETHERS

The RIVM in the Netherlands has derived a precautionary value for the allowed total daily intake for GenX of 21 ng/kg bw/day, this is in the same order of magnitude as the current TDI's for PFOS and PFOA. This results in a drinking water screening level of 150 ng/L (Smit 2016); the North Carolina Department of Health and Human Services has set a drinking water target level of 140 ng/L, based on this evaluation of the RIVM.

3.8 FATE AND TRANSPORT

Understanding the fate and transport of PFASs in the environment is fundamental to the investigation, risk assessment, and potential remediation of contaminated sites. The following sections provide a summary of migration considerations for PFASs and potential release scenarios that could result in PFASs' impacts to sediment, surface water, soil, groundwater, and through atmospheric deposition.

There is a growing awareness of the potential importance of polyfluorinated PFAA precursors as a source of PFAAs. Several studies have assessed the chemical composition of multiple types of Class B firefighting foams (Place and Field 2012; D'Agostino and Mabury 2014; Barzen-Hanson, Roberts et al., 2017; Backe et al. 2013). PFCAs may constitute a small fraction of the total PFASs identified in many fluorotelomer based Class B firefighting foams (though a portion of their precursors have been identified). In addition, these studies have identified sulfonamide-based PFAA precursors to PFSAs in firefighting foams manufactured using ECF (Place and Field 2012; D'Agostino and Mabury 2014; Barzen-Hanson, Roberts et al., 2017). As mentioned in section 3.3, there are reports that many PFAA precursors are transformed to PFAAs upon release to the environment (McGuire et al. 2014; Weber et al. 2017) and several transformation pathways have been elucidated (see Figures 3.11 and 3.12) (Weiner et al. 2013; Liu and Mejia Avendano 2013; D'Agostino and Mabury 2017a; Harding-Marjanovic et al. 2015). Some of the these PFAA precursors have recently been detected in groundwater at AFFF-impacted sites (Houtz et al. 2013; Backe et al. 2013), but their spatial distribution, particularly in relation to the PFAAs and co-contaminants has only recently begun to be elucidated as advanced analytical tools are increasingly used commercially (Ross, Houtz et al. 2018; Ross, Horneman et al. 2018).

The fate and transport properties of two groups of PFAAs, the PFCAs and PFSAs, have been the most extensively characterized of any of the PFASs to date, with most focus on PFOS and PFOA. Multiple sorption mechanisms may control the degree of PFAS sorption to sediments and soils during transport in water. An understanding of fate and transport of PFASs requires knowledge of both compound-specific factors (e.g., sorption and transformation potential) and environmental factors such as the aquifer properties and groundwater geochemistry. A summary of select physiochemical properties for commonly assesses PFASs is presented in Table 3.16.

Fate and transport processes will depend upon partitioning, transport, and transformation of PFASs in multiple environmental matrices. There are multiple partitioning processes which can affect PFASs fate and transport, including hydrophobic and lipophobic interactions (as a result of matrix interaction with the perfluoroalkyl groups), electrostatic interactions (with the polar, hydrophilic functional groups on many PFASs), and interfacial behaviors (as a result of the surfactant properties associated with the longer chain PFASs). Transport retardation is likely more significant for cationic and zwitterionic PFASs than it is for anionic PFASs because of the electrostatic interaction with the soil particle surface that is usually negatively charged (Brusseau 2018). In one study, for example, the positively charged perfluorohexane sulfonamide amine was detected in AFFF-impacted soils and in aquifer solids but not in groundwater (Houtz et al. 2013).

Once released to the environment, the potential mobility of various types of PFASs is somewhat dependent on their perfluoroalkyl chain length (also to some extent the degree of chain branching), the nature of the terminal and functional group(s) (i.e., anionic, cationic, zwitterionic or neutral—and thus potentially volatile), their pKa values and concentrations. However, many other environmental factors can influence the fate and transport of PFASs (Campos Pereira et al. 2018; Li et al. 2018; Guelfo et al. 2018). These may include multiple environmental factors such as the soil or aquifer fraction of organic carbon (f_{oc}), electrostatic potential and particle-size distribution of the matrix; the aqueous pH, presence of dissolved natural organic matter (NOM), nature of co-contaminants (which influence the biogeochemistry of water bodies), ionic strength of surface water or groundwater, the presence and nature of NAPL, plus the amount of rainfall and extent to which the groundwater table fluctuates seasonally will influence the extent of leaching of soil-borne contaminants to water bodies.

PFAA precursors can have differing physical and chemical properties compared to the perfluorinated daughter products to which they will transform. They may be more or less mobile depending on the size, hydrophobicity, and charge of the functional group (Xiao, Ling et al., 2017) or contain more complex anionic, cationic, or zwitterionic functional groups, leading to differences in their transport behavior.

As the PFAA precursors are susceptible to more rapid biotransformation to PFAAs under aerobic conditions, understanding the fate and transport processes of PFASs would potentially only be complete when an assessment of the total mass of PFASs in environmental matrices is complete. The biogeochemical status of the various environmental matrices vadose zone (i.e., dominantly aerobic or anaerobic) will influence the rate at which PFAAs can transform from PFAA precursors, so the site biogeochemistry may be an important factor to consider when evaluating the fate and transport of PFASs.

3.8.1 PFAS Distribution in Environmental Matrices

A study of the occurrence of select PFASs at non-FTA sites on active U.S. Air Force installations with historic AFFF use of varying magnitude reported measured concentrations from several hundred samples in multiple media (i.e., surface soil, subsurface soil, sediment, surface water, and groundwater), which demonstrate the relative distribution of the PFASs in differing matrices and partitioning behavior of long- and short-chain PFAAs (Anderson et al. 2016). Measurement of polyfluorinated PFAA precursors, such as fluorotelomers (i.e., 6:2 FTS) or known sulfonamide-based PFAA precursors, such as perfluorohexane sulfonamide amine (PFHxSaAm), were not attempted, although their transformation to PFAAs was mentioned as a potential factor in the distribution of PFAAs. A selection of the results are presented in Table 3.14. This indicates that soils and sediments represent a potential ongoing source of long-chain PFAAs to surface and groundwaters, such that surface and subsurface soils may still hold a significant mass of the PFASs, especially for longer chain PFASs, when considering the mass present in the soil source area versus the mass present in the groundwater plumes. The contribution that polyfluorinated PFAA precursors make to the perceived distribution of PFAAs has been investigated in only a few studies using advanced analytical tools that measure total PFASs, such as TOP Assay (McGuire et al. 2014; Weber et al. 2017; Dauchy et al. 2017b; Ross, Houtz et al., 2018; Ross, Horneman et al., 2018).

3.8.1.1 PFASs in Soils

Soils and sediments can represent a secondary source of PFASs to associated surface waters or groundwater through leaching percolation or the changes in elevation of surfaces waters or groundwater. The distribution of PFASs in soils can be complex and based upon the physical and chemical properties of differing PFASs in addition to site-specific factors. Lithological variations in particle size distribution can influence phase interface partitioning, the fraction of organic carbon influences hydrophobic interaction with the perfluoroalkyl chain and particle surface charge can influence binding to PFASs by ionic mechanisms.

The partitioning of longer chain and cationic, zwitterionic PFASs to solid phase matrices, means that source zone can retain significant concentrations of PFASs, such that source zones soils can potentially represent a long-term ongoing source of PFASs to groundwater (Weber et al. 2017; McGuire et al. 2014; Anderson et al. 2016). The continued leaching of PFASs from a fire training area impacted source area was reported to be over 18 years, with 20 years reported for infiltration beds where secondarily treated domestic wastewater was deposited from a WWTP (Weber et al. 2017).

On sites where the primary source is a fire training area, there may be secondary sources of PFASs, such as from emergency response activities, use of Class B foams in hangars and buildings, and for testing and maintenance of equipment (Anderson et al. 2016).

It should be noted that PFASs in the atmosphere can contribute to background soil levels, but the background levels from atmospheric deposition are typically significantly lower than those expected to be associated with use of PFASs that are released directly to ground. A description of detected levels of PFASs in multiple environmental matrices is provided in section 3.8.2 and in Tables 3.17a, 3.17b, and 3.17c.

3.8.1.2 Leaching

The unsaturated zone can be an ongoing source of PFASs to associated water bodies as PFASs located in unsaturated zone deposits can be subject to leaching downward

TABLE 3.14
Summary statistics for select PFASs measured by matrix.

PFAS	Parameter	Surface Soil	Subsurface Soil	Matrix Sediment	Surface Water	Groundwater
PFBA	DFc	38.46%	29.81%	24.24%	84.00%	85.51%
	Median	1	0.96	1.7	0.076	0.18
	Maximum	31	14	140	110	64
PFBS	DF	35.16%	34.62%	39.39%	80.00%	78.26%
	Median	0.775	1.3	0.71	0.106	0.2
	Maximum	52	79	340	317	110
PFPA	DF	53.85%	45.19%	45.45%	92.00%	87.68%
	Median	1.2	0.96	1.7	0.23	0.53
	Maximum	30	50	210	133	66
PFHxA	DF	70.33%	65.38%	63.64%	96.00%	94.20%
	Median	1.75	1.04	1.7	0.32	0.82
	Maximum	51	140	710	292	120
PFHxS	DF	76.92%	59.62%	72.73%	88.00%	94.93%
	Median	5.7	4.4	9.1	0.71	0.87
	Maximum	1300	520	2700	815	290
PFHpA	DF	59.34%	45.19%	48.48%	84.00%	85.51%
	Median	0.705	0.66	1.07	0.099	0.235
	Maximum	1.4	17	130	57	75
PFOA	DF	79.12%	48.08%	66.67%	88.00%	89.86%
	Median	1.45	1.55	2.45	0.382	0.405
	Maximum	58	140	950	210	250
PFOSA	DF	64.84%	29.81%	75.76%	52.00%	48.55%
	Median	1.2	0.47	1.3	0.014	0.032
	Maximum	620	160	380	15	12
PFOS	DF	98.90%	78.85%	93.94%	96.00%	84.06%
	Median	52.5	11.5	31	2.17	4.22
	Maximum	9700	1700	190000	8970	4300
PFNA	DF	71.43%	14.42%	12.12%	36.00%	46.38%
	Median	1.3	1.5	1.1	0.096	0.105
	Maximum	23	6.49	59	10	3
PFDA	DF	67.03%	12.50%	48.48%	52.00%	34.78%
	Median	0.98	1.4	1.9	0.067	0.023
	Maximum	15	9.4	59	3.2	1.8
PFDS	DF	48.35%	11.54%	33.33%	8.00%	20.29%
	Median	3.7	3.55	2	17.8	0.125
	Maximum	265	56	2200	35.6	2
PFUnA	DF	45.05	9.62	24.24	20	8.7
	Median	0.798	1.15	1.6	0.021	0.025
	Maximum	10	2	14	0.21	0.086
PFDoA	DF	21.98%	6.73%	45.45%	20.00%	4.35%
	Median	1.95	2.4	2.8	0.058	0.022
	Maximum	18	5.1	84	0.071	0.062
PFTriA	DF	15.38%	13.46%	24.24%	0.00%	1.45%
	Median	0.665	1.9	1.65	na	0.019
	Maximum	6.4	4.7	29	na	0.019
PFTeA	DF	10.99%	6.73%	15.15%	0.00%	1.45%
	Median	1.1	3.4	1.66	na	0.021
	Maximum	4.7	5.4	4.16	na	0.027

Notes:
Median values are reported using only detected concentrations.
All soil and sediment values reported in µg/kg and all water samples reported in µg/L.
DF = detection frequency

Source: Reprinted with permission from Anderson, H. R., G. C. Long, R. C. Porter, and J. K. Anderson. Occurence of select perfluoroalkyl substances at U.S. Air Force aqueous film-forming foam release sites other than fire-training areas: Field-validation of critical fate and transport properties. Chemosphere, 150 (2016): 678–685. Copyright 2016 Elsevier.

as a result of a water flux such as during rainfall (precipitation) or if the ground is being irrigated (Sepulvado et al. 2011; Ahrens and Bundschuh 2014). PFASs associated with vadose zone deposits can also leach from soils to groundwater as a result of seasonal changes in groundwater elevation, potentially in combination with increased seasonal precipitation events.

Leaching can drive soil-borne PFASs to underlying groundwater or toward surface waters as releases of PFASs often emanate from atmospheric deposition or release to the ground surface. The potential for PFASs to leach will be determined by a combination of environmental factors (soil particle size distribution, mineralogy, pH, etc.) and the migration potential for individual PFASs based on their structural properties (e.g., long- or short-chains, charge of functional group, etc.) (Gellrich et al. 2012). The air–water interface was recently described to be a significant source of retention of PFOA, indicating that it significantly affects the migration potential of PFASs (Lyu et al., 2018).

Leaching of PFASs from landfill is another potentially important process, especially for unlined landfills and when considering management of leachate (Benskin et al. 2012c; Lang et al. 2017; Lang et al. 2016; Yan 2015). The potential for PFASs leachate to impact plant uptake has been studied, with low uptake reported for PFOA and PFOS (Stahl et al. 2013), but uptake and concentration of the shorter-chain PFAAs into some plants has also been described (Blaine et al. 2013; Blaine et al. 2014a; Blaine et al. 2014b).

There are reports of both transport of PFASs by leaching and long-term sorption and retention of PFASs in vadose zone soils following long-term leaching (Hellsing 2016; Braunig et al. 2017; Lindstrom, Strynar, Delinsky et al., 2011; Filipovic, Laudon et al., 2015; Sepulvado et al. 2011; Ross, Horneman et al. 2018; Ross, McDonough et al. 2018). As cationic and zwitterionic PFAA precursors undergo potentially stronger sorption to soils and sediments via ion exchange processes, they are expected to be significantly less mobile than their anionic transformation products (Place and Field 2012; ESTCP 2017) and can be present in multiple types of fire-fighting foams (Backe et al., 2013). Cationic and zwitterionic PFAA precursors have been identified in ECF derived foams, such as PFHxSaAm, PFOSaAm, PFHxSaAmA and PFOSaAmA (Backe et al. 2013), as described in Table 3.3a. The impact of PFAA precursors, as potential soil-borne or sediment-borne source of PFAAs and their contribution to the leaching potential of PFAAs is still to be evaluated in detail for many PFAA precursors, but studies indicate that perfluorooctane sulfonamides are long-term sources of PFOS from sediments to surface waters (Benskin, Ikonomou, Gobas et al., 2012; Benskin et al. 2013). However, anionic precursors have been described to be potentially more mobile than the PFAAs they transform to create as they have been described to be less well retained via treatment using GAC (Xiao, Ulrich et al., 2017), so they have the potential to migrate from source zones in plumes, but may be transformed to PFAAs as they transit away from sources areas, especially in more aerobic aquifers.

A variety of anionic, cationic, and zwitterionic precursors have also been described in fire-fighting foams manufactured by FT, as shown in Table 3.3b, so are expected to exhibit variable fate and transport properties. The sorption of the FTSs (anionic), fluorotelomer sulfonamide amine (FTSaBs) (zwitterionic), and 6:2

fluorotelomer sulfonamidoamines (6:2 FTSaAm) (cationic) to six soils with varying organic carbon, effective cation-exchange capacity, and anion-exchange capacity was evaluated (Barzen-Hanson, Davis et al., 2017). The 6:2 FTSaAm was depleted from the aqueous phase in all but one soil, which is attributed to electrostatic and hydrophobic interactions. Sorption of the FTSs was driven by hydrophobic interactions, while the FTSaBs behave more like cations that strongly associate with the solid phase relative to groundwater. Zwitterionic PFASs may be mobile within the environment depending on site conditions and types of firefighting foams applied to the site.

The partitioning coefficient (K_d) values for a range of PFAAs and precursors on GAC were assessed by (Xiao, Ulrich et al., 2017) which found that all PFAA precursors exhibited lower Kd values (less sorption potential) than PFOS and PFHpS, indicating the potential for increased mobility within the aquifer and poorer removal via GAC. Using advanced analytical techniques, such as the TOP Assay (described in section 3.5.2), PFAA precursors have been identified migrating within groundwater from FTA source zones (Weber et al. 2017).

Increased precipitation and a rising groundwater table was observed to increase the mass flux of PFAAs away from source zones (Ross, Houtz et al. 2018; Ross, Horneman et al. 2018). The TOP Assay was used to investigate areas associated with fire training, which appeared to be dominantly impacted with PFOS using conventional analytical methods, however, and an increase of 127 percent of total PFAAs was detected following use of the TOP Assay (as compared to conventional analysis), which included PFNA, PFOA, PFHpA, PFHxA, PFPeA, and PFBA, as shown in Figure 3.22. As PFOS was the dominant PFAA detected, with much lower concentrations of 6:2 FTS, 8:2 FTS, and PFCAs measured prior to TOP Assay, the PFCAs identified following TOP Assay are considered representative of PFSAs, as TOP Assay converts sulfonamide precursors to PFCAs whereas biotransformation reactions create PFSAs, as detailed in section 3.5.2.2.1. The additional PFCAs detected using TOP assay likely emanate from the cationic, perfluoroalkyl sulfonamide amine and zwitterionic, perfluoroalkyl sulfonamide amino carboxylate precursors identified in multiple samples of ECF derived foams (Backe et al. 2013), as shown in Table 3.3a. The ratios of C4 to C8 PFCAs discovered in the soils at the site following TOP Assay, are very similar to the amounts of differing chain length perfluoroalkyl sulfonamide amines and perfluoroalkyl sulfonamide amino carboxylates, reported in Table 3.3a. The presence of soil-borne cationic or zwitterionic precursors to PFBS and PFPeS indicate an ongoing source of these shorter-chain PFAAs, which would otherwise be expected to be more mobile and not retained in soils. The major PFAAs detected after TOP Assay were PFHxA and PFOA, which indicate that PFHxSaAm and PFHxSaAmA are likely present as an ongoing source of PFHxS, and PFOSaAm and PFOSaAmA remain as an ongoing source of PFOS.

The seasonal fluctuations of groundwater at this site created an elevation change of 3.6 m, which caused an increased mass flux of PFAAs, in downgradient groundwater monitoring wells, as groundwater rises in the spring. A fourfold increase in the concentration of PFHxS in groundwater downgradient of the source area was observed in May versus January, which is likely due to a combination of increased

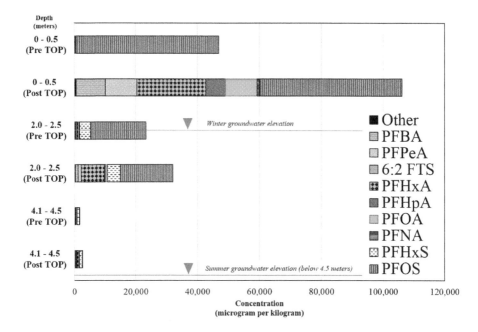

FIGURE 3.22 TOP Assay distribution of PFASs in soils impacted by PFASs by AFFF.

leaching from more frequent precipitation events, but the increase correlates more closely with the rise in the groundwater table and subsequent mobilization of PFASs from unsaturated soils. Similar studies have demonstrated PFAA-precursor transport with PFAAs within an aquifer using TOP Assay and identification of sources areas, based on the presence of precursors (McGuire et al. 2014; Weber et al. 2017). Use of TOP Assay for site investigations of sources areas can potentially enable better understanding of the total source mass and provide confidence that sources of both ECF and fluorotelomer foams have been identified. This enables the development of remedial strategies that are more cost-effective as targeting the most impacted locations on a site.

3.8.1.3 Transport and Retardation in Groundwater

The persistence of PFAAs, coupled with their relatively high aqueous solubility, dominant anionic charge, and low/moderate sorption to soils, makes many PFAAs increasingly mobile, as the length of the perfluoroalkyl chain diminishes, resulting in the potential for long groundwater plumes (sometimes multiple kilometers) which have the potential to impact groundwater abstraction wells and other receptors over a much wider area than conventional contaminants (such as hydrocarbons or chlorinated solvents). The total lack of biodegradation observed by all PFASs, with only biotransformation possible as a mechanism to convert polyfluoroalkyl substances to PFAAs means that physical transport processes are much more important when considering PFASs transport and exposure potential. As short-chain PFAAs are not effectively removed by many water treatment technologies, they have the potential to recirculate around the water cycle indefinitely.

Processes including advection, dispersion, and diffusion can cause migration of PFASs in the environment. Advection will drive PFASs' mobility in many groundwater plumes, but an understanding of sorption is required to estimate the retardation of PFASs in differing aquifer systems. In groundwater plumes, transverse dispersivity is generally limited, meaning plumes are usually relatively narrow as they migrate from a source zone with groundwater flow (Payne 2008). Diffusion of PFASs into lower permeability horizons and soils (such as clays) or concrete over time, may increase the residence time for PFASs in groundwater as the lower permeability zones represent an ongoing source of PFASs. An example is the penetration of PFASs by 12 cm into a concrete pad at a fire training area where diffusion was reported to contribute to distribution within the concrete (Baduel et al. 2015). During remedial treatment of groundwater for PFASs, back diffusion from low permeability deposits at certain sites may influence the ability to achieve low cleanup targets.

The partitioning of PFASs into groundwater from soils is somewhat dependent on the perfluoroalkyl chain length. Shorter-chain length PFASs are more water soluble than long chain, with generally lower K_d and organic carbon-water coefficient (K_{oc}) values, suggesting greater mobility in aquifers. An exception to this is encountered for K_{oc} values of some short-chain PFCAs, as shown in Table 3.15 (Guelfo and Higgins 2013), indicating multiple sorptive mechanisms may be influencing short-chain mobility, with sorption not explained by only f_{oc} of soil.

A study of the role of the fraction of organic carbon and pH on K_d identified that the sorption behavior of a range of PFASs could not be accounted for by single soil or sediment property (Li et al. 2018). A combination of pH and f_{oc} together was reported to explain a significant proportion of the variations in K_d values, but the pH, f_{oc}, and clay content were described to all have a significant effect on sorption. A decrease in pH has been associated with increased sorption and retardation of PFASs (Higgins and Luthy 2006a; McKenzie et al. 2015), likely as a result in a change in the surface charges of aquifer minerals at lower pH which tend to become increasingly cationic, causing greater binding of anionic PFAAs.

PFAS sorption increased with increasing perfluorocarbon chain length, with 0.60 and 0.83 log K_{OC} units per CF_2 moiety for C_3-C_{10} PFCAs and C_4, C_6, and C_8 PFSAs, respectively. Other studies reported the K_{OC} was found to increase approximately 0.5 log unit for every addition CF_2 group (Higgins and Luthy 2006a). Kd values for a range of PFAAs and precursors on GAC were recently assessed, which found that all PFAA precursors assessed exhibited lower Kd values (less sorption potential) than PFOS and PFHpS, indicating the potential for less attraction of the shorter-chain PFAAs of the basis of hydrophobic interaction with GAC (Xiao, Ulrich et al., 2017). However, although sorption generally increases with increasing chain length and with increasing solid-phase f_{oc}, the K_{oc} of PFBA and PFPeA did not follow the chain-length dependent trend with a log K_{oc} value for PFBA being similar to that of PFOA at 1.88. (Guelfo and Higgins 2013). PFSAs have been described to sorb more strongly than PFCAs (Campos Pereira et al. 2018) and branched isomers tend to exhibit less sorption (Kärrman 2011). Sorption of PFPAs and PFPiAs has been assessed which also showed chain dependent sorption with PFPiAs showing greater sorption than PFSAs, PFCA,s or PFPAs (Lee and Mabury 2017). The effects of the perfluoroalkyl chain length on hydrophobic sorption potential and water solubility also applies to PFAA

TABLE 3.15
Select properties of PFASs

Compound	Number of Fluorinated Carbons	Aqueous Solubility (20–25°C) [g/L]	Log K_{oc} [L/kg]	K_{oc} (L/kg)	K_d [L/kg at pH 7]	pKa
PFCAs						
PFBA	3	Miscible	1.88	76	–	−0.2 to 0.7
PFPeA	4	112.6	1.37	23	–	−0.06
PFHxA	5	21.7	1.31 to 1.91	20 to 81	–	−0.13
PFHpA	6	42	1.63 to 2.19	43 to 155	0.4 to 1.1	−0.15
PFOA	7	3.4 to 9.5	1.31 to 2.63	20 to 427	0 to 3.4	−0.16 to 3.8
PFNA	8	9.5	2.36 to 3.9	229 to 7,943	2.6 to 5.9	−0.17
PFDA	9	9.5	2.76 to 2.96	575 to 912	2.0 to 3.1	−0.17
PFUnA	10	0.004*	3.30	1,995	12 to 103	−0.17
PFDoA	11	0.0007*	–	–	24 to 269	−0.17 to 0.8
PFSAs						
PFBS	4	46.2 to 56.6	1.00 to 1.79	10 to 62	–	−6.0 to −5.0
PFHxS	6	2.3	1.78 to 3.1	60 to 1,259	0.6 to 3.2	−6.0 to −5.0
PFOS	8	0.52 to 0.57	2.4 to 3.7	251 to 5,012	0.1 to 97	−6.0 to −5.0
FOSA and FTSs						
FOSA	8	–	2.5 to 2.62	316 to 417	35 to 56	–
6:2 FTS	6	1.3*	–	–	–	1.31
8:2 FTS	8	0.06*	0.01	–	–	1.32

Notes:
This table is meant to supplement Appendix C.
* Indicates parameters estimated with published equations. Calculated parameters are based on the neutral form of the substances (and not the conjugate base, which predominates for some PFAS at neutral pH).
Sources: Data are from peer-reviewed literature as summarized by Pancras et al. (2016) and ITRC (2018a, 2018b).

precursors of the same physical structure; for example, 8:2 FTS is more hydrophobic and less soluble than 6:2 FTS. Chain-length dependent sorption appears evident for simpler precursor functional groups (e.g., FTS) but is less clear for precursors with more complex functional groups (Xiao, Ulrich et al., 2017).

It was suggested that the most long-chained PFASs, similar to other hydrophobic organic compounds, are preferentially sorbed to the highly condensed domains of the humic fraction, where hydrophobic interaction dominates as the sorptive process. However, shorter-chain PFASs are bound to a large extent to humic and fulvic acid, where cation exchange effects are more significant, as PFAAs can sorb to the surface of charged mineral surfaces by ion exchange processes. This indicates that the f_{oc} may dominate sorption for longer chain PFASs whereas ion exchange processes are more important when considering retardation mechanisms for shorter-chain PFASs (Campos Pereira et al. 2018). PFAAs generally have pKa values less than three and are usually anionic under almost all natural conditions they are, therefore, repelled and poorly sorbed by negatively charged mineral surfaces (e.g., clay particles) (Lipson 2013) which contributes to increased mobility.

As mentioned in section 3.8, cationic and zwitterionic PFAA precursors can undergo stronger sorption to soils and sediments via ion exchange processes so are expected to be significantly less mobile than the anionic PFAAs they ultimately form. Cationic precursors are anticipated to be less mobile than zwitterions, and some zwitterionic PFASs may still be mobile within the environment depending on their structure and site conditions. This is illustrated by the detection of 6:2 fluorotelomer sulfonamide alkylbetaine (6:2 FTAB) (a zwitterionic precursor) at significantly higher concentrations than any other PFAS (including PFOS and PFOA) within soils and surface water associated within a FTA (Boiteux et al. 2016). Although 6:2 FTAB is a zwitterion, its terminal charge is anionic, so this may be the determining factor when considering the ion exchange processes that influence its fate and transport. The terminal charge(s) of zwitterions may determine whether they can be transported in aquifers in a similar manner to anion or cations, or can exhibit a combination of potential properties as may be expected for PFHxSaAm.

As mentioned in section 3.3, as the longer-chain PFASs can have surfactant properties, there is the potential that they can form micelles above critical micelle concentrations (CMC); however, CMCs for PFOS and PFOA are relatively high (i.e., gram per liter [g/L]) compared with typical groundwater PFASs concentrations encountered. Hemimicelles can form, however, at 0.001 to 0.01 of the CMC, on solid surfaces (Guelfo and Higgins 2013). The tendency of some PFASs to form hemimicelles on solid surfaces may affect their partitioning in release scenarios to source zone surfaces such as concrete, surface soils, subsurface unsaturated and saturated zone soils. Relatively high concentrations of PFASs were generally released to ground in fire training activities, such that surface sorption could be an important sorption mechanism in source zones. The tendency of longer-chain PFASs to aggregate at surfaces at higher concentrations could influence adsorption on to environment matrices (Yu et al., 2009; Du et al. 2014). This surface sorption could be enhanced as a result of the fluorophilic "molecular brush" effect where perfluoroalkyl chains self-assemble at interfaces and pack together closely to create a dense layer (Zaggia et al. 2016). This effect may be more pronounced at higher PFAS concentrations as a result of a tendency to also form hemimicelles. The tendency of PFOA to migrate to the air-water interface and be a potentially significant source of retention, contributing to approximately 50 to 75 percent of total retention on two natural sands, has also been reported in laboratory test systems employed (Lyu et al. 2018). The potential for a PFAS's smear zone as the groundwater tables rises and falls into the vadose zone seems apparent, especially in sources zones.

The ionic strength of a water body can also create a dramatic change in solubility of PFOS, with solubility reported in pure water (570 to 680 mg/L) versus sea water (12.4 to 25 mg/L) (OECD 2002b; Giesy et al. 2006), which has been described as a "salting out" effect. This solution chemistry effect on increasing sorption and thus retardation of PFASs has also been reported for polyvalent cations (e.g., Ca^{2+}, Fe^{3+}, etc.) (McKenzie et al. 2015; Higgins and Luthy 2006a). At higher ionic strengths, the presence of cations may assist in stabilizing and promoting the molecular brush self-assembly mechanism described earlier. The potential effects of remedial approaches on PFASs partitioning and retardation, especially the use of *in situ* chemical oxidation, such as with persulfate of Fenton's reagent, where the pH decreases and the ionic strength increases, has been highlighted (McKenzie et al. 2015; McKenzie et al. 2016).

Sorption of PFAS can also be influenced by the presence of co-contaminants such as non-aqueous phase liquid (NAPL) and nonfluorinated surfactants with the effect on PFAA sorption being complex and dependent on the type of soil and co-contaminant as well as the type and concentration of PFAAs (McKenzie et al. 2016). For example, this study found the following:

- Whereas NAPL and nonfluorinated surfactants decreased the sorption of PFOS at lower PFOS concentrations (1 μg/L), they led to increases in sorption at higher PFOS concentrations (500 μg/L);
- Co-contaminants generally increased the sorption of shorter-chain PFAAs to all soils;
- While negatively charged nonfluorinated surfactants increased sorption in negatively charged soils, an amphoteric nonfluorinated surfactant increased sorption in positively charged soils; and
- While competitive effects among PFAAs may be a factor in low f_{oc} soils, overall, competitive effects were considered to be minor.

Retardation factors for differing PFAAs are presented in Table 3.15 using published K_{oc} values (Guelfo and Higgins 2013), as compared to variable f_{oc} values, which may be considered low, medium, and high (Interstate Technology & Regulatory Council [ITRC] 2015) with a bulk density of 1.5 and a porosity of 0.25 applied. The calculated data presented in Table 3.16 indicate a lack of retardation for the

TABLE 3.16
Retardation factors of Select PFAAs

Compound	Perfluoroalkyl Carbons	Average Log K_{oc} [L/kg]	K_{oc} [L/kg]	R_f Low F_{oc}(0.0004)	R_f Medium F_{oc}(0.0026)	R_f High F_{oc}(0.0210)
PFBA	3	1.88	76	1	2	11
PFPeA	4	1.37	23	1	1	4
PFHxA	5	1.31	20	1	1	4
PFHpA	6	1.63	43	1	2	6
PFOA	7	1.89	78	1	2	11
PFNA	8	2.36	229	2	5	30
PFDA	9	2.96	912	3	15	116
PFUnA	10	3.56	3631	10	58	458
PFBS	4	1.79	62	1	2	9
PFHxS	6	2.05	112	1	3	15
PFOS	8	2.80	631	3	11	81

Notes:
SDS = sodium decyl sulfate (an anionic surfactant)
AO = *n,n*-dimethyldodecylamine *n*-oxide (an amphoteric surfactant)
Source: Values calculated using average log K_{oc} values published in Guelfo and Higgins (2013).

shorter-chain PFAAs in aquifers with a low f_{oc}, however, as mentioned previously, multiple retardation mechanisms, such as ion exchange, may influence retardation at actual sites, which may depend on the local mineralogy and ambient pH. A modeling study undertaken after concomitant release of both PFOS and methyl tert-butyl ether (MTBE) into a dual porosity chalk aquifer system after the Buncefield fuel depot fire incident in 2005 (Lipson 2013) found that PFOS transport velocity was significantly lower than the average groundwater velocity (indicating some retardation), but that while MTBE moved 17 meters per year (m/year) with a calculated R_f of 378, PFOS traveled at 29 m/year with a calculated R_f of 221 in the same aquifer, indicating the relatively higher mobility of PFOS versus MTBE in this setting (Lipson 2013).

3.8.1.4 Surface Waters and Sediments

The oceans have been described as the main sink for PFASs, with an estimate that they contain the majority of PFCAs released to the environment (Armitage 2006). PFAAs in surface water can potentially impact groundwater through groundwater recharge or they can be transported to the oceans and transported by ocean currents (Liu et al. 2016; ATSDR 2008; Benskin et al. 2012).

In a study of the suspended particulate matter and sediment from Tokyo Bay, Japan, the more water-soluble shorter-chain PFCAs (C<7) have been reported to be exclusively detected in the dissolved phase, longer-chain PFCAs (C≥7), PFSAs, N-EtFOSAA, and perfluorooctane sulfonamide (PFOSA) appeared to bind more strongly to particles (Ahrens, Taniyasu, et al., 2010).

The sorption of varying chain-length PFAAs to sediments was investigated using natural sediments of varying iron oxide and organic carbon content (Higgins and Luthy 2006a). Sorptive potential was assessed for PFCA, PFSAs, and N-EtFOSAA. Sorption was reported to be influenced by both sediment-specific and solution-specific parameters, with sediment f_{oc}, rather than sediment iron oxide content being the dominant sediment-parameter affecting sorption, indicating the importance of hydrophobic interactions. However, sorption also increased with increasing dissolved concentrations of Ca^{2+} and decreasing pH, suggesting that electrostatic interactions play a role. The chain length of the perfluoroalkyl unit was the dominant structural feature influencing sorption, with each CF_2 unit contributing 0.50 to 0.60 log units to the measured distribution coefficients. A sulfonate terminal functional group contributed an additional 0.23 log units to the measured distribution coefficient, when compared to carboxylates. In addition, the perfluorooctyl sulfonamide acetic acids demonstrated substantially stronger sorption than PFOS (Higgins and Luthy 2006a).

An investigation of the influence of solution salinity, pH, and the sediment characteristics on the sorption and desorption of PFOS, showed that the sorption of PFOS onto sediment increased by a factor of three as the $CaCl_2$ concentration increased from 0.55 to 55 g/L at pH 7.0, and a factor of nearly six as pH increased to 8.0 (You et al. 2010). Desorption hysteresis was reported to occur at all salinities, with results suggesting that PFOS can be largely removed from the water with increasing salinity and gets trapped onto sediments irreversibly. As mentioned in sections 3.3 and 3.8.1.3, the solubility of PFOS decreases significantly between fresh and saline waters as a "salting-out "effect can occur, with enhancement of sorption of PFAAs to particulate matter at high salinity also reported (Jeon et al. 2011).

The ionic strength of a water body, such as in a halocline as brackish water develops in an estuary, before becoming saline, has also been reported to affect the partitioning of long-chain PFAAs (Hong et al. 2013). Field-based K_d for long-chain PFAAs (C≥8) were significantly correlated with salinity, with K_d values increased exponentially as a function of salinity. A "salting-out" effect was reported, where PFAAs were largely scavenged by adsorption onto suspended solids and/or sediments in estuarine environments. The effect on K_d, as in reports previously mentioned, was described as being directly proportional to the number of carbon atoms in the PFAAs. Concentrations of PFASs in sediments were dependent on the organic carbon content in sediment cores, with a large proportion of longer-chain PFASs observed in the upper layers of the sediment cores from industrialized coastal regions (Shen et al. 2018).

The presence of PFAA precursors potentially more strongly bound to sediments, then some PFAAs, is also a potential source of PFASs, with reports of the hydrophobic sulfonamido ethanol-based phosphate (SAmPAP) diesters impacting marine sediments as an ongoing source of PFOS (Benskin et al. 2013; Benskin, Ikonomou, Gobas et al., 2012).

3.8.1.5 Vapor Migration

Vapor migration plays only a minor role in assessing the mobility of most PFASs in the environment due to the low to very low vapor pressure of the majority of PFASs. FTOHs are reported as having varying vapor pressures, but compared with other PFASs, they have much higher vapor pressures and are therefore classified as volatile. FTOHs are also intermediates in the biotransformation of multiple fluorotelomer PFAA precursors (Liu and Mejia Avendano 2013), but they may be transient intermediates, before further transformation to less volatile PFAAs.

3.8.1.6 Atmospheric Deposition

Many PFASs are of relatively low volatility, as a result of charged terminal functional groups such as the PFAAs. However airborne transport of some PFASs, such as FTOHs, can occur through industrial releases (e.g., stack emissions) and may be a relevant migration pathway. Once airborne, some polyfluoroalkly substances, such as fluorotelomer alcohols, can be transformed to less volatile PFAAs via photooxidation, and they may be transported back to ground, so they can eventually accumulate to detectable levels in soil and surface water through atmospheric deposition (Young 2010; Ahrens and Bundschuh 2014; Rankin et al. 2016). The conversion of 8:2 FTOH to PFOA via atmospheric oxidation has been reported (Wallington 2006). Atmospheric deposition via both dry or wet deposition are potentially relevant for PFASs (Barton 2007; Barton 2010; Drey 2010; Taniyasu 2013). The PFAAs tend to suffer atmospheric transport via airborne particulate matter (Ge et al. 2017; Dreyer et al. 2015). Short-range atmospheric transport and deposition may result in impacts to soil and groundwater, with air emissions from the fluoropolymer manufacturing site reported to transport PFASs to a well field, as PFASs are deposited onto the soil, and then migrate downward with precipitation into the underlying aquifer (Davis et al. 2007). Atmospheric deposition was described as the sole source of PFASs to mountainous lakes outside of Grenoble, France, where fish liver samples were analyzed (Ahrens et al. 2010). The results of this study indicate that atmospheric deposition of PFOS or its precursors in the high mountain lakes is homogeneous, so potentially represents background, whereas the deposition of PFCAs depends strongly on more local contamination sources. Fate and transport modeling of PFOA concentrations in air, local to a fluoropolymer manufacturing site, surface water, groundwater, and municipal water systems has been performed to describe PFOA air dispersion, transit through the vadose zone, surface water transport, and groundwater flow and transport (Shin et al. 2011).

3.8.2 Detections and Background Levels in the Environment

There is an almost ubiquitous presence of PFASs in surface water, groundwater, air, soil, sediment, snow, biota including wildlife, and humans at low levels (Shoeib et al. 2005; Taniyasu et al. 2003; Yamashita et al. 2004; Yamashita et al. 2008; Higgins et al. 2005; Giesy and Kannan 2001; Loos 2013; Loos et al. 2013; Loos et al. 2009; Loos et al. 2010; Loos et al. 2017; Yeung and Mabury 2016; Kärrman et al. 2006; Rankin et al. 2016; D'Agostino and Mabury 2017b; Eschauzier, Voogt et al., 2012; Houde et al. 2011; Houtz et al. 2016; Llorca et al, 2012; Sammut et al. 2017).

PFASs are distributed through atmospheric deposition, diffusion/dispersion/advection, infiltration, and transformation processes (ITRC 2018b). The global distribution of POPs has been described as occurring via various processes involving evaporation and deposition leading to a "grasshopper" effect. Where POPs can migrate to higher latitudes in a series of relatively jumps whereby they migrate, rest, and migrate again in tune with seasonal temperature changes at mid-latitudes (Wania and Mackay 1996). Long-range atmospheric and oceanic transport has also been reported. This has resulted in the widespread occurrence of POPs in environmental matrices. Two major long-range transport models have been proposed for transport of PFASs (Lau et al. 2007; Butt et al. 2010) with the relative contribution of each pathway to overall transport uncertain. One pathway involves long-range atmospheric transport of volatile polyfluoroalkyl substances as PFAA precursors such as FTOHs followed by atmospheric oxidation to PFAAs which are then deposited onto the land or water, via precipitation events. The second pathway involves long-range aqueous transport of emitted PFAAs to remote locations by currents on the ocean surface. These two pathways, among others, can lead to the widespread background concentrations in various matrices that are summarized in Tables 3.17a, 3.17b, and 3.17c, which is not an exhaustive literature review.

Detectable levels of PFASs are observed in air, potentially as a result of being in the gaseous phase, associated with particles or in aerosols. Background levels of PFASs have been detected in air from a semiurban location in Toronto, which showed that the most abundant PFAS class in the gas phase was the FTOHs, representing about 80 to 84 percent of the ΣPFASs (Ahrens et al. 2012). This indicates that the volatile FTOHs may represent the dominant PFAS subject to atmospheric transport, which as described, suffer atmospheric oxidation to PFAAs, which can return to impact land and water. There is therefore a mechanism whereby volatile PFASs evaporate to the atmosphere and are converted to the more polar PFAAs which may then potentially be deposited back into rainfall. Polyfluoroalkyl substances detected in the atmosphere over the Atlantic and Southern Oceans have been suggested as

TABLE 3.17A
Summary of detections and background concentrations of PFAAs and total PFASs measured in the environment

Location	Source	Value[b]	ΣPFASs	PFOA	PFOS	PFBS	PFBA	PFHxA	PFHxS	PFHpA	PFNA	PFDA	PFUnDA
						Air (pg/m³)							
Albany, New York, USA	Kim and Kannan 2007	Mean (g)	7.29	3.16	1.7	–	–	–	0.31	0.26	0.21	0.63	–
		Median (g)	6.26	2.86	1.42	–	–	–	0.34	0.23	0.2	0.56	(<0.12)
		Range (g)	5.10 - 11.6	1.89 - 6.53	0.94 - 3.0	–	–	–	0.13 - 0.44	0.13 - 0.42	0.16 - 0.31	0.24 - 1.56	ND - 0.16
		Mean (p)	4.03	2.03	0.64	–	–	–	–	0.37	0.13	0.27	–
		Median (p)	3.96	1.57	0.66	–	–	–	–	0.29	<0.12	0.22	–
		Range (p)	2.05 - 6.04	0.76 - 4.19	0.35 - 1.16	–	–	–	<0.12	<0.12 - 0.81	<0.12 - 0.40	0.13 - 0.49	ND
Toronto, Canada	Ahrens et al. 2012	Mean (g)	127	0.48	1.06	0.11	14	0.7	0.11	0.99	0.32	0.36	0.2
		Median (g)	123	0.52	0.89	0.11	14.3	0.64	0.12	0.87	0.28	0.3	0.19
		Range (g)	67.1 - 209	0.07 - 0.98	0.33 - 2.62	0.05 - 0.2	6.84 - 22.2	ND - 1.73	ND - 0.3	0.08 - 2.11	ND - 0.7	0.05 - 0.61	ND - 0.41
		Mean (p)	6.23	0.01	0.5	0.01	ND	ND	ND	ND	0.03	0.04	0.08
		Median (p)	6.28	ND	0.37	0.0003	–	–	–	–	0.02	0.03	0.11
		Range (p)	2.86 - 11.66	ND - 0.06	ND - 2.29	ND - 0.06	–	–	–	–	ND - 0.12	ND - 0.12	ND - 0.2
Shenzhen, China	Liu et al. 2015	Mean	15 ± 8.8	5.4 ± 3.8	3.1 ± 1.2	–	1.9 ± 1.8	1.5 ± 1.5	0.31 ± 0.39	0.042 ± 0.1	0.49 ± 0.33	0.48 ± 0.38	0.018 ± 0.064
		Median	12	3.6	3.3	–	2	1.8	ND	ND	0.5	0.6	ND
		Range	3.4 - 34	1.5 - 15	ND - 4.3	–	ND - 5.0	ND - 3.6	ND - 1.2	ND - 0.30	ND - 1.0	ND - 1.2	ND - 0.22
Germany	Dreyer et al. 2015	Mean	–	0.7	0.65	–	–	–	–	–	–	–	–
		Range	–	0.1 - 4.8	0.2 - 3.5	–	–	–	–	–	–	–	–
Atlantic to Southern Ocean	Wang, Xi et al. 2015	Mean	23.8	–	–	–	–	–	–	–	–	–	–
		Median	22.2	–	–	–	–	–	–	–	–	–	–
		Range	2.78 - 68.8	–	–	–	–	–	–	–	–	–	–
North Greenland	Bossi et al. 2016	Mean	11.22	–	–	–	–	–	–	–	–	–	–
		Range	<0.11 - 67.34	–	–	–	–	–	–	–	–	–	–
Northern Part of South China Sea	Lai et al. 2016	Mean	54.5	–	–	–	–	–	–	–	–	–	–
		Median	55.6	–	–	–	–	–	–	–	–	–	–
		Range	18 - 109.9	–	–	–	–	–	–	–	–	–	–

Location	Source	Value[b]	ΣPFASs	PFOA	PFOS	PFBS	PFBA	PFHxA	PFHxS	PFHpA	PFNA	PFDA	PFUnDA
						Soil/Sediment (µg/kg)							
Canadian Arctic region	Stock et al. 2007	Mean	20.8	1.6	15.9	0.032	–	–	0.963	1.63	0.420	0.113	0.021
		Median	5.30	ND	0.06	ND	–	–	0.850	ND	ND	ND	ND
		Range	ND - 106	0.63 - 7.5	0.022 - 85	ND - 0.11	–	–	0.7 - 3.5	0.95 - 6.8	0.23 - 3.2	0.071 - 0.6	ND - 0.072
Global	Strynar et al. 2012	Mean	32.4	8.81	2.67	ND	–	3.39	–	16.2	0.61	0.63	0.24
		Median	12.85	2.42	1.88	ND	–	1.71	–	3.52	ND	0.28	ND
		Range	7.81 - 129	0.76 - 31.7	ND - 10.1	ND	–	ND - 12.4	–	ND - 79.1	ND - 0.61	ND - 2.03	ND - 1.31
North America	Rankin et al. 2016	Mean	2.11	0.54	0.39	–	–	0.32	0.01	0.24	0.25	0.17	0.18
		Median	1.56	0.36	0.23	–	–	0.23	0.01	0.17	0.10	0.05	0.06
		Range	0.18 - 5.97	0.02 - 1.84	0.02 - 1.96	–	–	0.03 - 1.99	ND - 0.04	0.02 - 1.30	0.01 - 1.06	0.005 - 0.10	ND - 1.51
Europe		Mean	1.77	0.49	0.58	–	–	0.08	0.02	0.12	0.08	0.04	0.03
		Median	0.35	0.13	0.06	–	–	0.05	0.004	0.06	0.06	0.02	0.02
		Range	0.05 - 5.54	0.02 - 2.67	ND - 3.13	–	–	0.02 - 0.25	ND - 0.10	0.01 - 0.38	ND - 0.22	ND - 0.58	ND - 0.08
Asia		Mean	4.83	1.26	0.18	–	–	1.51	0.01	1.18	0.35	0.12	0.26
		Median	3.02	1.10	0.15	–	–	0.27	0.01	0.33	0.33	0.11	0.15
		Range	0.21 - 14.32	0.04 - 3.44	0.07 - 0.41	–	–	0.03 - 7.54	0.003 - 0.02	0.03 - 3.75	0.01 - 0.72	ND - 0.22	0.005 - 1.01
Africa		Mean	0.61	0.11	0.07	–	–	0.26	0.01	0.13	0.02	0.02	0.01
		Median	0.52	0.13	0.07	–	–	0.09	ND	0.08	0.01	0.01	0.01
		Range	0.16 - 1.58	0.01 - 0.24	0.01 - 0.14	–	–	0.02 - 0.87	ND - 0.01	0.02 - 0.31	0.003 - 0.04	0.003 - 0.04	0.001 - 0.03
Australia, South America, and Antarctica		Mean	0.50	0.10	0.08	–	–	0.07	0.02	0.24	0.02	0.06	0.02
		Median	0.28	0.05	0.05	–	–	0.03	ND	0.05	0.01	0.01	0.003
		Range	0.08 - 1.32	0.02 - 0.49	0.01 - 0.26	–	–	0.004 - 0.22	ND - 0.04	ND - 1.16	ND - 0.05	ND - 0.24	ND - 0.04
Beijing, China	Wang, Vestergren et al. 2016	Mean	6.12	1.02	1.72	0.88	0.62	ND	0.14	ND	0.50	0.26	0.32
		Median	3.46	0.93	0.51	0.51	0.57	ND	ND	ND	ND	ND	ND
		Range	1.6 - 25.9	0.57 - 2.03	ND - 15.1	0.06 - 2.17	0.29 - 1.57	ND	ND - 0.95	ND	ND - 8.98	ND - 4.27	ND - 5.22

(Continued)

TABLE 3.17A (Continued)

Location	Source	Value[b]	ΣPFASs	PFOA	PFOS	PFBS	PFBA	PFHxA	PFHxS	PFHpA	PFNA	PFDA	PFUnDA
						Groundwater (ng/L)							
Minor Aquifer, United Kingdom	Environment Agency 2008	Mean	1180	—	—	—	—	—	—	—	—	—	—
		Median	390	—	—	—	—	—	—	—	—	—	—
		Range	120 - 6,560	—	—	—	—	—	—	—	—	—	—
Chalk Aquifer, United Kingdom	Environment Agency 2008	Mean	1350	—	—	—	—	—	—	—	—	—	—
		Median	220	—	—	—	—	—	—	—	—	—	—
		Range	100 - 8,100	—	—	—	—	—	—	—	—	—	—
Permotriassic Sandstone Aquifer, United Kingdom	Environment Agency 2008	Mean	1460	—	—	—	—	—	—	—	—	—	—
		Median	310	—	—	—	—	—	—	—	—	—	—
		Range	100 - 7,470	—	—	—	—	—	—	—	—	—	—
Europe	Loos et al. 2010	Mean	—	3	4	0	—	—	1	1	0	0	—
		Median	—	1	0	0	—	—	0	0	0	0	—
		Maximum	—	39	135	25	—	—	19	21	10	11	—
Spain	Llorca et al. 2012	Mean	17.4	11.7	4.5	ND	0.9	ND	0.25	ND	ND	ND	ND
		Range	1.8 - 33	1.3 - 22	ND - 9	ND	ND - 1.8	ND	ND - 0.5	ND	ND	ND	ND
New Jersey, USA	NJDEP 2014	Mean	33	8.6	2.1	0.3	0.3	3	1	1	10	ND	—
		Median	5	ND	ND	ND	ND	ND	ND	ND	ND	ND	—
		Range	ND - 145	ND - 57	ND - 12	ND - 6	ND - 6	ND - 36	ND - 10	ND - 22	ND - 96	ND	—
						Surface Water (ng/L)							
Tennessee River	Hansen et al. 2002	Mean	244	175	69.2	—	—	—	—	—	—	—	—
		Median	52.3	<25	52.3	—	—	—	—	—	—	—	—
		Range	16.8 - 598	<25 to 598	16.8 - 144	—	—	—	—	—	—	—	—
Industrial areas in Japan	Saito et al. 2004	Mean	20.49	16.43	4.06	—	—	—	—	—	—	—	—
		Median	3.53	2.46	1.15	—	—	—	—	—	—	—	—
		Range	1.00 - 494	0.10 - 456	0.24 - 37.3	—	—	—	—	—	—	—	—

Location	Source	Value[b]	ΣPFASs	PFOA	PFOS	PFBS	PFBA	PFHxA	PFHxS	PFHpA	PFNA	PFDA	PFUnDA
Canadian Arctic region	Stock et al. 2007	Mean	80.9	8.2	32.5	–	–	–	8.6	11.8	2.0	9.2	–
		Median	99.1	9.5	43	–	–	–	10	9.3	1.8	7.5	–
		Range	9.9 - 158.2	0.4 - 16	0.9 - 90	–	–	–	ND - 24	ND - 49	ND - 6.1	ND - 29	–
European Rivers	Loos et al. 2009	Mean	–	12	39	–	–	4	–	1	2	1	0
		Median	–	3	6	–	–	0	–	1	1	0	0
		Maximum	–	174	1371	–	–	109	–	27	57	7	3
Elbe River, Germany	Ahrens, Plassmann, et al. 2009	Mean	39.9	8.49	5.98	1.30	2.42	4.20	0.56	2.39	1.60	0.80	0.19
		Median	39.5	8.85	6.65	1.20	2.56	4.26	0.56	2.55	1.82	0.68	0.16
		Range	23.2 - 54	3.04 - 12.63	1.27 - 10.58	ND - 2.47	ND - 4.25	2.69 - 5.98	0.24 - 0.96	1.22 - 3.9	0.7 - 2.1	0.19 - 2.24	(0.05) - 0.6
Elbe River, Germany	Ahrens, Felizeter, et al. 2009	Mean	19.2	6.4	11.5	1.7	–	3.1	0.75	1.3	0.62	0.42	0.05
		Median	21.1	7.2	2.5	2.3	–	2.8	0.85	1.3	0.70	0.43	0.03
		Range	7.6 - 26.4	2.8 - 9.6	<0.06 - 82.2	(1.5) - 3.4	–	1.6 - 5	(0.3) - 1.3	0.8 - 2.4	0.2 - 1.1	(0.2) - 0.7	<0.004 - 0.1
Germany	Llorca et al. 2012	Mean	21	0.6	0.3	0.2	3.1	1.6	0.4	5	ND	0.06	ND
		Median	13	ND	ND	ND	ND	ND	ND	ND	ND	ND	ND
		Range	ND - 88	ND - 6.5	ND - 4.6	ND - 4.7	ND - 23	ND - 13	ND - 5.6	ND - 24	ND - 1.5	ND - 213	ND - 0.6
Spain		Mean	176	8.0	121	0.2	21	3.5	3.4	4.6	3.2	10	0.03
		Median	42	2.5	ND	ND	9.6	ND	ND	ND	ND	ND	ND
		Range	ND - 2,878	ND - 68	ND - 2708	ND - 4.1	ND - 125	ND - 31	ND - 37	ND - 27	ND - 52	ND - 213	ND - 0.6
Sweden Faroe Islands	Eriksson et al. 2013	Mean	1.44	0.29	0.22	ND	0.59	ND	ND	0.10	0.16	0.046	0.038
		Median	1.426	0.25	0.16	ND	0.59	ND	ND	0.13	0.15	0.034	0.037
		Range	0.82 - 2.07	0.23 - 0.42	ND - 0.57	ND	0.56 - 1.20	ND	ND	ND - 0.14	0.13 - 0.22	0.029 - 0.086	0.029 - 0.049
Oujiang River, China**	Wang, Huang et al. 2013	Mean	–	–	49.7	–	–	–	–	–	–	–	–
		Median	–	–	29.6	–	–	–	–	–	–	–	–
		Range	–	–	25.4 - 162	–	–	–	–	–	–	–	–
New Jersey, USA	NJDEP 2014	Mean	49	21	6.8	1.0	ND	3.8	7.5	2.3	3.9	ND	–
		Median	20.5	10.5	ND	ND	ND	ND	ND	ND	ND	ND	–
		Range	ND - 174	ND - 100	ND - 43	ND - 6	ND	ND - 17	ND - 46	ND - 10	ND - 19	ND	–

(Continued)

TABLE 3.17A (Continued)

Location	Source	Value[b]	ΣPFASs	PFOA	PFOS	PFBS	PFBA	PFHxA	PFHxS	PFHpA	PFNA	PFDA	PFUnDA
Elbe River, Germany	Heydebreck et al. 2015	Mean	21.2	1.45	2.45	2.84	1.25	3.61	0.96	0.52	0.83	(0.51)	0.40
		Median	19.4	2.43	ND	3.05	1.27	3.79	0.86	ND	0.815	-1.135	0.24
		Range	5.8 – 38.8	(1.87) – 4.48	ND – 13.1	0.96 – 4.55	ND – 2.34	1.88 – 5.48	0.35 – 1.73	ND – 2.22	0.35 – 1.31	ND – 3.34	ND – 1.91
Rhine River, Germany		Mean	52.4	6.05	1.50	19.5	2.08	3.35	2.12	2.41	0.44	0.21	ND
		Median	47.0	5.90	1.84	18.8	2.12	3.16	2.12	2.34	0.44	0.22	ND
		Range	19.8 – 139.6	4.38 – 8.88	ND – 3.36	1.75 – 50	0.27 – 3.32	2.22 – 6.8	1.04 – 4.44	1.55 – 3.68	0.09 – 0.76	0.02 – 0.38	ND
Xiaoqing River, China		Mean	91,156*	79,492*	0.1365	ND	917*	3,159	ND	5,470	9.19	0.21	0.12
		Median	75,400*	64,415*	ND	ND	910*	2,833	ND	3,803	5.93	(0.665)	0.09
		Range	53.0 – 824,862*	38.1 – 723,713*	ND – 4.01	ND	5.95 – 5,963*	3.06 – 25,243	ND	2.91 – 53,209	0.56 – 93.5	ND – 11.5	ND – 1.07
Sweden Faroe Islands	Filipovic, Laudon et al. 2015	Mean	0.692	0.216	0.070	–	–	0.223	0.024	0.073	0.084	0.021	0.005
		Median	0.523	0.201	0.057	–	–	ND	0.023	0.089	0.087	0.021	ND
		Range	0.164 – 2.037	0.098 – 0.434	0.022 – 0.256	–	–	ND – 1.195	0.006 – 0.057	ND – 0.166	0.05 – 0.151	ND – 0.07	ND – 0.037
Cape Fear River Watershed (A), North Carolina, USA	Sun et al. 2016	Mean	212	34.0	29.0	<10	26.0	48	10.0	39.0	<10	<25	–
		Median	355	46.0	44	<10	33.0	78	14	67	<10	<25	–
		Range	18 – 1,502	<10 – 137	<25 – 346	<10 – 80	<10 – 99	<10 – 318	<10 – 193	<10 – 324	<10 – 38	<25 – 35	–
Cape Fear River Watershed (B), North Carolina, USA		Mean	47	<10	<25	<10	12.0	<10	<10	<10	<10	<25	–
		Median	62	<10	<25	<10	12.0	11	<10	11	<10	<25	–
		Range	189 – 345	<10 – 32	<25 – 43	<10 – 11	12 – 38	<10 – 42	<10 – 14	<10 – 85	<10 – <10	<25 – <25	–
Cape Fear River Watershed (C), North Carolina, USA		Mean	710	<10	<25	<10	22.0	<10	<10	<10	<10	<25	–
		Median	0	<10	<25	<10	<10	<10	<10	<10	<10	<25	–
		Range	55 – 4,696	<10 – 17	<25 – 40	<10 – <10	<10 – 104	<10 – 24	<10 – 14	<10 – 24	<10 – <10	<25 – <25	–
Beijing, China	Wang, Vestergren et al. 2016	Mean	121	11.0	4.05	16.0	18.2	5.62	8.46	2.88	9.00	0.70	0.10
		Median	73.9	9.38	2.15	14.4	16.3	3.46	1.20	1.65	0.53	ND	ND
		Range	22.9 – 483	4.03 – 24	ND – 13.15	0.64 – 51.3	7.28 – 41.5	1.75 – 13.3	ND – 45	0.8 – 8.35	ND – 47	ND – 3.93	ND – 0.47

Per- and Polyfluoroalkyl Substances

Location	Source	Value[b]	ΣPFASs	PFOA	PFOS	PFBS	PBA	PFHxA	PFHxS	PFHpA	PFNA	PFDA	PFUnDA
Qiantang River Watershed, China	Lu et al. 2017	Mean	130	63.2	5.1	2.0	7.8	34.3	9.7	4.9	2.1	1.0	—
		Median	116	53.5	4.1	ND - 20.4	4.9	19.5	ND	3.9	1.9	1.0	—
		Range	26.9 - 323	2.4 - 166	ND - 20.6	ND - 20.4	ND - 33.3	1.1 - 123	ND - 203	ND - 13.6	ND - 12.4	ND - 10.4	—
Maltese Islands	Sammut et al. 2017	Mean	10.0	2.39	1.21	—	—	1.34	1.31	1.08	0.90	1.84	—
		Median	6.84	0.88	0.72	—	—	0.43	1.39	0.35	0.74	2.67	—
		Range	0.89 - 35.0	ND - 16.2	ND - 8.6	—	—	ND - 17.0	ND - 2.47	ND - 11.2	ND - 3.44	ND - 4.4	—
United Kingdom	Wilkinson et al. 2017	Range	—	<1.13 - 189	<1.52 - 119	<1.13 - 115	—	—	—	—	<0.75 - 209	—	—
Global	Pan et al. 2018	Mean	—	8.19	4.39	5.65	4.71	4.74	28.3	1.32	0.93	0.43	0.16
		Median	—	6.17	3.17	2.04	4.46	1.73	2.19	1.02	0.61	0.25	0.07
		Range	—	0.15 - 52.8	0.09 - 29.7	0.07 - 146	0.84 - 22.8	0.1 - 198	0.09 - 1434	0.02 - 5.7	0.05 - 5.73	ND - 5.75	ND - 3.06
						Rainwater, Snow (ng/L)							
Sweden Faroe Islands	Filipovic Laudon et al. 2015	Mean	1.997	0.540	0.068	—	—	0.652	0.008	0.305	0.214	0.129	0.057
		Median	1.715	0.464	0.067	—	—	0.552	ND	0.277	0.192	0.135	0.054
		Range	0.765 - 3.621	0.116 - 1.181	ND - 0.166	—	—	0.105 - 1.691	ND - 0.051	0.107 - 0.681	0.084 - 0.794	0.026 - 0.447	ND - 0.212
Antarctic Peninsula	Wang, Xi et al. 2015	Mean	0.2	—	—	—	—	—	—	—	—	—	—
		Range	0.125 - 0.303	—	—	—	—	—	—	—	—	—	—
Maltese Islands	Sammut et al. 2017	Mean	3.88	0.24	0.10	—	—	0.19	0.83	0.31	0.59	1.68	—
		Median	4.66	0.23	0.06	—	—	0.29	1.09	0.29	0.61	2.65	—
		Range	0.38 - 6.3	0 - 0.68	0 - 0.27	—	—	0 - 0.33	0 - 1.15	0.28 - 0.38	0 - 0.99	0 - 2.73	—
						Treatment Plant Effluent (ng/L)							
Northern Bavaria, Germany	Becker et al. 2008	Mean	83.45	38.45	45	—	—	—	—	—	—	—	—
		Median	64.5	34	30.5	—	—	—	—	—	—	—	—
		Range	20.7 - 190	8.7 - 93	12 - 140	—	—	—	—	—	—	—	—
Minnesota, USA	MDH 2008	Mean	—	23.1	162.1	27.2	65.9	18.6	23.2	8.4	9.9	5.1	ND
		Median	—	18.9	30.5	19.7	35.9	19.6	21.1	4.9	8.5	4.3	ND
		Range	—	3.38 - 64.1	<5.14 - 1510	<5.17 - 107	<2.6 - 565	<2.6 - 40.2	<5.2 - 53.1	<2.57 - 45.6	<2.59 - 31.4	<2.59 - 7.56	ND

(Continued)

TABLE 3.17A (Continued)

Location	Source	Value[b]	ΣPFASs	PFOA	PFOS	PFBS	PFBA	PFHxA	PFHxS	PFHpA	PFNA	PFDA	PFUnDA
Germany	Ahrens, Felizeter, et al. 2009	Mean	99.3	32.5	11.5	8.6	–	11.5	1.6	4.2	3.4	6.0	1.1
		Median	48.1	19	2.5	5.1	–	5.0	1.2	2.3	1.4	2.3	0.2
		Range	30.5 - 266	12.3 - 77.6	<0.06 - 82.2	1.8 - 25.9	–	3.7 - 57.4	(0.3) - 6.3	1.6 - 15.7	1.0 - 18.6	0.9 - 34.5	<0.004 - 8.8
Germany	Llorca et al. 2012	Range	31 - 62	ND - 1.8	ND	ND	7.6 - 9.8	ND - 1.1	ND	15 - 53	ND	ND - 5.3	ND
Spain	Loos 2013	Range	112 - 563	ND - 17	27 - 501	ND - 3.6	ND - 25	ND - 10	ND - 15	ND - 28	ND - 18	ND - 5.3	ND - 5.3
~20 European Countries		Mean	–	255	62.5	–	–	304	48.6	82.9	35.1	23.9	–
		Maximum	–	15,900	2,101	–	–	23,867	922	2,962	2,735	1,687	–
San Francisco, California, USA	Houtz et al. 2016	Mean	–	21 ± 13	13 ± 4.4	16 ± 5.8	16 ± 5.8	26 ± 5.1	4.8 ± 0.9	4.4 ± 2.2	8.4 ± 3.6	3.5 ± 1.7	–
United Kingdom	Wilkinson et al. 2016	Mean	157	31.7	32.2	51.3	–	–	–	–	41.8	–	–
		Median	172	21.7	26.8	44.5	–	–	–	–	17.2	–	–
		Range	80.1 - 229	3.9 - 87.1	3.5 - 60.1	9.6 - 96.0	–	–	–	–	5.5 - 113	–	–
Sea/Ocean Water													
Central to Western Pacific Ocean	Yamashita et al. 2004	Range	–	0.015 - 0.056	0.0011 - 0.0046	–	–	–	0.0004 - 0.0006	–	<0.0014	–	–
Eastern Pacific Ocean		Range	–	0.136 - 0.142	0.054 - 0.078	–	–	–	0.0022 - 0.0028	–	<0.0014	–	–
Mid Atlantic Ocean		Range	–	0.104 - 0.150	0.038 - 0.073	–	–	–	0.0026 - 0.0043	–	–	–	–
Sulu Sea		Range	–	0.076 - 0.51	<17 - 0.109	–	–	–	<0.2	–	–	–	–
South China Sea		Range	–	0.16 - 0.42	0.0088 - 0.113	–	–	–	<0.2	–	–	–	–
Tokyo Bay		Range	–	154.3 - 192.0	12.7 - 25.4	9.6 - 96.0	–	–	3.3 - 5.6	–	–	–	–

Per- and Polyfluoroalkyl Substances

Location	Source	Value[b]	ΣPFASs	PFOA	PFOS	PFBS	PFBA	PFHxA	PFHxS	PFHpA	PFNA	PFDA	PFUnDA
Eastern Pacific Ocean	Yamashita et al. 2008	Range	–	0.01 - 0.06	–	–	–	–	–	–	–	–	–
Mid Atlantic Ocean		Range	–	0.067 - 0.439	0.013 - 0.073	–	–	–	–	–	–	–	–
North Atlantic Ocean		Range	–	0.052 - 0.338	0.0086 - 0.036	–	–	–	–	–	–	–	–
South Pacific Ocean and Indian Ocean		Range	–	<0.005 - 0.011	<0.005 - 0.011	–	–	–	–	–	–	–	–
Western Pacific		Range	–	0.140 - 0.5	–	–	–	–	–	–	–	–	–
North Sea, Germany	Ahrens, Plassmann, et al. 2009	Mean	15.96	4.22	1.77	1.61	2.13	1.52	0.28	0.71	0.27	0.12	0.02
		Median	15.55	3.93	1.82	1.42	2.15	1.49	0.24	0.69	0.22	0.06	ND
		Range	13.6 - 18.6	3.58 - 5.47	0.18 - 4.16	(0.04) - 3.93	(0.42) - 5.02	1.07 - 1.96	0 - 0.72	0.48 - 1.05	0.09 - 0.71	(0.03) - 0.31	(0.07) - 0.13
Coastline of Germany	Heydebreck et al. 2015	Mean	12.1	1.72	1.51	1.34	0.74	2.39	0.56	(0.76)	0.21	0.01	ND
		Median	12.2	2.01	1.59	1.39	1.06	2.71	0.63	(1.00)	0.23	ND	ND
		Range	7.56 - 17.8	(1.04) - 6.37	ND - 3.29	0.84 - 2.02	(0.91) - 2.22	0.52 - 4.99	0.14 - 1.23	ND - (0.19)	ND - 0.43	ND - 0.09	ND
Coastline of the Netherlands		Mean	17.0	5.70	0.08	1.48	0.79	1.34	0.50	3.20	0.671	0.18	0.04
		Median	10.7	4.54	ND	1.05	0.04	0.83	0.27	1.19	0.275	(0.09)	ND
		Range	4.9 - 44.7	1.79 - 13.2	ND - 0.88	0.7 - 3.38	ND - 4.56	0.32 - 5.01	ND - 1.88	ND - 10.51	0.14 - 2.58	ND - 2.42	ND - 0.39
Laizhou Bay, China		Mean	13,251*	12,290*	0.98	ND	54.1	329	ND	383	1.39	ND	0.03
		Median	4,526*	4,380*	ND	ND	21.8	84	ND	82	1.09	ND	ND
		Range	1,125 - 58,718*	1,050 - 52,500*	ND - 4.97	ND	8.25 - 200	25.53 - 2,125	ND	16.19 - 2,730	0.59 - 3.48	ND	ND - 0.26

TABLE 3.17B
Summary of detections and background concentrations of other PFASs measured in air samples.

Location	Source	Value[b]	ΣFTOHs	12:2 FTOH	10:2 FTOH	8:2 FTOH	6:2 FTOH	MeFBSA	MeFOSA	EtFOSA	MeFBSE	MeFOSE	EtFOSE
						Air (pg/m³)							
Canadian Arctic region	Stock et al. 2007	Mean (g)	8.8	—	0.35	8.0	0.5	—	—	4.8	1.1	1.7	ND
		Median (g)	ND	—	ND	5.45	ND	—	—	ND	ND	ND	ND
		Range (g)	ND-34.8	—	ND-3.5	ND-18.6	ND-19.7	—	—	(7.4)-18.6	(22.7)-34	(31.2)-39.2	ND
		Mean (p)	0.41	—	0.41	ND	ND	—	—	(0.97)	1.1	3.9	4.56
		Median (p)	ND	—	ND	ND	ND	—	—	ND	ND	ND	ND
		Range (p)	ND-4.1	—	ND-4.1	ND	ND	—	—	ND	ND-5.4	ND-26.9	ND-27.9
Toronto, Canada	Ahrens et al. 2012	Mean (g)	101	—	11.4	41.7	48.3	—	ND	0.17	—	0.25	0.02
		Median (g)	98.3	—	9.99	34	54.3	—	ND	0.18	—	0.25	—
		Range (g)	59-153	—	8.17-16.7	25.7-74.6	25.1-62.1	—	ND	ND-0.29	—	0.12-0.46	ND-0.12
		Mean (p)	4.96	—	1.92	2.02	1.02	—	ND	0.02	—	0.41	0.06
		Median (p)	5.36	—	2.07	2.14	1.15	—	ND	0.03	—	0.33	—
		Range (p)	2.66-7.2	—	1.44-2.09	1.22-2.5	nd-2.62	—	ND	ND-0.03	—	0.21-0.66	ND-0.29
Germany	Dreyer et al. 2015	Mean	—	—	—	—	—	—	0.67	—	—	—	—
		Range	—	—	—	—	—	—	0.1-4.4	—	—	—	—
Atlantic to Southern Ocean	Wang, Xi et al. 2015	Mean	22.17	2.30	3.4	14.9	1.6	0.14	0.22	0.13	0.09	0.45	0.07
		Median	17.12	1.85	2.2	12.4	0.65	0.05	0.14	0.08	0.05	0.25	0.05
		Range	0.01-50.67	0.43-6.07	0.37-11.35	1.51-50.67	0.14-7.82	0.01-0.57	0.04-0.67	0.02-0.74	0.01-0.44	0.04-1.75	0.01-0.25
North Greenland	Bossi et al. 2016	Mean	—	—	1.59	4.93	2.82	—	0.44	0.33	—	0.61	0.5
		Range	—	—	<0.20-9.68	<0.45-22.4	<0.45-16.5	—	<0.20-3.41	<0.22-1.93	—	<0.15-7.46	<0.11-5.96
Northern Part of South China Sea	Lai et al. 2016	Mean	53	2.5	8.1	38	4.4	<0.9	0.3	0.4	<0.3	—	—
		Median	53.9	2.2	8.6	38.2	4	1.3	0.4	0.4	0.8	—	—
		Range	17.8-105.8	1.3-6	2-17.8	12.5-75.8	0.7-8.6	<0.9-1.3	<0.3-0.6	0.1-1.3	<0.3-1.2	—	—

TABLE 3.17C
Summary of detections and background concentrations of other PFASs measured in surf samples

Location	Source	Value[b]	8:2 FTS	6:2 FTS	4:2 FTS	FOSA	EtFOSE	HFPO-DA	HFPO-TA	ADONA	6:2 Cl-PFESA	sum PFPCPeS	sum PFECHS
						Surface Water (ng/L)							
Elbe River, Germany	Ahrens, Plassmann, et al. 2009	Mean	—	0.03	—	6.53	0.01	—	—	—	—	—	—
		Median	—	ND	—	6.59	ND	—	—	—	—	—	—
		Range	—	ND - 0.47	—	2.02 - 11.94	ND - 0.12	—	—	—	—	—	—
Elbe River, Germany	Ahrens, Felizeter, et al. 2009	Mean	—	0.87	—	—	—	—	—	—	—	—	—
		Median	—	0.8	—	—	—	—	—	—	—	—	—
		Range	—	<0.2 - 1.1	—	—	—	—	—	—	—	—	—
Oujiang River, China**	Wang et al. 2013	Mean	—	—	—	—	—	—	—	—	34.6	—	—
		Median	—	—	—	—	—	—	—	—	39.9	—	—
		Range	—	—	—	—	—	—	—	—	11.7 - 54.7	—	—
Elbe River, Germany	Heydebreck et al. 2015	Mean	—	0.93	—	0.06	—	ND	—	—	—	—	—
		Median	—	0.45	—	0.05	—	ND	—	—	—	—	—
		Range	—	ND - 2.86	—	ND - 0.33	—	ND	—	—	—	—	—
Rhine River, Germany		Mean	—	0.09	—	0.06	—	8.8	—	—	—	—	—
		Median	—	ND	—	0.07	—	ND	—	—	—	—	—
		Range	—	ND - 0.55	—	0.04 - 0.1	—	ND - 108	—	—	—	—	—
Xiaoqing River, China		Mean	—	ND	—	ND	—	484	—	—	—	—	—
		Median	—	ND	—	ND	—	98.75	—	—	—	—	—
		Range	—	ND	—	ND	—	3.3 - 3,825	—	—	—	—	—

(*Continued*)

TABLE 3.17C (Continued)

Location	Source	Value[b]	8:2 FTS	6:2 FTS	4:2 FTS	FOSA	EtFOSE	HFPO-DA	HFPO-TA	ADONA	6:2 Cl-PFESA	sum PFPCPeS	sum PFECHS
Cape Fear River Watershed (A), North Carolina, USA	Sun et al. 2016	Mean	–	–	–	–	–	<10	–	–	–	–	–
		Median	–	–	–	–	–	<10	–	–	–	–	–
		Range	–	–	–	–	–	<10 - <10	–	–	–	–	–
Cape Fear River Watershed (B), North Carolina, USA		Mean	–	–	–	–	–	<10	–	–	–	–	–
		Median	–	–	–	–	–	<10	–	–	–	–	–
		Range	–	–	–	–	–	304 - 10	–	–	–	–	–
Cape Fear River Watershed (C), North Carolina, USA		Mean	–	–	–	–	–	631	–	–	–	–	–
		Median	–	–	–	–	–	<10	–	–	–	–	–
		Range	–	–	–	–	–	55 - 4,560	–	–	–	–	–
Beijing, China	Wang et al. 2016	Mean	–	–	–	–	–	–	–	–	0.43	11.66	23.00
		Median	–	–	–	–	–	–	–	–	0.17	ND	2.28
		Range	–	–	–	–	–	–	–	–	ND - 5.32	ND - 128.5	ND - 195.1
Global	Pan et al. 2018	Mean	0.03	0.97	0.01	–	–	2.55	0.79	0.02	2.08	–	–
		Median	ND	0.21	ND	–	–	0.95	0.21	ND	0.31	–	–
		Range	ND - 0.28	ND - 13.9	ND - 0.11	–	–	0.18 - 144	ND - 34.8	ND - 1.55	ND - 52.2	–	–

Notes (for Tables 3.17a, 3.17b, and 3.17c):

ΣPFASs = total PFASs based on sum of all PFAS analytes reported

ND = indicates samples were either not detected or were below the level of detection and/or quantitation

(g) = gas phase air sample result

(p) = particulate phase air sample result

Mean and median are presented as reported in the original publication or calculated using reported sample results. For mean and median results calculated using sample results, the mean is calculated using zero for non-detect values and the median includes non-detect results.

* = estimated value reported and used in summary of mean, median, and range; reported value outside calibration range

** = results reported as an average of three samples taken from each location. Average sample concentrations used to summarize mean, median, and range.

evidence for global distribution (Dreyer et al. 2009). In this study the total gas-phase concentrations of PFASs in samples ranged from 4.5 pg m^{-3} in the Southern Ocean to 335 pg m^{-3} in European source regions. Concentrations of 8:2 FTOH was usually observed in the highest concentrations, and PFAS concentrations decreased from continental regions toward marine regions and from Central Europe toward the Arctic and Antarctica. Various studies have shown dominance of FTOHs in the gas phase from over open oceans and in remote regions, such as Greenland (Bossi et al. 2016), the South China Sea (Lai et al. 2016) and the Arctic (Butt et al. 2010).

Reports describing the analysis of PFASs in groundwater in the United Kingdom, from around 200 sampling locations, were published in 2007 and 2008 (Environment Agency 2007; 2008). Groundwater samples were taken from locations mostly in, or close to, industrial and/or urban areas and/or where firefighting foams may have been used. The results described concentrations of the sum of PFASs investigated in 36 chalk aquifer units containing a maximum of 8,100 ppt PFASs, with a mean of 1,350 ppt and a median of 220 ppt with 36 percent of sites were found to have detectable PFASs. In Permo-Triassic sandstone aquifer units 19.4 percent of the 72 monitored locations contained PFASs which had a mean concentration of 1,460 ppt and a median of 310 ppt with a maximum reported concentration of 7,470 ppt. Data from the 75 minor aquifers sampled revealed a mean PFASs concentrations of 1,180 ppt, a median of 390 ppt with a maximum of 6,560 ppt, with PFASs detected at 24 percent of the 75 locations sampled. The result indicates that widespread and relatively elevated concentrations of PFASs (as compared to the U.S. EPA HAL) are already apparent in many locations. The data suggest that many aquifers in industrial or urban areas globally will likely have PFASs present at levels well above the drinking water health advisories promulgated in many countries, which are usually in the range of 70 to 100 ppt, for long-chain PFASs. The detection of PFASs in groundwater may prove to be a common occurrence.

A global survey of 32 PFASs in surface soils at 62 locations representing all continents found quantifiable levels of PFCAs were observed in all samples with total concentrations ranging from up to 14.3 μg/kg while PFSAs were detected in all samples but one, ranging from up to 3.3 μg/kg, confirming the global distribution of PFASs in terrestrial settings (Rankin et al. 2016). The most commonly detected PFAAs in soils were PFOA and PFOS at concentrations up to 2.67 and 3.1 μg/kg. The geometric mean PFCA and PFSA background concentrations in soils were observed to be higher in the northern hemisphere (0.93 and 0.170 μg/kg) compared to the southern hemisphere (0.190 and 0.033 μg/kg) (Rankin et al. 2016). Because the soils collected in this study were from locations absent of direct human activity, it was suggested that the atmospheric long-range transport (LRT) of volatile PFASs, such as FTOHs, followed by oxidation and deposition are a significant source of PFCAs and PFSAs to soils.

Background levels of PFASs have been measured in fresh waters (Yang et al. 2011), seawater (Ahrens et al. 2009), as wells as ice cores in the remote Arctic regions (Cai et al. 2012). The occurrence of PFASs in deeper groundwater aquifers depends on PFAS water solubility and appears to be inversely related to perfluoroalkyl chain length (Meyer et al. 2011). Additionally, concentrations of PFOS and PFOS precursors in fish were similar between a reference lake and lakes near the source, with concentrations of PFCAs and PFCA precursors in fish being dependent on proximity to local industrial sources (Ahrens, Marusczak, et al., 2010).

The presence of PFAAs in rain and surface waters on the Maltese Islands was assessed in a recent study (Sammut et al. 2017). The ΣPFAA concentrations in rain water ranged between 0.38 ng/L and 6 ng/L. As the Maltese archipelago is disconnected from any other mainland, it was suggested that the presence of PFAAs in rainwater confirms that remote environments can become contaminated by PFASs from rain events depending on prevailing wind trajectories.

A survey of European surface waters for several PFAAs was performed in 2009 and showed elevated concentrations of PFOS in several major rivers, including River Severn, in the United Kingdom (238 ppt), River Seine in France, (97 ppt), River Scheldt, in Belgium (154 ppt), River Rhine in Germany (32 ppt), and River Krka in Slovenia (1,371 ppt) (Loos et al. 2009). The median PFOS concentration across European rivers was reported as 6 ppt, with the median for PFOA at 3 ppt. Given that the AA EQS for PFOS in European surface water is 0.65 ppt, the detections are significantly higher. One solution to the challenge of exceeding these low compliance criteria may be to risk-assess individual rivers and to develop river-specific PFAS targets.

A recent survey of the worldwide distribution of PFASs including PFECAs and PFESAs in surface waters revealed the presence of elevated concentrations of a combination of PFAAs in the River Thames in the United Kingdom (maximum 66 ppt sum PFAAs), with the levels in the Thames being the highest recorded from sampling 10 major international rivers (Pan et al. 2018). The presence of PFECAs and PFESAs was reported in multiple rivers, showing the adoption of these replacements.

Given these challenges, establishing site-specific background concentrations for PFASs are necessary steps for developing a robust site characterization.

3.8.3 SITES OF CONCERN

While there are a limited number of primary manufacturing sites for PFASs, secondary manufacturing (i.e., those that use PFASs or apply them as additives to other products) are also potential point sources of PFASs to the environment (Xiao 2017; Janberg et al. 2006). Additionally, the diverse application of PFASs in multiple commercial products, coupled with the often very low (ng/L) environmental compliance criteria means there are potentially a number of locations where PFASs have been lost to ground in quantities considered significant enough to warrant investigation. For example, users of PFASs products, such as those who do fire training, also have directly deposited PFASs in firefighting foams onto the ground in fire training areas, via repeat application of Class B foams, potentially for decades. Many commercial buildings (e.g., warehouses) that house flammable liquids or maintain machinery, vehicles, or aircraft (e.g., garages or aircraft hangars) use fire-prevention sprinkler systems, which could contain PFASs. The use of PFASs in surface coatings of papers, leather, and fabrics means sites of coatings manufacturing may be heavily impacted. The multitude of other industrial applications, such as in floor polishes, paints, cosmetics, car waxes, etc. may mean the product manufacturing sites are impacted. PFAS-containing products now in domestic or industrial landfills will be a long-term source of dissolved PFASs in leachates or potentially directly to groundwater if the landfills are unlined. Municipal and industrial WWTPs can be point sources for discharge of PFASs in effluents, with the biosolids harvested from processing wastes also containing PFASs, which are potentially used as fertilizers for arable crops.

It is important to differentiate these potential source areas from the background concentrations of PFASs measured in many locations, as described in section 3.8.2, as finding PFASs in soil or groundwater at low concentrations at a specific location does not necessarily mean it originated from a site local to the sampling point, as they may originate from diffuse sources of PFASs (Muller et al. 2011).

The principal types of sites of concern regarding potential PFASs contamination may include:

(1) industrial facilities undertaking primary and secondary manufacturing, as well as facilities that use PFAS-containing materials in their industrial process (Lyons 2007); (2) sites handling firefighting foams; (3) WWTPs and biosolids application areas; and (4) landfills. A generalized conceptual site model (CSM) for each of these sites of concern is provided below.

3.8.3.1 CSM for Industrial Facilities

Industrial facilities where PFASs have been used may include fluoropolymer manufacturing, metal plating facilities (chromium, zinc, gold, copper, nickel, tin, brass, etc.), textile, fabric and carpet manufacturing or processing facilities, paper mills, leather tanneries, and in the manufacture of some paints, rubber, inks, cosmetics, detergents, polishes, dyes, adhesives, lubricants, photographic paper, cleaning agents, pesticides, hydraulic fluids, etc. Sites associated with these activities may have used PFASs in bulk quantities. In fact, the use of PFCAs for fluoropolymer production processes is understood to be the single largest known source of PFCA emissions to the environment and has been the single largest direct use of PFOA and perfluorononanoic acid (PFNA) (Prevedouros et al. 2006).

A generic CSM considering the potential fate and transport pathways of PFASs from an industrial site is shown in Figure 3.23. As shown, industrial facilities may release PFASs to the environment via wastewater discharges to surface water or via WWTPs, on- and off-site disposal of wastes, accidental releases such as leaks and spills, and stack emissions (ITRC 2018b). PFASs emissions to air can also result in land and groundwater impacts and potential human exposure in the surrounding area (Beekman et al. 2016; Liu et al. 2016).

3.8.3.2 CSM for Fire Training Areas and Class B Fire Response Areas

Compared to other industrial and commercial applications of PFASs, the use of Class B firefighting foams, such as AFFF, for fire training and in emergency response situations is considered one of the most dispersive activities. Particularly for training purposes they are likely to have been applied repeatedly and in large quantities to the ground surface sometimes over several decades and often without adequate containment or wastewater management.

Firefighting foams containing PFASs continue to be used at airports, military facilities, FTAs, petrochemical refineries, bulk fuel storage terminals, ports, and other facilities handling large volumes of flammable liquid hydrocarbon (Thalheimer et al. 2017; Weber et al. 2017; McGuire et al. 2014; Hu et al. 2016b).

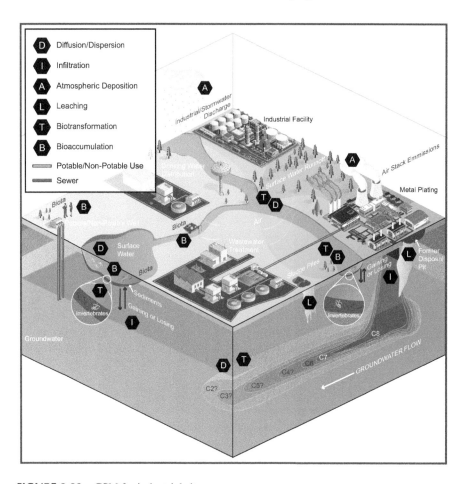

FIGURE 3.23 CSM for industrial sites.

The use of firefighting foams is commonly associated with FTAs, which typically comprise an area (concrete, paved and unpaved) on which liquid hydrocarbon or solvent fuel fires that are extinguished during fire training exercises. Waste fire water containing residual fuels and foam are often directed through site drainage to basic effluent treatment centers primarily designed to remove hydrocarbons rather than PFASs with water, and are then discharged to third-party WWTPs, surface waters or groundwater via soakaways. Such practices, as well as the inefficient capture of firewater within the FTA, stormwater runoff, and/or leaks through cracks in the concrete pad and associated drainage, have resulted in PFASs released to the environment. As a result, PFASs are often found within shallow surface soils as well as groundwater and surface waters around and beneath FTAs (e.g., Weber et al. 2017; McGuire 2013) with soil impacts presenting a long-term potential source to groundwater (Filipovic, Laudon et al., 2015; Filipovic, Woldegiorgis et al., 2015). A range of PFASs have also been identified at a depth of up to 12 cm within exposed concrete pads (Baduel et al. 2015) with modeling suggesting that PFASs' impact within concrete pads or training

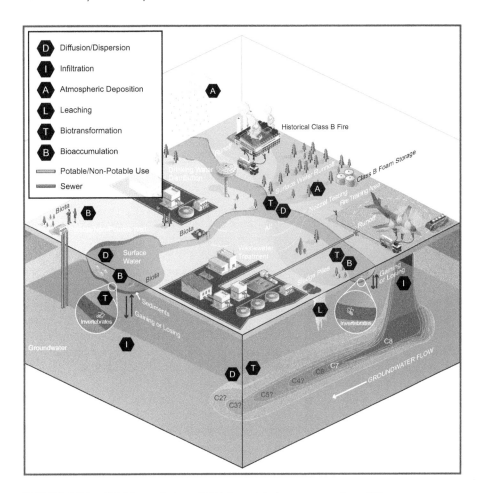

FIGURE 3.24 CSM for evolution of PFAAs and their precursors from FTA source areas.

pads will likely remain a source of PFASs for many decades (Baduel et al. 2015). Accordingly, examples of modern FTA design include improved water handling systems (e.g., rainwater harvesting), lining systems, control systems, and construction materials to effectively contain and reuse firewater (Knapton 2016).

A generic CSM considering the potential fate and transport pathways of PFASs from and FTA and other associated areas is shown in Figure 3.24. In addition to usage within FTAs, releases of firefighting foams to the environment may have occurred at fire stations, fire truck washdown and maintenance areas, emergency response incidents, maintenance and episodic discharge from fire suppression systems including within large aircraft hangars and buildings (Anderson et al. 2016; Thalheimer et al. 2017), or tank-based systems at hydrocarbon storage facilities.

3.8.3.2.1 PFASs Associated with Firefighting Activities
Groundwater that is impacted with PFASs from fire training activities is not necessarily dominated by PFSAs and PFCAs, with fluorotelomer sulfonates such as 6:2 FTS

sometimes reported at higher concentrations. For example in groundwater associated with fire training activities, high levels of PFOS (1 mg/L), PFOA (6.6 mg/L) were exceeded by the concentration of 6:2 FTS at 14.6 mg/L (Schultz et al. 2004). Depending on the products used and PFASs released at each site, a number of differing PFAAs and their PFAA precursors are anticipated to be detected in associated environmental matrices. For example, use of differing Class B foam types (i.e., ECF-based versus FT-based) will result in the detection of specific PFAAs and PFAA precursors. The widespread use of many differing foam types since the mid 1970s means a mixture of PFSAs, PFCAs, and fluorotelomers are likely at many fire training areas (Place and Field 2012).

The ECF-based foams were generally comprised of PFSAs, and PFOS is anticipated to be a dominant PFAA found in environmental matrices, whereas the fluorotelomer-based foams will likely have contained numerous differing polyfluorinated PFAA precursors, which require advanced analytical techniques (such as PIGE, TOP Assay, AOF, or LC-QTOF-MS) to be applied for them to be detected (Barzen-Hanson, Roberts et al., 2017; Place and Field 2012; Houtz et al. 2013; Harding-Marjanovic et al. 2015). However, PFAA precursors containing a sulfur bridge or sulfate group terminal to the fluorotelomer alkyl group can transform to create and fluorotelomer sulfonate, such as 10:2 FTS, 8:2 FTS, and 6:2 FTS. Detection of 6:2 FTS as a "meta-stable" transformation intermediate generated from these PFAA precursors is an indicator that they are present. 6:2 FTS is a transformation intermediate of these PFAA precursors, which is stable under anoxic conditions but amenable to aerobic biotransformation to ultimately create PFAAs including PFHxA, PFPeA, and PFBA as persistent terminal daughter products.

A study of surface water and sediments near four sites where AFFF was historically used showed a variety of PFAAs were detected, including short- and long-chain PFCAs (i.e., PFBA, PFPeA, PFHxA, PFHpA, PFOA, and PFNA) and PFSAs (i.e., PFBS, PFHxS, PFHpS, and PFOS), as well as PFAA precursors (including 6:2 FTS, 8:2 GTS, 6:2 FTAB, and 6:2 FtSaAM, and 8:2 FTS) (Dauchy et al. 2017b). The profiles of PFASs detected were reported to be influenced by parameters such as the route of PFAS transport after use (runoff, seepage, direct discharge), time elapsed since the cessation of firefighting activities, and firefighting foam composition (Filipovic, Woldegiorgis et al., 2015; Filipovic, Laudon et al., 2015). Reports of an AFFF release near a surface water body indicated the presence of up to 2.21 mg/L of PFOS immediately after an AFFF release from an incident on an airport in 2000 (Moody et al. 2002) with 290 ng/L of PFOS in surface water collected from a nearby down-gradient creek sample location nine years later, and between 45 and 65 ng/L of PFOS in surface water samples collected from the receiving lake ten years later (D'Agostino and Mabury 2017a).

Advanced analytical tools to attempt to detect PFAA precursors were not applied in any of these studies; however, there are two reports of the application of TOP Assay where more detailed site investigations attempting to measure total PFASs have been detailed (McGuire et al. 2014; Weber et al. 2017).

3.8.3.2.2 FTA Source Area Considerations

A generic CSM showing the potential fate and transport processes which could at an FTA source area be a result of the use of Class B firefighting foams is presented in Figure 3.25. As shown in the Figure and described in section 3.8.1, cationic PFAA

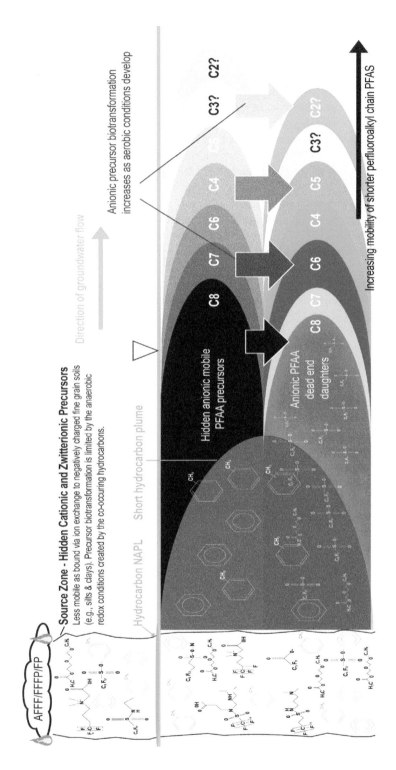

FIGURE 3.25 CSM of class B fire response, fire training, nozzle testing, and storage areas.

precursors (and some zwitterions) are expected to mainly be retained in the soils at the source zone via ion exchange processes (sorbed to negatively charged soil particles). The source zones are likely to be anaerobic as a result of the presence of residual hydrocarbons used in fire training, so PFAA precursors will biotransform extremely slowly due to the prevailing biogeochemical conditions and may present an ongoing source of anionic PFAAs, slowly released from the source. Anionic PFAA precursors will migrate away from the source and enter the redox recharge zone where conditions become increasingly aerobic thus promoting *in situ* generation of detectable PFAAs from the, often undetected, anionic PFAA precursors. The use of advanced analytical tools such as the TOP Assay to characterize sources and plumes of PFASs associated with FTAs has been described in only a limited number of reports (McGuire et al. 2014; Weber et al. 2017).

For some site investigations it may be useful to employ advanced analytical techniques (such as the TOP Assay or AOF) to characterize PFAS impacts and to accurately locate source areas and heavily impacted horizons. Identifying the sources of PFAAs emanating from FTAs where FT-derived AFFF have been used may be more challenging without the use of TOP Assay, as a significant proportion of the PFASs present may not be identified due to the limited number of PFASs for which there are analytical standards (i.e., use of U.S. EPA Method 537) or similar methods. The interpretation of TOP Assay data and application to site characterization is described in section 3.5.2.2.

3.8.3.3 CSM for WWTPs and Biosolid Application Areas

Most conventional and advanced wastewater treatment processes are not effective in removing PFASs (Ross, McDonough et al. 2018), and may cause polyfluorinated precursors to transform into PFAAs (Boulanger et al. 2005). PFASs can also be concentrated in the sewage sludge solid waste (biosolids), which is sometime applied as a fertilizer for crops, with the biosolids potentially acting as a source of PFASs to crops, groundwater or surface water via run-off (Lindstrom, Stryar, Delinsky et al., 2011).

In a recent European study of PFAS concentrations in 90 WWTP effluents, PFOA, PFHpA, and PFOS were detected in more than 90 percent of the waters, with PFOA at the highest median concentration (0.0129 µg/l) (Loos et al., 2013). The study stated (page 6483 of Loos et al., 2013): "Often PFAS concentrations increase in wastewater treatment plants as a result of biodegradation of PFAA precursors during the activated sludge process. PFOA is generally fully discharged into receiving rivers, while about half of PFOS is retained in the sewage sludge."

A generic CSM considering the potential fate and transport pathways of PFASs from WWTPs and biosolid application areas is shown in Figure 3.26. Treated wastewater effluent is typically discharged to surface waters, so the transition over this period from long-chain PFASs to the more mobile short-chain replacements, has been observed in a study of the patterns of PFASs detected in WWTP discharges. It was shown that from 2009 to 2014, concentrations of short-chain PFASs increased in effluent from municipal wastewater treatment plants (Houtz et al. 2016).

PFASs within biosolids that are used to amended soils have also been shown to have the potential for uptake by certain plants (Blaine et al. 2013; Vierke 2017) with

Per- and Polyfluoroalkyl Substances

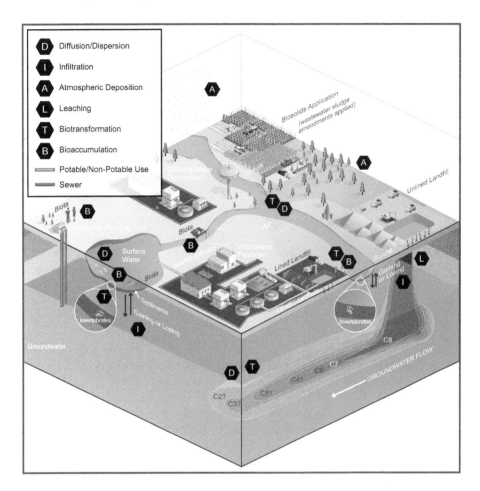

FIGURE 3.26 CSM of waste management processes.

short-chain PFASs found to accumulate within fruits and shoots and longer-chain PFAS within roots (Blaine et al. 2014a; Blaine et al. 2014b). WWTPs have also been described to be potential sources of PFASs to the atmosphere (Weinberg et al. 2011; Ahrens 2011; Ahrens et al. 2011).

3.8.3.4 CSM for Landfills

A generic CSM considering the potential fate and transport pathways of PFASs from a landfill site is shown in Figure 3.26. Many landfills are expected to be potential long-term sources of PFASs as a result of the receipt of multiple waste products containing PFASs, such as fabrics, leather goods, and more (Lang et al. 2016), with some fluoropolymers also reported to transform creating PFASs (Russell 2008; Washington et al. 2014; Washington et al. 2015). The concentrations of PFASs in landfill leachates are reported to be relatively high; however, the volume of leachate generated is estimated to be relatively low, in comparison to WWTPs (Busch et al. 2010).

The distribution of PFASs in landfill leachate tends to be dominated by shorter-chain PFASs as they are more mobile than longer chains and less hydrophobic (Allred 2015; Lang et al. 2016). The role of PFAA precursor transformation in municipal landfill leachate on changes in PFAA concentrations over time was investigated, with PFPeA and PFHxA typically dominated, but seasonal changes caused an increased flux of PFOS and numerous PFAA precursors (Benskin et al. 2012c). A more detailed assessment of the dominant PFASs revealed the presence of the 5:3 FTCA (or 5:3 acid) as the dominant PFAS in landfill leachate (Lang et al. 2017). Anaerobic transformation of 6:2 FTOH creates 5:3 FTCA, with 6:2 FTOH being a common intermediate in the transformation of C6 fluorotelomers (Zhang et al. 2013b; Liu and Mejia Avendano 2013). The slow clearance of 5:3 FTCA in higher organisms was recently reported, suggesting the potential for biopersistence (Kabadi et al. 2018; Liu and Mejia Avendano 2013). Volatile PFASs such as FTOHs may also be generated and released to the air by landfills (Ahrens et al. 2011).

3.9 PFAS-RELEVANT TREATMENT TECHNOLOGIES

The current state of the practice of remediation of PFASs is the sequential application of multiple technologies (i.e., a "treatment train" concept), primarily focused around reducing the treatment volume and concentrating the PFASs to be destroyed. In the complete absence of microbial and with significant challenges to practical field-scale chemical degradation, the viability of destruction-based treatment (e.g., combustion-based incineration) for PFASs increases with smaller concentrated waste streams. For soil or sediment treatment, PFASs may be concentrated through temperature-induced volatilization or through soil/sediment washing with extractants (e.g., methanol or hydroxide). Soil/sediment may also be stabilized in place to prevent leachability to water. For water treatment, PFASs may be concentrated through adsorption- or separation-based technologies; which are those technologies that exploit electrostatic and/or hydrophobic adsorption or partitioning to the gas-liquid interface. To begin the discussion, a specific section on the complications and status of biological treatment for PFASs is provided to justify why this technology is not further discussed in subsequent soil/sediment and water treatment subsections.

Due to the extreme recalcitrance, diversity, and ubiquity of the PFASs throughout the environment, PFAS remediation requires a treatment train to transition from large-volume and low-concentration PFASs to low-volume and higher-concentration PFASs. This section will discuss various soil/sediment and water (e.g., municipal waste and drinking water, stormwater, groundwater, surface water) technologies that are relevant to PFAS treatment. Relevant treatment technologies for soil/sediment and water are presented in Figures 3.27 and 3.28, respectively. Additionally, a table ranking the relative performance of the PFAS-relevant treatment technologies to address PFOA/PFOS, short-chain PFAAs, and PFAA precursors as well

Per- and Polyfluoroalkyl Substances 189

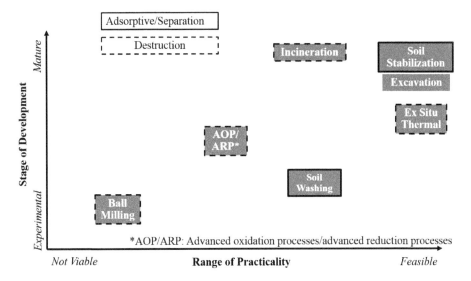

FIGURE 3.27 PFAS treatment technologies for soil, range of practicality, and stage of development.

as the availability and current cost of the technology is presented in Table 3.18. Many technologies are discussed from a practical perspective herein, but the complete list of PFAS-relevant treatment technologies is expected to increase as novel treatment technologies progress from the laboratory-scale to the commercial-scale. Additionally, proprietary reagents are not explicitly discussed, but rather the underlying scientific mechanism is discussed to present an unbiased account of the available PFAS-relevant treatment technologies.

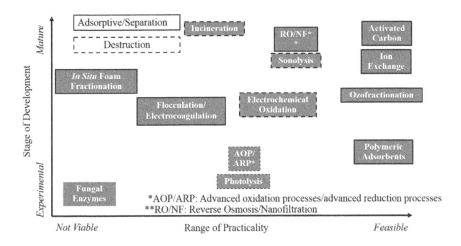

FIGURE 3.28 PFAS treatment technologies for water, range of practicality, and stage of development.

TABLE 3.18
Technology screening table for water

Treatment Technology	Effective on PFOA/PFOS?	Effective on Short Chain PFAAs?	Effective on Precursors?	Commercially Available?	Relative Capital Cost	Relative Operating Cost
Mature Technologies						
Granular activated carbon (adsorption)	Demonstrated	Unlikely/not demonstrated	Assumed likely	Yes	$	$$
Non-ion/anion-exchange resin (adsorption)	Demonstrated	Limited	Assumed likely	Yes	$	$$
Reverse osmosis/nanofiltration (separation)	Demonstrated	Demonstrated	Demonstrated	Yes	$$$	$$$
Incineration (destructive)	Demonstrated	Demonstrated	Assumed likely	Yes	$$$$	$$$$
In-Development Technologies						
Organically modified silica adsorbents (adsorption)	Demonstrated	Assumed likely	Assumed likely	Limited	$$	$$$
Ozofractionation (separation)	Demonstrated	Demonstrated	Demonstrated	Limited	$$$	$$$
Precipitation and sedimentation (separation)	Demonstrated	Limited	Assumed likely	Limited	$$	$$
Emerging Technologies						
Other polymeric adsorbents (adsorption)	Demonstrated	Assumed likely	Assumed likely	No	$$$	$$$
Advanced oxidation processes (destructive)	Limited	Unlikely/not demonstrated	Demonstrated	Yes	$$	$$
Advanced reduction processes (destructive)	Demonstrated	Assumed likely	Assumed likely	No	$$$	$$$
Electrochemical (destructive)	Demonstrated	Assumed likely	Assumed likely	No	$$$$	$$$$
Sonolysis (destructive)	Demonstrated	Demonstrated	Assumed likely	No	$$$$	$$$$
Photolysis (destructive)	Demonstrated	Assumed likely	Assumed likely	No	$$$	$$

3.9.1 BIOLOGICAL TREATMENT

Biodegradation is defined as the process by which organic substances are decomposed by microorganisms into substances such as carbon dioxide, water, and ammonia (OECD 2002a), also termed mineralization. The perfluoroalkyl moiety is not biodegradable, and mineralization of any PFAS to stoichiometric quantities of fluoride, carbon dioxide, and, potentially, sulfate for the PFSAs and their precursors, has not been demonstrated (Colosi et al. 2009; Liu and Mejia Avendano 2013; Luo et al. 2015; Ochoa-Herrera et al. 2016). PFAA precursors biotransform in the environment to form PFAAs, although the rate of this transformation may be slow and some of the transient intermediates formed have yet to be defined (Dinglasan et al. 2004; Wang et al. 2005; Wang et al. 2009; Fromel and Knepper 2010; Lui, Wang, Buck et al., 2010; Russell et al. 2010; Dasu et al. 2012; Dasu et al. 2013; Benskin et al. 2013; Weiner et al. 2013; Liu and Mejia Avendano 2013; Tseng et al. 2014; Harding-Marjanovic et al. 2015; D'Agostino and Mabury 2017a). Although some reports of biological attack on PFASs have been described as biodegradation (Dasu et al. 2012; D'Agostino and Mabury 2017a), none of the approximately 4,700 identified PFASs have been demonstrated to biodegrade.

A detailed account of the fundamental challenges associated with biodegradation of the perfluoroalkyl moiety is provided in section 3.3 and elsewhere (Ross, McDonough et al. 2018) and briefly include the strength of the C-F bonds (Goldman 1965; Matheson L. J. 1995; Kissa 2001) the stability of sequential C-F bonds (Kissa 2001), the intrinsic biodiversity and evolution of the natural consortia (Gribble 2002; Harper 2005), and thermodynamic competition. Laboratory-scale trials have evaluated the very slow biotransformation of PFOA using fungal enzymes, such as horseradish peroxidase and laccase (Colosi et al. 2009; Luo et al. 2015) under idealized laboratory conditions, and will continue to evaluate the efficacy of biological defluorination. However, minimal PFOA defluorination (30 percent) over long residence times (157 days) with short-chain PFAA generation demonstrate the persistence of PFAAs. Fungal enzymatic attack on PFOS, has recently been reported using a laccase enzyme with an artificial mediator added (Luo et al. 2018). In this study a maximum of 32 percent fluoride release from PFOS was reported over 162 days, with no data presented showing terminal products formed. The very slow reported kinetics for microbial attack on PFAAs under idealized laboratory conditions with artificial mediators added and the lack of evidence demonstrating biodegradation (mineralization) indicates that commercial bioremediation of PFASs is highly unlikely to succeed.

Anaerobic respiration of PFAAs will be challenging as sequential removal of 17 fluorine atoms will be required for PFOA and PFOS with multiple, potentially toxic, intermediate breakdown products being formed. Then in real-world aquifers there will likely be a preference for alternative, more thermodynamically favorable, terminal electron acceptors such as carbon dioxide over PFAA respiration.

3.9.2 SOIL AND SEDIMENT TREATMENT

The mechanisms describing adsorption of PFASs (long- and short-chain PFAA precursors and PFAAs) onto soil/sediment are described in section 3.8. It is these mechanisms that may render PFAS-impacted soil/sediment as a continued source of PFASs to water. Therefore, developing soil/sediment treatment technologies for PFASs that either prevent impacted soil/sediment from leaching PFASs to water or separating the PFASs from the soil/sediment matrix, will usually involve transferring PFASs into a liquid concentrate, for subsequent treatment. Additional treatment approaches for soils/sediments include thermal processes with desorption and thermal destruction elements. The treatment technologies to be discussed in this section have been arranged on a stage of development versus range of practicality plot in relation to one another (Figure 3.27). Excavation (with offsite disposal in a landfill or offsite incineration) is the current state of the practice for holistic PFASs treatment for soil/sediment; however, there are many promising alternatives that are actively being developed at the field pilot-scale. These include capping/covering with long-term monitoring, soil/sediment washing, soil/sediment stabilization and/or solidification, onsite thermal desorption with subsequent incineration, high-energy electron beam (eBeam), and mechanochemical destruction (MCD or ball milling).

The modes of implementation of the soil/sediment technologies to be discussed are generally excavation or soil/sediment mixing (either *ex situ* in above ground vessels or *in situ* with conventional construction equipment). This section will focus on the relevance of multiple technologies to PFASs, and for more fundamental information on these modes of implementation, the reader is referred elsewhere (Suthersan 2017). Excavation, with offsite disposal to landfill is relevant for PFAS-impacted source zones; however, the associated long-term liability should be carefully considered given the extreme persistence of PFASs and increased regulatory attention to PFAS treatment or monitoring in landfill leachates (Grieco 2018). Capping/covering of PFAS-impacted soil/sediment *in situ* or *ex situ* within engineered stockpiles may be viable option. Capping/covering of PFAS-impacted soils/sediment requires an appropriate risk-based regulatory framework considering the proximity of the nearest drinking water receptor with monitoring to perpetuity to prevent precipitation infiltration from leaching to groundwater. The remaining soil/sediment treatment technologies are detailed in the next subsections. It is noteworthy that although the conceptual mechanisms of these technologies appear valid, there are few (if any) field-scale demonstrations of these technologies specific to PFAS treatment.

3.9.2.1 Incineration

Incineration is a combustion-based process that involves heating a substance to 1,000°C for two seconds, and the applicability for a chemical is if 99 percent destruction can be achieved within two seconds at temperatures less than 1,000°C (Vecitis et al. 2009). Incineration-based thermal destruction of individual PFASs has been demonstrated in the literature at temperatures of approximately 300°C to 1,000°C (Burgess 1995; Tsang 1998; U.S. EPA 2003b; U.S. EPA 2003a; Krusic and Roe 2004; Krusic et al. 2005; Yamada et al. 2005; Loganathan et al. 2007; Vecitis et al. 2009; Wang, Lu, et al., 2011). During incineration, carbon to carbon (C-C) bonds tend to break before carbon to fluorine (C-F) bonds, which can create fluoroalkyl radicals (Burgess 1995). Therefore, incomplete thermal destruction could form short-chain fluorocarbons (e.g., tetrafluoromethane [CF_4], hexafluoroethane [C_2F_6], fluoroform [CHF_3], various difluoroethane isomers [1,1-$C_2H_2F_2$; 1,2-$C_2H_2F_2$]), which may be subject to air emission regulations. While these fluorocarbons have been detected in combustion effluent (3M Company 2003; Yamada et al. 2005; Wang, Shih et al., 2013; Merino et al. 2016), it has been reported that under combustion conditions, these radicals are expected to be readily transformed to carbon monoxide (CO), carbon dioxide (CO_2), and hydrogen fluoride (HF). This implies incomplete destruction of a full suite of PFASs at lower temperatures and the generation of incineration by-products, which are regulated greenhouse gases with long atmospheric half-lives (Wang, Lu, et al., 2011; Merino et al. 2016). Preliminary research has shown that certain additives may help reduce the formation of these by-products (Wang, Lu, et al., 2011; Wang, Shih et al., 2013), and this could increase the already high cost of incineration correspondingly. An incineration study conducted by the Ministry of the Environment of Japan (2013) indicates that incineration temperatures of 1,100°C in the primary furnace and 900°C in the secondary furnace of a rotary kiln did not result in significant production of fluorocarbons in the emission due to destruction of treated PFASs (PFOS specifically). This underscores the importance of the type of incineration process used to facilitate complete destruction of the full suite of PFASs with minimal by-product formation. A primary (1,100°C) furnace followed by a secondary furnace (900°C) is more commonly associated with a U.S. hazardous waste incineration process as opposed to municipal solid waste incinerator processes (National Research Council 2000).

3.9.2.2 Stabilization/Solidification

Soil/sediment stabilization/solidification (S/S) involves mixing the PFAS-containing soil/sediment with reagents to prevent subsequent leaching into the surrounding environment and/or discourage infiltrating water from accessing adsorbed PFASs. Solidification attempts to impart a load-bearing capacity (i.e., unconfined compressive strength [UCS]) to the monolith while also reducing the permeability and uses Portland cement and bentonite, while stabilization targets chemical reactions that reduce the propensity of adsorbed PFASs to dissolve into water. The reagents are designed to exploit both the hydrophobic and electrostatic properties of PFASs to "fix" PFAS mass in place (i.e., fixation) while also reduce the permeability of the mixed monolith. The reduction in permeability mitigates the potential for infiltrating water to facilitate dissolution of the stabilized PFASs. S/S for excavated waste to be disposed of offsite in a landfill and large-scale *in situ* applications for non-PFAS contaminants are well-established throughout the industry; however, commercially available PFAS-specific demonstrations are not yet available. The mixing can occur either *in situ* or *ex situ* within sealed containers using conventional construction equipment. From a practical perspective, the PFAS-containing soil/sediment can be mixed onsite using backhoes, augers, and/or front-end loaders or placed in machines called "pug mills," which can grind and mix materials at the same time. Water may be added to the mixed monolith to improve the workability of the reagent and soil/sediment to facilitate homogenous mixing depending on the degree of saturation.

Though S/S technology is mature, an understanding of the adsorption mechanisms of various PFASs and identification/development of effective adsorbents that can be used as reagents for S/S are developing in the literature. Commercially available products tested for S/S include activated carbon, organo-modified clays, pyrolyzed cellulose, and proprietary blends of activated carbon/clay/aluminum hydroxides

(Du et al. 2014; Storch 2016). Reagents under development for S/S of PFASs include graphene derivatives, iron oxides, polydiallyldimethylammonium chloride (PDM), polyamine (PA), and layered double hydroxides (Rattanaoudom et al. 2012; Hu et al. 2017; Lath 2017; Pennell 2017; Aly 2018). Short-chain PFASs have different physical/chemical properties than long-chain PFASs, as they are more water soluble and hence mobile, and interact differently with stabilization reagents. PFAA precursors interact differently with S/S reagents due to the presence of cationic, anionic, and zwitterionic functional groups. Incorporating TOP Assay into leachability and adsorption evaluations better informs S/S reagent selection to target total PFASs. As regulations develop, it is advisable that the permanence of S/S consider both short- and long-chain PFASs as well as PFAA precursors. Typically, the fixant and Portland cement concentrations are determined during predesign treatability testing. Treatability testing in support of an S/S design is important because it can save cost on reagents (often paying for itself by justifying a lower-than-anticipated reagent concentration), but also can identify adverse interactions between the fixants and Portland cement.

It is important to note that S/S does not destroy PFASs, and therefore a potential for long-term PFAS-leaching exists. It is imperative to understand the permanence associated with S/S of PFAS-containing soil/sediment managed onsite or offsite. To evaluate the permanence of S/S, long-term leachability test work at environmentally relevant "worst-case" scenarios is required; however, these data are not yet commercially available. Although some commercial entities have simulated long-term S/S leachability evaluations, they have done so at acidic pH conditions which represent a "best case" scenario for anionic PFAA adsorption (Ziltek 2017; Ross, McDonough et al. 2018). As the dominant surface charge is positive under acidic conditions, electrostatic adsorption of the anionic PFAAs are enhanced. In a bench-scale leachability test performed at neutral pH, impractical quantities of S/S reagent were required to effectively diminish PFAS leachate concentrations demonstrating the important influence of pH on S/S permanence (Bräunig et al. 2017).

The applicability of S/S for PFAS-containing soil/sediment may be site-specific. If regulators are amenable to onsite repurposed use of PFAS-containing soil/sediment (e.g., surface grading or excavation backfill), S/S may be a more cost-effective management strategy than offsite incineration. As S/S significantly reduces the leachability of PFASs, pretreatment of PFAS-containing soil/sediment using S/S prior to disposal in a landfill may reduce future liability. If the S/S materials are transported to landfills, specific landfill disposal requirements must be reviewed and understood as they relate to PFASs, co-contaminants, and the S/S reagents added. With the emerging nature of PFASs, landfill disposal requirements and state regulations are developing and typically vary.

3.9.2.3 Vapor Energy Generator Technology

The Vapor Energy Generator (VEG) system is a mobile *ex situ* thermal desorption treatment system (with subsequent chemical and thermal PFAS destruction) that uses a patented steam generator. The VEG system initially desorbs PFASs from soil/sediment by using steam to concentrate PFASs for more efficient thermal destruction at temperatures of 1,100°C to 1,700°C. Desorption of PFASs from the soil/sediment occurs in a fully enclosed *ex situ* chamber (chamber) with soil temperatures ranging between 650°C to 815°C. The concentrated PFAS vapor stream is conveyed through a modular filtration system (filter) containing an engineered mixture of lime, water, and zero-valent iron. Off gas from the filter generates synthesis gas (syngas), which is primarily hydrogen and carbon monoxide and used as a fuel source to generate additional steam in chamber. The purpose of the engineered mixture is to facilitate chemical reduction of the PFASs; however, only about 20 percent of concentrated PFASs are chemically reduced in the vapor stream. Following filtration, the residual steam concentrate is conveyed through the chamber where the remaining PFASs are thermally destroyed at 1,100°C to 1,700°C. Thermal destruction generates HF gas, which is neutralized with potassium hydroxide to produce a vapor stream that is circulated through the chamber.

A bench-scale study was performed to assess soil treatment in an FTA where AFFF was historically used (Endpoint 2016). The bench-scale study achieved 99 percent removal of 10 PFAAs (including PFOA, PFOS, and PFHxS) to below the limits of quantitation during a 30-minute treatment time at 950°C. The results suggest that the optimal conditions for complete PFAS removal from soil may be between 600°C and 950°C for 15 to 30 minutes, although the number of treatment tests run in this study was insufficient to fully optimize treatment conditions. Further, soil treatment may be influenced by soil types and moisture content. Consistent with other thermal desorption estimates, slight increases in short-chain PFAAs, such as PFBA and PFPeA, were observed after treatment for the lowest temperatures tested (480°C to 600°C). The generation of short-chain PFAAs may reflect transformation products of long-chain PFAAs or PFAA precursors.

The typical throughput rate depends largely on the necessary residence time within the VEG system. While field-scale data for PFAS-containing soil/sediment is not yet commercially available, an achievable throughput of approximately 260 cubic yards per day may be practical based on existing bench-scale data (Ross, McDonough et al. 2018). The VEG system is currently being optimized to incorporate PFAS-impacted liquid rather than potable water to improve total PFAS treatment efficiency (Javaherian 2016). A waste stream of approximately six 55-gallon drums for every 10,000 to 15,000 cubic yards of soil treated is produced. The profile for this waste is expected to be driven by the pH and fluorine concentrations.

While no full-scale PFAS-specific field-scale implementation has been completed, the VEG system has been used for 45 projects throughout the U.S. The combination of actual field-scale implementation and promising bench-scale study data positions the VEG system within a small group of commercially available onsite PFAS-relevant treatment technologies with the potential for complete destruction. The benefits of the VEG system include comparatively lower energy costs and minimized emissions (versus incineration), a relatively small operational footprint, and a comparatively lower mobilization cost (versus nationwide transportation costs to fixed incineration facilities). Potential drawbacks of the VEG system include the lack of actual performance data from a field-scale PFAS-containing soil/sediment site, specific testing on PFAA precursor treatment efficacy, and continual refinement to find the optimal treatment conditions (Ross, McDonough et al. 2018).

3.9.2.4 Soil/Sediment Washing

Soil/sediment washing is an *ex situ* separation-based technology. Contaminants that are adsorbed onto soil/sediment are physically separated from the soil/sediment by physical and/or chemical processes with a liquid extractant resulting in a higher concentration, smaller volume liquid waste. This technology was primarily used from the late 1980s to the early 2000s for soil remediation of metals, gasoline, fuel oils, and pesticides (U.S. EPA 1996; U.S. EPA 2017).

The applicability of soil washing to PFAS-containing soil/sediment involves taking large quantities of low-concentration PFAS-containing soil/sediment and concentrating the waste into a smaller volume of PFAS-containing liquid. Due to the primary concern of PFASs impacting drinking water, PFAS-relevant liquid treatment technologies are more commercially available and developing faster than solid treatment technologies. The concept of soil/sediment washing represents the industry state of the practice for PFOS and PFOA treatment, however its efficacy on PFASs such as cationic PFAA precursors, which may represent a significant mass of sorbed PFASs in AFFF impacted soils, has not been proven. Soil washing can be applied concentrate PFAS impacts via water treatment technologies, for energy-intensive destructive technologies (Ross, McDonough et al. 2018).

The soil/sediment washing process is more effective on soils that are primarily coarse grained because it can settle out of the soil/sediment wash bath faster than finer grained soil. Soil/sediment washing liquid may be recycled for further washing or treated for further PFAS removal using liquid treatment technologies. Fine solids that cannot settle from a soil/sediment wash bath within a reasonable time will require further management, and it may not be practical to soil wash PFAS-containing soil/sediment that is contains fine grained particles above 25 percent of the total mass.

3.9.2.5 High-Energy Electron Beam

A *high-energy electron beam (eBeam)* is a high-efficiency, flow-through, non-thermal, technology that uses electron accelerators to generate large numbers of highly energetic electrons from electricity (Cleland 2012; Pillai 2017). The technology has been used commercially for pasteurizing foods, sterilizing medical devices, cross-linking polymers, and eliminating insects and pests from fresh produce (Cleland 2012; Pillai 2016; Zembouai 2016; Pillai 2017). eBeam is applicable for use on solid and liquid matrices for many purposes (U.S. EPA 1997; Waite 1998; Praveen 2013; Briggs 2015). The amount of energy from eBeam that is absorbed by an irradiated material per unit mass is called dose. The absorbed dose during eBeam treatment depends on the type and thickness of the material, the beam power, and the length of time the material is exposed to the electron beam (Waite 1998).

During the irradiation of water, three primary reactive species are formed: solvated electrons and hydrogen radicals, which are strong reducing species; and hydroxyl radicals, which are strong oxidizing species. These species form as the water molecule disproportionates, creating both advanced reduction and oxidation processes without the addition of any chemicals. The absolute concentration of radicals formed during irradiation is dose and water-quality dependent, but it has been measured at greater than millimolar levels in potable, raw, and secondary wastewater effluent (Waite 1998).

The eBeam technology has been demonstrated to defluorinate PFOA/PFOS in aqueous samples at the bench-scale (Wang, Batchelor et al., 2016). During these trials, radical scavenging experiments indicated that the chemical reductants (solvated electrons and hydroxyl radical) were important in the eBeam degradation of PFOA/PFOS (Ma et al. 2017). Further evaluation of this technology for treating other PFASs (PFAA precursors and other long- and short-chain PFAAs) in soil/sediment, as well as testing over a range of concentrations, will be necessary to further understand treatment performance potential and to identify any deleterious by-products. Based on eBeam's applicability to heavy hydrocarbon-impacted soil (Briggs 2015), it is anticipated that eBeam will effectively treat PFAS-impacted soil/sediment; however, proof of concept testing is planned. The concept behind eBeam is for treatment of PFAS-containing soil/sediment *ex situ* treatment through the development of a modular skid that could be mobilized to specific sites or a large-scale system that would serve as a centralized treatment location.

3.9.2.6 Mechanochemical Destruction

MCD (or ball milling) is a destructive-based technology that uses 5 to 10 millimeter (mm) stainless-steel balls in conventional planetary ball mills. MCD refers to chemical surface reactions of transient temperature increase or the generation of triboplasmas (i.e., highly ionized neutral gas) (Heinicke et al. 1984) caused by a mechanical force. By facilitating collisions between the balls and contaminated soil/sediment through centripetal acceleration (at hundreds of revolutions per minute), the nondeformable steel balls deform the surface of the contaminated soil/sediment. The deformation at the surface of the soil/sediment results in a temperature increase. Additionally, conceptualized elastic and plastic deformations because of ball-to-ball and ball-to-soil/sediment collisions may facilitate the generation of the solvated electron, which facilitates chemical reduction (Environmental Decontamination Europe Ltd. 2018). The efficiency of the collisions may be optimized through the addition of silica sands. Presently, MCD is a technology that has been demonstrated at field scale for several organochlorine POPs but use on PFAS-impacted soil/sediment has not been proven and laboratory testing on impacted soil/sediment is required. Full-scale implementation for soil/sediment treatment for PFASs requires significantly more research.

The influence of co-milling agents such as potassium hydroxide (KOH), lime (CaO), silicon dioxide (SiO_2), and sodium hydroxide (NaOH) were evaluated to facilitate the generation of the hydroxyl radical (Zhang et al. 2013a). This work demonstrated greater than 90 percent reduction of PFOA/PFOS with an appreciable fluoride mass balance. Some of the defluorination was attributed to chemical destruction because of strong chemical oxidizing and reducing conditions. Given that PFSA's are very resistant to chemical oxidationand simultaneous propagation of advanced oxidizing and reducing species presents increased scavenging of radicals, it is possible that PFOA/PFOS destruction is via a thermal process. The viability of MCD has not been studied in detail with respect to the spectrum of compounds that PFASs comprise.

3.9.3 WATER TREATMENT

This section discusses a selection of water treatment technologies potentially relevant for the treatment of PFASs. Due to the ubiquity of PFAS detections in water throughout the environment (e.g., municipal waste, drinking water, stormwater, groundwater and surface water), PFAS treatment technologies will be discussed with respect to the water matrix in general. Where appropriate, caveats specific to a type of water will be specified. The relevant water treatment mechanisms for PFASs are adsorption, separation, and destruction. Some of these forms of treatment mechanisms are currently commercially available, while others are growing in their commercial application or are more experimental. Therefore, this section categorizes the available PFAS-relevant water treatment technologies into Mature, In Development, and Emerging technologies sections. Regardless of the treatment mechanism(s) selected for an application, the current state of the practice is a combination of water treatment technologies in a treatment train to complement one another to address the challenging physical and chemical properties of PFASs with the goal of concentrating the PFASs into a smaller volume for further destructive treatment. Additionally, combined technologies may be required to optimize PFAS treatment as common co-contaminants may inhibit PFAS treatment.

The mode of implementation of the water treatment technologies to be discussed are generally *ex situ* and involve water collection systems conveyed to a water treatment plant, which could potentially be enhanced using dynamic groundwater recirculation (DGR). Some *in situ* technologies will be discussed, such as using permeable reactive barriers (PRBs) as part of experimental attempts to intercept PFAS groundwater plumes and injected adsorbents, but the current state of the practice does not include a PFAS-relevant treatment technology that is holistic. Similarly to the soil/sediment section, this section will focus on the technology relevance to PFASs, and for more fundamental information on these modes of implementation, the reader is referred elsewhere (Suthersan 2017).

3.9.3.1 Mature Water Treatment Technologies

The water treatment technologies discussed in this section are currently commercially available and are being used at the practical water treatment scale. They include activated carbon, ion exchange, reverse osmosis and nanofiltration, and combustion-based incineration.

3.9.3.1.1 Activated Carbon (Granular, Powdered, Super-Fine Powdered)

Activated carbon (AC) is a common water treatment technology used to improve taste and odor, and to remove NOM and various synthetic organic chemicals. AC is a highly porous particulate material made from carbon-rich raw materials (e.g., wood, bamboo, coal, coconut shell). Its high specific surface area allows for effective removal of contaminants from water through both physical and chemical adsorption mechanisms. Adsorptive removal is described by sorption isotherms that show the distribution of the contaminant adsorbed (q_e; e.g., mass of contaminant per mass of AC) versus the liquid (C_e; e.g., mass contaminant per volume of liquid treated) at equilibrium. Empirically fitting the experimental isotherm data, typically with a Langmuir or Freundlich model, enables the determination of the AC adsorption capacity (q_m) under the experimental conditions. Adsorption kinetic studies supplement isotherm data to provide a baseline affinity of AC for a specific contaminant. Additionally, column studies or pilot testing predict a "breakthrough time," which is operationally defined as the volume of impacted water treated before the effluent concentration reaches a specified percentage of the influent concentration (Eddy 2003). Breakthrough indicates that the adsorption capacity of the AC has been met by contaminants (and/or co-contaminants) occupying all available adsorption sites. Treatment efficacy will continue to decrease following breakthrough unless the AC is reactivated/replaced. Breakthrough is commonly quantified in terms of "bed volumes (BV)," which is equivalent to the total treatment volume divided by the volume of the vessel containing the AC.

The granular form of activated carbon (GAC) is currently widely used for the removal of PFOA and PFOS and, to a lesser extent, other PFAAs, from water. The efficacy of PFAA treatment using GAC is well documented in the literature with results from bench-scale studies to full-scale application available to inform water treatment (Ochoa-Herrera and Sierra-Alvarez 2008; Yu et al. 2009; Vecitis et al. 2009; Senevirathna et al. 2010a; Senevirathna et al. 2010b; Ahrens, Taniyasu, et al., 2010; Takagi 2011; Eschauzier, Beerendonk et al., 2012; Appleman et al. 2013; Flores et al. 2013; Appleman et al. 2014; Cummings 2015; Higgins and Dickenson 2016; Merino et al. 2016; Nowack 2017; McCleaf et al. 2017; Inyang and Dickenson 2017; Ross, McDonough et al. 2018; Szabo 2017; McNamara et al. 2018). Currently, GAC is the treatment technology most utilities install to immediately address PFASs in drinking water (Higgins 2017) and it has become the basis of comparison for all new adsorbent technologies targeting PFAS removal (Ross, McDonough et al. 2018). Isotherm parameters for various types of GAC are summarized in Table 3.19, and may be used as a guide for choosing the most appropriate GAC for desired PFAA treatment. However, application-specific pre-design testing is necessary to inform the most efficient GAC selection (Nowack 2017).

GAC removal efficiency varies with PFAS characteristics (i.e., chain length and functional group), water chemistry (e.g., NOM, co-contaminants), and GAC properties (e.g., GAC type, specific surface area, pore size, and surface chemistry). Specifically, the adsorption capacity and breakthrough times decrease as chain length decreases and linear chain PFAAs have greater removal efficiencies compared to branched chain PFAAs (Hansen 2010; Zhao et al. 2011; Eschauzier, Beerendonk et al., 2012; Appleman et al. 2014; McCleaf et al. 2017). Desorption of short-chain PFAA compounds (e.g., PFBA, PFHxA, PFPeA) and increasing long-chain PFAA removal was observed at the end of GAC column experiments close to 50,000 bed volumes (McCleaf et al. 2017). This observation exacerbates the concern of holistic PFAS removal using GAC as the sustained influent concentration of long-chain PFAA demonstrate "rollover," or desorption of previously sorbed contaminants, due to displacing the previously adsorbed short-chain PFAAs. There are currently no published studies on the effectiveness of GAC in removing PFAA precursors; however, a recent theoretical study predicted that newly discovered PFAA precursors from an AFFF-impacted drinking water supply would break through GAC systems before both PFOA and PFOS (Xiao, Ulrich et al., 2017).

TABLE 3.19
Isotherm parameters for various types of GAC

Activated Carbon	Adsorption Capacity (mg/g)	Langmuir $C_{s,max}$ (mg/g)	Langmuir K_L (L/mg)	Freundlich K_f	Freundlich n	Reference
PFBA						
Calgon Filtrasorb 400 (F400)	–	–	–	130 $(\mu g/g)(\mu g/L)^{-n}$	0.56	Qui 2007
Calgon Filtrasorb 600 (F600)	–	15.3; 6.60	340.1; 315.7	–	–	Zhao et al. 2011*
PFBS						
Calgon Filtrasorb 400 (F400)	98.70	98.7	0.034	9.3 $(mg/g)(mg/L)^{-n}$	0.463	Ochoa-Herrera and Sierra-Alvarez 2008
Calgon Filtrasorb 400 (F400)	–	–	–	240 $(\mu g/g)(\mu g/L)^{-n}$	0.55	Qui 2007
Calgon Filtrasorb 600 (F600)	–	40.0; 32.3	134.2; 80.2	–	–	Zhao et al. 2011*
PFPeA						
Calgon Filtrasorb 400 (F400)	–	–	–	290 $(\mu g/g)(\mu g/L)^{-n}$	0.85	Qui 2007
PFHxA						
Calgon Filtrasorb 400 (F400)	–	–	–	490 $(\mu g/g)(\mu g/L)^{-n}$	0.81	Qui 2007
PFHxS						
Calgon Filtrasorb 400 (F400)	–	–	–	2,970 $(\mu g/g)(\mu g/L)^{-n}$	0.94	Qui 2007
PFHpA						
Calgon Filtrasorb 400 (F400)	–	–	–	300 $(\mu g/g)(ug/L)^{-n}$	0.93	Qui 2007
PFOA						
Calgon Filtrasorb 400 (F400)	112.1	112.1	0.038	11.8 $(mg/g)(mg/L)^{-n}$	0.443	Ochoa-Herrera and Sierra-Alvarez 2008
Calgon Filtrasorb 400 (F400)	–	–	–	680 $(\mu g/g)(\mu g/L)^{-n}$	1.10	Qui 2007
Calgon Filtrasorb 400 (F400)	–	–	–	3.06 $(\mu g/mg)(\mu g/L)^{-n}$	1.98	Zhi and Liu 2016

Activated Carbon	Adsorption Capacity (mg/g)	Langmuir $C_{s,max}$ (mg/g)	Langmuir K_L (L/mg)	Freundlich K_f	Freundlich n	Reference
Calgon Filtrasorb 600 (F600)	—	76.1; 63.6	62.7	—	—	Zhao et al. 2011*
Activated Carbon Fiber (ACF20)	—	—	—	0.51 (μg/mg)(μg/L)$^{-n}$	3.86	Zhi and Liu 2016
Pengcheng Activated Carbon Co. GAC	—	161	0.043	0.47 mol/g)(mmol/L)$^{-n}$	3.6	Yu et al. 2009
MeadWestvaco Co. BioNuchar (wood-based)	—	—	—	0.18 (μg/mg)(μg/L)$^{-n}$	0.71	Zhi and Liu 2016
MeadWestvaco Co. WVB 14x35 (wood-based)	—	—	—	0.14 (μg/mg)(μg/L)$^{-n}$	0.75	Zhi and Liu 2016
Siemens Inc. AquaCarb 1240C (coconut-shell based)	—	—	—	1.92 (μg/mg)(μg/L)$^{-n}$	1.13	Zhi and Liu 2016
Bamboo-derived GAC	476	426	0.000379	1.34 (mmol/g)(mmol/L)$^{-n}$	5.28	Deng et al. 2015
Xinhua Activated Carbon Co. PAC	—	16.502	0.606	24.238 (mg/g)(mg/L)$^{-n}$	0.450	Qu et al. 2009
Fluka 72343 PAC	—	426	0.05023	1.35 (mmol/g)(mmol/L)$^{-n}$	2.10	Rattanaoudom et al. 2012
Pengcheng Activated Carbon Co. PAC	—	277	0.00014	0.83 (mmol/g)(mmol/L)$^{-n}$	5.0	Yu et al. 2009

PFOS

Activated Carbon	Adsorption Capacity (mg/g)	Langmuir $C_{s,max}$ (mg/g)	Langmuir K_L (L/mg)	Freundlich K_f	Freundlich n	Reference
Calgon Filtrasorb 300 (F300)	196.2	196.2	0.068	38.5 (mg/g)(mg/L)$^{-n}$	0.332	Ochoa-Herrera and Sierra-Alvarez 2008
Calgon Filtrasorb 400 (F400)	—	−0.245	−29.71	25.9 (mg/g)(mg/L)$^{-n}$	1.123	Ochoa-Herrera and Sierra-Alvarez 2008
	236.4	236.4	0.124	60.9 (mg/g)(mg/L)$^{-n}$	0.289	Ochoa-Herrera and Sierra-Alvarez 2008
	—	—	—	660 (μg/g)(μg/L)$^{-n}$	1.41	Qui 2007
	—	—	—	28.4 (μg/g)(μg/L)$^{-n}$	2.20	Senevirathna et al. 2010
	—	—	—	1.11 (μg/mg)(μg/L)$^{-n}$	1.17	Zhi and Liu 2016

(*Continued*)

TABLE 3.19 (Continued)

Activated Carbon	Adsorption Capacity (mg/g)	Langmuir $C_{s,max}$ (mg/g)	Langmuir K_L (L/mg)	Freundlich K_f	Freundlich n	Reference
Calgon Filtrasorb 600 (F600)	—	81.3	42.9	—	—	Zhao et al. 2011*
	—	66.7	112.8	—	—	Zhao et al. 2011*
Calgon URV-MOD 1	211.6	211.6	0.080	36.7 $(mg/g)(mg/L)^{-n}$	0.371	Ochoa-Herrera and Sierra-Alvarez 2008
Activated Carbon Fiber (ACF20)	—	—	—	3.54 $(\mu g/mg)(\mu g/L)^{-n}$	0.81	Zhi and Liu 2016
Pengcheng Activated Carbon Co. GAC	—	185	0.078	0.43 $(mmol/g)(mmol/L)^{-n}$	5.5	Yu et al. 2009
MeadWestvaco Co. BioNuchar (wood-based)	—	—	—	0.97 $(\mu g/mg)(\mu g/L)^{-n}$	0.59	Zhi and Liu 2016
MeadWestvaco Co. WVB 14x35 (wood-based)	—	—	—	1.27 $(\mu g/mg)(\mu g/L)^{-n}$	0.54	Zhi and Liu 2016
Siemens Inc. AquaCarb 1240C (coconut-shell based)	—	—	—	1.72 $(\mu g/mg)(\mu g/L)^{-n}$	0.61	Zhi and Liu 2016
Bamboo-derived GAC	1160	1,100	0.000104	3.20 $(mmol/g)(mmol/L)^{-n}$	3.41	Deng et al. 2015
Norit 1240W PAC	46.49	—	—	—	—	Kinshott 2008
Fluka 72343 PAC	—	440	0.06218	1.19 $(mmol/g)(mmol/L)^{-n}$	2.14	Rattanaoudom et al. 2012
Pengcheng Activated Carbon Co. PAC	—	520	0.00011	1.27 $(mmol/g)(mmol/L)^{-n}$	5.5	Yu et al. 2009
PFNA						
Calgon Filtrasorb 400 (F400)	—	—	—	1,550 $(\mu g/g)(\mu g/L)^{-n}$	0.97	Qui 2007
PFDA						
Calgon Filtrasorb 400 (F400)	—	—	—	1,940 $(\mu g/g)(\mu g/L)^{-n}$	0.92	Qui 2007
PFHxDA						
Calgon Filtrasorb 400 (F400)	—	—	—	2,640 $(\mu g/g)(\mu g/L)^{-n}$	1.81	Qui 2007

Notes:
* Note that isotherm values were derived from the Brunauer–Emmett–Teller isotherm equation, which is an extension of the Langmuir isotherm.

High concentrations of NOM/co-contaminants may adversely influence PFAA adsorption through competition for adsorption sites (Higgins and Luthy 2006a; Ahrens, Taniyasu, et al., 2010; Zhao et al. 2011; Appleman et al. 2013; Hong et al. 2013; Pramanik et al. 2015; McCleaf et al. 2017). Due to the competition for adsorption sites between PFAA and NOM, a greater percentage of mesoporosity (i.e., slightly larger porosity) has been observed to be more favorable for GAC-based PFAA removal (Nassi et al. 2014; Nowack 2017). An example of this is the enhanced coconut shell carbon, which has been optimized for PFAA removal by overactivation to reduce the microporosity and increase the mesoporosity. Another form of GAC optimization to combat NOM competing with PFAAs for adsorption sites is pH adjustment. The organic acids that comprise NOM typically have pKa values on the order of four to five compared to PFAAs which are typically significantly less. This means that increasing the influent pH to circumneutral will dissociate organic acids, making them less hydrophobic, decreasing their likelihood to adsorb onto the GAC, and decreasing the competition with PFAAs. The type of GAC selected can also influence the efficiency and economics of a GAC water treatment system for PFASs. Wood- and bamboo-based GAC have been shown to outperform coal-based material with adsorption decreasing at both acidic and alkaline pH (Deng et al. 2015; Zhi and Liu 2015). While somewhat intuitive, a less dense GAC that achieves comparable PFAS removal as a denser form of GAC is advantageous from a cost perspective as GAC is purchased on a per-pound basis (such as sub-bituminous coal-based versus bituminous coal-based GAC).

A significant advantage of GAC is the capability to destroy PFAAs during the reactivation process. GAC can be reactivated under extreme thermal conditions have been reported to destroy adsorbed PFAAs, disintegrate approximately 15 percent of the GAC, and return a product that may be capable of improved PFAS removal. Complete destruction of PFOS requires a temperature range of approximately 1000° to 1,200°C (Schultz 2003; Yamada et al. 2003), and thermal reactivation kilns normally include afterburners for air pollution control that operate at temperatures above 1,100°C. Therefore, a typical thermal reactivation process (800°C to 1,000°C reactivation temperature, plus an afterburner) may be suited for PFOS destruction during GAC reactivation, which is required to reuse GAC in a treatment system.

The implementation of GAC for water treatment is feasible as GAC vessels are available in a variety of vessel sizes and conservative costs for GAC range from US$1 to US$2 per pound (lb). GAC can be mobilized for small-scale or large-scale applications, with vessels designed to handle a range of flow rates to avoid channelizing and inefficient removal. Common empty-bed contact times or residence times (i.e., the amount of time the liquid needs to be in contact with the GAC) ranges from 7 to 15 minutes. Although GAC is being implemented throughout the industry as an immediate response to concerns over impacts from PFASs wo water, the current drawbacks associated with GAC are the ineffective removal of short-chain PFAAs, the moderate efficiency of PFOA and PFOS removal, and the uncertainties regarding PFAA precursor removal. With the potential for regulations to expand to include a broader range of PFASs (including PFAA precursors) (Kotthoff and Bücking 2018), other forms of water treatment that complement the drawbacks of GAC will be necessary for complete and comprehensive PFAS removal.

GAC has a larger mean particle size (approximately 0.3 to 2.4 millimeters [mm]) than powdered activated carbon (PAC; approximately 10s to 100s microns [μm]) (Potwara 2012), and therefore PAC has a greater external surface area which can greatly increase its adsorptive capacity. The increased external surface area of PAC is anticipated to greatly improve the overall adsorption capacity; however, considerably longer residence times are necessary to fully utilize the capacity of the PAC. At the municipal water treatment scale of millions of gallons per day (MGD), the corresponding tank size to accommodate a residence time of greater than 30 minutes is impractical. A potential optimization is the generation of super-fine PAC (SPAC), which has demonstrated an ability to greatly increase adsorption kinetics and reduce the required residence time of PAC (Weber et al. 1983; Matsui et al. 2007, Matsui et al. 2012; Matsui et al. 2013; Bakkaloglu 2014). SPAC is a micro-ground PAC, with a mean particle size of less than 1 μm. Numerous literature studies examine the propensity of SPAC to adsorb NOM, with varying hypotheses as to why SPAC adsorptive kinetics and capacity outperforms PAC, such as the increased external surface area or the increased macro- and mesoporosity. There are conflicting reports that PAC outperforms GAC for PFAA removal (Qu et al. 2009; Rattanaoudom et al. 2012; Zhi and Liu 2015; Sun et al. 2016); however, PAC cannot be reactivated like GAC and is typically a single-use form of activated carbon. There is a data gap in the literature as to whether the higher throughput performance of PAC/SPAC (approximately 200 to 2,000 times greater sorption capacity) offsets the increased unit cost and need for disposal. PAC/SPAC application for PFAS removal is being developed at the laboratory scale, but SPAC has been implemented at the commercial scale with capacities in the MGD range. Another limitation of PAC/SPAC is that it is deployed as a slurry and separated from treated liquid by using a filtration unit.

3.9.3.1.2 Anion-Exchange Resins

Anion exchange (AIX) resins contain charged functional groups bound to a polyacrylic or polystyrene bead. Contaminant removal is usually accomplished by exchanging the counter ion exploiting electrostatic adsorption. AIX resins are an established technology for many common contaminants in both municipal water and groundwater treatment. Sulfate, chromate, nitrate, chloride, radionuclides (e.g., rhenium and technetium), and perchlorate are removed using resins that have been engineered to enhance their polyatomic ion selectivity. High concentrations of total dissolved solids (TDS) exert a considerable ionic strength influence, which may confound electrostatic adsorption onto the resins or greatly diminish the efficiency. High concentration of common anions such as chloride, nitrate, and sulfate may also affect performance.

Research on non-ion/AIX resins for PFASs has primarily focused on the removal of anionic PFAAs (Yu et al. 2009; Senevirathna et al. 2010a; Senevirathna et al. 2010b; Deng et al. 2010; Zaggia et al. 2016). Removal of cationic and zwitteronic PFAA precursors using resin technology has not been investigated or widely demonstrated. PFAA precursors are unlikely to be appreciably removed by AIX resins, and resins combined with another technology may be necessary if PFAA precursor removal is required. Available isotherm constants and calculated adsorption capacity values based on published literature are summarized in Table 3.20. There is considerable evidence in the literature that AIX resins outperform non-ion exchange resins, and therefore the focus of the discussion here is on AIX (Carter and Farrell 2010; Senevirathna et al. 2010b; Du et al. 2014; Chularueangaksorn et al. 2013).

TABLE 3.20
Freundlich isotherm constants for resins

Resin Name	Type and Functional Group	Concentration of PFAS (mg/L, unless noted)	pH (s.u.)	Freundlich coefficient [(µg/g) (µg/L)$^{(1/n)}$, unless noted]	1/n	Adsorption Capacity (µg/g, unless noted)	Reference
			PFBA				
Purolite A520E	(ion) TEQA	1,000	7.5	—	—	29.5 mg/g	Zaggia et al. 2016
Purolite A600E	(ion) TMQA	1,000	7.5	—	—	19.1 mg/g	Zaggia et al. 2016
Purolite A532E	(ion) BQA	1,000	7.5	—	—	52.3 mg/g	Zaggia et al. 2016
*Purolite A520E	(ion) TEQA	Drinking water (see note)	7.6	—	—	1.5 µg/g	Zaggia et al. 2016
*Purolite A600E	(ion) TMQA	Drinking water (see note)	7.6	—	—	0.6 µg/g	Zaggia et al. 2016
*Purolite A532E	(ion) BQA	Drinking water (see note)	7.6	—	—	3.3 µg/g	Zaggia et al. 2016
			PFBS				
Purolite A520E	(ion) TEQA	1,000	7.5	—	—	53.8 mg/g	Zaggia et al. 2016
Purolite A600E	(ion) TMQA	1,000	7.5	—	—	34.6 mg/g	Zaggia et al. 2016
Purolite A532E	(ion) BQA	1,000	7.5	—	—	109.2 mg/g	Zaggia et al. 2016
*Purolite A520E	(ion) TEQA	Drinking water (see note)	7.6	—	—	4.3 µg/g	Zaggia et al. 2016
*Purolite A600E	(ion) TMQA	Drinking water (see note)	7.6	—	—	2.8 µg/g	Zaggia et al. 2016
*Purolite A532E	(ion) BQA	Drinking water (see note)	7.6	—	—	5.9 µg/g	Zaggia et al. 2016

(Continued)

TABLE 3.20 (Continued)

Resin Name	Type and Functional Group	Concentration of PFAS (mg/L, unless noted)	pH (s.u.)	Freundlich coefficient [(µg/g) (µg/L)$^{(1/n)}$, unless noted]	1/n	Adsorption Capacity (µg/g, unless noted)	Reference
			PFOA				
Purolite A520E	(ion) TEQA	1,000	7.5	—	—	134.7 mg/g	Zaggia et al. 2016
Purolite A600E	(ion) TMQA	1,000	7.5	—	—	125.2 mg/g	Zaggia et al. 2016
Purolite A532E	(ion) BQA	1,000	7.5	—	—	142.1 mg/g	Zaggia et al. 2016
*Purolite A520E	(ion) TEQA	Drinking water (see note)	7.6	—	—	37.2 µg/g	Zaggia et al. 2016
*Purolite A600E	(ion) TMQA	Drinking water (see note)	7.6	—	—	28.5 µg/g	Zaggia et al. 2016
*Purolite A532E	(ion) BQA	Drinking water (see note)	7.6	—	—	45.1 µg/g	Zaggia et al. 2016
Amberlite A1400	(ion) PSDVB	20–250	3 and 7	3.35 (mmol$^{(1-1/n)}$ L$^{1/n}$ g^{-1})	0.13	3.39 and 1.81 mmol/g	Yu et al. 2009
US Filter A-714	(ion) None provided	4,320	2.5	—	—	686 mg/g	Lampert et al. 2007
US Filter A-399	(ion) None provided	4,320	2.5	—	—	—	Lampert et al. 2007
Amberlite XAD-7HP	(non) AAP	160–3,100 ng/L	7.8	1,831	1.0	—	Xiao et al. 2012
			PFOS				
Dow L493	(non) SDVB	100	6.4	54.6	1.2	7.93	Senevirathna et al. 2010
Amb XAD 4	(non) MCAP	100	6.4	79.1	0.6	1.95	Senevirathna et al. 2010
Dow V493	(non) SDVB	100	6.4	81.3	1.1	9.29	Senevirathna et al. 2010
Amb IRA-400	(ion) SDVB	100	6.4	95.9	0.5	2.01	Senevirathna et al. 2010
Dow MarathonA	(ion) SDVB	100	6.4	28.4	0.6	0.18	Senevirathna et al. 2010
Purolite A520E	(ion) TEQA	1,000	7.5	—	—	210.4 mg/g	Zaggia et al. 2016
Purolite A600E	(ion) TMQA	1,000	7.5	—	—	186.2 mg/g	Zaggia et al. 2016
Purolite A532E	(ion) BQA	1,000	7.5	—	—	260.5 mg/g	Zaggia et al. 2016

Resin Name	Type and Functional Group	Concentration of PFAS (mg/L, unless noted)	pH (s.u.)	Freundlich coefficient [(μg/g) (μg/L)$^{(1/n)}$, unless noted]	1/n	Adsorption Capacity (μg/g, unless noted)	Reference
*Purolite A520E	(ion) TEQA	Drinking water (see note)	7.6	—	—	3.7 μg/g	Zaggia et al. 2016
*Purolite A600E	(ion) TMQA	Drinking water (see note)	7.6	—	—	2.2 μg/g	Zaggia et al. 2016
*Purolite A532E	(ion) BQA	Drinking water (see note)	7.6	—	—	4.7 μg/g	Zaggia et al. 2016
Amberlite AI400	(ion) PSDVB	20–250	3 and 7	0.52 (mmol$^{(1-1/n)}$ L$^{1/n}$ g^{-1})	0.17	0.36 and 0.34 mmol/g	Yu et al. 2009
US Filter A-714	(ion) None provided	950	2.5	—	—	151 mg/g	Lampert et al. 2007
IRA67	(ion) Polyacrylic resin	400	3	—	0.072	5.70 mmol/g	Deng et al. 2010
IRA958	(ion) Polystyrene resin	400	3	—	0.076	4.86 mmol/g	Deng et al. 2010
Amberlite XAD-7HP	(non) AAP	80–1,800 ng/L	7.8	3,095	1.0	—	Xiao et al. 2012
Amberlite XAD-2	(non) None provided	100–800 ng/L	7.8	440	1.7	—	Xiao et al. 2012

Notes:

* Continuous flow pilot scale; all others are lab-based tests
(ion) indicates an ion-exchange polymer
(non) indicates a non-ion-exchange polymer
Drinking water sample contained PFOA = 430 ng/L, PFBA = 212 ng/L, PFBS = 171 ng/L, PFOS = 27 ng/L
BQA—Bifunctional quaternary amine
PSDVB—Polystyrene divinylbenzene
SDVB—Styrene divinylbenzene
TEQA—Triethyl quaternary amine
TMQA—Trimethyl quaternary amine
AAP—Aliphatic acrylic polymer
MCAP—Macroreticular cross-linked aromatic polymer
SDVB—Styrene divinylbenzene

Several AIX resins with a range of functional groups that enable different types of selectivity have been assessed for the removal of a small number of PFAAs from water using electrostatic adsorption (Du et al. 2014), although other interactions such as hydrophobic interactions within the AIX resins due to agglomerated PFAAs can also be a significant removal mechanism (Zaggia et al. 2016). Theoretically AIX resins that predominantly rely on electrostatic interactions would require regeneration with a brine to displace the PFAA-electrostatic adsorption whereas AIX resins that promote hydrophobic agglomeration between resins may require a methanol extraction for regeneration. There are commercial applications of AIX resin onsite regeneration with solvent/brine extractants (ECT2 2016). While onsite/offsite regeneration of spent resins can be achieved, current pricing and health and safety issues support single-use AIX resins with offsite incineration. Typical bed volume throughput projections for AIX resins greatly exceed those of GAC, and managing an organic solvent (e.g., methanol) onsite may be difficult and represents a health and safety concern. The use of ethanol as an alternative for regeneration processes has been successfully demonstrated (Punyapalakul et al. 2013).

AIX removal efficiency of PFAAs has been demonstrated to depend on the functional group (sulfonates are more effectively removed than carboxylates), the chain length (long-chain PFAAs are more effectively removed than short-chain PFAAs), isomer type (linear chain PFAAs are more effectively removed than branched chain PFAAs), and the geochemistry of the water (high TDS adversely influences PFAA removal) (Appleman et al. 2013; Higgins and Dickenson 2016; McCleaf et al. 2017). Many co-contaminants may be present in waste streams and are present at concentrations that are orders of magnitude greater than PFAAs, resulting in significant competition for adsorption sites.

The implementation of AIX for water treatment is feasible as AIX vessels are available in a variety of vessel sizes and conservative costs for AIX resin range from $300 to $400 per cubic foot. AIX vessels are typically smaller than GAC vessels with EBCT for resins of approximately three to five minutes, and data in the literature suggest higher treated throughput per AIX bed volume. AIX resins can be mobilized for small-scale or large-scale applications. AIX is increasingly being implemented throughout the industry due to an increased regulatory awareness of the number of PFAAs (including short-chain PFAAs) which can be assessed and removed from water. The current drawbacks associated with AIX resins are the significant influence of TDS, the potential to create a radioactive waste if the treated water has radionuclides (similar concern for any adsorbent treating groundwater and using electrostatic adsorption), and the unknown performance for PFAA precursor removal.

3.9.3.1.3 Reverse Osmosis/Nanofiltration

Reverse osmosis (RO) and nanofiltration (NF) are physical removal processes that reject aqueous phase contaminants by hindered diffusion through semipermeable (i.e., not microporous or nanoporous) membranes. Contaminant rejection is influenced by characteristic properties of individual compounds, such as electrostatic charge, size, shape, and mass. Rejection is also influenced by site- and system-specific factors, such as feed water temperature and operating flux. Warmer water reduces contaminant rejection somewhat via both thermal expansion of the membrane material and increased diffusive transport of solutes across the membrane barrier. By contrast, operating an RO/NF system at higher flux yields more throughput of water across the membrane barrier without affecting diffusive transport of contaminants, thus increasing rejection. The occurrence of inorganic scaling can also affect rejection, as the higher concentration at the membrane surface can create a localized osmotic gradient that induces increased solute passage—an effect known as "concentration polarization".

RO/NF systems may have comparatively high unit costs of treated water production due largely to energy demand. Pretreatment requirements can also increase RO/NF costs, as the use of an antiscalant chemicals and/or acid may be necessary to control scaling, depending on the feed water quality. Likewise, more turbid feed water may require the use of membrane filtration (e.g., microfiltration or ultrafiltration) to protect the RO/NF membranes from particulate fouling, and elevated iron and manganese concentrations may necessitate additional pretreatment, such as greensand filters.

Many studies throughout the literature demonstrate that RO/NF is one of the most effective water treatment technologies for removal of PFASs (Tang et al. 2006; Quinones and Snyder 2009; Thompson et al. 2011; Appleman et al. 2013; Flores et al. 2013; Higgins and Dickenson 2016; Steinle-Darling 2008). The PFAS removal effectiveness has been observed to be chain length and functional group insensitive, which implies that RO/NF is expected to be effective at removing the broad class of PFAS compounds (Higgins and Dickenson 2016). Additionally, concentration polarization is not expected to influence the rejection of PFAS compounds, which are organic, and typically present a low concentration (ppt to ppb); they are highly soluble and thus do not scale.

Concentrate management is also an important consideration for RO/NF treatment. The concentrate flow is typically 15 to 25 percent of the influent, although percentages can vary widely with case-specific water quality. Depending on the available management options, additional treatment of the concentrate may be necessary prior to disposal or repurposing for beneficial use.

3.9.3.1.4 Incineration

Incineration is considered a mature technology as part of a treatment train process for water treatment systems that use adsorption-based or separation-based technologies to consolidate PFASs into a concentrated waste stream. The concentrated waste streams are contained, transported offsite, and processed at an appropriate incineration facility to destroy PFASs. The details associated with PFAS-specific incineration are discussed in section 3.9.2.

The incineration of liquid represents a relatively energy inefficient and nonsustainable disposal technique. The amount of energy to incinerate 1 gallon of PFAS-containing liquid (assuming water as the base solvent) is approximately 260 kW-hr, which is equivalent to ~US$30 per hour of operation just in energy costs using an energy rate of US$0.11 per kW-hr. For perspective, in 2016 the average monthly energy consumption for a U.S. residence was approximately 897 kW-hr, implying that approximately 30 percent of the monthly energy usage of an American family is necessary to incinerate 1 gallon of liquid PFAS-containing waste (U.S. Energy Information Administration 2018). From a cost perspective, incineration is typically the benchmark against which other forms of PFAS destruction are measured. Another challenge with the incineration of a liquid PFAS waste is the phase change associated with heating a liquid. To manage the phase change from liquid to water vapor to superheated steam, the rate at which liquid PFAS waste can be incinerated is typically limited to small incremental volumes. As such, incineration is a high-cost treatment approach and due to the inherent energy requirements associated with the incineration of liquid waste streams, incineration is best suited for high-concentration waste streams.

3.9.3.2 Developing Treatment Technologies

The PFAS-relevant water treatment technologies discussed in this section may be currently commercially available but require more field-verification for practical water treatment scale application. They include organically modified silica adsorbents, ozofractionation, flocculation and coagulation.

3.9.3.2.1 Organically Modified Silica Adsorbents

The viability of mesoporous organosilica (MPOS) to adsorb organic compounds has been investigated for the past 15 years and is documented throughout the literature (Edmiston and Underwood 2009; Yang 2014; Edmiston 2014). The MPOS adsorbent is a silica-based polymeric structure and consists of crosslinked alkoxysilanes (Edmiston 2013). MPOS adsorbents generally have a greater percentage of mesoporosity than other commercially available adsorbents, which is expected to be more favorable for adsorption of PFASs in the presence of NOM.

MPOS is a developing adsorbent that targets a purely hydrophobic adsorption mechanism to remove PFASs from water, and it has shown the capability to adsorb the perfluoroalkyl moiety common to all PFASs (Edmiston 2017). The hypothesis of MPOS adsorbents is that the PFASs adsorb strongly via purely hydrophobic interactions related to the fluorinated carbon chain. This would imply broadly applicable PFAS removal. A feature of this adsorbent is that surface hydrophobicity and pore size may be customizable, mitigating competitive adsorption between PFAA and co-contaminants (Edmiston 2017). As MPOS adsorbents target purely hydrophobic adsorption, it is plausible that regeneration of the adsorbent could be achieved with a solvent rinse, but there are operational concerns of onsite management of a PFAS-enriched extract that needs further treatment. Optimizing throughput MPOS may make single use with incineration more economical and safe than regeneration.

MPOS adsorbents are commercially available in two forms: (1) as a bulk adsorbent, and (2) deposited as a film onto sand filtration media and porous silica media. The bulk adsorbent is described in the literature as "swellable" because the flexible pore structure expands when placed in solvents and is more appropriate for higher influent concentrations of PFASs. The MPOS fixed on media does not swell and is more appropriate for lower influent concentrations of PFASs. The current indication of cost for the MPOS adsorbent is approximately US$4 to US$5 per pound, and the material can be implemented in variable vessel sizes to accommodate small-scale and large-scale applications.

3.9.3.2.2 Ozofractionation

Ozofractionation (OZF) exploits the amphiphilic nature of many PFASs. While the functional group is hydrophilic, the perfluoroalkyl chain is hydrophobic, which makes the gas-liquid interface of a bubble ideal for agglomeration of PFASs. Bubbling ozone gas through tanks as PFAS-containing wastewater passes through creates a PFAS-enriched foam that can be collected. The waste stream generated by OZF can include sediment/solids, the PFAS-enriched foam concentrate (potentially with fine-grained solids) and spent adsorbent media or filtered rejectate (if necessary). Therefore, the OZF technology is a separation technology and represents the treatment train concept.

OZF is a patented process available commercially as Ozofractionative Catalyzed Reagent Addition (OCRA) (Dickson 2013; Dickson 2014). The OCRA system includes a series of water tanks in which micron-sized (less than 200 micrometers) ozone gas bubbles are passed through PFAS-impacted wastewater to facilitate gas-liquid interface partitioning in foam, which is collected for further treatment. Smaller bubble size is intentional as it maximizes the surface area for collecting PFASs. The commonly associated co-contaminants (i.e., NOM and petroleum hydrocarbons) can be oxidized by the ozone during the PFAS separation process.

OZF is a separation technology, meaning it generates a PFAS-enriched waste stream that needs further treatment. Treated wastewater with a high concentration of total suspended solids will require some form of sediment management, including leach testing to evaluate PFASs potentially adsorbed to the sediment (e.g., the oxidation of iron may co-precipitate PFASs). A general estimate of the concentrated waste generated is approximately 1 to 5 percent of the influent flow, however, reconcentration of waste has recently been demonstrated. This performance contrasted with RO/NF, which, depending on the background water quality, may exhibit concentrate of up to 30 percent, brings relevance to OZF suggesting it may pair well with more energy intensive, destruction-based technologies.

3.9.3.2.3 Flocculation and Coagulation

Commercial products are available to attempt to coagulate and sediment PFASs from wastewater. One form of a coagulant that is available commercially has demonstrated 40 percent to greater than 95 percent removal of PFAAs (short-chain to long-chain; Buhl 2017). This form of developing PFAS water treatment is considered a separation technology as the concentrated precipitate requires further management. The process includes the addition of a reagent that creates PFAS-enriched precipitate (through both hydrophobic and electrostatic interactions), which is subsequently removed as a sludge and concentrated via filtration and disposed of (Buhl 2017). This form of pretreatment may be combined with a conventional adsorption-based technology to prolong the operational life of the adsorbent. Typical doses of the coagulant range from 25 mg/L to 2 grams per liter. This form of treatment is designed to prolong the operational life of conventional adsorbents as (to date) the process cannot achieve low ppt concentrations alone.

PFAS separation can also be achieved using electrocoagulation (Lin et al. 2015). Electrocoagulation relies on an electric current to facilitate metal hydroxide precipitation, which has been demonstrated to co-precipitate PFAS via the electrostatic adsorption mechanism and potential inclusion. Electrocoagulation has resulted in 71 to 77 percent reductions in fluorinated surfactant concentrations (Baudequin et al. 2011).

3.9.3.2.4 Phytoremediation

The applicability of phytoremediation to manage the mass flux of PFASs in an aquifer has been suggested in some recent studies. Short-chain PFAAs have been observed to concentrate in fruit and long-chain PFAAs have been observed to concentrate in roots and shoots (Blaine et al. 2013; Blaine et al. 2014a; Blaine et al. 2014b). Additionally, an evaluation of a variety of trees and plant species to concentrate 26 PFASs was conducted (Gobelius et al. 2017), and attempted phytoremediation to address PFAAs using an established wetland has been reported (Plumlee et al. 2008). In general, the bioconcentration factors of PFASs appear to be significantly less than those of historically studied metals (Jaffré et al. 1979), with no volatilization or mineralization mechanism once PFASs are concentrated within the vegetation. For PFASs, the reported total whole tree burden of the sum of 26 PFASs, per tree, was up to 11 mg for birch and 1.8 mg for spruce (Gobelius et al., 2017). Therefore, phytoremediation seems unlikely to be applicable, as is very inefficient passive management of shallow groundwater PFAS impacts there would be a need for perpetual maintenance to harvest and incinerate the vegetation once it has reached PFAS saturation. Site-specific validation of PFAS uptake is recommended as well as a stidy of the effect of the local ecology on the PFASs impacting the trees and upgradient and downgradient groundwater monitoring to inform scheduled O&M.

3.9.3.3 Experimental Treatment Technologies

The PFAS-relevant water treatment technologies discussed in this section are generally not currently commercially available and require more laboratory-scale and field-verification for practical water treatment scale application. They include injectable particulate carbon, experimental adsorbents, advanced oxidation processes, advanced reduction processes, electrochemical treatment, sonolysis, and photolysis. Most of the experimental treatment technologies are destruction based.

3.9.3.3.1 Injectable Particulate Carbon

The use of particulate AC for injection to aquifers has been proposed to address multiple types of dissolved phase contaminants (Regenesis 2017; RPI 2017). For contaminants that can biodegrade, the concept suggests biodegradation is enhanced while the contaminant mass flux is diminished. The success of this technology is predicated on achieving homogenous distribution of the AC throughout the aquifer, which is significantly challenged by hydraulic fracturing and straining of particle sizes above 1 μm in pore spaces. The ongoing delivery of a terminal electron acceptor or electron donor is essential for contaminant biodegradation and this will not be facilitated when pore spaces strain out carbon particles and so potentially become blocked.

As PFAAs do not biodegrade (section 3.9.1), the emplaced AC will saturate with these contaminants. Additionally, the injected AC is expected to have the same limited effectiveness for removing short-chain PFAAs and PFAA precursors from water (Xiao, Golovko et al., 2017). Once the AC has saturated with PFASs, it will form a secondary source zone, subsequently releasing PFASs. The limited availability of pore space after particulate AC injection may severely inhibit attempted replenishment. GAC used in *ex situ* treatment systems can be easily replaced when contaminant breakthrough is observed, or the remedy can be modified as necessary. Injectable particulate AC cannot be easily replaced or amended with additional treatment technology. As activated carbon has demonstrated poor performance for removal of short chain PFAAs and modeling has shown diminished performance for treatment of PFAA precursors, the same limitations are anticipated considering injected particulate carbon. To summarize, significantly more understanding of the distribution and efficacy of injectable particulate AC for treating a range of PFASs is needed.

3.9.3.3.2 Experimental Adsorbents

Considerable research and development (R&D) is ongoing within the industry to develop engineered adsorbents that exploit hydrophobic and electrostatic adsorption mechanisms for holistic PFAS removal more efficiently than currently available media. Laboratory-scale studies have been conducted on the following adsorbents: carbon nano tubes (Li 2011; Deng et al. 2012), chitosan (Yu 2009; Zhang et al. 2011), cationic coagulants (Aly 2018), biochar and ash (Chen et al. 2011), cyclodextrin (Xiao, Ling et al., 2017; Kawano 2013), modified cellulose (Reeve 2018), organo-clays (Das et al. 2013), and non-ion exchange resins (Senevirathna et al. 2010b; Xiao et al. 2012).

These experimental adsorbents are limited by one or several of the following:

- Currently available R&D proof of concept testing has focused on demonstrating effective removal of PFOA/PFOS, rather than the broader class of PFAS (short-chain and PFAA precursors). With academic university involvement, this data gap is slowly being addressed.
- Currently available R&D proof of concept testing has not evaluated co-contaminant influence of PFAS adsorption effectiveness.
- Manufacturing on a large scale is a limitation, and it may take weeks of synthesis to generate grams of an adsorbent, for example. This will drive the commercial unit cost of the adsorbents considerably higher than GAC and AIX.
- Without demonstration data at the practical water treatment scale, it is difficult to predict actual adsorption capacity thresholds and longer-term performance.
- Regeneration of adsorbents that have reached their adsorption capacity may have an influence on subsequent performance. Long-term practical scale PFAS-treatment data with subsequent regeneration is required to understand the adsorbent lifetime.

3.9.3.3.3 Advanced Oxidation Processes

Chemical oxidation that uses advanced oxidation processes (AOP) utilizes radicals (such as the hydroxyl radical, the ozone radical, and the sulfate radical) to chemically attack organic compounds potentially leading to mineralization. Numerous combinations of chemical oxidants and activation chemistries can enable manipulation of reactivity to control kinetics, oxidation potential, and secondary water quality concerns. Radical oxidizing species generally have a stronger oxidizing potential than stoichiometric oxidants. With respect to PFASs, the carbon fluorine bond has a very high bond dissociation energy (BDE), therefore oxidative attack is thermodynamically unfavorable. In addition, the reactive kinetics of the generated radicals have orders of magnitude more affinity to react with common co-contaminants and other water chemistry scavengers (Buxton 1988). Lastly, the theorized oxidative pathway appears to be sequential defluorination, which is expected to generate increasingly recalcitrant short-chain PFAAs. AOP may have a role as part of a treatment train to facilitate the transformation of PFAA precursors to PFAAs or as a pretreatment mechanism to consume NOM that may optimize the overall PFAS treatment. In general, chemical oxidation processes have demonstrated conversion of PFAA precursors to PFAAs (Plumlee et al. 2008; Houtz 2013; Bruton and Sedlak 2017). However, the complete mineralization of PFASs by conventional AOPs remains debatable. It had been demonstrated that chemical oxidation of PFCAs (i.e., PFOA) is more likely achievable oxidation than that of PFSAs (i.e., PFOS) (Hori et al. 2005; Hori et al. 2007a; Hori et al. 2008; Schröder and Meesters 2005; Qiu 2006; Chen et al. 2006; Kutsuna 2006; Dillert 2007; Vecitis et al. 2009; Lee et al. 2009; Lee 2010; Ahmad 2012; Kerfoot 2013; Mitchell et al. 2014).

3.9.3.3.4 Advanced Reduction Processes

Whereas AOP treatment of PFAS-impacted water is challenged by the strongly electronegative fluorine atoms closely bound to the carbon atoms resisting oxidative attack (Ross, McDonough et al. 2018), the electronegative fluorine atoms are susceptible to reductive attack (Song et al. 2013). The use of advanced reducing processes (ARP), or the use of strong reducing radicals, has been demonstrated to be effective for the destruction of PFAAs at the laboratory scale. This includes plasma, eBeam, ultra violet- (UV-) irradiated-sulfite, -iodide, and -dithionite, and electrochemical treatment (Huang et al. 2007; Ochoa-Herrera et al. 2008; Park et al. 2009; Qu et al. 2010; Li et al. 2014; Mitchell et al. 2014; Hayashi et al. 2015; Wang, Batchelor et al., 2016; Merino et al. 2016; Ma et al. 2017; Stratton et al. 2017; Ross, McDonough et al. 2018; Mededovic Thagard 2018).

The hypothesized functional reductant credited with attack on PFASs is the solvated electron (i.e., hydrated electron or aquated electron) (Park et al. 2009; Song et al. 2013; Mitchell et al. 2014). The solvated electron is typically generated through the disproportionation of the water molecule into hydrogen radical, the solvated electron, and the hydroxyl radical. The large standard reduction potential (–2.9 volts) makes the solvated electron a powerful reductant (Buxton 1988). The solvated electrons attack the α- position C-F bonds instead of C-C bonds, to initiate the defluorination process (Qu et al. 2010; Song et al. 2013). Solvated electrons are nonselective powerful reductants and can be generated through UV-irradiation of reductants, but are readily scavenged by dissolved oxygen and nitrate, suggesting that its application for *ex situ* water treatment will be challenging, and oxygen and anions, such as nitrate, which will consume the reductants, could impact the treatment efficacy (Schaefer et al. 2017; Ross, McDonough et al. 2018).

3.9.3.3.5 Electrochemical Treatment

Electrochemical treatment refers to direct electron transfer from an anode to a molecule within an electrochemical cell designed with an anode, cathode, and electrolyte (Horst et al. 2018). Electrochemical treatment processes are generally described to effect contaminant destruction via two mechanisms: (1) indirect electrochemical oxidation, where strong oxidants, such as hydroxyl radicals, are produced within the electrolyte; and (2) direct electron transfer, where chemical reduction takes place directly at the anode. These processes may offer an opportunity to simultaneously convert PFAA precursors to PFAAs and then chemically reduce the PFAAs. Therefore, electrochemical treatment represents a destructive-based treatment mechanism.

The first known reporting of electrochemical treatment of PFASs (specific to PFOA/PFOS) was in 2015 (Schaefer et al. 2015), and this form of PFAS water treatment is under development. The effectiveness of electrochemical treatment to mineralize long-chain PFAA requires further study as, over typical residence times of 4 to 10 hours, the persistent generation of short-chain PFAA has been observed (Gomez-Ruiz et al. 2017). Further, greater than 4,000 mg/L perchlorate was generated after 10 hours of electrochemical treatment (Gomez-Ruiz et al. 2017). The production of toxic by-products (e.g., hydrogen fluoride, chlorine gas, bromate, perchlorate, and adsorbable organic halides) has been reported and may form when treating PFAS-contaminated wastewater with co-contaminants via electrochemical treatment (Trautmann et al. 2015). Water treatment techniques exist for perchlorate, such as AIX resins and anaerobic bioreactors, but the additional cost of these technologies may limit the application of electrochemical treatment for PFAS destruction in many cases where waters contain chloride or bromide. Consistent with other literature on the destruction of PFOS, it is comparatively more challenging to destroy than PFOA under similar conditions. For example, electrochemical treatment was recently demonstrated to be approximately 90 percent effective for PFOA and 65 percent effective for PFOS at environmental relevant concentrations of 300 to 600 µg/L using a boron-doped diamond anode (Schaefer et al. 2017).

Of importance to electrochemical treatment is the current density and the materials of the anode. Literature values for relevant current densities range from 1 to 50 milliamps per centimeter squared (Schaefer et al. 2015; Schaefer et al. 2017; Urtiaga et al. 2015; Gomez-Ruiz et al. 2017), and boron-doped diamond anodes appear to be emerging as the most effective from an operational standpoint (Urtiaga et al. 2015), though R&D continues to identify new options such as the titanium suboxide anode (Huang 2017). The energy usage of electrochemical treatment will require optimization (currently 0.1 to 0.5 kW-hr per liter of water treated), and electrochemical treatment is currently suited to low volumes of high-concentration PFASs.

3.9.3.3.6 Sonolysis

When sound waves are applied to a liquid at frequencies of between 20 kilohertz (kHz) to 1,500 kHz, the subsequent generation of microbubbles undergoes successive rarefaction and compression. This repeated elongation and compression of the microbubbles intensifies and eventually results in cavitation because of the negative pressure. When the microbubbles collapse, significant energy in the form of heat is released and literature observations suggest achievable temperatures of 5,000 degrees Kelvin along the surface of the collapsing bubbles (Campbell et al. 2009).

This temperature is significantly greater than the temperatures demonstrating thermal destruction of PFASs and occurs along the gas-liquid interface, which is a preferential aggregation site for dissolved PFASs. The demonstration of sonolysis for the destruction of PFASs is widespread in the literature (Mason 2000; Drees 2005; Moriwaki et al. 2005; Vecitis et al. 2008; Vecitis et al. 2010; Rayne and Forest 2009; Cheng et al. 2008; Cheng et al. 2010; Hao et al. 2014; Rodriguez-Freire et al. 2015; Rodriguez-Freire et al. 2016; Wood et al. 2017; Gole et al. 2017; Gole et al. 2018; Fernandez et al. 2016).

Generally, lower frequency sonolysis creates larger bubbles with higher energy output and higher frequency sonolysis creates smaller bubbles with significantly more surface area and less energy output (Drees 2005). The energy input to achieve high-frequency sonolysis generally precludes its use in conventional sonolysis applications to treat high volumes of low concentrations of PFASs implications. However, for efficient PFAS destruction in the presence of a broad range of organic contaminants, a frequency range of 500 to 1,100 kHz seems appropriate based on demonstrations in the literature (Rodriguez-Freire et al. 2015; Fernandez et al. 2016; Rodriguez-Freire et al. 2016). Sonolysis is anticipated to destroy a wide range of PFAS compounds (long-chain, short-chain, and PFAA precursors). Literature demonstrations consistently exhibit pseudo-first order kinetics with faster kinetics for larger PFASs. Some studies have evaluated sonolysis for PFAS destruction using a landfill leachate to simulate co-contaminants, and effective destruction of PFOA/PFOS was demonstrated (Cheng et al. 2010).

The drawbacks associated with sonolysis for water treatment are the energy requirements (generally 0.1 to 0.5 kW-hr per liter of water treated) and no practical field-scale demonstrations of PFAS destruction using sonolysis. Sonolysis has been used on a full-scale basis in the chemical and water treatment industries. The technology offers modular design of reaction units for scale-up, but this has yet to be achieved for PFAS treatment. Sound is generated and transferred to water through a transducer, which also concentrates the energy. Different types of transducers are available commercially, and selection depends on several factors, including the required energy intensity, frequency, and reactor size and geometry. The number of modular units or transducers needed in a large tank would depend upon the reaction kinetics and flow rate to be treated, but also on the achievable sound field to ensure uniform cavitation. This may have an advantage for small volumes of concentrated PFAS waste because the treatment volumes can be controlled. For scale-up, there are many optimization factors that can be explored (sound field distribution, bubbling gas in-line, pH changes, and changing the external temperature and pressure) as well as the generation of the hydroxyl radical, which may have varying degrees of success for different PFASs and conversion of PFAA precursors to PFAAs. Such activity depends on the number and location of transducers, frequency, the geometry of reactor and power dissipation, and larger scale sonolytic reactors can suffer from dead zones (Gole et al. 2017, Wood et al. 2017).

3.9.3.3.7 Photolysis

Photolysis is a chemical transformation, commonly observed for a variety of organic compounds, that results in cleavage of chemical bonds and subsequent compound degradation when exposed to UV or visible light (i.e., 290 to 600 nanometer wavelength). Direct photolysis occurs when the energy from light adsorption by the compound directly causes its degradation. Indirect photolysis occurs when a compound is transformed by a transfer of energy from another species (excited or formed in the presence of light), such as hydroxyl radicals. Due to the high BDE of the C-F bond, direct photolysis proceeds slowly to produce carbon dioxide (CO_2), fluoride (F-), and short-chain intermediates (Hori et al. 2004; Chen et al. 2006; Yamamoto et al. 2007). As such, photocatalysts are used to facilitate indirect photodegradation under a wide range of light wavelengths (Hoffman 1995; Qui 2007; Li et al. 2012; Xu et al. 2017).

Photolysis of long-chain PFASs in the natural environment has been observed in both field and laboratory studies; however, results indicate short-chain intermediates are more resistant to photodegradation (Huang et al. 2007; Taniyasu et al. 2013; Liu et al. 2017). Photolysis has been demonstrated both with and without photocatalysts; however, photocatalysts increase the defluorination rate and efficiency (Hori et al. 2004; Chen et al. 2006; Hori et al. 2007b; Qu et al. 2010; Giri et al. 2011; Li et al. 2012; Song et al. 2013; Zhang et al. 2015; Xu et al. 2017). Additional considerations for the use of photolytic decomposition of PFASs include aqueous solution characteristics, such as co-contaminants, solution acidity, temperature, and ionic strength. The presence of NOM may decrease decomposition by photolysis due to competition for light (Qui 2007). Increasing solution acidity, temperature, and ionic strength, however, may increase photodegradation rates (Zhang et al. 2015; Liu et al. 2017).

Photolysis may provide a practical destruction-based mechanism for concentrated PFAS waste but requires further research as to the fate of sequential defluorination products and various geochemical inhibitors. Irradiation of liquid with UV light is implementable on a large scale (i.e., disinfection), but the generated reactive species may be quickly scavenged. The addition of photocatalysts appears to have promise but is also highly specialized and not well understood. Due to the general recalcitrance of PFAAs for the majority of AOPs, the primary treatment mechanism is believed to be the solvated electron, which has been demonstrated to be highly reactive with other geochemical constituents present at greater concentrations than the PFAAs.

3.10 CONCLUSIONS

As understanding of the types, properties, and environmental behavior of PFASs increases, so do the tools and techniques available for their management. The recent dramatically increased understanding and focus on PFASs allows informed assessment of potential liabilities, development of robust CSMs, and provision of pragmatic, risk-based management solutions relevant to an evolving global regulatory climate.

Many PFASs can potentially travel from their point of release forming large plumes, as they are mobile and persistent. So the use of a CSM to define risks posed by specific PFASs to potential receptors is considered essential. A robust, site-specific CSM is the cornerstone for assessment of potential environmental and human health risks within the contaminated land management sector.

Environmental site investigations have been initiated at hundreds of locations globally as a preliminary evaluation of the perceived risks posed by PFASs to human health and/or the environment. Some of these PFASs' impacted sites are likely to require a combination of rapid, comprehensive, cost-effective, and/or aggressive remediation options. The transparent definition of linkages between sources of contamination, the potential pathways PFASs transit (such as in groundwater plumes) and the exposure to receptors is considered essential to pragmatically manage the potential environmental and human health risks that PFASs may pose at individual sites.

From a regulatory perspective, the general trend appears to be enforcement of lower standards and inclusion of additional PFASs beyond PFOS, PFOA, and PFHxS. In 2018 both ATSDR and EFSA drafted lower but similar tolerable intake levels for PFOS and PFOA, which, if adopted will further diminish advisory levels for PFOS and PFOA in drinking water to well below the 70 to 100 ppt levels already seen in many countries. As even lower standards are proposed, however, they start to approach background concentrations in many matrices and there needs to be an appreciation of the existing global background distribution of many PFASs. A description of background concentrations is included in section 3.8.2 to assist with rationalizing whether a detection of a specific PFAS is likely to emanate from a local source or be a background detection. The frequency and concentrations of PFASs detected in aquifers sampled in the United Kingdom, as reported in section 3.8.2 suggests that they have widespread distribution in groundwater underlying urban and industrialized areas. The further concern, considering the diminishing ppt level standards, is the volume of soil/sediment and water that may be considered out of compliance as very low standards are adopted. For example, the concept that aerial emissions, local to or far from an impacted area may be a major contributor to PFASs levels in surface water needs to be considered. Alternatively, given the very widespread use of PFASs in an immense range of consumer products, there may be multiple diffuse plumes from several sources, such as unlined landfills, car washing/waxing operations, disposal of floor polish waste, clothes washing waste, etc. via drains that leak to groundwater, rainwater run-off from buildings coated with waterproofing containing PFASs, and so forth, which can all combine to create detectable levels of PFASs in groundwater or surface waters.

There is ongoing uncertainty regarding the environmental risk posed by fluorotelomers and some short-chain PFAAs. Fluorotelomers transform in the environment to ultimately create PFAAs, via intermediates such as 6:2 FTS. Under anaerobic conditions 6:2 fluorotelomers may be transformed to lesser well-characterized intermediates such as 5:3 FTCA, which has recently been highlighted as potentially biopersistent (i.e., showing slow clearance from organisms, so having potential for bioaccumulation) (Kabadi et al. 2018). The short-chain PFAAs have been identified as concentrating in the edible portion of crops (Blaine et al. 2013; Blaine et al. 2014a; Blaine et al. 2014b), so this could open a differing exposure pathway for PFASs. There are ongoing concerns regarding human toxicity that are still to be assessed for many PFASs (Vierke 2017). However, the difference in compliance concentrations for the long- and short-chain PFAAs should be a consideration. Long-chain PFAAs tend to be regulated at ppt levels, whereas to date, when the toxicity of short-chain PFAAs are evaluated individually by regulators, they tend to be regulated to ppb levels. Examples of compliance concentrations for short-chain PFAAs (PFBA and PFBS) from Minnesota are 7 ppb. In Bavaria, where the concentration range for drinking water standards varies between 6 and 10 ppb for the short-chain PFAAs, the regulators consider there to be adequate toxicological information (i.e., PFHxA 6 ppb; PFBA 10 ppb; PFBS 6 ppb) (Bavaria State Office for Environment 2017). This report also describes that for other short-chain PFAAs and known polyfluoroalkyl precursors (6:2 FTS, PFPeA, PFHpA, PFOSA) where insufficient toxicological data are considered to be available, the drinking water standards range from 100 ppt to 3 ppb.

The criteria used to assess the safety of chemicals evolve, seemingly, as a result of detections of PFASs in drinking water. The Stockholm Convention on Persistent Organic Pollutants aims to eliminate or restrict the production and use of persistent organic pollutants (POPs) but uses methods to classify potential chemicals of concern on the basis that they are persistent (P), meet criteria demonstrating significant potential bioaccumulation (B), and are considered toxic (T)—termed PBT. However, additional emerging methods are used to evaluate environmental hazards posed by PFASs. These include assessing persistence, mobility (M), and toxicity (PMT) and the assessment of very bioaccumulative and very persistent (vPvB) compounds (Vierke 2017). The protection of drinking water resources using a PBT framework, initially focused on bioaccumulative, non polar compounds such as PCBS has been described as only being marginally effective for protection of drinking water (Reemtsma et al 2016). A focus on the mobility of chemicals in combination with their persistence has led to a preliminary assessment of substances registered under REACH that could fulfill new PMT and very persistent and very mobile (vPvM) criteria (UBA and NGI 2018). The suggestion that mobility is relevant to the assessment of chemicals in the environment is not new, as several authors have suggested approaches where mobility is included (Giger et al. 2005; Reemtsma & Jekel 2006; Reemtsma et al. 2016). The proposal to regulate substances under REACH as PMT substances (Neumann 2017) and vPvM substances (Neumann & Schliebner 2017) may indicate the direction of travel for chemical regulations in many countries.

The fate and transport of PFASs has been described in section 3.9, which identified the unsaturated zone as often an ongoing source of PFASs to groundwater via leaching as a result of seasonal precipitation events and fluctuations in groundwater elevations. In addition to these factors, the fate and transport of PFASs, from FTAs as well as other potential source areas (e.g., landfills), can be influenced by the presence of PFAA precursors as described in Figure 3.25. This shows that the biotransformation of PFAA precursors can be important for conceptualizing fate and transport for the large family of PFASs. Polyfluorinated precursors may form a reservoir of future PFAA-contaminant mass within the environment as variable rates of their transformation, which can depend on the geochemical conditions, will be observed. Therefore, the use of TOP Assay at an impacted site to find and characterize source zones may be essential to develop effective CSMs to understand the location and concentrations of the contaminant mass, such that substantial funds on remediation are not committed before a robust CSM is established. One concern may be that without an adequately comprehensive chemical analysis of PFASs, remedial works may need to be repeated in the future, thus incurring additional expense.

The use of TOP Assay by regulators in Queensland, Australia, for the analysis of multiple environmental matrices may be the first example of this analytical tool being used for remedial compliance (Storch 2018). As PFASs comprise thousands of individual compounds, which may be analogous to other contaminant classes such as petroleum hydrocarbons and PCBSs, some rationalization of the analysis of PFASs seems necessary. This could be conducted in a way similar to how spectated total petroleum hydrocarbon (TPH) analysis is done based on aliphatic and aromatic fractions of varying chain lengths, and PCBs are classified as being "dioxin like." It is to be determined if for PFASs use of the TOP Assay will evolve to become a tool that allows estimation of long- and short-chain PFASs in multiple environmental matrices, considering the complexity of this contaminant class.

An objective, systematic assessment of risk, considering the many PFASs detected, before defining which remedies to use and thus target specific contaminants for treatment, provides a more comprehensive approach. Rationalizing specific exposure pathways will be essential to assess the need for remediation and to avoid extracting groundwater until it meets very low drinking water targets.

The opportunity to potentially future-proof risk assessments by distillation of analytes, to define specific PFASs to consider for treatment, may offer many advantages. However, as the understanding of PFASs toxicology evolves there is always a risk that this approach does not adequately protect against the potential that currently acceptable standards will diminish in future. However, this approach should pragmatically determine the "risk driver" PFASs for specific sites considering the source pathway receptor linkages associated with each site. Defining the "risk driver" indicator compounds PFASs could allow transparent decision-making to define which PFASs are in need of treatment at each site and, thus, focus the remedial objectives. This type of pragmatic approach should balance the risks PFASs pose with the resources applied to their management, considering the need to address stakeholder expectations now, while regulations evolve, considering that PFASs are not biodegradable, and so persist indefinitely.

The concept of extracting and treating groundwater because it exceeds an extremely low health advisory, MCL, or drinking water standard may prove to be unsustainable in many cases. The costs and resources applied to managing PFASs seem to be best spent applied to sources which are identified to be impacting receptors. The costs of PFAS remediation at a German airport were reported to be approaching €100M (Weber 2016), this may be an indication of costs involved to achieve drinking water standards (or MCLs). For PFASs the used of risk based contaminated site management will be important to balance the costs and benefits of PFAS remediation, especially considering the challenges associated with remediation of PFASs.

Risk-based management of PFASs seems essential when considering many wider environmental perspectives and stakeholder concerns. These could include several factors that are relevant to a more holistic and sustainable management approach, such as (1) the amount of time it may take to reach a ppt or ppb standard for differing PFASs; (2) the reality that low levels of background PFAS concentrations are fairly ubiquitously spread in multiple matrices via aerial deposition; (3) the time, resources, and environmental impact (CO_2 footprint, energy usage, etc.) applied to manage PFASs versus the environmental benefit. Ideally these factors should be considered to assess the sustainability of remedial activities. If no identified receptor is established for a PFAS's impacted groundwater body, for example, then the question of what harm it poses should be asked, as PFASs will attenuate by dilution in a groundwater plume. Advice for application of monitored natural attenuation for PFASs would be very similar to that applied to metals (ITRC 2010). Monitored natural attenuation (MNA) has already been applied to a PFOS plume in the United Kingdom (Miles et al. 2017).

The presence of PFAA precursors has important implications for PFAS treatment as described in section 3.9. Specifically, due to the extremely recalcitrant nature of the PFAS chemistry, few destructive treatment technologies are currently commercially viable, and the state of the practice for PFAS treatment involves a series of treatment technologies designed to reduce the volume of impacted matrices to be disposed of/incinerated, though a long list of developing and innovative treatment technologies are on the horizon and with it should come improvements in treatment efficiency and efficacy for the broad class of PFASs present in the environment. Testing whether these emerging remedial technologies are effective to treat a comprehensive range of PFASs seems likely to become increasingly important.

As PFASs comprise both PFAAs and polyfluorinated PFAA precursors, with an increasing number of PFASs becoming regulated, it appears that groundwater remediation technologies need to be adaptable to cope with the expanding range of individual PFASs. Treatment adaption to remove additional PFASs may be achieved more easily via use of *ex situ* treatment technologies. Future-proofing *in situ* technologies is likely to be problematic as they may be designed to address only a limited number of PFASs. For example, treating a range of PFASs *in situ* using injected activated carbon will be problematic as short-chain PFASs and many precursors are either not retained or are poorly removed using activated carbon generally and ultimately the injected carbon particles will act as a secondary source of PFASs.

Many water treatment technologies are available that claim to be able to treat PFASs, but most have not been assessed to treat a broader array of PFAAs or polyfluorinated precursors. Most water treatment technologies involve adsorption of PFASs to a support matrix, which then needs to be disposed of or regenerated. So no single technologies are currently available that can both remove and destroy PFASs simultaneously. Therefore, immense opportunities are available to develop more effective and sustainable remediation solution for PFASs, which seem likely to involve a treatment train of technologies.

Soil remediation technologies are evolving, but there are currently no technologies commercially available that are proven to remove and destroy a comprehensive range of PFASs, but some of the thermal treatments are showing promise and may be applicable. From the perspective of a pragmatic and sustainable solution to PFAS's impacts in soil both above and below the water table in source areas such as FTAs, soil stabilization appears to be an approach that is viable but longer-term testing of performance is required.

To conclude, the challenges that PFAS's management pose to the contaminated land and remediation community may seem significant, but new analytical tools and innovative remedial technologies are rapidly developing. The focused ingenuity of thousands of professional scientists and engineers will develop an array of solutions to mitigate the risks PFASs may pose to human health and the environment.

REFERENCES

3M Company. 2018. 3M(TM) Fluorosurfactants [Online]. Available: https://www.3m.com/3M/en_US/company-us/all-3m-products/~/All-3M-Products/Chemi/. Accessed November 11.

3M Company. 2003. Perfluorooctanic acid (PFOA) Physiochemical Properties and Environmental Fate Data. August 1.

3M Company. 1999. Perfluorooctane Sulfonate: Current Summary of Human Sera, Health and Toxicology Data. January 21.

Agency for Toxic Substances and Disease Registry (ATSDR). 2018. Toxicological profile for perfluoroalkyls, draft for public comment [Online]. June. Available: www.atsdr.cdc.gov/toxprofiles/tp200.pdf. Accessed September 15, 2018.

Agency for Toxic Substances and Disease Registry (ATSDR). 2017. An overview of perfluoroalkyl and polyfluoroalkyl substances and interim guidance for clinicians responding to patient exposure concerns. interim guidance. June 7, 2017 [Online]. Accessed September 15, 2018. www.atsdr.cdc.gov/pfas/docs/pfas_clinician_fact_sheet_508.pdf.

Agency for Toxic Substances and Disease Registry (ATSDR). 2016. Health consultation, exposure investigation, biological sampling of per- and polyfluoroalkyl substances (PFAS) in the vicinity of Lawrence, Morgan, and Limestone Counties, Alabama. *In:* SERVICES, U.S. D.O.H. A.H. (ed.). Atlanta, Georgia.

Agency for Toxic Substances and Disease Registry (ATSDR). 2009. Draft toxicological profile for perfluoroalkyls draft toxicological profile for perfluoroalkyls. U.S. Department of Health and Human Services, Public Health Service Agency for Toxic Substances and Disease Registry.

Agency for Toxic Substances and Disease Registry (ATSDR). 2008. Health consultation: PFOS detections in the City of Brainerd, Minnesota—City of Brainerd, Crow Wing County, Minnesota, St. Paul, Minnesota: Minnesota Department of Health.

Ahmad, M. 2012. *Innovative Oxidation Pathways for the Treatment of Traditional and Emerging Contaminants.* Environmental Engineering Doctor of Philosophy, Washington State University.

Ahrens, L. & Bundschuh, M. 2014. Fate and effects of poly- and perfluoroalkyl substances in the aquatic environment: A review. *Environ Toxicol Chem, 33*, 1921.

Ahrens, L., Harner, T., Shoeib, M., Lane, D. A. and Murphy, J. G., 2012. Improved characterization of gas–particle partitioning for per-and polyfluoroalkyl substances in the atmosphere using annular diffusion denuder samplers. *Environmental science & technology, 46*(13), pp.7199–7206.

Ahrens, L. 2011. Polyfluoroalkyl compounds in the aquatic environment: A review of their occurrence and fate. *J. Environ. Monit.*, 13, 20–31.

Ahrens, L., Shoeib, M., Harner, T., Lee, S. C., Guo, R. & Reiner, E. J. 2011. Wastewater treatment plant and landfills as sources of polyfluoroalkyl compounds to the atmosphere. *Environ Sci Technol, 45*, 8098–8105.

Ahrens, L., Marusczak, N., Rubarth, J., Dommergue, A., Nedjai, R., Ferrari, C. & Ebinghaus, R. 2010. Distribution of perfluoroalkyl compounds and mercury in fish liver from high-mountain lakes in France originating from atmospheric deposition. *Environmental Chemistry, 7.*

Ahrens, L., Taniyasu, S., Yeung, L. W., Yamashita, N., Lam, P. K. & Ebinghaus, R. 2010. Distribution of polyfluoroalkyl compounds in water, suspended particulate matter and sediment from Tokyo Bay, Japan. *Chemosphere, 79*, 266–272.

Ahrens, L., Barber, J. L., Xie, Z. & Ebinghaus, R. 2009. Longitudinal and latitudinal distribution of perfluoroalkyl compounds in the surface water of the Atlantic Ocean. *Environ Sci Technol, 43*, 3122–3127.

Ahrens, L., Felizeter, S., Sturm, R., Xie, Z. and Ebinghaus, R. 2009. Polyfluorinated compounds in waste water treatment plant effluents and surface waters along the River Elbe, Germany. *Marine pollution bulletin*, 58(9), pp. 1326–1333.

Ahrens, L., Plassmann, M., Xie, Z. and Ebinghaus, R. 2009. Determination of polyfluoroalkyl compounds in water and suspended particulate matter in the river Elbe and North Sea, Germany. *Frontiers of Environmental Science & Engineering in China, 3*(2), pp. 152–170.

Allen, J. G. 2018. These toxic chemicals are everywhere—even in your body. And they won't ever go away. [Online]. *Washington Post.* Available: www.washingtonpost.com/opinions/these-toxic-chemicals-are-everywhere-and-they-wont-ever-go-away/2018/01/02/82e7e48a-e4ee-11e7-a65d-1ac0fd7f097e_story.html?noredirect=on&utm_term=.b16300a449cd. Accessed September 24, 2018.

Allred, B. M., J. R. Lang, M. A. Barlaz, & J. A. Field. 2015. Physical and biological release of poly-and perfluoroalkyl substances (PFAS) from municipal solid waste in anaerobic model landfill reactors. *Environmental Science and Technology, 49,* 7648–7656.

Aly, Y. H. 2018. In situ remediation method for enhanced sorption of perfluoro-alkyl substances onto Ottawa Sand. *Journal of Environmental Engineering, 144,* 04018086-1 to 04018086-9.

Anderson, R. H., Long, G. C., Porter, R. C. & Anderson, J. K. 2016. Occurrence of select perfluoroalkyl substances at U.S. Air Force aqueous film-forming foam release sites other than fire-training areas: Field-validation of critical fate and transport properties. *Chemosphere, 150,* 678–685.

Appleman, T. D., Higgins, C. P., Quinones, O., Vanderford, B. J., Kolstad, C., Zeigler-Holady, J. C. & Dickenson, E. R. 2014. Treatment of poly- and perfluoroalkyl substances in U.S. full-scale water treatment systems. *Water Research, 51,* 246–255.

Appleman, T. D., Dickenson, E. R., Bellona, C. & Higgins, C. P. 2013. Nanofiltration and granular activated carbon treatment of perfluoroalkyl acids. *J Hazard Mater, 260,* 740–746.

Arkema. 2018. Paints, coatings and adhesives [Online]. Available: https://www.arkema.com.cn/en/products/markets-overview/paint-coatings-and-adhesives/. Accessed November 10.

Armitage, J. M., M. Macleod, & I. T. Cousins. 2009. Modeling the global fate and transport of perfluorooctanoic acid (PFOA) and perfluorooctanoate (PFO) emitted from direct sources using a multispecies mass balance model. *Environmental Science and Technology, 43,* 1134–1140.

Armitage, J. M., I. T. Cousins, R. C. Buck, K. Prevedouros, M. H. Russell, M. Macleod, & S. H. Korzeniowski. 2006. Modeling global-scale fate and transport of perfluorooctanoate emitted from direct sources. *Environmental Science and Technology, 40,* 6969–6975.

Asher, B. J., Wang, Y., De Silva, A. O., Backus, S., Muir, D. C., Wong, C. S. & Martin, J. W. 2012. Enantiospecific perfluorooctane sulfonate (PFOS) analysis reveals evidence for the source contribution of PFOS-precursors to the Lake Ontario foodweb. *Environmental Science & Technology, 46,* 7653–7660.

Awad, E., Zhang, X., Bhavsar, S. P., Petro, S., Crozier, P. W., Reiner, E. J., Fletcher, R., Tittlemier, S. A. & Braekevelt, E. 2011. Long-term environmental fate of perfluorinated compounds after accidental release at Toronto airport. *Environmental Science & Technology, 45,* 8081–8089.

Backe, W. J., Day, T. C. & Field, J. A. 2013. Zwitterionic, cationic, and anionic fluorinated chemicals in aqueous film forming foam formulations and groundwater from U.S. military bases by nonaqueous large-volume injection HPLC-MS/MS. *Environmental Science & Technology, 47,* 5226–5234.

Baduel, C., Paxman, C. J. & Mueller, J. F. 2015. Perfluoroalkyl substances in a firefighting training ground (FTG), distribution and potential future release. *J Hazard Mater, 296,* 46–53.

Bakkaloglu, S. 2014. *Adsorption of synthetic organic chemicals: A comparison of superfine powdered activated carbon with powdered activated carbon.* PhD, Clemson University.

Banks, R. E. 2000. *Fluorine Chemistry at the Millennium: Fascinated by Fluorine*, Elsevier Science.

Banks, R. E., Smart, B. E. & Tatlow, J. C. (Eds.) 1994. *Organofluorine Chemistry Principles and Commercial Applications*. Springer Science+Business Media New York. ISBN 978-1-4899-1204-6; DOI 10.1007/978-1-4899-1202-2.

Banzhaf, S., Filipovic, M., Lewis, J., Sparrenbom, C. J. & Barthel, R. 2017. A review of contamination of surface-, ground-, and drinking water in Sweden by perfluoroalkyl and polyfluoroalkyl substances (PFASs). *Ambio, 46*, 335–346.

Barry, V., Winquist, A. & Steenland, K. 2013. Perfluorooctanoic acid (PFOA) exposures and incident cancers among adults living near a chemical plant. *Environ Health Perspect, 121*, 1313–1318.

Barton, C. A., Zarzecki, C. J. & Russell M. H. 2010. A site-specific screening comparison of modeled and monitored air dispersion and deposition for perfluorooctanoate. *Journal of the Air and Waste Management Association, 60*, 402–411.

Barton, C. A., Kaiser M. A. & Russell M. H. 2007. Partitioning and removal of perfluorooctanoate during rain events: The importance of physical-chemical properties. *Journal of Environmental Monitoring, 8*, 839–846.

Barzen-Hanson, K. A., Davis, S. E., Kleber, M. & Field, J. A. 2017. Sorption of fluorotelomer sulfonates, fluorotelomer sulfonamido betaines, and a fluorotelomer sulfonamido amine in national foam aqueous film-forming foam to soil. *Environmental Science & Technology, 51*, 12394–12404.

Barzen-Hanson, K. A., Roberts, S. C., Choyke, S., Oetjen, K., Mcalees, A., Riddell, N., Mccrindle, R., Ferguson, P. L., Higgins, C. P. & Field, J. A. 2017. Discovery of 40 classes of per- and polyfluoroalkyl substances in historical aqueous film-forming foams (AFFFs) and AFFF-impacted groundwater. *Environmental Science & Technology, 51*, 2047–2057.

Barzen-Hanson, K. A. & Field, J. A. 2015. Discovery and implications of C2 and C3 perfluoroalkyl sulfonates in aqueous film-forming foams and groundwater. *Environmental Science & Technology Letters, 2*, 95–99.

Baudequin, C., Couallier, E., Rakib, M., Deguerry, I., Severac, R. & Pabon, M. 2011. Purification of firefighting water containing a fluorinated surfactant by reverse osmosis coupled to electrocoagulation–filtration. *Separation and Purification Technology, 76*, 275–282.

Bavarian State Office for the Environment. 2017. *Preliminary evaluation guidelines of PFC impurities in water and soil* [Online]. Available: www.lfu.bayern.de/analytik_stoffe/doc/leitlinien_vorlaufbewertung_pfc_verunreinigungen.pdf. Accessed September 24, 2018.

Becker, A. M., Gerstmann, S. and Frank, H. 2008. Perfluorooctane surfactants in waste waters, the major source of river pollution. *Chemosphere, 72*(1), pp. 115–121.

Beekman, M., Zweers, P., Muller, A., De Vries, W., Janssen, P. & Zeilmaker, M. 2016. Evaluation of substances used in the GenX technology by Chemours, Dordrecht. *RIVM rapport 2016-0174*. The Netherlands.

Beesoon, S., Webster, G. M., Shoeib, M., Harner, T., Benskin, J. P. & Martin, J. W. 2011. Isomer profiles of perfluorochemicals in matched maternal, cord, and house dust samples: manufacturing sources and transplacental transfer. *Environ Health Perspect, 119*, 1659–1664.

Begley, T. H., Hsu, W., Noonan, G. and Diachenko, G., 2008. Migration of fluorochemical paper additives from food-contact paper into foods and food simulants. *Food additives and contaminants, 25*(3), pp.384–390.

Begley, T. H., White, K., Honigfort, P., Twaroski, M. L., Neches, R. and Walker, R. A., 2005. Perfluorochemicals: potential sources of and migration from food packaging. *Food additives and contaminants, 22*(10), pp.1023–1031.

Benskin, J. P., Ikonomou, M. G., Gobas, F. A., Begley, T. H., Woudneh, M. B. & Cosgrove, J. R. 2013. Biodegradation of N-ethyl perfluorooctane sulfonamido ethanol (EtFOSE) and EtFOSE-based phosphate diester (SAmPAP diester) in marine sediments. *Environ Sci Technol, 47*, 1381–1389.

Benskin, J. P., Ikonomou, M. G., Gobas, F. A., Woudneh, M. B. & Cosgrove, J. R. 2012. Observation of a novel PFOS-precursor, the perfluorooctane sulfonamido ethanol-based phosphate (SAmPAP) diester, in marine sediments. *Environ Sci Technol, 46*, 6505–6514.

Benskin, J. P., Ikonomou, M. G., Woudneh, M. B. & Cosgrove, J. R. 2012. Rapid characterization of perfluoroalkyl carboxylate, sulfonate, and sulfonamide isomers by high-performance liquid chromatography-tandem mass spectrometry. *J Chromatogr A, 1247*, 165–170.

Benskin, J. P., Li, B., Ikonomou, M. G., Grace, J. R. & Li, L. Y. 2012c. Per- and polyfluoroalkyl substances in landfill leachate: Patterns, time trends, and sources. *Environ Sci Technol, 46*, 11532–11540.

Benskin, J. P., Muir, D. C., Scott, B. F., Spencer, C., De Silva, A. O., Kylin, H., Martin, J. W., Morris, A., Lohmann, R., Tomy, G., Rosenberg, B., Taniyasu, S. & Yamashita, N. 2012d. Perfluoroalkyl acids in the Atlantic and Canadian Arctic Oceans. *Environ Sci Technol, 46*, 5815–5823.

Benskin, J. P., De Silva, A. O. & Martin, J. W. 2010. Isomer profiling of perfluorinated substances as a tool for source tracking: A review of early findings and future applications. *Rev Environ Contam Toxicol, 208*, 111–160.

Bhhatarai, B. & Gramatica, P. 2011. Prediction of aqueous solubility, vapor pressure and critical micelle concentration for aquatic partitioning of perfluorinated chemicals. *Environ Sci Technol, 45*, 8120–8128.

Björnsdotter M. K., Yeung L. W. Y., Kärrman A., Ericson Jogsten I. 2018. Ultra-short-chain perfluoroalkyl substances (PFASs) including trifluoromethanesulfonic acid (TFMS) in environmental waters. *Presentation*, Dioxin 2018, 38th International Symposium on Halogenated Persistent Organic Pollutants & 10th International PCB Workshop. August 26-31, Krakow, Poland.

Blaine, A. C., Rich, C. D., Sedlacko, E. M., Hundal, L. S., Kumar, K., Lau, C., Mills, M. A., Harris, K. M. & Higgins, C. P. 2014a. Perfluoroalkyl acid distribution in various plant compartments of edible crops grown in biosolids-amended soils. *Environmental Science & Technology, 48*, 7858–7865.

Blaine, A. C., Rich, C. D., Sedlacko, E. M., Hyland, K. C., Stushnoff, C., Dickenson, E. R. & Higgins, C. P. 2014b. Perfluoroalkyl acid uptake in lettuce (Lactuca sativa) and strawberry (Fragaria ananassa) irrigated with reclaimed water. *Environmental Science & Technology, 48*, 14361–14368.

Blaine, A. C., Rich, C. D., Hundal, L. S., Lau, C., Mills, M. A., Harris, K. M. & Higgins, C. P. 2013. Uptake of perfluoroalkyl acids into edible crops via land applied biosolids: field and greenhouse studies. *Environmental Science & Technology, 47*, 14062–14069.

Board, G. 2016. Toxic legacy Teflon chemical sticks around in water supplies. *Ohio Valley Resource*, Media article.

Boiteux, V., Bach, C., Sagres, V., Hemard, J., Colin, A., Rosin, C., Munoz, J.-F. & Dauchy, X. 2016. Analysis of 29 per- and polyfluorinated compounds in water, sediment, soil and sludge by liquid chromatography–tandem mass spectrometry. *International Journal of Environmental Analytical Chemistry, 96*, 705–728.

Bossi, R., Vorkamp, K. and Skov, H. 2016. Concentrations of organochlorine pesticides, polybrominated diphenyl ethers and perfluorinated compounds in the atmosphere of North Greenland. *Environmental pollution, 217*, pp.4–10.

Boulanger, B., Vargo, J. D., Schnoor, J. L. & Hornbuckle, K. C. 2005. Evaluation of Perfluorooctane Surfactants in a Wastewater Treatment System and in a Commercial Surface Protection Product. *Environmental Science & Technology, 39*, 5524–5530.

Braunig, J., Baduel, C., Heffernan, A., Rotander, A., Donaldson, E. & Mueller, J. F. 2017. Fate and redistribution of perfluoroalkyl acids through AFFF-impacted groundwater. *Sci Total Environ, 596–597,* 360–368.

Bräunig, J., Baduel, C. & Mueller, J. 2017. Influence of a commercial sorbent on the leaching behaviour and bioavailability of selected perfluoroalkyl acids (PFAAs) from soil impacted by AFFF. *Presentation.*

Brice, T. J. T., P. W. 1956. *Fluorocarbon sulfonic acids and derivatives.* 2,732,298.

Briggs, K. and D. Staack. 2015. Enhancement of bioremediation of soils contaminated with organic hydrocarbon using an electron beam. *Third International Symposium on Bioremediation and Sustainable Environmental Technologies.* Miami, FL.

Brusseau, M. G. 2018. Assessing the potential contributions of additional retention processes to PFAS retardation in the subsurface. *Science of the Total Environment, 613,* 176–185.

Bruton, T. A. & Sedlak, D. L. 2017. Treatment of aqueous film-forming foam by heat-activated persulfate under conditions representative of in situ chemical oxidation. *Environmental Science & Technology, 51,* 13878–13885.

Buck, R. C., Franklin, J., Berger, U., Conder, J. M., Cousins, I. T., De Voogt, P., Jensen, A. A., Kannan, K., Mabury, S. A. & Van Leeuwen, S. P. J. 2011. Perfluoroalkyl and polyfluoroalkyl substances in the environment: Terminology, classification, and origins. *Integr. Environ. Assess. Manage, 7,* 513.

Buhl, J. 2017. Removing PFASs from water. *Remediation Solutions,* XXVI.

Bull, S., Burnett, K., Vassaux, K., Ashdown, L., Brown, T. & Rushton, L. 2014. Extensive literature search and provision of summaries of studies related to the oral toxicity of perfluoroalkylated substances (PFASs), their precursors and potential replacements in experimental animals and humans Area 1: Data on toxicokinetics (absorption, distribution, metabolism, excretion) in in vitro studies, experimental animals and humans Area 2: Data on toxicity in experimental animals Area 3: Data on observations in humans. *EFSA.*

Burgess, D. R., Zachariah, M. R., Tsang, W., Westmoreland, P. R. 1995. Thermochemical and Chemical Kinetic Data for Fluorinated Hydrocarbons. *Progress in Energy and Combustion Science, 21,* 453–529.

Busch, J., Ahrens, L., Sturm, R. & Ebinghaus, R. 2010. Polyfluoroalkyl compounds in landfill leachates. *Environ Pollut, 158,* 1467–1471.

Butenhoff, J. L., S. C. Chang, G. W. Olsen, and P. J. Thomford. 2012. Chronic dietary toxicity and carcinogenicity study with potassium perfluorooctane sulfonate in Sprague Dawley rats. *Toxicology* 293:1–15.

Butenhoff, J. L., Kennedy, G. L., Jr., Chang, S. C. & Olsen, G. W. 2012. Chronic dietary toxicity and carcinogenicity study with ammonium perfluorooctanoate in Sprague-Dawley rats. *Toxicology, 298,* 1–13.

Butenhoff J. L., Ehresman D. J., Chang S.-C., Parker G. A., Stump D. G. 2009. Gestational and lactational exposure to potassium perfluorooctanesulfonate (K+PFOS) in rats: developmental neurotoxicity. *Reprod. Toxicol.* 27:319–330

Butenhoff, J. L; Kennedy, G. L, Jr; Frame, S. R., O'Connor, J. C., York, R. G. 2004. The reproductive toxicology of ammonium perfluorooctanoate (APFO) in the rat. Toxicology 196: 95–116

Butenhoff, J. L., Costa, G., Elcombe, C., Farrar, D., Hansen, K., Iwai, H., Jung, R., Kennedy, G., Leider, P. & Thomford, P. 2002. Toxicity of ammonium perfluorooctanoate in male cynomolgus monkeys after oral dosing for 6 months. *Toxicol Sci, 69,* 244–257.

Butt, C.M., Berger, U., Bossi, R., Tomy, G.T., 2010. Review: levels and trends of poly- and perfluorinated compounds in the arctic environment. *Sci Total Environ, 408,* 2936–2965.

Buxton, G. V., Greenstock, C. L., Helman, W. P., & Ross, A. B. 1988. Critical review of rate constants for reactions of hydrated electrons, hydrogen atoms and hydroxyl radicals (OH/O) in aqueous solution. *Journal of Physical and Chemical Reference Data, 17,* 513–886.

C8 Science Panel. 2017. *C8 Science Panel Website* [Online]. Available: www.c8sciencepanel. org/ [Accessed July 20, 2018].

Cai, M., Zhao, Z., Yin, Z., Ahrens, L., Huang, P., Cai, M., Yang, H., He, J., Sturm, R., Ebinghaus, R. & Xie, Z. 2012. Occurrence of perfluoroalkyl compounds in surface waters from the North Pacific to the Arctic Ocean. *Environ Sci Technol, 46*, 661–668.

California State Water Quality Control Board. 2018. *Perfluorooctanoic acid (PFOA) and perfluorooctanesulfonic acid (PFOS)* [Online]. Available: www.waterboards.ca.gov/ drinking_water/certlic/drinkingwater/PFOA_PFOS.html. [Accessed July 16, 2018].

Campbell, T. Y., Vecitis, C. D., Mader, B. T. & Hoffmann, M. R. 2009. Perfluorinated surfactant chain-length effects on sonochemical kinetics. *Journal of Physical Chemistry A, 113*, 9834–9842.

Campos Pereira, H., Ullberg, M., Kleja, D. B., Gustafsson, J. P. & Ahrens, L. 2018. Sorption of perfluoroalkyl substances (PFASs) to an organic soil horizon: Effect of cation composition and pH. *Chemosphere, 207*, 183–191.

Carter, K. E. & Farrell, J. 2010. Removal of perfluorooctane and perfluorobutane sulfonate from water via carbon adsorption and ion exchange. *Separation Science and Technology, 45*, 762–767.

Caporiccio, G. 1996. United States Patent Number 5,493,049, Fluorinated Compounds Containing Hetero Atoms and Polymers Thereof. February 20.

Caporiccio, G. 1994. United States Patent Number 5,350,878, Fluorinated Compounds Containing Hetero Atoms and Polymers Thereof. September 27.

Case, M. T., York, R. G., Christian, M. S. 2001. Rat and rabbit oral developmental toxicology studies with two perfluorinated compounds. *Int J Toxicol.* 20, 101–109.

Casson, R. & Chiang, S.-Y. D. 2018. Integrating total oxidizable precursor assay data to evaluate fate and transport of PFASs. *Remediation Journal, 28*, 71–87.

Caverly Rae, J. M., Craig, L., Slone, T. W., Frame, S. R., Buxton, L. W. & Kennedy, G. L. 2015. Evaluation of chronic toxicity and carcinogenicity of ammonium 2,3,3,3-tetrafluoro-2-(heptafluoropropoxy)-propanoate in Sprague–Dawley rats. *Toxicology Reports, 2*, 939–949.

Costello, M. G. & Moore, G. G. I. 1994. *Direct fluorination process for making perfluorinated organic substances*. U.S. patent application.

Chang, S. C., Noker, P. E., Gorman, G. S., Gibson, S. J., Hart, J. A., Ehresman, D. J. & Butenhoff, J. L. 2012. Comparative pharmacokinetics of perfluorooctanesulfonate (PFOS) in rats, mice, and monkeys. *Reprod Toxicol, 33*, 428–440.

Chang, S. C., Das, K., Ehresman, D. J., Ellefson, M. E., Gorman, G. S., Hart, J. A., Noker, P. E., Tan, Y. M., Lieder, P. H., Lau, C., Olsen, G. W. & Butenhoff, J. L. 2008. Comparative pharmacokinetics of perfluorobutyrate in rats, mice, monkeys, and humans and relevance to human exposure via drinking water. *Toxicol Sci, 104*, 40–53.

Chen, H., Yao, Y., Zhao, Z., Wang, Y., Wang, Q., Ren, C., Wang, B., Sun, H., Alder, A. C. & Kannan, K. 2018. Multi-media distribution and transfer of per- and polyfluoroalkyl substances (PFASs) surrounding two fluorochemical manufacturing facilities in Fuxin, China. *Environ Sci Technol.*

Chen, J., Zhang, P. & Zhang, L. 2006. Photocatalytic decomposition of environmentally persistent perfluorooctanoic acid. *Chemistry Letters, 35*, 230–231.

Chen, X., Xia, X., Wang, X., Qiao, J. & Chen, H. 2011. A comparative study on sorption of perfluorooctane sulfonate (PFOS) by chars, ash and carbon nanotubes. *Chemosphere, 83*, 1313–1319.

Cheng, J., Vecitis, C. D., Park, H., Mader, B. T. & Hoffmann, M. R. 2010. Sonochemical degradation of perfluorooctane sulfonate (PFOS) and perfluorooctanoate (PFOA) in groundwater: Kinetic effects of matrix inorganics. *Environmental Science & Technology, 44*, 445–450.

Cheng, J., Vecitis, C. D., Park, H., Mader, B. T. & Hoffmann, M. R. 2008. Sonochemical degradation of perfluorooctane sulfonate (PFOS) and perfluorooctanoate (PFOA) in landfill groundwater: Environmental matrix effects. *Environmental Science & Technology, 42*, 8057–8063.

Cheremisinoff, N. P. 2017. Perfluorinated chemicals: Contaminants of concern. Scrivener Publishing LLC. ISBN 978-1-119-36353-8.

Chularueangaksorn, P., Tanaka, S., Fujii, S. & Kunacheva, C. 2013. Regeneration and reusability of anion exchange resin used in perfluorooctane sulfonate removal by batch experiments. *Journal of Applied Polymer Science, 130*, 884–890.

Cleland, M. R. 2012. *Electron beam materials irradiators in industrial accelerators and their applications*, Singapore, World Scientific Publishing Company.

Colosi, L. M., Pinto, R. A., Huang, Q. & Weber, W. J., Jr. 2009. Peroxidase-mediated degradation of perfluorooctanoic acid. *Environmental Toxicology and Chemistry, 28*, 264–271.

Committee on Toxicity of Chemicals in Food, C. P. A. T. E. C. 2006. Statement on the tolerable daily intake for perfluorooctane sulfonate.

Conder, J. M., Hoke, R. A., Wolf, W. D., Russell, M. H. & Buck, R. C. 2008. Are PFCAs bioaccumulative? A Critical review and comparison with regulatory criteria and persistent lipophilic compounds. *Environmental Science & Technology, 42*, 995–1003.

Cui, L., Zhou Q.-f., Liao C.-y., Fu J.-j., Jiang G.-b. 2009. Studies on the toxicological effects of PFOA and PFOS on rats using histological observation and chemical analysis. *Arch. Environ. Contam. Toxicol.* 56:338–349

Cummings, L., A. Matarazzo, N. Nelson, F. Sickels, & C. T. Storms 2015. Recommendation on perfluorinated compound treatment options for drinking water. *In:* Subcommittee, N. J. D. W. Q. I. T. (ed.).

D'Agostino, L. A. & Mabury, S. A. 2017a. Aerobic biodegradation of 2 fluorotelomer sulfonamide-based aqueous film-forming foam components produces perfluoroalkyl carboxylates. *Environmental Toxicology and Chemistry.*

D'Agostino, L. A. & Mabury, S. A. 2017b. Certain perfluoroalkyl and polyfluoroalkyl substances associated with aqueous film forming foam are widespread in canadian surface waters. *Environ Sci Technol, 51*, 13603–13613.

D'Agostino, L. A. & Mabury, S. A. 2014. Identification of novel fluorinated surfactants in aqueous film forming foams and commercial surfactant concentrates. *Environ Sci Technol, 48*, 121–129.

D'Eon, J. C., Hurley, M. D., Wallington, T. J. & Mabury, S. A. 2006. Atmospheric chemistry ofN-methyl Perfluorobutane Sulfonamidoethanol, C4F9SO2N(CH3)CH2CH2OH: Kinetics and mechanism of reaction with OH. *Environmental Science & Technology, 40*, 1862–1868.

Danish Ministry of the Environment, Environmental Protection Agency (DEPA). 2015a. Perfluoroalkylated substances: PFOA, PFOS and PFOSA: Evaluation of health hazards and proposal of a health-based quality criterion for drinking water, soil and ground water. *Report 1665.*

Danish Ministry of the Environment, Environmental Protection Agency (DEPA). 2015b. *Perfluorinated alkyl acid compounds (PFAS compounds) incl. PFOA, PFOS and PFOSA* [Online]. Available: http://mst.dk/media/92446/pfoa-pfos-pfosa-datablad-final-27-april-2015.pdf. Accessed September 15, 2018.

Danish Ministry of the Environment, Environmental Protection Agency (DEPA). 2015c. Alternatives to perfluoroalkyl and polyfluoroalkyl substances (PFAS) in textiles. Survey of chemical substances in consumer products No. 137, 2015. ISBN no. 978-87-93352-16-2.

Danish Ministry of the Environment, Environmental Protection Agency (DEPA). 2013. Survey of PFOS, PFOA and other perfluoroalkyl and polyfluoroalkyl substances: Part of the LOUS-review. Copenhagen: Danish Environmental Protection Agency.

Dainkin Global. 2018. Oil and Gas [Online]. Available: www.daikinchemicals.com/solutions/industries/oil-and-gas.html. Accessed November 11.

Danner, B. 2014. Polymerisation of hexafluoropropylene oxde. *United States Patent,* US008653229B2.

Das, K., P. Rosen, B. Mitchell, C. R. Wood, B. D. Abbott, & C. Lau. 2011. *Perfluorophosphonic acid activates peroxisome proliferator-activated receptor-alpha but not constitutive androstane receptor in the murine liver* [Online]. Available: https://cfpub.epa.gov/si/si_public_record_report.cfm?dirEntryId=230824 Accessed September 24, 2018.

Das, P., Arias E, V. A., Kambala, V., Mallavarapu, M. & Naidu, R. 2013. Remediation of perfluorooctane sulfonate in contaminated soils by modified clay adsorbent—A risk-based approach. *Water, Air, & Soil Pollution,* 224.

Dasu, K., Lee, L. S., Turco, R. F. & Nies, L. F. 2013. Aerobic biodegradation of 8:2 fluorotelomer stearate monoester and 8:2 fluorotelomer citrate triester in forest soil. *Chemosphere, 91,* 399–405.

Dasu, K., Liu, J. & Lee, L. S. 2012. Aerobic soil biodegradation of 8:2 fluorotelomer stearate monoester. *Environmental Science & Technology, 46,* 3831–3836.

Dauchy, X., Boiteux, V., Bach, C., Colin, A., Hemard, J., Rosin, C. & Munoz, J.-F. 2017a. Mass flows and fate of per- and polyfluoroalkyl substances (PFASs) in the wastewater treatment plant of a fluorochemical manufacturing facility. *Science of The Total Environment, 576,* 549–558.

Dauchy, X., Boiteux, V., Bach, C., Rosin, C. & Munoz, J. F. 2017b. Per- and polyfluoroalkyl substances in firefighting foam concentrates and water samples collected near sites impacted by the use of these foams. *Chemosphere, 183,* 53–61.

Davis, K. L., Aucoin, M.D., Larsen, B.S., Kaiser, M.A., & Hartten, A. S. 2007. Transport of ammonium perfluorooctanoate in environmental media near a fluoropolymer manufacturing facility. *Chemosphere, 67,* 2011–2019.

De Silva, A. O., Spencer, C., Scott, B. F., Backus, S. & Muir, D. C. 2011. Detection of a cyclic perfluorinated acid, perfluoroethylcyclohexane sulfonate, in the Great Lakes of North America. *Environmental Science & Technology, 45,* 8060–8066.

De Voogt, P. 2010. *Reviews of environmental contamination and toxicology perfluorinated_ alkylated substances.* Springer Science+Business Media, LLC. ISSN 0179-5953, ISBN 978-1-4419-6879-1, DOI 10.1007/978-1-4419-6880-7.

Deem, W. R. 1973. United States Patent No. 3,758,618. Production of Tetrafluoroethylene Oligomers. September 11.

Deng, S., Nie, Y., Du, Z., Huang, Q., Meng, P., Wang, B., Huang, J. & Yu, G. 2015. Enhanced adsorption of perfluorooctane sulfonate and perfluorooctanoate by bamboo-derived granular activated carbon. *J Hazard Mater, 282,* 150–157.

Deng, S., Zheng, Y. Q., Xu, F. J., Wang, B., Huang, J. & Yu, G. 2012. Highly efficient sorption of perfluorooctane sulfonate and perfluorooctanoate on a quaternized cotton prepared by atom transfer radical polymerization. *Chemical Engineering Journal, 193–194,* 154–160.

Deng, S., Yu, Q., Huang, J. & Yu, G. 2010. Removal of perfluorooctane sulfonate from wastewater by anion exchange resins: Effects of resin properties and solution chemistry. *Water Research, 44,* 5188–5195.

Department of Environment Regulation (DER). Government of Western Australia. 2016. Interim guideline on the assessment and management of perfluoroalkyl and polyfluoroalkyl substances (PFAS). *Contaminated Sites Guidelines.* February.

DeWitt, J. C. 2015. Toxicological effects of perfluoroalkyl and polyfluoroalkyl substances. *Springer International Publishing Switzerland.* ISSN 2168-4219, ISBN 978-3-319-15517-3, DOI 10.1007/978-3-319-15518-0.

Dickson, M. D. 2013. *Method for treating industrial waste.* Australia patent application 2012289835.

Dickson, M. D. 2014. United States US 2014O190896A1 Patent Application Publication.

Dillert. R., D. B., & H. Hidaka 2007. Light-induced degradation of perfluorocarboxylic acids in the presence of titanium dioxide. *Chemosphere, 67*, 785–792.

Ding, G. & Peijnenburg, W. J. G. M. 2013. Physicochemical properties and aquatic toxicity of poly- and perfluorinated compounds. *Crit. Rev. Environ. Sci. Technol., 43*, 598.

Dinglasan, M. J. A., Ye, Y., Edwards, E. A. & Mabury, S. A. 2004. Fluorotelomer alcohol biodegradation yields poly- and perfluorinated acids. *Environmental Science & Technology, 38*, 2857–2864.

Dong, G.-H., Y.-H. Zhang, L. Zheng, W. Liu, Y.-H. Jin, and Q.-C. He. 2009. Chronic effects of perfluorooctane sulfonate exposure on immunotoxicity in adult male C57BL/6 mice. *Archives of Toxicology* 83:805–815.

Drees, C. W. 2005. *Sonochemical Degradation of PFOS*. Chemical Engineering Masters of Science, The Ohio State University.

Dreyer A., I. Weinberg, C. Temme, and R. Ebinghaus. 2009. "Polyfluorinated compounds in the atmosphere of the Atlantic and Southern Oceans: evidence for a global distribution." *Environmental Science and Technology* 43(17):6507–6514.

Dreyer A., Matthias, V., Weinberg, I., & Ebinghaus, R. 2010. Wet deposition of poly- and perfluorinated compounds in Northern Germany. *Environmental Pollution, 158*, 1221–1227.

Department of Toxic Substances Control (DTSC). 2018. *Product–chemical profile for perfluoroalkyl and polyfluoroalkyl substances (PFASs) in carpets and rugs* [Online]. Available: http://dtsc.ca.gov/SCP/upload/Product-Chemical-Profile-PFAS-Carpets-and-Rugs.PDF [Accessed] September 24, 2018.

Du, Z., Deng, S., Bei, Y., Huang, Q., Wang, B., Huang, J. & Yu, G. 2014. Adsorption behavior and mechanism of perfluorinated compounds on various adsorbents—a review. *Journal of Hazardous Materials, 274*, 443–454.

Dutch Institute for National Health (RIVM). 2017a. *Derivation of a lifetime drinking-water guideline for 2,3,3,3-tetrafluoro-2-(heptafluoropropoxy)propanoic acid (FRD-903)— Revised version January 2017* [Online]. Available: www.rivm.nl/Onderwerpen/G/GenX/ Downloaden/148_2016_M_V_bijlage_afleiding_richtwaarde_FRD903_in_drinkwater_ Definitief_revisie_jan_2017.org Accessed: September 24, 2018.

Dutch Institute for National Health (RIVM) 2017b. Water quality standards for PFOA A proposal in accordance with the methodology of the Water Framework Directive.

ECT2. 2016. Sustainable removal of poly and perfluorinated Alkyl substances (PFAS) from groundwater using synthetic media—resins. *Factsheet*.

Eddy, M. 2003. *Wastewater engineering: Treatment and reuse*, New York: McGraw-Hill.

Environmental Decontamination Europe Ltd. 2018. Mechano-chemical destruction (MCD) of PFAS in soil. Presentation. May.

Edmiston, P. L. 2017. College of Wooster, Ohio. *personal communication*.

Edmiston, P. L., Jolly, S. & Spoonamore, S. 2014. Nanoengineered organosilica materials for the treatment of produced water. *Aquananotechnology: Global Prospects*.

Edmiston, P. L. 2013. *Modified sol-gel derived sorbent material and method for using the same*. U.S. Patent, US 2013/0012379 A1.

Edmiston, P. L. & Underwood, L. A. 2009. Absorption of dissolved organic species from water using organically modified silica that swells. *Separation and Purification Technology, 66*, 532–540.

Eleuterio, H. S. 1971. Polymerization of perfluoro epoxides. *Journal of Macromolecular Science: Part A, Chemistry*, 6.

Ellis, D. A., Martin, J. W., De Silva, A. O., Mabury, S. A., Hurley, M. D., Sulbaek-Andersen, M. P. & Wallington, T. J. 2004. Degradation of fluorotelomer alcohols: A likely atmospheric source of perfluorinated carboxylic acids. *Environmental Science & Technology, 38*, 3316–3321.

Endpoint. 2016. Bench-scale VEG research & development study: Implementation memorandum for ex-situ thermal desorption of perfluoroalkyl compounds (PFCs) in soils. *Technical Note*. www.endpoint-inc.com/wp-content/uploads/2016/05/VEG-Bench-Scale-PFCs-Soil.pdf.
England, P. H. 2009. PFOS and PFOA toxicological overview. *Public Health England*.
Environment Agency. 2008. Incidence and attenuation of PFOS in groundwater.
Environment Agency. 2007. Investigation of PFOS and other perfluorochemicals in groundwater and surface water in England and Wales.
Environment Protection Authority, Victoria. 2018. *Interim position statement on PFAS* [Online]. Available: www.epa.vic.gov.au/~/media/Publications/1669%202.pdf Accessed September 24, 2018.
Eriksson, U., Kärrman, A., Rotander, A., Mikkelsen, B. and Dam, M. 2013. Perfluoroalkyl substances (PFASs) in food and water from Faroe Islands. *Environmental science and pollution research*, 20(11), pp. 7940–7948.
Eschauzier, C., Beerendonk, E., Scholte-Veenendaal, P. & De Voogt, P. 2012. Impact of treatment processes on the removal of perfluoroalkyl acids from the drinking water production chain. *Environmental Science & Technology, 46*, 1708–1715.
Eschauzier, C., De Voogt, P., Brauch, H.-J. & Lange, F. T. 2012. Polyfluorinated chemicals in European surface waters, ground- and drinking waters. *Polyfluorinated Chemicals and Transformation Products*.
ESTCP, S. 2017. SERDP and ESTCP workshop on research and demonstration needs for management of afff-impacted sites.
European Chemicals Agency (ECHA). 2015. Background Document to the Opinion on the Annex XV dossier proposing restrictions on Perfluorooctanoic acid (PFOA), PFOA salts and PFOA-related substances. Committee for Risk Assessment (RAC), Committee for Socio-economic Analysis (SEAC). ECHA/RAC/RES-O-0000006229-70-02/F. September 11.
European Chemicals Agency (ECHA). 2014. Germany and Norway propose a restriction on Perfluorooctanoic acid (PFOA), its salts and PFOA-related substances. Information note. (On-line). Available: https://echa.europa.eu/documents/10162/3b6926a2-64cb-4849-b9be-c226b56ae7fe. Accessed September 24, 2018.
European Chemicals Agency (ECHA). 2013. *Registered substances*. [Online]. Available: http://echa.europa.eu/information-on-chemicals/registered-substances Accessed September 24, 2018.
European Food Safety Authority (EFSA). 2018. Risk to human health related to the presence of perfluorooctane sulfonic acid and perfluorooctanoic acid in food. [Online]. Available: www.documentcloud.org/documents/4620589-EFSA-PFOS-and-PFOA-Draft.html. Accessed September 15, 2018.
European Food Safety Authority (EFSA). 2017. Draft Scientific opinion on the health risks related to the presence of perluoroalkylated substances in food (EFSA -Q-2017-00549). [Online]. Available: http://registerofquestions.efsa.europa.eu/roqFrontend/questionLoader?question=EFSA-Q-2017-00549. Accessed September 24, 2018.
European Food Safety Authority (EFSA). 2015. Draft Scientific opinion on the health risks related to the presence of perluoroalkylated substances in food (EFSA-Q-2015-00526). [Online]. Available: http://registerofquestions.efsa.europa.eu/roqFrontend/questionLoader?question=EFSA-Q-2015-00526. Accessed September 24, 2018.
European Food Safety Authority (EFSA). 2008. Perfluorooctane sulfonate (PFOS), perfluorooctanoic acid (PFOA) and their salts, Scientific Opinion of the Panel on Contaminants in the Food Chain. *EFSA Journal, 653,* 1–131. February.
EuropeanUnion. 2011. Commission Regulation (EU) No 253/2011 of 15 March 2011 amending Regulation (EC) No 1907/2006 of the European Parliament and of the Council on the Registration, Evaluation, Authorisation and Restriction of Chemicals (REACH) as regards Annex XIII. *Official Journal of the European Union.* http://eur-lex.europa.eu/eli/reg/2011/253/oj/eng.

Fei, C., Mclaughlin, J. K., Lipworth, L. & Olsen, J. 2008. Prenatal exposure to perfluoroocatanoate (PFOA) and perfluorooctanesulfonate (PFOS) and maternally reported developmental milestones in infancy. *Environ Health Perspect, 116*(10), 1391–1395.

Fei, C., Mclaughlin, J. K., Tarone, R. E. & Olsen, J. 2007. Perfluorinated chemicals and fetal growth: A study within the Danish National Birth Cohort. *Environ Health Perspect, 115*, 1677–1682.

Fernandez, N. A., Rodriguez-Freire, L., Keswani, M. & Sierra-Alvarez, R. 2016. Effect of chemical structure on the sonochemical degradation of perfluoroalkyl and polyfluoroalkyl substances (PFASs). The Royal Society of Chemistry Environmental Science: Water Research and Technology.

Field, J. A., Schultz, M. & Barofsky, D. 2003. Identifying hydrocarbon and fluorocarbon surfactants in specialty chemical formulations of environmental interest by fast atom bombardment/mass spectrometry. *CHIMIA International Journal for Chemistry, 57*, 556–560.

Filipovic, M., Laudon, H., Mclachlan, M. S. & Berger, U. 2015. Mass balance of perfluorinated alkyl acids in a pristine boreal catchment. *Environ. Sci. Technol., 49*, 12127.

Filipovic, M., Woldegiorgis, A., Norstrom, K., Bibi, M., Lindberg, M. & Osteras, A. H. 2015. Historical usage of aqueous film forming foam: A case study of the widespread distribution of perfluoroalkyl acids from a military airport to groundwater, lakes, soils and fish. *Chemosphere, 129*, 39–45.

Flores, C., Ventura, F., Martin-Alonso, J. & Caixach, J. 2013. Occurrence of perfluorooctane sulfonate (PFOS) and perfluorooctanoate (PFOA) in N.E. Spanish surface waters and their removal in a drinking water treatment plant that combines conventional and advanced treatments in parallel lines. *Sci Total Environ, 461–462*, 618–626.

Flynn, R. M., Vitcak, D. R., Fuckanin, R. S. & Elsbernd, C. L. S. 2003. *Fluorinated polyether compositions*. U.S. patent application WO 2004060964 A1.

Frömel, T., Gremmel, C. Dimzon, I. & De Voogt, P. 2016. Investigations on the presence and behavior of precursors to perfluoroalkyl substances in the environment as a preparation of regulatory measures.

Frömel, T. & Knepper, T. P. 2010. *Biodegradation of fluorinated alkyl substances*.

Fromme, H., Tittlemier, S. A., Volkel, W., Wilhelm, M. & Twardella, D. 2009. Perfluorinated compounds—exposure assessment for the general population in Western countries. *Int J Hyg Environ Health, 212*, 239–270.

Food Standards Australia New Zealand (FSANZ). 2017. Final health-based guidance values Australia. *Food Standards Australia New Zealand*.

Fukushi, T., P. J. Scott, C. N. Ferguson,. M. A. Grootaert, D. Duchesne, Y. Moua, L. A. Last, T. A. Shefelbine, & L. M. B. Rodgers. 2014. United States Patent No: US 8,835,551 B2, Ultra Low Viscosity Iodine Containing Amorphous Fluoropolymers. September 16.

Full-Scale Treatment of PFAS-Impacted Wastewater Using Ozofractionation. P. J. Storch, J. Lagowski, M. Dickson, D. Solomon, and I. Ross. Peter Storch (ARCADIS/Australia) 2018 Battelle Conference Proceedings, Palm Springs, CA

Funkhouser, M. 2014. *The toxicologic effects of perfluorooctane sulfonate (PFOS) on a freshwater gastropod, Physa pomilia, and a parthenogenetic decapod, Procambarus fallax f. virginali*. Thesis. Texas Tech University for Masters of Science. May.

Gannon, S. A., Fasano, W. J., Mawn, M. P., Nabb, D. L., Buck, R. C., Buxton, L. W., Jepson, G. W. & Frame, S. R. 2016. Absorption, distribution, metabolism, excretion, and kinetics of 2,3,3,3-tetrafluoro-2-(heptafluoropropoxy) propanoic acid ammonium salt following a single dose in rat, mouse, and cynomolgus monkey. *Toxicology, 340*, 1–9.

Gauthier, S. A. & Mabury, S. A. 2005. Aqueous photolysis of 8:2 fluorotelomer alcohol. *Environmental Toxicology and Chemistry, 24*, 1837.

Gebbink, W. A., Berger, U. & Cousins, I. T. 2015. Estimating human exposure to PFOS isomers and PFCA homologues: The relative importance of direct and indirect (precursor) exposure. *Environ Int, 74*, 160–169.

Gebbink, W. A., Ullah, S., Sandblom, O. & Berger, U. 2013. Polyfluoroalkyl phosphate esters and perfluoroalkyl carboxylic acids in target food samples and packaging—method development and screening. *Environ Sci Pollut Res Int, 20,* 7949–7958.

Gellrich, V., Stahl, T. & Knepper, T. P. 2012. Behavior of perfluorinated compounds in soils during leaching experiments. *Chemosphere, 87,* 1052.

German Environment Agency (UBA) and Norwegian Geotechnical Institute (NGI). 2018. *PMT and vPvM substances under REACH* [Online]. Available: www.umweltbundesamt. de/sites/default/files/medien/421/dokumente/workshop_program_05032018.pdf. Accessed September 24, 2018.

Giesy, J. P., Naile, J. E., Khim, J. S., Jones, P. D. & Newsted, J. L. 2010. Aquatic toxicology of perfluorinated chemicals. *Rev Environ Contam Toxicol, 202,* 1–52.

Giesy, J. P. Maybury, S. A., Kannan, K., Jones, P. D., Newsted, J. L., & Coady, K. 2006. In: R. Hites (ed.) *Handbook of Environmental Chemistry.* Heidleberg, Germany: Springer-Verlag.

Giesy, J. P., Naile, J. E., Khim, J. S., Jones, P. D. & Newsted, J. L. 2010. Aquatic toxicology of perfluorinated chemicals. *Rev Environ Contam Toxicol, 202,* 1–52.

Giger W, Alder AA, Göbel A, Golet E, McArdell C, Molnar E, Schaffner C. 2005. Polar POPS ante portas - Antibiotics and Anticorrosives as Contaminants in Wastewaters, Ambient Water and Drinking Water; Lecture at CEEC Workshop "Classical and Environmental Contaminants: from Lakes to Ocean" Dübendorf CH.

Giri, R. R., Ozaki, H., Morigaki, T., Taniguchi, S. & Takanami, R. 2011. UV photolysis of perfluorooctanoic acid (PFOA) in dilute aqueous solution. *Water Sci Technol, 63,* 276–282.

Glynn, A., Berger, U., Bignert, A., Ullah, S., Aune, M., Lignell, S. & Darnerud, P. O. 2012. Perfluorinated alkyl acids in blood serum from primiparous women in Sweden: Serial sampling during pregnancy and nursing, and temporal trends 1996–2010. *Environ. Sci. Technol., 46,* 9071.

Gobelius, L., Hedlund, J., Durig, W., Troger, R., Lilja, K., Wiberg, K. & Ahrens, L. 2018. Per- and polyfluoroalkyl substances in Swedish groundwater and surface water: implications for environmental quality standards and drinking water guidelines. *Environ Sci Technol.*

Gobelius, L., Lewis, J. & Ahrens, L. 2017. Plant uptake of per- and polyfluoroalkyl substances at a contaminated fire training facility to evaluate the phytoremediation potential of various plant species. *Environmental Science & Technology, 51,* 12602–12610.

Goldenthal, E. I. 1978. Final Report, Ninety Day Subacute Rhesus Monkey Toxicity Study. International Research and Development Corporation, Study No. 137-090, November 10, 1978. U.S. Environmental Protection Agency Administrative Record 226-0447.

Goldenthal, E. I., D. C. Jessup, R. G. Geil, N. D. Jefferson, and R. J. Arceo. 1978. Ninety-Day Subacute Rat Study. Study No. 137-085. International Research and Development Corporation, Mattawan, MI.

Goldenthal, E. I., D. C. Jessup, R. G. Geil, and J. S. Mehring. 1979. Ninety-Day Subacute Rhesus Monkey Toxicity Study. Study No. 137-087. International Research and Development Corporation, Mattawan, MI.

Goldman, P. 1965. The enzymatic cleavage of the carbon-fluorine bond in fluoroacetate. *Journal of Biological Chemistry, 240*(8), 3434–3438.

Gole, V. L., Sierra-Alvarez, R., Peng, H., Giesy, J. P., Deymier, P. & Keswani, M. 2018. Sonochemical treatment of per- and poly-fluoroalkyl compounds in aqueous film-forming foams by use of a large-scale multi-transducer dual-frequency based acoustic reactor. *Ultrasonics Sonochemistry, 45,* 213–222.

Gole, V. L., Fishgold, A., Sierra-Alvarez, R., Deymier, P. & Keswani, M. 2017. Treatment of (per) fluorooctanesulfonic acid (PFOS) using a large-scale sonochemical reactor. *Separation and Purification Technology.*

Gomez-Ruiz, B., Gómez-Lavín, S., Diban, N., Boiteux, V., Colin, A., Dauchy, X. & Urtiaga, A. 2017. Efficient electrochemical degradation of poly- and perfluoroalkyl substances (PFASs) from the effluents of an industrial wastewater treatment plant. *Chemical Engineering Journal, 322,* 196–204.

Gomis, M. I., Vestergren, R., Borg, D. & Cousins, I. T. 2018. Comparing the toxic potency in vivo of long-chain perfluoroalkyl acids and fluorinated alternatives. *Environ Int, 113,* 1–9.

Gomis, M. I., Wang, Z., Scheringer, M. & Cousins, I. T. 2015. A modeling assessment of the physicochemical properties and environmental fate of emerging and novel per- and polyfluoroalkyl substances. *Science of the Total Environment, 505,* 981–991.

Gordon, S. C. 2011. Toxicological evaluation of ammonium 4,8-dioxa-3H-perfluorononanoate, a new emulsifier to replace ammonium perfluorooctanoate in fluoropolymer manufacturing. *Regul Toxicol Pharmacol, 59,* 64–80.

Gorrochategui, E., Perez-Albaladejo, E., Casas, J., Lacorte, S. & Porte, C. 2014. Perfluorinated chemicals: Differential toxicity, inhibition of aromatase activity and alteration of cellular lipids in human placental cells. *Toxicol Appl Pharmacol, 277,* 124–130.

Gribble, G. 2002. *Naturally occurring organofluorines,* Berlin, Heidelberg: Springer.

Grieco, S. A. Lang, J., 2018. Per- and polyfluoroalkyl substances (PFAS) in municipal solid waste and landfill leachate. National Waste & Recycling Association and Environmental Research and Education Foundation.

Guelfo, J. L., Marlow, T., Klein, D. M., Savitz, D. A., Frickel, S., Crimi, M. & Suuberg, E. M. 2018. Evaluation and management strategies for per- and polyfluoroalkyl substances (PFASs) in drinking water aquifers: Perspectives from impacted U.S. Northeast communities. *Environ Health Perspect, 126,* 065001.

Guelfo, J. L. & Higgins, C. P. 2013. Subsurface transport potential of perfluoroalkyl acids at aqueous film-forming foam (AFFF)-impacted sites. *Environmental Science & Technology, 47,* 4164–4171.

Guo, Z., Liu, X., Krebs, K. A. & Roache, N. F. 2009. Perfluorocarboxylic acid content in 116 articles of commerce. *In:* AGENCY, U. E. P. (ed.). www.oecd.org/env/48125746.pdf: Research Triangle Park, NC.

Hagen, D. F., Belisle, J., Johnsson, J. D., & Venkaterswarlu, P., 1981. Characterization of fluorinated metabolites by gas chromatographic helium microwave plasma detector–The biotransformation of 1H,1H,2H,2H-perfluorodecanol to perfluorooctanoate. *Anal. Biochem., 118,* 336–343.

Hamm, M. P., Cherry, N. M., Chan, E., Martin, J. W. & Burstyn, I. 2010. Maternal exposure to perfluorinated acids and fetal growth. *Journal of Exposure Science and Environmental Epidemiology 20,* 589–597.

Hamrock, S. J. & P. T. Pham. 2000. United States Patent Number 6,063,522, Electrolyes Containing Mixed Fluorochemical/Hydrocarbon Imide and Methide Salts. May 16.

Hansen, K. J., Johnson, H. O., Eldridge, J. S., Butenhoff, J. L. and Dick, L. A. 2002. Quantitative characterization of trace levels of PFOS and PFOA in the Tennessee River. *Environmental Science & Technology, 36*(8), pp. 1681–1685.

Hansen, M. C., Børresen, M. H., Schlabach, M. & Cornelissen, G. 2010. Sorption of perfluorinated compounds from contaminated water to activated carbon. *Journal of Soils and Sediments, 10,* 179–185.

Hao, F., Guo, W., Wang, A., Leng, Y. & Li, H. 2014. Intensification of sonochemical degradation of ammonium perfluorooctanoate by persulfate oxidant. *Ultrasonics Sonochemistry, 21,* 554–558.

Harding-Marjanovic, K. C., Houtz, E. F., Yi, S., Field, J. A., Sedlak, D. L. & Alvarez-Cohen, L. 2015. Aerobic biotransformation of fluorotelomer thioether amido sulfonate (Lodyne) in AFFF-amended microcosms. *Environmental Science & Technology, 49,* 7666–7674.

Harper D. B., O. H. D., Murphy, C. 2005. Fluorinated natural products: Occurrence and biosynthesis. *Natural Production of Organohalogen Compounds,* 56.

Hayashi, R., Obo, H., Takeuchi, N. & Yasuoka, K. 2015. Decomposition of perfluorinated compounds in water by DC plasma within oxygen bubbles. *Electrical Engineering in Japan, 190*, 9–16.

Held, T., and M. Reinhard. 2016. Analyzed PFAS—the tip of the iceberg? *Contaminated Sites Spectrum, 25*, 173–186., doi: 0942-3818.

Heinicke, G., Hennig, H. P., Linke, E., Steinike, U., Thiessen, K. P. & Meyer, K. 1984. Tribochemistry Akademie, Verlag, Berlin 1984 495 S., 329 Abb., 106 Tab. Preis:98,-M. *Crystal Research and Technology, 19*, 1424–1424.

Hellsing, M. S., S. Josefsson, A. V. Hughes, & L. Ahrens. 2016. Sorption of perfluoroalkyl substances to two types of minerals. *Chemosphere, 159*, 385–391.

Herzke, D., Olsson, E. & Posner, S. 2012. Perfluoroalkyl and polyfluoroalkyl substances (PFASs) in consumer products in Norway: A pilot study. *Chemosphere, 88*, 980–987.

Herzke, D., Schlabach, M. & Mariussen, E. 2007. Literature survey of polyfluorinated organic compounds, phosphor containing flame retardants, 3-nitrobenzanthrone, organic tin compounds, platinum and silver.

Heydebreck, F., Tang, J., Xie, Z. and Ebinghaus, R. 2015. Correction to Alternative and Legacy Perfluoroalkyl Substances: Differences between European and Chinese River/Estuary Systems. *Environmental science & technology, 49*(24), pp. 14742–14743.

Higgins, C. 2017. Sorption of poly- and perfluoroalkyl substances (PFASs) relevant to aqueous film forming foam (AFFF) impacted groundwater by biochars and activated carbon. *Presentation*.

Higgins, C. & Dickenson, E. R. 2016. Treatment and mitigation strategies for poly and perfluoroalkyl substances. *Water Research Foundation,* WRF report 4322, 1–123.

Higgins, C. P. & Luthy, R. G. 2006a. Sorption of perfluorinated surfactants on sediments. *Environ Sci Technol, 40*, 7251–7256.

Higgins, C. P. & Luthy, R. G. 2006b. Sorption of perfluorinated surfactants on sediments. *Environmental Science & Technology, 40*, 7251–7256.

Higgins, C. P., Field, J. A., Criddle, C. S. & Luthy, R. G. 2005. Quantitative determination of perfluorochemicals in sediments and domestic sludge. *Environmental Science & Technology, 39*, 3946–3956.

Hill, J. T., Erdman, J.P. 1977. Anionic polymerization of fluorocarbon epoxides. *ACS Symposium Series, 59*, 269–284.

Hintzer, K. 2006. Polymerization of hexafluoropropylene oxide. *European Patent,* EP 1 698 650 A1.

Hoffman, M. R., Martin, S. T., Choi, W. & Bahnemann, D. W. 1995. Environmental applications of semiconductor photocatalysis. *Chemical Reviews, 95*, 69–96.

Hoke, R. A., Ferrell, B. D., Sloman, T. L., Buck, R. C. & Buxton, L. W. 2016. Aquatic hazard, bioaccumulation and screening risk assessment for ammonium 2,3,3,3-tetrafluoro-2-(heptafluoropropoxy)-propanoate. *Chemosphere, 149*, 336–342.

Holmes, N. 2018. *Fluorotelomer-based perfluoroalkyl acids detected pre- and post-TOP assay.* Data transmittal via email communication from N. Holmes to I. Ross. March.

Holmquist, H., Schellenberger, S., Van Der Veen, I., Peters, G. M., Leonards, P. E. & Cousins, I. T. 2016. Properties, performance and associated hazards of state-of-the-art durable water repellent (DWR) chemistry for textile finishing. *Environ Int, 91*, 251–264.

Hong, S., Khim, J. S., Park, J., Kim, M., Kim, W. K., Jung, J., Hyun, S., Kim, J. G., Lee, H., Choi, H. J., Codling, G. & Giesy, J. P. 2013. In situ fate and partitioning of waterborne perfluoroalkyl acids (PFAAs) in the Youngsan and Nakdong River Estuaries of South Korea. *Sci Total Environ, 445–446*, 136–145.

Hori, H., Nagaoka, Y., Murayama, M. & Kutsuna, S. 2008. Efficient decomposition of perfluorocarboxylic acids and alternative fluorochemical surfactants in hot water. *Environmental Science & Technology, 42*, 7438–7443.

Hori, H., Yamamoto, A., Koike, K., Kutsuna, S., Osaka, I. & Arakawa, R. 2007a. Persulfateinduced photochemical decomposition of a fluorotelomer unsaturated carboxylic acid in water. *Water Research, 41,* 2962–2968.

Hori, H., Yamamoto, A., Koike, K., Kutsuna, S., Osaka, I. & Arakawa, R. 2007b. Photochemical decomposition of environmentally persistent short-chain perfluorocarboxylic acids in water mediated by iron(II)/(III) redox reactions. *Chemosphere, 68,* 572–578.

Hori, H., Nagaoka, Y., Yamamoto, A., Sano, T., Yamashita, N., Taniyasu, S., Kutsuna, S., Osaka, I. & Arakawa, R. 2006. Efficient decomposition of environmentally persistent perfluorooctanesulfonate and related fluorochemicals using zerovalent iron in subcritical water. *Environmental Science & Technology, 40,* 1049–1054.

Hori, H., Hayakawa, E., Einaga, H., Kutsuna, S., Koike, K., Ibusuki, T., Kiatagawa, H. & Arakawa, R. 2004. Decomposition of environmentally persistent perfluorooctanoic acid in water by photochemical approaches. *Environmental Science & Technology, 38,* 6118–6124.

Hori, H., Yamamoto, A., Hayakawa, E., Taniyasu, S., Yamashita, N., Kutsuna, S., Kiatagawa, H. & ArakawA, R. 2005. Efficient decomposition of environmentally persistent perfluorocarboxylic acids by use of persulfate as a photochemical oxidant. *Environmental Science & Technology, 39,* 2383–2388.

Horst, J., J. Mcdonough, I. Ross, M. Dickson, J. Miles, J. Hurst, & P. Storch. 2018. Water treatment technologies: The next generation. *Groundwater Monitoring and Remediation.*

Houde, M., De Silva, A. O., Muir, D. C. & Letcher, R. J. 2011. Monitoring of perfluorinated compounds in aquatic biota: An updated review. *Environ Sci Technol, 45,* 7962–7973.

Houtz, E. F., Sutton, R., Park, J. S. & Sedlak, M. 2016. Poly- and perfluoroalkyl substances in wastewater: Significance of unknown precursors, manufacturing shifts, and likely AFFF impacts. *Water Res, 95,* 142–149.

Houtz, E. 2013. *Oxidative measurement of perfluoroalkyl acid precursors: Implications for urban runoff management and remediation of AFFF-contaminated groundwater and soil.* Doctor of Philosophy, University of California, Berkeley.

Houtz, E. F., Higgins, C. P., Field, J. A. & Sedlak, D. L. 2013. Persistence of perfluoroalkyl acid precursors in AFFF-impacted groundwater and soil. *Environmental Science & Technology, 47,* 8187–8195.

Houtz, E. F. & Sedlak, D. L. 2012. Oxidative conversion as a means of detecting precursors to perfluoroalkyl acids in urban runoff. *Environmental Science & Technology, 46,* 9342–9349.

Hu, X. C., Andrews, D. Q., Lindstrom, A. B., Bruton, T. A., Schaider, L. A., Grandjean, P., Lohmann, R., Carignan, C. C., Blum, A., Balan, S. A., Higgins, C. P. & Sunderland, E. M. 2016a. Detection of poly- and perfluoroalkyl substances (PFASs) in U.S. drinking water linked to industrial sites, military fire training areas, and wastewater treatment plants. *Environmental Science & Technology Letters.*

Hu, X. C., Andrews, D. Q., Lindstrom, A. B., Bruton, T. A., Schaider, L. A., Grandjean, P., Lohmann, R., Carignan, C. C., Blum, A., Balan, S. A., Higgins, C. P. & Sunderland, E. M. 2016b. Detection of poly- and perfluoroalkyl substances (PFASs) in U.S. drinking water linked to industrial sites, military fire training areas, and wastewater treatment plants. *Environ. Sci. Technol. Lett., 3,* 344.

Hu, Z., Song, X., Wei, C. & Liu, J. 2017. Behavior and mechanisms for sorptive removal of perfluorooctane sulfonate by layered double hydroxides. *Chemosphere, 187,* 196–205.

Huang, L., Dong, W. & Hou, H. 2007. Investigation of the reactivity of hydrated electron toward perfluorinated carboxylates by laser flash photolysis. *Chemical Physics Letters, 436,* 124–128.

Huang, Q. 2017. Electrochemical degradation of perfluoroalkyl acids by macroporous titanium suboxide anode. *SERDP ESTCP Symposium Per- and Polyfluoroalkyl Substances (PFAS): Treatments to Replacements,* November 28, 2017.

Integral Consulting Inc. 2016. *Compendium of State Regulatory Activities on Emerging Contaminants* [Online]. Available: http://anr.vermont.gov/sites/anr/files/specialtopics/Act154ChemicalUse/REF%20Article%20-%20Integral%2C%20Compendium%20of%20State%20Regulatory%20Authorities%202016.pdf. AccessedSeptember 24, 2018.

Ingelido, A. M., Abballe, A., Gemma, S., Dellatte, E., Iacovella, N., De Angelis, G., Zampaglioni, F., Marra, V., Miniero, R., Valentini, S., Russo, F., Vazzoler, M., Testai, E. & De Felip, E. 2018. Biomonitoring of perfluorinated compounds in adults exposed to contaminated drinking water in the Veneto Region, Italy. *Environ Int, 110*, 149–159.

International Standards Organization (ISO). 2009. Water quality—Determination of perfluorooctanesulfonate (PFOS) and perfluorooctanoate (PFOA)—Method for unfiltered samples using solid phase extraction and liquid chromatography/mass spectrometry. ISO 25101:2009(E). March 1.

Inyang, M. & Dickenson, E. R. V. 2017. The use of carbon adsorbents for the removal of perfluoroalkyl acids from potable reuse systems. *Chemosphere, 184*, 168–175.

Interstate Technology & Regulatory Council (ITRC). 2018a. Naming Conventions and Physical and Chemical Properties of Per- and Polyfluoroalkyl Substances (PFAS). [Online]. Available: https://pfas-1.itrcweb.org/wp-content/uploads/2018/03/pfas_fact_sheet_naming_conventions__3_16_18.pdf. [Accessed September 24, 2018].

Interstate Technology & Regulatory Council (ITRC 2018b). PFAS Fact Sheet fate and Transport. [Online]. Available: https://pfas-1.itrcweb.org/wp-content/uploads/2018/03/pfas_fact_sheet_fate_and_transport__3_16_18.pdf. Accessed September 24, 2018.

Interstate Technology & Regulatory Council (ITRC). 2015. *Integrated DNAPL site characterization and tools selection (ISC-1)*. Washington, D.C.: Interstate technology & regulatory council, DNAPL Site Characterization Team. *Appendix I. Representative Values for Foc* [Online]. Available: www.itrcweb.org/DNAPL-ISC_tools-selection/Content/Appendix%20I.%20Foc%20Tables.htm. Accessed September 24, 2018.

Interstate Technology & Regulatory Council (ITRC). 2010. A decision framework for applying monitored natural attenuation processes to metals and radionuclides in groundwater. (Online). Available: www.itrcweb.org/GuidanceDocuments/APMR1.pdf. Accessed September 24, 2018.

Jaffré, T., Kersten, W. J., Brooks, R. R., & Reeves, R. D. 1979. Nickel uptake by the Flacourtiaceae of New Caledonia. *Proceedings of the Royal Society London, 205*, 385–394.

Jarnberg, U., Holmstrom, K., Van Bavel, B. & Kärrman, A. 2006. Perfluoroalkylated acids and related compounds (PFAS) in the Swedish environment—Chemistry, sources, exposures. Stockholm University and Orebro University.

Javaherian 2016. VEG Technology in-situ and ex-situ thermal treatment of soils enhanced oil & NAPL recovery. *Data Sheet*.

Jeon, J., Kannan, K., Lim, B. J., An, K. G. & Kim, S. D. 2011. Effects of salinity and organic matter on the partitioning of perfluoroalkyl acid (PFAs) to clay particles. *J Environ Monit, 13*, 1803–1810.

Ji, K., Kim, Y., Oh, S., Ahn, B., Jo, H. & Choi, K. 2008. Toxicity of perfluorooctane sulfonic acid and perfluorooctanoic acid on freshwater macroinvertebrates (Daphnia Magna and Moina Macrocopa) and fish (Oryzias Latipes). *Environ. Toxicol. Chem., 27*, 2159.

Kabadi, S. V., Fisher, J., Aungst, J. & Rice, P. 2018. Internal exposure-based pharmacokinetic evaluation of potential for biopersistence of 6:2 fluorotelomer alcohol (FTOH) and its metabolites. *Food Chem Toxicol, 112*, 375–382.

Kannan, K., S. Corsolini, J. Falandysz, G. Fillmann, K. S. Kumar, B. G. Loganathan, M. Ali M. et al. 2004. Perfluorooctanesulfonate and related fluorochemicals in human blood from several countries. *Environmental science & technology* 38, no. 17 (2004): 4489–4495.

Kärrman A., Lafossas, C. & Møskeland, T. 2011. Environmental levels and distribution of structural isomers of perfluoroalkyl acids after aqueous fire-fighting foam (AFFF) contamination. *Environmental Chemistry, 8*, 372–380.

Kärrman, A., Van Bavel, B., Jarnberg, U., Hardell, L. & Lindstrom, G. 2006. Perfluorinated chemicals in relation to other persistent organic pollutants in human blood. *Chemosphere, 64*, 1582–1591.

Kawano, S., Kida, T., Takemine, S., Matsumura, C., Nakano, T., Kuramitsu, M., Adachi, K., Akashi, M. 2013. Efficient removal and recovery of perfluorinated compounds from water by surface-tethered β-cyclodextrins on polystyrene particles. *Chemistry Letters, 42*, 392–394.

Kerfoot, W. B. 2013. Method and apparatus for treating perfluoroalkyl compounds. *In:* Office, U. S. P. A. T. (ed.) Washington, D.C. www.google.com/patents/US9694401.

Kim, S. K. & K. Kannan. 2007. Perfluorinated acids in air, rain, snow, surface runoff, and lakes: relative importance of pathways to contamination of urban lakes. *Environmental science & technology, 41*(24), pp. 8328–8334.

Kirsch, P. 2004. *Modern Fluoroorganic Chemistry*, Weinheim, WILEY-VCH Verlag GmbH & Co. KGaA.

Kissa, E. ed. 2001. *Fluorinated* surfactants and repellents. Vol. 97. CRC Press, 2001.

Kjølholt, J., Jensen, A. A., & Warming, M., 2015. Short-chain polyfluoroalkyl substances (PFAS): A literature review of information on human health effects and environmental fate and effect aspects of short-chain PFAS. *Environmental Project No. 1707, 2015*. www2.mst.dk/Udgiv/publications/2015/05/978-87-93352-15-5.pdf: DEPA.

Klein, D. 2017. Toxicity of perfluoroalkyl substances. *Presentation.*

Knapton, J. 2016. Jersey FTA design.

Kotthoff, M. & Bücking, M. 2018. Four chemical trends will shape the next decade's directions in perfluoroalkyl and polyfluoroalkyl substances research. *Frontiers in Chemistry*, 6.

Krafft, M. P. & Riess, J. G. 2015. Per- and polyfluorinated substances (PFASs): Environmental challenges. *Current Opinion in Colloid & Interface Science, 20*, 192–212.

Krusic, P. J., Marchione, A. A. & Roe, D. C. 2005. Gas-phase NMR studies of the thermolysis of perfluorooctanoic acid. *Journal of Fluorine Chemistry, 126*, 1510–1516.

Krusic, P. J. & Roe, D. C. 2004. Gas-phase NMR technique for studying the thermolysis of materials: thermal decomposition of ammonium perfluorooctanoate. *Anal Chem, 76*, 3800–3803.

Kutsuna, S., Nagaoka, Y., Takeuchi, K., & Hori, H. 2006. TiO2-induced heterogeneous photodegradation of a fluorotelomer alcohol in air. *Environ. Sci. Technol., 40*, 6824–6829.

Lagow, R. J. & Inoue, S. 1978. Method for producing perfluoroether oligomers having terminal carboxylic acid groups. *Patent*, US4113772 A.

Lai, S., Song, J., Song, T., Huang, Z., Zhang, Y., Zhao, Y., Liu, G., Zheng, J., Mi, W., Tang, J. and Zou, S. 2016. Neutral polyfluoroalkyl substances in the atmosphere over the northern South China Sea. *Environmental Pollution, 214*, pp.449–455.

Lang, J. R., Allred, B. M., Field, J. A., Levis, J. W. & Barlaz, M. A. 2017. National estimate of per- and polyfluoroalkyl substance (PFAS) release to U.S. municipal landfill leachate. *Environ Sci Technol, 51*, 2197–2205.

Lang, J. R., Allred, B. M., Peaslee, G. F., Field, J. A. & Barlaz, M. A. 2016. Release of per- and polyfluoroalkyl substances (PFASs) from carpet and clothing in model anaerobic landfill reactors. *Environ Sci Technol, 50*, 5024–5032.

Lath, S., Navarro, Divina A., Kumar, Anu, Losic, Dusan, Mclaughlin, Michael J. 2017. Adsorption of Perfluorooctanoic acid (PFOA) using graphene-based materials. http://ziltek.com.au/pdf/Z914-01-RemBind-Platform%20Presentation-AdsorptionofPFOAusingGraphene-CleanUp2017.pdf.

Lau, C. 2016. Perfluoroalkyl substances: Highlights of toxicological findings—EPA. *Presentation.*

Lau, C., Anitole, K., Hodes, C., Lai, D., Pfahles-Hutchens, A. & Seed, J. 2007. Perfluoroalkyl acids: A review of monitoring and toxicological findings. *Toxicol Sci, 99*, 366–394.

Lau C., J. R. Thibodeaux, R.G. Hanson, M. G. Narotsky, J. M. Rogers, A. B. Lindstrom, M. J. Strynar. 2006. Effects of perfluorooctanoic acid exposure during pregnancy in the mouse. *Toxicol.Sci.* 90, 510–518.

Lau, C., Butenhoff, J. L. & Rogers, J. M. 2004. The developmental toxicity of perfluoroalkyl acids and their derivatives. *Toxicol Appl Pharmacol, 198*, 231–241.

Lau C., J. R. Thibodeaux, R. G. Hanson, J. M. Rogers, B. E. Grey, M. E. Stanton, J. L. Butenhoff, L. A. Stevenson. 2003. Exposure to perfluorooctane sulfonate during pregnancy in rat and mouse. II: postnatal evaluation. *Toxicol. Sci.* 74:382–392

Lee, H. & Mabury, S. A. 2017. Sorption of perfluoroalkyl phosphonates and perfluoroalkyl phosphinates in soils. *Environ Sci Technol, 51*, 3197–3205.

Lee, H., Tevlin, A. G., & Mabury, S. A. 2013. "Fate of polyfluoroalkyl phosphate diesters and their metabolites in biosolids-applied soil: biodegradation and plant uptake in greenhouse and field experiments." *Environmental science & technology* 48, no. 1: 340–349.

Lee, H., D'eon, J. & Mabury, S. A. 2010. Biodegradation of polyfluoroalkyl phosphates as a source of perfluorinated acids to the environment. *Environmental Science & Technology, 44*, 3305–3310.

Lee, Y. C., Lo, S. L., Chiueh, P. T. & Chang, D. G. 2009. Efficient decomposition of perfluorocarboxylic acids in aqueous solution using microwave-induced persulfate. *Water Research, 43*, 2811–2816.

Lee, Y. C., Lo, S. L., Chiueh, P. T., Liou, Y. H. & Chen, M. L. 2010. Microwave-hydrothermal decomposition of perfluorooctanoic acid in water by iron-activated persulfate oxidation. *Water Research, 44*, 886–892.

Li, X., Fang, J., Liu, G., Zhang, S., Pan, B. & Ma, J. 2014. Kinetics and efficiency of the hydrated electron-induced dehalogenation by the sulfite/UV process. *Water Research, 62*, 220–228.

Li, X., Zhang, P., Jin, L., Shao, T., Li, Z. & Cao, J. 2012. Efficient photocatalytic decomposition of perfluorooctanoic acid by indium oxide and its mechanism. *Environ Sci Technol, 46*, 5528–5534.

Li, X., et al. 2011. Enhanced adsorption of PFOA and PFOS on multiwalled carbon nanotubes under electrochemical assistance. *Environmental Science and Technology, 45*, 8498–8505.

Li, Y., Oliver, D. P. & Kookana, R. S. 2018. A critical analysis of published data to discern the role of soil and sediment properties in determining sorption of per and polyfluoroalkyl substances (PFASs). *Sci Total Environ, 628–629*, 110–120.

Lin, H., Wang, Y., Niu, J., Yue, Z. & Huang, Q. 2015. Efficient sorption and removal of perfluoroalkyl acids (PFAAs) from aqueous solution by metal hydroxides generated in situ by electrocoagulation. *Environ Sci Technol, 49*, 10562–10569.

Lindstrom, A. B., Strynar, M. J., Delinsky, A. D., Nakayama, S. F., Mcmillan, L., Libelo, E. L., Neill, M. & Thomas, L. 2011. Application of WWTP biosolids and resulting perfluorinated compound contamination of surface and well water in Decatur, Alabama, USA. *Environ Sci Technol, 45*, 8015–8021.

Lindstrom, A. B., Strynar, M. J. & Libelo, E. L. 2011. Polyfluorinated compounds: Past, present, and future. *Environ Sci Technol, 45*, 7954–7961.

Lipson, D., Raine, B., & Webb, M., 2013. Transport of perfluorooctane sulfonate (PFOS) fractured bedrock at a well-characterized site. *Proceedings of SETAC-EU Conference, 2013*.

Lui, B., Zhang, H., Yao, D., Li, J., Xie, L., Wang, X., Wang, Y., Liu, G., and Yang, B. 2015. Perfluorinated compounds (PFCs) in the atmosphere of Shenzhen, China: Spatial distribution, sources and health risk assessment. *Chemosphere* 138: 511–518.

Liu, C., Gin, K. Y., Chang, V. W., Goh, B. P. & Reinhard, M. 2011. Novel perspectives on the bioaccumulation of PFCs: The concentration dependency. *Environmental Science & Technology, 45*, 9758–9764.

Liu, G., Dhana, K., Furtado, J. D., Rood, J., Zong, G., Liang, L., Qi, L., Bray, G. A., Dejonge, L., Coull, B., Grandjean, P. & Sun, Q. 2018. Perfluoroalkyl substances and changes in body weight and resting metabolic rate in response to weight-loss diets: A prospective study. *PLoS Med, 15*, e1002502.

Liu, J. & Mejia Avendano, S. 2013. Microbial degradation of polyfluoroalkyl chemicals in the environment: A review. *Environment International, 61*, 98–114.

Liu, J., Qu, R., Wang, Z., Mendoza-Sanchez, I. & Sharma, V. K. 2017. Thermal- and photo-induced degradation of perfluorinated carboxylic acids: Kinetics and mechanism. *Water Res, 126*, 12–18.

Liu, J., Wang, N., Buck, R. C., Wolstenholme, B. W., Folsom, P. W., Sulecki, L. M. & Bellin, C. A. 2010. Aerobic biodegradation of [14C] 6:2 fluorotelomer alcohol in a flow-through soil incubation system. *Chemosphere, 80*, 716–723.

Liu, J., Wang, N., Szostek, B., Buck, R. C., Panciroli, P. K., Folsom, P. W., Sulecki, L. M. & Bellin, C. A. 2010. 6-2 Fluorotelomer alcohol aerobic biodegradation in soil and mixed bacterial culture. *Chemosphere, 78*, 437–444.

Liu, Z., Lu, Y., Wang, T., Wang, P., Li, Q., Johnson, A. C., Sarvajayakesavalu, S. & Sweetman, A. J. 2016. Risk-assessment and source identification of perfluoroalkyl acids in surface and ground water: Spatial distribution around a mega-fluorochemical industrial park, China. *Environ Int, 91*, 69–77.

Llorca, M., Farre, M., Pico, Y., Muller, J., Knepper, T. P. & Barcelo, D. 2012. Analysis of perfluoroalkyl substances in waters from Germany and Spain. *Sci Total Environ, 431*, 139–150.

Loganathan, B. G. & Kwan-Sing Lam, P. 2011. *Global contamination trends of persistent organic chemicals.* Boca Raton, FL: CRC Press.

Loganathan, B. G., Sajwan, K. S., Sinclair, E., Senthil Kumar, K. & Kannan, K. 2007. Perfluoroalkyl sulfonates and perfluorocarboxylates in two wastewater treatment facilities in Kentucky and Georgia. *Water Res, 41*, 4611–4620.

Logeshwaran, P., Sivaram, A. K., Megharaj, M. & Naidu, R. 2016. Evaluation of cyto- and gentoxic effects of Class B firefighting foam products: Tridol-S 3% AFFF and Tridol-S 6% AFFF to Allium cepa. *Environmental Technology & Innovation.*

Long, Y., Y. Wang, G. Ji, L. Yan, F. Hu, and A. Gu. 2013. Neurotoxicity of perfluorooctane sulfonate to hippocampal cells in adult mice. *PLOS ONE* 8(1):e54176.

Loos, R. 2013. EU wide monitoring survey on waste water treatment plant effluents. *European Commission, JSC Scientific and Policy Reports.*

Loos, R., Carvalho, R., António, D. C., Comero, S., Locoro, G., Tavazzi, S., Paracchini, B., Ghiani, M., Lettieri, T., Blaha, L., Jarosova, B., Voorspoels, S., Servaes, K., Haglund, P., Fick, J., Lindberg, R. H., Schwesig, D. & Gawlik, B. M. 2013. EU-wide monitoring survey on emerging polar organic contaminants in wastewater treatment plant effluents. *Water Research, 47*, 6475–6487.

Loos, R., Gawlik, B. M., Locoro, G., Rimaviciute, E., Contini, S. & Bidoglio, G. 2009. EU-wide survey of polar organic persistent pollutants in European river waters. *Environ Pollut, 157*, 561–568.

Loos, R., Locoro, G., Comero, S., Contini, S., Schwesig, D., Werres, F., Balsaa, P., Gans, O., Weiss, S., Blaha, L., Bolchi, M. & Gawlik, B. M. 2010. Pan-European survey on the occurrence of selected polar organic persistent pollutants in ground water. *Water Res, 44*, 4115–4126.

Loos, R., Tavazzi, S., Mariani, G., Suurkuusk, G., Paracchini, B. & Umlauf, G. 2017. Analysis of emerging organic contaminants in water, fish and suspended particulate matter (SPM) in the Joint Danube Survey using solid-phase extraction followed by UHPLC-MS-MS and GC-MS analysis. *Sci Total Environ, 607–608*, 1201–1212.

Loveless, S. E., Finlay, C., Everds, N. E., Frame, S. R., Gillies,P. J., O'Connor, J. C., Powley, C. R., Kennedy, G. L. 2006. Comparative responses of rats and mice exposed to linear/branched, linear, or branched ammonium perfluorooctanoate (APFO). *Toxicology* 220, 203–217.

Lu, G. H., N. Gai, P. Zhang, H. T. Piao, S. Chen, X. C. Wang, X. C. Jiao, X. C. Yin, K. Y. Tan, & Y. L. Yang. 2017. Perfluoroalkyl acids in surface waters and tapwater in the Qiantang River watershed—Influences from paper, textile, and leather industries. *Chemosphere, 185*, pp. 610–617.

Luebker, D. J., M. T. Case, R. G. York, J. A. Moore, K. J. Hansen, and J. L. Butenhoff. 2005. Two-generation reproduction and cross-foster studies of perfluorooctanesulfonate (PFOS) in rats. *Toxicology* 215:126–148.

Luebker, D. J., R. G. York, K. J. Hansen, J. A. Moore, and J. L. Butenhoff. 2005. Neonatal mortality from in utero exposure to perfluorooctanesulfonate (PFOS) in Sprague-Dawley rats: Dose-response and biochemical and pharmacokinetic parameters. *Toxicology* 215:149–169.

Luo, Q., Lu, J., Zhang, H., Wang, Z., Feng, M., Chiang, S.-Y. D., Woodward, D. & Huang, Q. 2015. Laccase-catalyzed degradation of perfluorooctanoic acid. *Environmental Science & Technology Letters, 2*, 198–203.

Lyons, C. 2007. *Stain-resistant, Nonstick, Waterproof, and Lethal: The Hidden Dangers of C8*, Praeger Publications.

Lyu, Y., Brusseau, M. L., Chen, W., Yan, N., Fu, X. & Lin, X. 2018. Adsorption of PFOA at the air-water interface during transport in unsaturated porous media. *Environ Sci Technol.*

Ma, S.-H., Wu, M.-H., Tang, L., Sun, R., Zang, C., Xiang, J.-J., Yang, X.-X., Li, X. & Xu, G. 2017. EB degradation of perfluorooctanoic acid and perfluorooctane sulfonate in aqueous solution. *Nuclear Science and Techniques, 28*.

Maisonet, M., Terrell, M. L., Mcgeehin, M. A., Christensen, K. Y., Holmes, A., Calafat, A. M. & Marcus, M. 2012. Maternal concentrations of polyfluoroalkyl compounds during pregnancy and fetal and postnatal growth in British girls. *Environ Health Perspect, 120*.

Martin, J. W., Asher, B. J., Beesoon, S., Benskin, J. P. & Ross, M. S. 2010. PFOS or PreFOS? Are perfluorooctane sulfonate precursors (PreFOS) important determinants of human and environmental perfluorooctane sulfonate (PFOS) exposure? *J Environ Monit, 12*, 1979–2004.

Martin, J. W., Ellis, D. A., Mabury, S. A., Hurley, M. D. & Wallington, T. J. 2006. Atmospheric chemistry of perfluoroalkanesulfonamides: Kinetic and product studies of the OH radical and Cl atom initiated oxidation of N-Ethyl Perfluorobutanesulfonamide. *Environmental Science & Technology, 40*, 864–872.

Mason, T. J. 2000. Large-scale sonochemical processing: Aspiration and actuality. *Ultrasonics Sonochemistry, 7*, 145–149.

Massachusetts Department of Environmental Protection (MassDEP). 2017. DRAFT Fact Sheet, Guidance on Sampling and Analysis for PFAS at Disposal Sites Regulated under the Massachusetts Contingency Plan. January.

Mastrantonio, M., Bai, E., Uccelli, R., Cordiano, V., Screpanti, A. & Crosignani, P. 2017. Drinking water contamination from perfluoroalkyl substances (PFAS): An ecological mortality study in the Veneto Region, Italy. *Eur J Public Health.*

Matheson L. J., Visscher P. T., Schaefer, J. K., & Oremland, R. S. 1995. Summary of research results on bacterial degradation of Trifluoroacetate (TFA). *U.S. Geological Survey*, Open File Report, 96–219.

Matsui, Y., Aizawa, T., Kanda, F., Nigorikawa, N., Mima, S. & Kawase Y. 2007. Adsorptive removal of geosmin by ceramic membrane filtration with superpowdered activated carbon. *Journal of Water Supply: Research and Technology—AQUA, 56*, 411–418.

Matsui, Y. N., Taniguchi, T. & Matsushita, T. 2013. Geosmin and 2-methylisoborneol removal using superfine powdered activated carbon: Shell adsorption and branched pore kinetic model analysis and optimal particle size. *Water Research, 47*, 2873–2880.

Matsui, Y. T., Nakao, S., Knappe, D. & Matsushita, T. 2012. Characteristics of competitive adsorption between 2-methylisoborneol and natural organic matter on superfine and conventionally sized powdered activated carbon. *Water Research, 46*, 4741–4749.

McCleaf, P., Englund, S., Ostlund, A., Lindegren, K., Wiberg, K. & Ahrens, L. 2017. Removal efficiency of multiple poly- and perfluoroalkyl substances (PFASs) in drinking water using granular activated carbon (GAC) and anion exchange (AE) column tests. *Water Research, 120*, 77–87.

McGuire, M. E., Schaefer, C., Richards, T., Backe, W. J., Field, J. A., Houtz, E., Sedlak, D. L., Guelfo, J. L., Wunsch, A. & Higgins, C. P. 2014. Evidence of remediation-induced alteration of subsurface poly- and perfluoroalkyl substance distribution at a former firefighter training area. *Environmental Science & Technology, 48*, 6644–6652.

McGuire, M. E. 2013. An in-depth site characterization of poly- and perfluoroalkyl substances at an abandoned fire training area. *Thesis.* Colorado School of Mines, Master of Science in Environmental Science and Engineering.

McKenzie, E. R., Siegrist, R. L., McCray, J. E. & Higgins, C. P. 2016. The influence of a nonaqueous phase liquid (NAPL) and chemical oxidant application on perfluoroalkyl acid (PFAA) fate and transport. *Water Res, 92*, 199–207.

McKenzie, E. R., Siegrist, R. L., McCray, J. E. & Higgins, C. P. 2015. Effects of chemical oxidants on perfluoroalkyl acid transport in one-dimensional porous media columns. *Environmental Science & Technology, 49*, 1681–1689.

McNamara, J. D., Franco, R., Mimna, R. & Zappa, L. 2018. Comparison of activated carbons for removal of perfluorinated compounds from drinking water. *Journal of American Water Works Association, 110*, E2–E14.

Mededovic Thagard, S., T. Holsen, S. Richardson, P. Kulkami, C. Newell, M. Nixon, A. Bodour, & C. Varley. An enhanced contact electrical discharge plasma reactor: An effective technology to degrade per- and polyfluoroalkyl substances (PFAS). 11th International Conference on the Remediation of Chlorinated and Recalcitrant Compounds, 2018 Palm Springs, CA.

Meissner, E. A. W. A. 2007. Oligomerization of hexafluoropropylene oxide in the presence of alkali metal. *Polish Journal of Chemical Technology, 9*, 95–97.

Mejia-Avendano, S., Vo Duy, S., Sauve, S. & Liu, J. 2016. Generation of perfluoroalkyl acids from aerobic biotransformation of quaternary ammonium polyfluoroalkyl surfactants. *Environ Sci Technol, 50*, 9923–9932.

Merino, N., Qu, Y., Deeb, R. A., Hawley, E. L., Hoffmann, M. R. & Mahendra, S. 2016. Degradation and removal methods for perfluoroalkyl and polyfluoroalkyl substances in water. *Environmental Engineering Science, 33*, 615–649.

Meyer, T., De Silva, A. O., Spencer, C. & Wania, F. 2011. Fate of perfluorinated carboxylates and sulfonates during snowmelt within an urban watershed. *Environ Sci Technol, 45*, 8113–8119.

Miles, J. Ross, I., Horneman, A., Hurst, J., and Burdick, J. 2017. Fate and Transport Modelling of PFOS in a Fractured Chalk Aquifer towards a Large Scale Drinking Water Abstraction. ACS 253rd Meeting, San Francisco, 2nd-6th April 2017.

Ministry of the Environment of Japan. 2013. Summary of the guideline on the treatment of wastes containing perfluorooctane sulfonic acid (PFOS) and its salts in Japan.

Minnesota Department of Health (MDH). 2008. Health Consultation, PFOS Detections in the City of Brainerd, Minnesota, City of Brainerd, Crow Wing County, Minnesota. Prepared under cooperative agreement with the U.S. Department of Health and Human Services, Agency for Toxic Substances and Disease Registry.

Mitchell, S. M., Ahmad, M., Teel, A. L. & Watts, R. J. 2014. Degradation of perfluorooctanoic acid by reactive species generated through catalyzed H_2O_2 propagation reactions. *Environmental Science & Technology Letters, 1*, 117–121.

Mogensen, U. B., Grandjean, P., Heilmann, C., Nielsen, F., Weihe, P. & Budtz-Jorgensen, E. 2015. Structural equation modeling of immunotoxicity associated with exposure to perfluorinated alkylates. *Environ Health, 14*, 47.

Moller, A., Ahrens, L., Surm, R., Westerveld, J., Van Der Wielen, F., Ebinghaus, R. & De Voogt, P. 2010. Distribution and sources of polyfluoroalkyl substances (PFAS) in the River Rhine watershed. *Environ Pollut, 158*, 3243–3250.

Moody, C. A., Martin, J. W., Kwan, W. C., Muir, D. C. & Mabury, S. A. 2002. Monitoring perfluorinated surfactants in biota and surface water samples following an accidental release of fire-fighting foam into Etobicoke Creek. *Environ Sci Technol, 36*, 545–551.

Moriwaki, H., Takagi, Y., Tanaka, M., Tsuruho, K., Okitsu, K. & Maeda, Y. 2005. Sonochemical decomposition of perfluorooctane sulfonate and perfluorooctanoic acid. *Environmental Science & Technology, 39*, 3388–3392.

Muller, C. E., Spiess, N., Gerecke, A. C., Scheringer, M. & Hungerbuhler, K. 2011. Quantifying diffuse and point inputs of perfluoroalkyl acids in a nonindustrial river catchment. *Environ Sci Technol, 45*, 9901–9909.

Nabb, D. L., Szostek, B., Himmelstein, M. W., Mawn, M. P., Gargas, M. L., Sweeney, L. M., Stadler, J. C., Buck, R. C. & Fasano, W. J. 2007. In vitro metabolism of 8-2 fluorotelomer alcohol: Interspecies comparisons and metabolic pathway refinement. *Toxicol Sci, 100*, 333–344.

Nassi, M., Sarti, E., Pasti, L., Martucci, A., Marchetti, N., Cavazzini, A., Di Renzo, F. & Galarneau, A. 2014. Removal of perfluorooctanoic acid from water by adsorption on high surface area mesoporous materials. *Journal of Porous Materials*.

National Research Council. 2000. *Waste incineration and public health*. National Academies Press.

Neumann, M., D. Sattler, L. Vierke, and I. Schliebner. 2017. *A proposal for criteria and an assessment procedure to identify Persistent, Mobile and Toxic (PM or PMT) substances registered under REACH*. Presentation, 16th International Conference on Chemistry and the Environment in 2017. Oslo, Norway. June 22.

New Hampshire Department of Environmental Services (NHDES). 2016. Perfluorinated Compound (PFC) Sample Collection Guidance. November.

New Jersey Department of Environmental Protection (NJDEP). 2014. Occurence of Perfluorinated Chemicals in Untreated New Jersey Drinking Water Sources, Final Report. April.

Newton, S., Mcmahen, R., Stoeckel, J. A., Chislock, M., Lindstrom, A. & Strynar, M. 2017. Novel polyfluorinated compounds identified using high resolution mass spectrometry downstream of manufacturing facilities near Decatur, Alabama. *Environ Sci Technol, 51*, 1544–1552.

Ng, C. A. & Hungerbuhler, K. 2013. Bioconcentration of perfluorinated alkyl acids: How important is specific binding? *Environ Sci Technol, 47*, 7214–7223.

Nilsson, H., Kärrman, A., Rotander, A., Van Bavel, B., Lindstrom, G. & Westberg, H. 2013. Professional ski waxers' exposure to PFAS and aerosol concentrations in gas phase and different particle size fractions. *Environ Sci Process Impacts, 15*, 814–822.

North Carolina Department of Health and Human Services. 2017. Questions and Answers Regarding North Carolina Department of Health and Human Services Updated Risk Assessment for GenX (Perfluoro-2-propoxypropanoic acid). Dated July 14. Available: https://ncdenr.s3.amazonaws.com/s3fs-public/GenX/NC%20DHHS%20Risk%20 Assessment%20FAQ%20Final%20Clean%20071417%20PM.pdf. Accessed September 15, 2018.

Nowack, K. 2017. GAC treatment for PFC removal. Presentation.

O'Brien, J. M., Austin, A. J., Williams, A., Yauk, C. L., Crump, D. & Kennedy, S. W. 2011. Technical-grade perfluorooctane sulfonate alters the expression of more transcripts in cultured chicken embryonic hepatocytes than linear perfluorooctane sulfonate. *Environ. Toxicol. Chem., 30*, 2846.

Oakes, K. D., Benskin, J. P., Martin, J. W., Ings, J. S., Heinrichs, J. Y., Dixon, D. G. & Servos, M. R. 2010. Biomonitoring of perfluorochemicals and toxicity to the downstream fish community of Etobicoke Creek following deployment of aqueous film-forming foam. *Aquat Toxicol, 98*, 120–129.

Ochoa-Herrera, V., Field, J. A., Luna-Velasco, A. & Sierra-Alvarez, R. 2016. Microbial toxicity and biodegradability of perfluorooctane sulfonate (PFOS) and shorter chain perfluoroalkyl and polyfluoroalkyl substances (PFASs). *Environmental Science: Processes & Impacts, 18*, 1236–1246.

Ochoa-Herrera, V. & Sierra-Alvarez, R. 2008. Removal of perfluorinated surfactants by sorption onto granular activated carbon, zeolite and sludge. *Chemosphere, 72*, 1588–1593.

Ochoa-Herrera, V., Sierra-Alvarez, R., Somogyi, A., Jacobsen, N. E., Wysocki, V. H. & Field, J. A. 2008. Reductive defluorination of perfluorooctane sulfonate. *Environmental Science & Technology, 42*, 3260–3264.

Organisation for Economic Cooperation and Development (OECD). 2018a. *Comprehensive global database of per-and polyfluoroalkyl substances (PFASs)* [Online]. Available: https://urldefense.proofpoint.com/v2/url?u=http-3A__www.oecd.org_chemicalsafety_risk-2Dmanagement_global-2Ddatabase-2Dof-2Dper-2Dand-2Dpolyfluoroalkyl-2Dsubstances.xlsx&d=DwMFAg&c=tpTxelpKGw9ZbZ5Dlo0lybSxHDHIiYjksG4icXfalgk&r=vdRSWKVbQgcVIUfImGQWVd5-426hjlO1zHhzz_48Z2Y&m=BF1JbwFkvEq0MZO7EFXfLimutJ6URbVpHGHL6QovNT4&s=xT5RZuTaOUUt_8RnCuanl_NHUDPwepavFMZ9GqEHrwY&e= Accessed September 24, 2018.

Organisation for Economic Cooperation and Development (OECD). 2018b. *Portal on Per and Poly Fluorinated Chemicals* [Online]. Available: www.oecd.org/chemicalsafety/portal-perfluorinated-chemicals/. Accessed September 24, 2018.

Organisation for Economic Cooperation and Development (OECD). 2015. Report 30: Working towards a global emission inventory of PFAS: Focus on PFCAs status quo and the way forward. *OECD*.

Organisation for Economic Cooperation and Development (OECD). 2013. Report 27: Synthesis paper on per- and polyfluorinated chemicals (PFCs). Environment, Health and Safety, Environment Directorate, *OECD*.

Organisation for Economic Cooperation and Development (OECD). 2007a. Report 21: Lists of PFOS, PFAS, PFOA, PFCA, related compounds and chemicals that may degrade to PFCA. *OECD*.

Organisation for Economic Cooperation and Development (OECD. 2007b). Report 23: Report of an OECD Workshop on perfluorocarboxylic acids (PFCAs) and precursors. 2007.

Organisation for Economic Cooperation and Development (OECD). 2005. Report No. 19: Results of survey on production and use of PFOS, PFAS and PFOA, related substances and products/mixtures containing these substances. *OECD*.

Organisation for Economic Cooperation and Development (OECD). 2002a. Glossary of statistical terms: Biodegradation definition. https://stats.oecd.org/glossary/detail.asp?ID=203.

Organisation for Economic Cooperation and Development (OECD). 2002b. Hazard assessment of perfluorooctane sulfonate (PFOS) and its salts. *OECD*.

Ohmori K., Katayama K., & Kawashima, Y. 2003. Comparison of the toxicokinetics between perfluorocarboxylic acids with different carbon chain length. *Toxicology, 184*, 135–140.

Oliaei, F., Kessler, K. & Kriens, D. 2006. Investigation of perfluorochemical (PFC) contamination in Minnesota Phase One. *Report to Senate Environment Committee*.

Oliaei, F., Kriens, D., Weber, R., & Watson, A. 2013. PFOS and PFC releases and associated pollution from a PFC production plant in Minnesota (USA). *Environ Sci Pollut Res* (2013) 20:1977-1992. DOI 10.1007/s11356-012-1275-4.

Olsen, G. W., Burris, J. M., Ehresman, D. J., Froehlich, J. W., Seacat, A. M., Butenhoff, J. L. & Zobel, L. R. 2007. Half-life of serum elimination of perfluorooctanesulfonate, perfluorohexanesulfonate, and perfluorooctanoate in retired fluorochemical production workers. *Environ Health Perspect, 115*, 1298–1305.

Olsen, G. W., Chang, S. C., Noker, P. E., Gorman, G. S., Ehresman, D. J., Lieder, P. H. & Butenhoff, J. L. 2009. A comparison of the pharmacokinetics of perfluorobutanesulfonate (PFBS) in rats, monkeys, and humans. *Toxicology, 256*, 65–74.

Olson, A. D. 2017. An investigation into the toxicity, bioconcentration, and risk of perfluoroalkyl substances in aquatic taxa. *Thesis*. Texas Tech University, PhD in Environmental Toxicology.

Pan, Y., Zhang, H., Cui, Q., Sheng, N., Yeung, L. W. Y., Sun, Y., Guo, Y. & Dai, J. 2018. Worldwide distribution of novel perfluoroether carboxylic and sulfonic acids in surface water. *Environ Sci Technol*.

Pan, Y., Zhang, H., Cui, Q., Sheng, N., Yeung, L. W. Y., Guo, Y., Sun, Y. & Dai, J. 2017. First report on the occurrence and bioaccumulation of hexafluoropropylene oxide trimer acid: An emerging concern. *Environ Sci Technol, 51*, 9553–9560.

Pancras, T., Schrauwen, G., Held, T., Baker, K., Ross, I. & Slenders, H. 2016. Environmental fate and effects of poly-and perfluoroalkyl substances (PFAS). *CONCAWE*. www.concawe.eu/wp-content/uploads/2016/06/Rpt_16-8.pdf.

Park, H., Vecitis, C. D., Cheng, J., Choi, W., Mader, B. T. & Hoffmann, M. R. 2009. Reductive defluorination of aqueous perfluorinated alkyl surfactants: Effects of ionic headgroup and chain length. *Journal of Physical Chemistry A, 113*, 690–696.

Parsons, J. R., Sáez, M., Dolfing, J. & Voogt, P. D. 2008. Biodegradation of perfluorinated compounds. D. M. Whitacre (ed.) *Reviews of Environmental Contamination and Toxicology, Vol 196*, doi: 10.1007/978-0-387-78444-1_2, Springer Science+Business Media, LLC.

Payne, F. C., Quinnan, J. A. & Potter S. T. 2008. *Remediation hydraulics*. Boca Raton, FL: CRC Press.

Van Pee, K., and S. Unversucht. "Biological dehalogenation and halogenation reactions." *chemosphere* 52, no. 2 (2003): 299–312.

Pée, K-H, U. S. 2003. Biological dehalogenation and halogenation reactions. *Chemosphere, 52*, 299–312.

Pennell, K. D., Capiro, Natalie L., Fortner, John D., Arnold, William A., Simcik, Matt F., Hatton, James. In situ treatment options for PFAS-contaminated groundwater. *In:* University, T., ed. CleanUp 2017, September 12, 2017, Melbourne, Australia. www.cleanupconference.com/wp-content/uploads/2017/09/4A1.pdf.

Perkins, R. G., Butenhoff, J. L., Kennedy, G. L. and Palazzolo, M. (2004). 13-week dietary toxicity study of ammonium perfluorooctanoate (APFO) in male rats. *Drug Chem Toxicol*. 27, 361–378

Public Health England (PHE). 2009. PFOS and PFOA: *properties, incident management and toxicology*. Prepared by the Toxicology Department, CRCE, PHE, 2009, Version 1.

Phillips, M. M., Dinglasan-Panlilio, M. J. A., Mabury, S. A., Solomon, K. R. & Sibley, P. K. 2010. Chronic toxicity of fluorotelomer acids to Daphnia magna and Chironomus dilutus. *Environmental Toxicology and Chemistry, 29*, 1123–1131.

Phillips, M. M., Dinglasan, M. J. A., Mabury, S. A., Solomon, K. R. & Sibley, P. K. 2007. Fluorotelomer acids are more toxic than perfluorinated acids. *Environmental Science & Technology, 41*, 7159–7163.

Pillai, S. D., & Shima Shayanfar. 2017. *Electron beam technology and other irradiation technology applications in the food industry*. Berlin, Germany: Springer Verlag.

Pillai, S. D. 2016. Introduction to electron-beam food irradiation. *Chemical Engineering Progress, 112*, 36–44.

Place, B. J. & Field, J. A. 2012. Identification of novel fluorochemicals in aqueous film-forming foams used by the US military. *Environmental Science and Technology, 46*, 7120–7127.

Plassmann, M. M. & Berger, U. 2013. Perfluoroalkyl carboxylic acids with up to 22 carbon atoms in snow and soil samples from a ski area. *Chemosphere, 91*, 832.

Plassmann, M. M. & Berger, U. 2010. Trace analytical methods for semifluorinated N-alkanes in snow, soil, and air. *Anal. Chem., 82,* 4551.

Plumlee, M. H., Larabee, J. & Reinhard, M. 2008. Perfluorochemicals in water reuse. *Chemosphere, 72,* 1541–1547.

Post, G. B., Gleason, J. A. & Cooper, K. R. 2017. Key scientific issues in developing drinking water guidelines for perfluoroalkyl acids: Contaminants of emerging concern. *PLoS Biol, 15,* e2002855.

Potwara, R. 2012. The ABCs of activated carbon. *Water Quality Products,* 14–15.

Pramanik, B. K., Pramanik, S. K. & Suja, F. 2015. A comparative study of coagulation, granular- and powdered-activated carbon for the removal of perfluorooctane sulfonate and perfluorooctanoate in drinking water treatment. *Environ Technol, 36,* 2610–2617.

Praveen, C., Palmy R. Jesudhasan, Robert S. Reimers, & Suresh D. PILLAI 2013. Electron beam inactivation of selected microbial pathogens and indicator organisms in aerobically and anaerobically digested sewage sludge. *Bioresource Technology, 144,* 6.

Prevedouros, K., Cousins, I. T., Buck, R. C. & Korzeniowski, S. H. 2006. Sources, Fate and Transport of Perfluorocarboxylates. *Environmental Science & Technology, 40,* 32–44.

Program, N. T. 2016. For the systematic review of immunotoxicity associated with exposure to perfluorooctanoic acid (PFOA) or perfluorooctane sulfonate (PFOS).

Punyapalakul, P., Suksomboon, K., Prarat, P. & Khaodhiar, S. 2013. Effects of surface functional groups and porous structures on adsorption and recovery of perfluorinated compounds by inorganic porous silicas. *Separation Science and Technology, 48,* 775–788.

Qui, Y. 2007. *Study on treatment technologies for perfluorochemicals in wastewater.* PhD, Kyoto University.

Qiu, Y., Fujii, S., Tanaka, S. & Koizumi, A. 2006. Preliminary study on the treatment of perfluorinated chemicals (PFCs) by advanced oxidation processes (AOPs). *Proceedings of 21th Seminar of JSPS-MOE Core University Program, Kyoto, Japan.*

Qu, Y., Zhang, C., Li, F., Chen, J. & Zhou, Q. 2010. Photo-reductive defluorination of perfluorooctanoic acid in water. *Water Research, 44,* 2939–2947.

Qu, Y., Zhang, C., Li, F., Bo, X., Liu, G. & Zhou, Q. 2009. Equilibrium and kinetics study on the adsorption of perfluorooctanoic acid from aqueous solution onto powdered activated carbon. *J Hazard Mater, 169,* 146–152.

Queensland Department of Environment and Heritige Protection. 2016. Operational policy, environmental management of firefighting foams. *In:* Protection, D. O. E. A. H. (ed.) *Handout.*

Quinones, O. & Snyder, S. A. 2009. Occurrence of perfluoroalkyl carboxylates and sulfonates in drinking water utilities and related waters from the United States. *Environ Sci Technol, 43,* 9089–9095.

Rand, A. A. & Mabury, S. A. 2017. Is there a human health risk associated with indirect exposure to perfluoroalkyl carboxylates (PFCAs)? *Toxicology, 375,* 28–36.

Rankin, K., Mabury, S. A., Jenkins, T. M. & Washington, J. W. 2016. A North American and global survey of perfluoroalkyl substances in surface soils: Distribution patterns and mode of occurrence. *Chemosphere, 161,* 333–341.

Rappazzo, K. M., Coffman, E. & Hines, E. P. 2017. Exposure to perfluorinated alkyl substances and health outcomes in children: A systematic review of the epidemiologic literature. *Int J Environ Res Public Health,* 14.

Rattanaoudom, R., Visvanathan, C. & Boontanon, S. K. 2012. Removal of concentrated PFOS and PFOA in synthetic industrial wastewater by powder activated carbon and hydrotalcite. *International Conference on the Challenges in Environmental Science and Engineering.* Melbourne, Australia.

Rayne, S. & Forest, K. 2009. Perfluoroalkyl sulfonic and carboxylic acids: a critical review of physicochemical properties, levels and patterns in waters and wastewaters, and treatment methods. *Journal of Environmental Science and Health. Part A, Toxic/hazardous Substances & Environmental Engineering, 44,* 1145–1199.

Rayne, S. F. & Friesen, K. J. 2008. Congener-specific numbering systems for the environmentally relevant C4 through C8 perfluorinated homologue groups of alkyl sulfonates, carboxylates, telomer alcohols, olefins, and acids, and their derivatives. *J. Environ. Sci. Health Pt. A* 43, 1391–1401.

Reemtsma, T., Berger U., Arp, H., Gallard, H., Knepper, T., Neumann, M., Quintana, J. B., and de Voogt, P. 2016. Mind the Gap: Persistent and Mobile Organic Compounds Water Contaminants That Slip Through. 10308–10315.

Reemtsma, T., and Jekel, M., eds. 2006. *Organic pollutants in the water cycle: properties, occurrence, analysis and environmental relevance of polar compounds.* John Wiley & Sons.

Reeve, B., & You, A. 2018. CustoMem PFAS Data Sheet.

Regenesis, 2017. PlumeStop® Liquid Activated Carbon™. https://regenesis.com/eur/remediation-products/plumestop-liquid-activated-carbon/.

Rhoads, K. R., Janssen, E. M. L., Luthy, R. G. & Criddle, C. S. 2008. Aerobic biotransformation and fate of N-ethyl perfluorooctane sulfonamidoethanol (N-EtFOSE) in activated sludge. *Environmental Science & Technology, 42*, 2873–2878.

Riddell, N., Arsenault, G., Benskin J. P., Chittim, B., Martin J. W., McAlees A., & Mccrindle, R. 2009. Branched perfluorooctane sulfonate isomer quantification and characterization in blood serum samples by HPLC/ESI-MS(/MS). *Environ Sci Technol 43*, 7902–7908.

Ritter, E. E., Dickinson, M. E., Harron, J. P., Lunderberg, D. M., Deyoung, P. A., Robel, A. E., Field, J. A. & Peaslee, G. F. 2017. PIGE as a screening tool for per- and polyfluorinated substances in papers and textiles. *Nuclear Instruments and Methods in Physics Research Section B: Beam Interactions with Materials and Atoms, 407*, 47–54.

Ritter, S. K. 2010. Fluorochemicals go short. *Chemical and Engineering News.* 88(5), pp. 12–17.

Rodriguez-Freire, L., Abad-Fernandez, N., Sierra-Alvarez, R., Hoppe-Jones, C., Peng, H., Giesy, J. P., Snyder, S. & Keswani, M. 2016. Sonochemical degradation of perfluorinated chemicals in aqueous film-forming foams. *Journal of Hazardous Materials, 317*, 275–283.

Rodriguez-Freire, L., Balachandran, R., Sierra-Alvarez, R. & Keswani, M. 2015. Effect of sound frequency and initial concentration on the sonochemical degradation of perfluorooctane sulfonate (PFOS). *Journal of Hazardous Materials, 300*, 662–669.

Ross, I., Houtz, E., Burdick, J., McDonough, J., & Quinnan, J. 2018. Characterization and treatment of poly- and perfluoroalkyl substances (PFASs). *Advances in Remediation, A Better Way Forward.* Arcadis.

Ross, I., Horneman, A. Houtz, E. F., Christensen, A. A. G., Burdick, J., Hurst, J., & Miles, J. 2018. Detailed site investigation of unsaturated and saturated zones of a fire training area for per- and polyfluoroalkyl substances (PFASs) using advanced analytical tools. *The Eleventh International Conference on Remediation of Chlorinated and Recalcitrant Compounds.* Palm Springs, California.

Ross, I., McDonough, J., Miles, J., Storch, P., Thelakkat Kochunarayanan, P., Kalve, E., Hurst, J., S. Dasgupta, S. & Burdick, J. 2018. A review of emerging technologies for remediation of PFASs. *Remediation Journal, 28*, 101–126.

Royer, L. A., Lee, L. S., Russell, M. H., Nies, L. F. and Turco, R. F. 2015. Microbial transformation of 8: 2 fluorotelomer acrylate and methacrylate in aerobic soils. *Chemosphere, 129*, pp. 54–61.

RPI 2017. Trap & Treat® BOS 100®. www.trapandtreat.com/products/bos-100/.

Russell, M. H., Szostek, B., & Buck, R. C. 2008. Investigation of the biodegradation potential of a fluoroacrylate polymer product in aerobic soils. *Environ Sci Technol, 42*, 800–807.

Russell, M. H., Berti, W. R., Szostek, B., Wang, N. & Buck, R. C. 2010. Evaluation of PFO formation from the biodegradation of a fluorotelomer-based urethane polymer product in aerobic soils. *Polymer Degradation and Stability, 95*, 79–85.

Saito, N., Harada, K., Inoue, K., Sasaki, K., Yoshinaga, T. and Koizumi, A. 2004. Perfluorooctanoate and perfluorooctane sulfonate concentrations in surface water in Japan. *Journal of occupational health*, 46(1), pp. 49–59.

Sammut, G., Sinagra, E., Helmus, R. & De Voogt, P. 2017. Perfluoroalkyl substances in the Maltese environment: Surface water and rain water. *Sci Total Environ, 589*, 182–190.

Santoro, M. A. 2008. *Brief history of perfluorochemical production, products and environmental presence*. Presentation.at ASTSWMO Mid-Year Meeting, April 23–24.

Schaefer, C. E., Andaya, C., Burant, A., Condee, C. W., Urtiaga, A., Strathmann, T. J. & Higgins, C. P. 2017. Electrochemical treatment of perfluorooctanoic acid and perfluorooctane sulfonate: Insights into mechanisms and application to groundwater treatment. *Chemical Engineering Journal, 317*, 424–432.

Schaefer, C. E., Andaya, C., Urtiaga, A., Mckenzie, E. R. & Higgins, C. P. 2015. Electrochemical treatment of perfluorooctanoic acid (PFOA) and perfluorooctane sulfonic acid (PFOS) in groundwater impacted by aqueous film forming foams (AFFFs). *Journal of Hazardous Materials, 295*, 170–175.

Schaider, L. A., Balan, S. A., Blum, A., Andrews, D. Q., Strynar, M. J., Dickinson, M. E., Lunderberg, D. M., Lang, J. R. & Peaslee, G. F. 2017. Fluorinated compounds in U.S. fast food packaging. *Environmental Science & Technology Letters*.

Scheirs, J., 1997. *Modern fluoropolymers*, New York: John Wiley & Sons.

Schellenberger, S., P. Gillgard, A. Stare, A. Hanning, O. Levenstam, S. Roos, & I. Tl. Cousins. 2018. Facing the rain after the phase out: Performance evaluation of alternative fluorinated and non-fluorinated durable water repellents for outdoor fabrics. *Chemosphere, 193*, pp. 675–684.

Schröder, H. F. & Meesters, R. J. W. 2005. Stability of fluorinated surfactants in advanced oxidation processes—A follow-up of degradation products using flow injection–mass spectrometry, liquid chromatography–mass spectrometry and liquid chromatography–multiple stage mass spectrometry. *Journal of Chromatography A, 1082*, 110–119.

Schultz, M. et al. (2003). Fluorinated alkyl surfactants. *Environ. Eng. Sci.* 20(5) 487–501.

Schultz, M. M., Barofsky, D. F. & Field, J. A. 2004. Quantitative determination of fluorotelomer sulfonates in groundwater by LC MS/MS. *Environmental Science & Technology, 38*, 1828–1835.

Schwarzenbach, R. P. G., Imboden, D. M. 2003. *Environmental organic chemistry*, 2nd ed., New York: John Wiley & Sons.

Schwerdtle T. 2018. PFOS/PFOA in food: Main conclusions of the EFSA risk assessment. Presentation at the 38th International Symposium on Halogenated Persistent Organic Pollutants, 26 – 31 August 2018, Kraków, Poland.

Seacat, A. M., P. J. Thomford, K. J. Hansen, L. A. Clemen, S. R. Eldridge, C. R. Elcombe, and J. L. Butenhoff. 2003. Sub-chronic dietary toxicity of potassium perfluorooctanesulfonate in rats. *Toxicology* 183:117–131.

Seacat, A. M., Thomford, P. J., Hansen, K. J., Olsen, G. W., Case, M. T. & Butenhoff, J. L. 2002. Subchronic toxicity studies on perfluorooctanesulfonate potassium salt in cynomolgus monkeys. *Toxicol Sci, 68*, 249–264.

Senevirathna, S. T., Tanaka, S., Fujii, S., Kunacheva, C., Harada, H., Shivakoti, B. R. & Okamoto, R. 2010a. A comparative study of adsorption of perfluorooctane sulfonate (PFOS) onto granular activated carbon, ion-exchange polymers and non-ion-exchange polymers. *Chemosphere, 80*, 647–651.

Senevirathna, S. T., Tanaka, S., Fujii, S., Kunacheva, C., Harada, H., Ariyadasa, B. H. A. K. T. & Shivakoti, B. R. 2010b. Adsorption of perfluorooctane sulfonate (n-PFOS) onto non-ion-exchange polymers and granular activated carbon: Batch and column test. *Desalination, 260*, 29–33.

Sepulvado, J. G., Blaine, A. C., Hundal, L. S. & Higgins, C. P. 2011. Occurrence and fate of perfluorochemicals in soil following the land application of municipal biosolids. *Environ Sci Technol, 45*, 8106–8112.

Serex, T., Anand, S., Munley, S., Donner, E. M., Frame, S. R., Buck, R. C. & Loveless, S. E. 2014. Toxicological evaluation of 6:2 fluorotelomer alcohol. *Toxicology, 319,* 1–9.

Sergei V., Kostjuk, E. O., Ganachaud, F., Ame´Duri, B. & Boutevin, B. 2009. Anionic ring-opening polymerization of hexafluoropropylene oxide using alkali metal fluorides as catalysts: A mechanistic study. *Macromolecules, 42,* 612–619.

Shen, A., Lee, S., Ra, K., Suk, D. & Moon, H.-B. 2018. Historical trends of perfluoroalkyl substances (PFASs) in dated sediments from semi-enclosed bays of Korea. *Marine Pollution Bulletin, 128,* 287–294.

Sheng, N., Pan, Y., Guo, Y., Sun, Y. and Dai, J., 2018. Hepatotoxic Effects of Hexafluoropropylene Oxide Trimer Acid (HFPO-TA), A Novel Perfluorooctanoic Acid (PFOA) Alternative, on Mice. *Environmental science & technology.*

Shin, H-M., V. M. Vieira, P. B. Ryan, R. Detwiler, B. Sanders, K. Steenland, and S. M. Bartell. 2011. "Environmental Fate and Transport Modeling for Perfluorooctanoic Acid Emitted from the Washington Works Facility in West Virginia" *Environmental Science and Technology* 45: 1435–1442.

Shinoda, K. H., & Hayashi, T. 1972. The physicochemical properties of aqueous solutions of fluorinated surfactants. *J. Phys. Chem. C, 76,* 909–914.

Shoeib, M., Harner, T., Wilford, B. H., Jones, K. C. & Zhu, J. 2005. Perfluorinated sulfonamides in indoor and outdoor air and indoor dust: Occurrence, partitioning, and human exposure. *Environmental Science & Technology, 39,* 6599–6606.

Simcik, M. F. 2005. Aquatic processes and systems in perspective. Global transport and fate of perfluorochemicals. *J Environ Monit, 7,* 759–763.

Solvey Specialty Polymers. 2016. Fomblin(R) HC, Transformative Specialty Ingredient in Cosmetic Formulations. R 09/2016, Version 1.5.

Song, Z., Tang, H., Wang, N. & Zhu, L. 2013. Reductive defluorination of perfluorooctanoic acid by hydrated electrons in a sulfite-mediated UV photochemical system. *Journal of Hazardous Materials, 262,* 332–338.

Stahl, T. & Brunn, H. 2011. Toxicology of perfluorinated compounds. *Environmental Sciences Europe, 23,* 38.

Stahl, T., Failing, K., Berger, J., Georgii, S., & Brunn, H. 2012. Perfluorooctanoic acid and perfluorooctane sulfonate in liver and muscle tissue from wild boar in Hesse, Germany. *Arch Environ Contam Toxicol., 62,* 696–703.

Stahl, T., Riebe, R. A., Falk, S., Failing, K. & Brunn, H. 2013. Long-term lysimeter experiment to investigate the leaching of perfluoroalkyl substances (PFASs) and the carry-over from soil to plants: results of a pilot study. *J Agric Food Chem, 61,* 1784–1793.

Stedingk, V. & Bergman Å. 2004. *En miljökemisk översikt av polyfluorerade kemikalier (PFCs).* Stockholm University, Stockholm.

Steinle-Darling, E. & Reinhard, M. 2008. Nanofiltration for trace organic contaminant removal: Structure, solution, and membrane fouling effects on the rejection of perfluorochemicals. *Environ. Sci. Technol., 42,* 5292–5297.

Stock, N. L., Furdui, V. I., Muir, D. C. and Mabury, S. A. 2007. Perfluoroalkyl contaminants in the Canadian Arctic: evidence of atmospheric transport and local contamination. *Environmental science & technology, 41*(10), pp. 3529–3536.

P. J. Storch, J. Lagowski, M. Dickson, D. Solomon, and I. Ross. 2018. Full-Scale Treatment of PFAS-Impacted Wastewater Using Ozofractionation. Presentation, 2018 Battelle Conference Proceedings, Palm Springs, CA

Storch, P. 2016. Re: PFAS in situ soil mixing treatability testing results. *Personal communication.*

Stratton, G. R., Dai, F., Bellona, C. L., Holsen, T. M., Dickenson, E. R. & Mededovic Thagard, S. 2017. Plasma-based water treatment: Efficient transformation of perfluoroalkyl substances in prepared solutions and contaminated groundwater. *Environ Sci Technol.*

Strynar, M. J., Lindstrom, A. B., Nakayama, S. F., Egeghy, P. P. and Helfant, L. J. 2012. Pilot scale application of a method for the analysis of perfluorinated compounds in surface soils. *Chemosphere*, 86(3), pp. 252–257.

Sun, M., Arevalo, E., Strynar, M., Lindstrom, A., Richardson, M., Kearns, B., Pickett, A., Smith, C. & Knappe, D. R. U. 2016. Legacy and emerging perfluoroalkyl substances are important drinking water contaminants in the Cape Fear River watershed of North Carolina. *Environmental Science & Technology Letters, 3*, 415–419.

Suthersan, S., Horst, J., Schnobrich, M., Welty, N., McDonough, J. 2017. *Remediation Engineering: Design Concepts*, Boca Raton, FL, Taylor & Francis Group.

Swedish Chemicals Agency. 2015. Occurrence and use of highly fluorinated substances and alternatives. Report 7/15. ISSN 0284-1185.

Szabo., Jeff., Hall, J., Magnuson, M., 2017. Treatment of perfluorinated alkyl substances in wash water using granular activated carbon and mixed media. *In:* Program, O. O. R. A. D. H. S. R. (ed.).

Takagi, S. 2011. Fate of perfluorooctanesulfonate and perfluorooctanoate in drinking water treatment processes. *Water Research, 45*, 3925–3932.

Tang, C. Y., Fu, Q. S., Robertson, A. P., Criddle, C. S. & Leckie, J. O. 2006. Use of reverse osmosis membranes to remove perfluorooctane sulfonate (PFOS) from semiconductor wastewater. *Environmental Science & Technology, 40*, 7343–7349.

Taniyasu, S., Kannan, K., Horii, Y., Hanari, N. & Yamashita, N. 2003. A survey of perfluorooctane sulfonate and related perfluorinated organic compounds in water, fish, birds, and humans from Japan. *Environ Sci Technol, 12*, 2634–2639.

Taniyasu, S., Yamashita, N., Moon, H. B., Kwok, K. Y., Lam, P. K., Horii, Y., Petrick, G. & Kannan, K. 2013. Does wet precipitation represent local and regional atmospheric transportation by perfluorinated alkyl substances? *Environment International, 55*, 25–32.

Taniyasu, S., Yamashita, N., Yamazaki, E., Petrick, G. & Kannan, K. 2013. The environmental photolysis of perfluorooctanesulfonate, perfluorooctanoate, and related fluorochemicals. *Chemosphere, 90*, 1686–1692.

Thalheimer, A. H., Mcconney, L. B., Kalinovich, I. K., Pigott, A. V., Franz, J. D., Holbert, H. T., Mericas, D. & Puchacz, Z. J. 2017. *Use and potential impacts of AFFF containing PFASs at airports.*

Thomford, P. J. 2002. 104-Week Dietary Chronic Toxicity and Carcinogenicity Study with Perfluorooctane Sulfonic Acid Potassium Salt (PFOS; T-6295) in Rats. Final Report. Volumes I-IX. Covance Study No. 6329-183. 3M Company, St. Paul, MN.

Thompson, J., Eaglesham, G., Reungoat, J., Poussade, Y., Bartkow, M., Lawrence, M. & Mueller, J. F. 2011. Removal of PFOS, PFOA and other perfluoroalkyl acids at water reclamation plants in South East Queensland Australia. *Chemosphere, 82*, 9–17.

Trautmann, A. M., Schell, H., Schmidt, K. R., Mangold, K. M. & Tiehm, A. 2015. Electrochemical degradation of perfluoroalkyl and polyfluoroalkyl substances (PFASs) in groundwater. *Water Sci Technol, 71*, 1569–1575.

Trier, Xenia, C. Taxvig, A. K. Rosenmai, & G. A. Pedersen. 2018. PFAS in Paper and Board for Food Contact, Options for Risk Management of Poly- and Perfluorinated Substances. *Nordic Council of Ministers*, 2018:573. ISSN 0908-6692.

Tsang W. & Babushok V. 1998. On the incinerability of highly fluorinated organic compounds. *Combustion Science and Technology, 139*, 385–402.

Tsao, M. S. 1959. The aqueous chemistry of inorganic free radicals. The mechanism of the photolytic decomposition of aqueous persulphate ion and evidence regarding the sulphate-hydroxyl radical interconversion equilibrium. *Journal Weiss Experientia, 63*, 8.

Tseng, N., Wang, N., Szostek, B. & Mahendra, S. 2014. Biotransformation of 6:2 fluorotelomer alcohol (6:2 FTOH) by a wood-rotting fungus. *Environmental Science & Technology, 48*, 4012–4020.

Urtiaga, A., Fernandez-Gonzalez, C., Gomez-Lavin, S. & Ortiz, I. 2015. Kinetics of the electrochemical mineralization of perfluorooctanoic acid on ultrananocrystalline boron-doped conductive diamond electrodes. *Chemosphere, 129*, 20–26.

United States Army Corps of Engineers. 2016. DRAFT Standard Operating Procedure 047: Per/Poly Fluorinated Substances (PFAS) Field Sampling. March.

United States Energy Information Administration. 2018. Frequently Asked Questions: How much electricity does an American home use? [Online]. Available: www.eia.gov/tools/faqs/faq.php?id=97&t=3. Accessed September 24.

United Nations Environment Programme (UNEP). 2017. Stockholm Convention. *Chemicals proposed for listing under the Convention* [Online]. Available: http://chm.pops.int/TheConvention/ThePOPs/ChemicalsProposedforListing/tabid/2510/Default.aspx. Accessed September 15, 2018.

United Nations Environment Programme (UNEP). 2015. PFAS analysis in water for the Global Monitoring Plan of the Stockholm Convention, Set-up and guidelines for monitoring. Jana Weiss, Jacob de Boer, Urs Berger, Derek Muir, Ting Ruan, Alejandra Torre, Foppe Smedes, Branislav Vrana, Fabrice Clavient, and Heidelore Fiedler. Division of Technology, Industry and Economics. Geneva. April.

United Nations Environment Programme (UNEP). 2014. Stockholm Convention on Persistent Organic Pollutants. UNEP/POPS/POPRC.10/INF/7*. July 28.

United States Army Corps of Engineers (U.S. ACE). 2016. DRAFT – Standard Operating Procedure 047: Per/Poly Fluorinated Alkyl Substances (PFAS) Field Sampling. Revision: 1. March.

United States Environmental Protection Agency (U.S. EPA). 1996. A citizen's guide to soil washing. *In:* Office, T. I. (ed.).

United States Environmental Protection Agency (U.S. EPA). 1997. High Voltage Environmental Applications, Inc. Electron Beam Technology. *In:* Development, O. O. R. A. (ed.). https://cfpub.epa.gov/si/si_public_record_Report.cfm?dirEntryId=96217: Washington D.C.

United States Environmental Protection Agency (U.S. EPA). 2003a. Final report: Laboratory-scale thermal degradation of perfluorooctanyl sulfonate and related substances. *In:* Office of Pollution Prevention & Toxics, D. A. (ed.). Washington D.C: U.S. Environmental Protection Agency.

United States Environmental Protection Agency (U.S. EPA). 2003b. Laboratory-scale thermal degradation of perfluorooctanyl sulfonate and related substances. *In:* Office Of Pollution Prevention & Toxics, D. A. (ed.). Washington D.C.: US Environmental Protection Agency.

United States Environmental Protection Agency (U.S. EPA). 2005. *Draft Risk Assessment of the Potential Human Health Effects Associated with Exposure to Perfluorooctanoic Acid and Its Salts.* Available: at the National Service Center for Environmental Publications: https://nepis.epa.gov/. Accessed September 24, 2018.

United States Environmental Protection Agency (U.S. EPA). 2009a. Determination of selected perfluorinated alkyl acids in drinking water by solid phase extraction and liquid chromatography/tandem mass spectrometry (LC/MS/MS).

United States Environmental Protection Agency (U.S. EPA). 2009b. Long-chain perfluorinated chemicals (PFCs) action plan. *In:* Agency, U. S. E. P. (ed.).

United States Environmental Protection Agency (U.S. EPA). 2014a. *Health effects document for perfluorooctane sulfonate (PFOS)* [Online]. Available at https://peerreview.versar.com/epa/pfoa/pdf/Health-Effects-Document-for-Perfluorooctane-Sulfonate-(PFOS).pdf Accessed September 24, 2018.

United States Environmental Protection Agency (U.S. EPA). 2014b. Health effects document for perfluorooctanoic acid (PFOA) [Online]. Available at www.epa.gov/sites/production/files/2016-05/documents/pfoa_hesd_final-plain.pdf. Accessed September 24, 2018.

United States Environmental Protection Agency (U. S. EPA). 2016. *Fact Sheet, PFOA & PFOS Drinking Water Health Advisories* [Online]. Available: www.epa.gov/ground-water-and-drinking-water/drinking-water-health-advisories-pfoa-and-pfos Accessed September 24, 2018.

United States Environmental Protection Agency (U.S. EPA). 2017. Superfund Remedy Report, 15th ed. *In: Management*, O. O. L. A. E. (ed.).

Vecitis, C. D., Park, H., Cheng, J., Mader, B. T. & Hoffmann, M. R. 2008. Enhancement of perfluorooctanoate and perfluorooctanesulfonate activity at acoustic cavitation bubble interfaces. *The Journal of Physical Chemistry C, 112*, 16850–16857.

Vecitis, C. D., Park, H., Cheng, J., Mader, B. T. & Hoffmann, M. R. 2009. Treatment technologies for aqueous perfluorooctanesulfonate (PFOS) and perfluorooctanoate (PFOA). *Frontiers of Environmental Science & Engineering in China, 3*, 129–151.

Vecitis, C. D., Wang, Y., Cheng, J., Park, H., Mader, B. T. & Hoffmann, M. R. 2010. Sonochemical degradation of perfluorooctanesulfonate in aqueous film-forming foams. *Environ Sci Technol, 44*, 432–438.

Vestergren, R., Cousins, I. T., Trudel, D., Wormuth, M. & Scheringer, M. 2008. Estimating the contribution of precursor compounds in consumer exposure to PFOS and PFOA. *Chemosphere, 73*, 1617–1624.

Vierke, L. 2017. *Regulation needs support from research: Short-chain PFASs under REACH* [Online]. Available: www.mn.uio.no/kjemi/english/research/projects/ICCE2017/wednesday-21.06/helga-eng-auditorium-1/hr.-16:30/1715-vierke.pdf [Accessed September 14, 2018].

Rickard, R. W. 2009. Toxicology perfluorocarboxylates. *Presentation*.

Waite, T. D., Kurucz, C. N., Cooper, W. J. & Brown D. 1998. Full-scale electron beam systems for treatment of water, wastewater and medical waste. *International Atomic Energy Agency*. www.iaea.org/inis/collection/NCLCollectionStore/_Public/29/050/29050419.pdf.

Wallington, T. J., Hurley, M. D., Zia, J., Wuebbles, D. J., Sillman, S., Ito, A., Penner, J. E., Ellis, D. A., Martin, J., Mabury, S. A., Nielsen, O. J. & Sulbaek Andersen, M. P. 2006. Formation of C7F15COOH (PFOA) and other perfluorocarboxylic acids during the atmospheric oxidation of 8:2 fluorotelomer alcohol. *Environmental Science and Technology, 40*, 924–930.

Wang, F., Lu, X., Shih, K. & Liu, C. 2011. Influence of calcium hydroxide on the fate of perfluorooctanesulfonate under thermal conditions. *J Hazard Mater, 192*, 1067–1071.

Wang, F., Shih, K., Lu, X. & Liu, C. 2013. Mineralization behavior of fluorine in perfluorooctanesulfonate (PFOS) during thermal treatment of lime-conditioned sludge. *Environ Sci Technol, 47*, 2621–2627.

Wang, I. J., Hsieh, W. S., Chen, C. Y., Fletcher, T., Lien, G. W., Chiang, H. L., Chiang, C. F., Wu, T. N. & Chen, P. C. 2011. The effect of prenatal perfluorinated chemicals expsoures on pediatric atopy. *Environmental Research, 111*, 785–791.

Wang, L., Batchelor, B., Pillai, S. D. & Botlaguduru, V. S. V. 2016. Electron beam treatment for potable water reuse: Removal of bromate and perfluorooctanoic acid. *Chemical Engineering Journal, 302*, 58–68.

Wang, N., Buck, R. C., Szostek, B., Sulecki, L. M. & Wolstenholme, B. W. 2012. 5:3 Polyfluorinated acid aerobic biotransformation in activated sludge via novel "one-carbon removal pathways." *Chemosphere, 87*, 527–534.

Wang, N., Liu, J., Buck, R. C., Korzeniowski, S. H., Wolstenholme, B. W., Folsom, P. W. & Sulecki, L. M. 2011. 6:2 fluorotelomer sulfonate aerobic biotransformation in activated sludge of waste water treatment plants. *Chemosphere, 82*, 853–858.

Wang, N., Szostek, B., Buck, R. C., Folsom, P. W., Sulecki, L. M., Capka, V., Berti, W. R. & Gannon, J. T. 2005. Fluorotelomer alcohol biodegradationdirect evidence that perfluorinated carbon chains breakdown. *Environmental Science & Technology, 39*, 7516–7528.

Wang, N., Szostek, B., Buck, R. C., Folsom, P. W., Sulecki, L. M. & Gannon, J. T. 2009. 8-2 fluorotelomer alcohol aerobic soil biodegradation: Pathways, metabolites, and metabolite yields. *Chemosphere, 75*, 1089–1096.

Wang, S., T. Okazoe, E. Murotani, K. Watanabe, D. Shirakawa, K. Oharu. 2006. United States Patent Application Publication No. US 2006/0030733 A1, Process for Producing Perfluorodiacyl Fluorinated Compounds. February 9.

Wang, S., Huang, J., Yang, Y., Hui, Y., Ge, Y., Larssen, T., Yu, G., Deng, S., Wang, B. and Harman, C., 2013. First report of a Chinese PFOS alternative overlooked for 30 years: its toxicity, persistence, and presence in the environment. *Environmental science & technology, 47*(18), pp.10163-10170.

Wang, Y., Vestergren, R., Shi, Y., Cao, D., Xu, L., Cai, Y., Zhao, X. and Wu, F. 2016. Identification, tissue distribution, and bioaccumulation potential of cyclic perfluorinated sulfonic acids isomers in an airport impacted ecosystem. *Environmental science & technology, 50*(20), pp. 10923–10932.

Wang, Z., Cousins, I. T., Berger, U., Hungerbuhler, K. & Scheringer, M. 2016. Comparative assessment of the environmental hazards of and exposure to perfluoroalkyl phosphonic and phosphinic acids (PFPAs and PFPiAs): Current knowledge, gaps, challenges and research needs. *Environ Int, 89–90*, 235–247.

Wang, Z., Cousins, I. T., Scheringer, M., Buck, R. C. & Hungerbuhler, K. 2014a. Global emission inventories for C4-C14 perfluoroalkyl carboxylic acid (PFCA) homologues from 1951 to 2030, Part I: Production and emissions from quantifiable sources. *Environ Int, 70*, 62–75.

Wang, Z., Cousins, I. T., Scheringer, M., Buck, R. C. & Hungerbuhler, K. 2014b. Global emission inventories for C4-C14 perfluoroalkyl carboxylic acid (PFCA) homologues from 1951 to 2030, Part II: The remaining pieces of the puzzle. *Environ Int, 69*, 166–176.

Wang, Z., Cousins, I. T., Scheringer, M. & Hungerbuehler, K. 2015. Hazard assessment of fluorinated alternatives to long-chain perfluoroalkyl acids (PFAAs) and their precursors: Status quo, ongoing challenges and possible solutions. *Environ Int, 75*, 172–179.

Wang, Z., Cousins, I. T., Scheringer, M. & Hungerbühler, K. 2013. Fluorinated alternatives to long-chain perfluoroalkyl carboxylic acids (PFCAs), perfluoroalkane sulfonic acids (PFSAs) and their potential precursors. *Environment International, 60*, 242–248.

Wang, Z., Dewitt, J. C., Higgins, C. P. & Cousins, I. T. 2017. A never-ending story of per- and polyfluoroalkyl substances (PFASs)? *Environmental Science & Technology, 51*, 2508–2518.

Wang, Z., Xie, Z., Mi, W., Moöller, A., Wolschke, H. and Ebinghaus, R. 2015. Neutral poly/per-fluoroalkyl substances in air from the Atlantic to the Southern Ocean and in Antarctic snow. *Environmental science & technology, 49*(13), pp. 7770–7775.

Washington, J. W., Jenkins, T. M., Rankin, K. & Naile, J. E. 2015. Decades-scale degradation of commercial, side-chain, fluorotelomer-based polymers in soils and water. *Environ Sci Technol, 49*, 915–923.

Washington, J. W., Naile, J. E., Jenkins, T. M. & Lynch, D. G. 2014. Characterizing fluorotelomer and polyfluoroalkyl substances in new and aged fluorotelomer-based polymers for degradation studies with GC/MS and LC/MS/MS. *Environ Sci Technol, 48*, 5762–5769.

Weber, A. K., Barber, L. B., Leblanc, D. R., Sunderland, E. M. & Vecitis, C. D. 2017. Geochemical and hydrologic factors controlling subsurface transport of poly- and perfluoroalkyl substances, Cape Cod, Massachusetts. *Environ Sci Technol, 51*, 4269–4279.

Weber Jr, W. J., Voice, T.C. & Jodellah, A. 1983. Adsorption of humic substances: The effects of heterogeneity and system characteristics. *Journal of American Water Works Association, 75*, 612–619.

Weinberg, I., Dreyer, A. & Ebinghaus, R. 2011. Waste water treatment plants as sources of polyfluorinated compounds, polybrominated diphenyl ethers and musk fragrances to ambient air. *Environ Pollut, 159*, 125–132.

Weiner, B., Yeung, L. W. Y., Marchington, E. B., D'agostino, L. A. & Mabury, S. A. 2013. Organic fluorine content in aqueous film forming foams (AFFFs) and biodegradation of the foam component 6:2 fluorotelomermercaptoalkylamido sulfonate (6: 2 FTSAS). *Environmental Chemistry, 10*, 486.

Wilkinson, J. L., Hooda, P. S., Swinden, J., Barker, J. and Barton, S. 2017. Spatial distribution of organic contaminants in three rivers of Southern England bound to suspended particulate material and dissolved in water. *Science of the Total Environment, 593*, pp. 487–497.

Wilson, A. 2016a. *Emerging contaminants—Challenges for consultants and laboratories* [Online]. Available: https://forumcourt.sharepoint.com/_layouts/15/guestaccess.aspx?guestaccesstoken=6VkP6zf592L6IA3Ka8cR29sw1IiPMVygHk9nu5PazK8%3d&docid=0535161364da541258b1924549c8e0636 [Accessed September 24, 2018]. >

Wilson, A. 2016b. Emerging contaminants challenges for consultants and laboratories. *Presentation.*

Winquist, A., Lally, C., Shin, H. M. & Steenland, K. 2013. Design, methods, and population for a study of PFOA health effects among highly exposed mid-Ohio valley community residents and workers. *Environ Health Perspect, 121*, 893–899.

Wong, F., Macleod, M., Mueller, J. F. & Cousins, I. T. 2014. Enhanced elimination of perfluorooctane sulfonic acid by menstruating women: Evidence from population-based pharmacokinetic modeling. *Environ Sci Technol, 48*, 8807–8814.

Wong, F., Macleod, M., Mueller, J. F. & Cousins, I. T. 2015. Response to comment on "enhanced elimination of perfluorooctane sulfonic acid by menstruating women: evidence from population-based pharmacokinetic modeling." *Environ Sci Technol, 49*, 5838–5839.

Wood, R. J., Lee, J. & Bussemaker, M. J. 2017. A parametric review of sonochemistry: Control and augmentation of sonochemical activity in aqueous solutions. *Ultrason Sonochem, 38*, 351–370.

World Health Organisation. 2014. *IARC Interim Annual Report 2014.* Report No. SC/51/2, GC/57/2. December 10.

Xia, J. & Niu, C. 2017. Acute toxicity effects of perfluorooctane sulfonate on sperm vitality, kinematics and fertilization success in zebrafish. *Chin. J. Oceanol. Limnol., 35*, 723.

Xiao, F. 2017. Emerging poly- and perfluoroalkyl substances in the aquatic environment: A review of current literature. *Water Res, 124*, 482–495.

Xiao, F., Davidsavor, K. J., Park, S., Nakayama, M. & Phillips, B. R. 2012. Batch and column study: Sorption of perfluorinated surfactants from water and cosolvent systems by Amberlite XAD resins. *J Colloid Interface Sci, 368*, 505–511.

Xiao, F., Golovko, S. A. & Golovko, M. Y. 2017. Identification of novel non-ionic, cationic, zwitterionic, and anionic polyfluoroalkyl substances using UPLC-TOF-MSE high-resolution parent ion search. *Anal Chim Acta, 988*, 41–49.

Xiao, L., Ling, Y., Alsbaiee, A., Li, C., Helbling, D. E. & Dichtel, W. R. 2017. Beta-cyclodextrin polymer network sequesters perfluorooctanoic acid at environmentally relevant concentrations. *Journal of the American Chemical Society.*

Xiao, X., Ulrich, B. A., Chen, B. & Higgins, C. P. 2017. Sorption of poly- and perfluoroalkyl substances (PFASs) relevant to aqueous film-forming foam (AFFF)-impacted groundwater by biochars and activated carbon. *Environmental Science & Technology, 51*, 6342–6351.

Xiao, F., Hanson, R. A., Golovko, S. A., Golovko, M. Y., & Arnold, W. A. 2018. PFOA and PFOS are generated from zwitterionic and cationic precursor compounds during water disinfection with chlorine or ozone. *Environmental Science & Technology Letters.*

Xu, B., Ahmed, M. B., Zhou, J. L., Altaee, A., Wu, M. & Xu, G. 2017. Photocatalytic removal of perfluoroalkyl substances from water and wastewater: Mechanism, kinetics and controlling factors. *Chemosphere, 189*, 717–729.

Yamada, T., P.H. Taylor (2003). Laboratory scale thermal degradation of perfluoro-octanyl sulfonate and related precursors. Final Report, 3M Company.
Yamada, T., Taylor, P. H., Buck, R. C., Kaiser, M. A. & Giraud, R. J. 2005. Thermal degradation of fluorotelomer treated articles and related materials. *Chemosphere, 61*, 974–984.
Yamamoto, T., Noma, Y., Sakai, S.-I. & Shibata, Y. 2007. Photodegradation of perfluorooctane sulfonate by UV irradiation in water and alkaline 2-propanol. *Environmental Science & Technology, 41*, 5660–5665.
Yamashita, N., Kannan, K., Taniyasu, S., Horii, Y., Okazawa, T., Petrick, G. & Gamo, T. 2004. Analysis of perfluorinated acids at parts-per-quadrillion levels in seawater using liquid chromatography-tandem mass spectrometry. *Environmental Science & Technology, 38*, 5522–5528.
Yamashita, N., Taniyasu, S., Petrick, G., Wei, S., Gamo, T., Lam, P. K. & Kannan, K. 2008. Perfluorinated acids as novel chemical tracers of global circulation of ocean waters. *Chemosphere, 70*, 1247–1255.
Yan, H., Cousins, I. T., Zhang, C. & Zhou, Q. 2015. Perfluoroalkyl acids in municipal landfill leachates from China: Occurrence, fate during leachate treatment and potential impact on groundwater. *Science of the Total Environment, 524*, 23–31.
Yang, H., Spoonamore, S. 2014. Stormwater runoff treatment using bioswales augmented with advanced nanoengineered materials. *In:* D. E. Reisner, T. P. (ed.) *Aquananotechnology: Global Prospects.* Boca Raton, FL: CRC Press.
Yeung, L. W., Miyake, Y., Taniyasu, S., Wang, Y., Yu, H., So, M. K., Jiang, G., Wu, Y., Li, J., Giesy, J. P., Yamashita, N. & Lam, P. K. 2008. Perfluorinated compounds and total and extractable organic fluorine in human blood samples from China. *Environ Sci Technol, 42*, 8140–8145.
Yeung, L. W., Robinson, S. J., Koschorreck, J. & Mabury, S. A. 2013a. Part I. A temporal study of PFCAs and their precursors in human plasma from two German cities 1982–2009. *Environ Sci Technol, 47*, 3865–3874.
Yeung, L. W., Robinson, S. J., Koschorreck, J. & Mabury, S. A. 2013b. Part II. A temporal study of PFOS and its precursors in human plasma from two German cities in 1982–2009. *Environ Sci Technol, 47*, 3875–3882.
Yeung, L. W. Y. & Mabury, S. A. 2016. Are humans exposed to increasing amounts of unidentified organofluorine? *Environmental Chemistry*, 13.
Yi, S., Harding-Marjanovic, K. C., Houtz, E. F., Gao, Y., Lawrence, J. E., Nichiporuk, R. V., Iavarone, A. T., Zhuang, W.-Q., Hansen, M., Field, J. A., Sedlak, D. L. & Alvarez-Cohen, L. 2018. Biotransformation of AFFF component 6:2 fluorotelomer thioether amido sulfonate generates 6:2 fluorotelomer thioether carboxylate under sulfate-reducing conditions. *Environmental Science & Technology Letters, 5*, 283–288.
York, R. G., G. L. Kennedy, G. W. Olsen, and J. L. Butenhoff. 2010. Male reproductive system parameters in a two-generation reproduction study of ammonium perfluorooctanoate in rats and human relevance. *Toxicology* 271:64–72.
You, C., Jia, C. & Pan, G. 2010. Effect of salinity and sediment characteristics on the sorption and desorption of perfluorooctane sulfonate at sediment-water interface. *Environmental Pollution, 158*, 1343–1347.
Young, C. J., Mabury, S. A., 2010. Atmospheric perfluorinated acid precursors: Chemistry, occurrence, and impacts. *Reviews of Environmental Contamination and Toxicology, 208*, 1–109.
Yu, Q., Zhang, R., Deng, S., Huang, J. & Yu, G. 2009. Sorption of perfluorooctane sulfonate and perfluorooctanoate on activated carbons and resin: Kinetic and isotherm study. *Water Res, 43*, 1150–1158.
Yuan, G., Peng, H., Huang, C. & Hu, J. 2016. Ubiquitous occurrence of fluorotelomer alcohols in eco-friendly paper-made food-contact materials and their implication for human exposure. *Environ Sci Technol, 50*, 942–950.

Zaggia, A. & Ameduri, B. 2012. Recent advances on synthesis of potentially non-bioaccumulable fluorinated surfactants. *Current Opinion in Colloid & Interface Science, 17,* 188–195.

Zaggia, A., Conte, L., Falletti, L., Fant, M. & Chiorboli, A. 2016. Use of strong anion exchange resins for the removal of perfluoroalkylated substances from contaminated drinking water in batch and continuous pilot plants. *Water Research, 91,* 137–146.

Zembouai, I., Kaci, M., Bruzaud, S., Pillin, I., Audic, J-L., Shayanfar, S., Pillai, S. D. 2016. Electron beam radiation effects on properties and ecotoxicity of PHBV/PLA blends in presence of organo-modified montmorillonite. *Polymer Degradation and Stability, 132,* 10.

Zhang, C., Qu, Y., Zhao, X. & Zhou, Q. 2015. Photoinduced reductive decomposition of perfluorooctanoic acid in water: effect of temperature and ionic strength. *CLEAN—Soil, Air, Water, 43,* 223–228.

Zhang, K., Huang, J., Yu, G., Zhang, Q., Deng, S. & Wang, B. 2013a. Destruction of perfluorooctane sulfonate (PFOS) and perfluorooctanoic acid (PFOA) by ball milling. *Environmental Science Technology, 47,* 6471–6477.

Zhang, Q., Deng, S., Yu, G. & Huang, J. 2011. Removal of perfluorooctane sulfonate from aqueous solution by crosslinked chitosan beads: Sorption kinetics and uptake mechanism. *Bioresour Technol, 102,* 2265–2271.

Zhang, S., Lu, X., Wang, N. & Buck, R. C. 2016. Biotransformation potential of 6:2 fluorotelomer sulfonate (6:2 FTSA) in aerobic and anaerobic sediment. *Chemosphere, 154,* 224–230.

Zhang, S., Szostek, B., McCausland, P. K., Wolstenholme, B. W., Lu, X., Wang, N. & Buck, R. C. 2013b. 6:2 and 8:2 fluorotelomer alcohol anaerobic biotransformation in digester sludge from a WWTP under methanogenic conditions. *Environ Sci Technol, 47,* 4227–4235.

Zhang, X., Chen, L., Fei, X. C., Ma, Y. S. & Gao, H. W. 2009. Binding of PFOS to serum albumin and DNA: Insight into the molecular toxicity of perfluorochemicals. *BMC Mol Biol, 10,* 16.

Zhao, D., Cheng, J., Vecitis, C. D. & Hoffmann, M. R. 2011. Sorption of perfluorochemicals to granular activated carbon in the presence of ultrasound. *J Phys Chem A, 115,* 2250–2257.

Zhi, Y. & Liu, J. 2015. Adsorption of perfluoroalkyl acids by carbonaceous adsorbents: Effect of carbon surface chemistry. *Environmental Pollution, 202,* 168–176.

Ziltek 2017. RemBind®. http://ziltek.com.au/rembind.html.

Acronyms

PFAS Acronyms

4:2 Cl-PFESA	4:2 chlorinated perfluoroalkyl ether sulfonate
6:2 Cl-PFESA	6:2 chlorinated perfluoroalkyl ether sulfonate
8:2 Cl-PFESA	8:2 chlorinated perfluoroalkyl ether sulfonate
4:2 FTOH	4:2 fluorotelomer alcohol
4:2 FTS	4:2 fluorotelomer sulfonate
4:2 FTTAoS	4:2 fluorotelomer thioamido sulfonate
4:2 PAP	4:2 polyfluoroalkyl phosphate
5:1:2 FTB	5:1:2 fluorotelomer betaine
5:3 FTB	5:3 fluorotelomer betaine
5:3 FTCA	5:3 fluorotelomer carboxylate / carboxylic acid
6:2 diPAP	6:2 di-substituted polyfluoroalkyl phosphate
6:2 FTAB	6:2 fluorotelomer sulfonamide alkylbetaine
6:2 FTOH	6:2 fluorotelomer alcohol
6:2 FTS	6:2 fluorotelomer sulfonate / sulfonic acid
6:2 FTSaAm	6:2 fluorotelomer sulfonamidoamine
6:2 FTSaB	6:2 fluorotelomer sulfonamide amine
6:2 FTTAoS	6:2 fluorotelomer thioamido sulfonate
6:2 FTTHN	6:2 fluorotelomer thiohydroxy ammonium
6:2 PAP	6:2 polyfluoroalkyl phosphate
7:1:2 FTB	7:1:2 fluorotelomer betaine
7:3 FTB	7:3 fluorotelomer betaine
8:2 FTOH	8:2 fluorotelomer alcohol
8:2 FTS	8:2 fluorotelomer sulfonate / sulfonic acid
8:2 FTSaAm	8:2 fluorotelomer sulfonamido amine
8:2 FTSaB	8:2 fluorotelomer sulfonamide amine
8:2 FTTAoS	8:2 fluorotelomer thioamido sulfonate
8:2 diPAP	8:2 di-substituted polyfluoroalkyl phosphate
C_8/C_8 PFPiA	8:8 perfluoroalky phosphinic acid
9:1:2 FTB	9:1:2 fluorotelomer betaine
9:3 FTB	9:3 fluorotelomer betaine
10:2 FTSaB	10:2 fluorotelomer sulfonamide amine
12:2 FTSaB	12:2 fluorotelomer sulfonamide amine
ADONA	trade name for 4,8-dioxa-3H-perfluorononanoate
APFO	ammonium perfluorooctanoate
***N*-BuFASA**	*N*-butyl perfluoroalkane sulfonamide
***N*-BuFASAA**	*N*-butyl perfluoroalkane sulfonamido acetic acid
***N*-BuFOSA**	*N*-butyl perfluorooctane sulfonamide
CTFE	chlorotrifluoroethylene
diPAPs	disubstituted polyfluoroalkyl phosphates
diSAmPAP	bis-[2-(*N*-ethylperfluorooctane-1-sulfonamido)ethyl] phosphate
***N*-EtFASA**	*N*-ethyl perfluoroalkane sulfonamide

N-EtFASAA	*N*-ethyl perfluoroalkane sulfonamido acetic acid
N-EtFBSA	*N*-ethyl perfluorobutane sulfonamide
N-EtFOSA	*N*-ethyl perfluorooctane sulfonamide
FASA	perfluoroalkane sulfonamide, also referred to as PFASA
X-FASA/E	(*N*-methyl/ethyl) perfluoroalkyl sulfonamide/ sulfonamidoethanol
FASAA	perfluoroalkane sulfonamido acetic acid, also referred to as PFASAA
FASE	perfluoroalkane sulfonamido ethanols
FEP	fluorinated ethylene propylene
FOSA	perfluorooctane sulfonamide also referred to as PFOSA
FOSAA	perfluorooctanesulfonamidoacetate also referred to as PFOSAA
FOSE	perfluorooctanesulfonamidoethanol
FTASs	fluorotelomer thioether amdo sulfonic acids
FTBs	fluorotelomer betaines
FTCAs	fluorotelomer carboxylic acids
FTOs	fluorotelomer olefins
FTOHs	fluorotelomer alcohols
FTSaBs	fluorotelomer sulfonamide amines
FTSB	fluorotelomer sulfonamide betaine
FTSs	fluorotelomer sulfonic acids/ sulfonate
HFPO-DA	hexafluoropropylene oxide-dimer acid; also sometimes referred to as GenX
HFPO-TA	hexafluoropropylene oxide trimer acid
N-MeFASA	N-methyl polyfluoroalkane sulfonamide
N-MeFASAA	N-methyl perfluoroalkane sulfonamido acetic acid
N-EtFOSA	N-ethyl perfluorooctane perfluororooctane sulphonamide
N-EtFOSAA	N-ethyl perfluororooctane sulfonamidoacetate
N-EtFOSE	N-ethyl perfluororooctane sulfonamide ethanol
N-MeFOSA	N-methyl perfluorooctane sulfonamide
N-MeFOSAA	N-methyl perfluorooctane sulfonamidoacetate
N-MeFOSE	N-methyl perfluorooctane sulfonamidoethanol
PAPs	polyfluoroalkyl phosphates
PACF	perfluoroalkyl carbonyl fluoride
PASFs	perfluoroalkane sulfonyl fluorides
PFAAs	perfluoroalkyl acids
PFAIs	perfluoroalkyl iodides
PFASs	per- and polyfluoroalkyl substances
PFBA	perfluorobutanoic acid
PFBS	perfluorobutane sulfonate
PFBSaAm	perfluorobutane sulfonamido amine
PFBSaAmA	perfluorobuptane sulfonamide amino carboxylate
PFCAs	perfluoroalkyl carboxylic acids
PFCs	perfluorinated compounds
PFDA	perfluorodecanoic acid
PFDoA	perfluorododecanoic acid
PFDS	perfluorodecane sulfonic acid/sulfonate

PFPDA	perfluorodecyl phophonate
PFECAs	perfluoroalkyl ether carboxylic acids
PFECH	perfluoroethylenecyclohexane sulfonate
PFESAs	perfluoroalkyl ether sulphonic acids
PFEtA	perfluoroethanoic acid
PFEtS	perfluoroethane sulfonate
PFHpA	perfluoroheptanoic acid
PFHpS	perfluoroheptane sulfonate
PFHpSaAm	perfluoroheptane sulfonamido amine
PFHpSaAmA	perfluoroheptane sulfonamide amino carboxylate
PFHxPA	perfluorohexyl phosphonic acid
PFHxA	perfluorohexanoic acid
PFHxS	perfluorohexane sulfonic acid
PFHxSaAm	perfluorohexane sulfonamide amine; N,N-dimethyl-3-{[(trideca-perfluorohexyl)sulfonyl]amino}propan-1-aminium
PFHxSaAmA	perfluorohexane sulfonamide amino carboxylate; 3-(N-(2-carboxyethyl)-trideca-perfluorohexylsulfonamido)-N,N-dimethylpropan-1-aminium
PFHxPA	perfluorohexyl phosphonate
PFNA	perfluorononanoic acid
PFNS	perfluorononane sulfonic acid
PFOA	perfluorooctanoic acid/ perfluorooctanoate
PFOPA	perfluorooctyl phosphonic acid
PFOS	perfluorooctane sulfonic acid/ perfluorooctane sulfonate
PFOSA	perfluorooctane sulfonamide
PFOSaAm	N,N-dimethyl-3-{[(trideca-perfluorooctyl)sulfonyl]amino}propan-1-aminium
PFOSaAmA	3-(N-(2-carboxyethyl)-trideca-perfluorohoctylsulfonamido)-N,N-dimethylpropan-1-aminium
PFPAs	perfluoroalkyl phosphonic acids
PFPE	perfluoropolyether
PFPeA	perfluoropentanoic acid
PFPeS	perfluoropentanoic sulfonate
PFPeSaAm	perfluoropentane sulfonamido amine
PFPeSaAmA	perfluoropentane sulfonamide amino carboxylate
PFPrA	perfluoropropane carboxylate
PFPiAs	perfluoroalkyl phosphinic acids
PFPrS	perfluoropropane sulfonate
PFSAs	perfluoroalkyl sulfonic acids
PFUdA	perfluoroundecanoic acid
POSF	perfluorooctane sulfonyl fluoride
PTFE	polytetrafluorethylene
PVDF	polyvinlidene difluoride
SAmPAP	sulfonamido ethanol-based phosphate esters
TFE	tetrafluoroethylene
TFMS	trifluoromethane sulfonate

Non-PFAS Acronyms

<LOQ	less than the level of quantification
µg/g	micrograms per gram
µg/kg	microgram per kilogram
µg/L	microgram per liter
µm	microns
AAP	aliphatic acrylic polymer
AC	activated carbon
AFFF	aqueous film forming foam
AIX	anion exchange
AO	n,n-dimethyldodecylamine n-oxide (an amphoteric surfactant)
AOF analysis	adsorbable organofluorine analysis
AOP	advanced oxidation processes
ARP	advanced reducing processes
ATSDR	Agency for Toxic Substances and Disease Registry
BCF	bioconcentration factor
BDE	bond dissociation energy
BQA	bifunctional quaternary amine
br-	used to indicate branched isomers
BV	bed volumes
CaO	lime
CDC	Centers for Disease Control and Prevention
Ce	mass contaminant per volume of liquid treated by activated carbon (AC)
cm	centimeter
CMC	critical micelle concentrations
CO	carbon monoxide
CO_2	carbon dioxide
CSM	conceptual site model
DEPA	Danish Ministry of the Environment, Environmental Protection Agency
DTSC	Department of Toxic Substances Control
DF	detection frequency
eBeam	high-energy electron beam
ECF	electrochemical fluorination process
EFSA	European Chemicals Agency
EFSA	European Food Standards Agency
EOF	extractable organofluorine
EQS	environmental quality standards
EU	European Union
foc	fraction of organic carbon
FSANZ	Food Standards Australia New Zealand
FT	fluorotelomerization
FTA	fire training area
g/L	gram per liter
GAC	granular activated carbon

HDPE	high-density polyethylene
HF	hydrogen fluoride
IARC	International Agency for Research on Cancer
ISO	International Standards Organization
Kd	partitioning coefficient
kHz	kilohertz
Koc	organic carbon-water coefficient
KOH	potassium hydroxide
Kow	octanol and water partition coefficient
kW-hr	kilowatt-hours
L/kg	liter/kilogram
L/mg	liter/milligram
LC-MS/MS	liquid chromatography with tandem mass spectrometry
LOAEL	lowest observed adverse effect level
LRT	long-range transport
MCAP	macroreticular crosslinked aromatic polymer
MCD	mechanochemical destruction
mg	milligram
mg/g	milligram per gram
mg/kg	milligram per kilogram
mg/kg bw/d	milligrams per kilogram of body weight per day
mg/L	milligram per liter
MGD	millions of gallons per day
mm	millimeters
mmol/L	micromoles per liter
mol/g	moles per gram
MPOS	mesoporous organosilica
MS	mass spectrometry
MTBE	methyl tert-butyl ether
n-	used to indicate linear isomer
NaOH	sodium hydroxide
NAPL	non-aqueous phase liquid
ND	not detected
NF	nanofiltration
ng/kg bw/day	nanogram per kilogram of body weight per day
ng/L	nanogram per liter
ng/mL	nanogram per milliliter
NHANES	National Health and Nutrition Examination Survey
NOAEL	no observed adverse effect level
NOM	natural organic matter
oC	degrees Celsius
OCRA	Ozofractionative Catalyzed Reagent Addition
OECD	Oganisation for Economic Co-operation and Development
OZF	ozofractionation
PA	polyamine
PAC	powdered activated carbon

PCE	perchloroethene
PDM	polydiallyldimethylammonium chloride
pH	negative log base 10 of the hydrogen ion concentration in moles per liter
PHE	Public Health England
PIGE	particle-induced gamma emission
pKa	acid dissociation constant
PMT	persistence, mobility, and toxicity
POD	point of departure
POPs	persistent organic pollutants
ppb	parts per billion
ppt	parts per trillion
PRB	permeable reactive barriers
PSDVB	polystyrene divinylbenzene
qe	mass of contaminant per mass of activated carbon (AC)
qm	activated carbon (AC) adsorption capacity
QTOF-MS	quadrupole time-of-flight—mass spectroscopy
R&D	research and development
Rf	retardation factor
RfDs	reference doses
RIVM	National Institute for Public Health and the Environment
RMR	resting metabolic rate
RO	reverse osmosis
S/S	stabilization/solidification
SDS	sodium decyl sulfate (an anionic surfactant)
SDVB	styrene divinylbenzene
SiO_2	silicon dioxide
SPAC	super-fine powdered activated carbon
Swedish EPA	Swedish Environmental Protection Agency
TDIs	tolerable daily intakes
TDS	total dissolved solids
TEA	terminal electron acceptor
TEQA	triethyl quaternary amine
TMQA	trimethyl quaternary amine
TOP Assay	total oxidizable precursor assay
UCS	unconfined compressive strength
UKCOT	United Kingdom Committee on Toxicology
UNEP	United Nations Environment Programme
U.S. EPA	U.S. Environmental Protection Agency
UV	ultraviolet
VEG	vapor energy generator
vPvB	very persistent and very bioaccumulative
vPvM	very persistent and very mobile
WWTP	wastewater treatment plant
μg/L	micrograms per liter

4 Hexavalent Chromium

Margaret Gentile
(Arcadis U.S. Inc., Margaret.Gentile@arcadis.com)

Greg Imamura
(Arcadis U.S. Inc., Greg.Imamura@arcadis.com)

Brent Alspach
(Arcadis U.S. Inc., Brent. Alspach@arcadis.com)

4.1 BASIC INFORMATION

Chromium is a common groundwater contaminant that has been the focus of regulation and treatment for many years. Because it has several forms that can be present in groundwater, most of the regulations to date have focused on total chromium. Despite the maturity in treatment of chromium as a contaminant, the understanding of the toxicology of the hexavalent form of chromium has been evolving and, consequently, the related regulations are currently being revisited. In the California, a new 10 micrograms per liter (µg/L) standard was in effect from 2014 to 2017 and a new Maximum Contaminant Level (MCL) is pending. In addition, the United States Environmental Protection Agency (USEPA) is currently conducting a risk assessment that will eventually be used to consider adoption of a federal MCL specific to hexavalent chromium (Cr[VI]). Given the lower levels that will need to be achieved in treatment, not just in environmental media but also in water supply, this regulatory activity (discussed in detail later in this chapter) has resurfaced attention on Cr(VI).

4.1.1 Geochemistry of Chromium

Chromium is a transition element. It is the 21st most abundant element in the earth's crust with an average natural concentration of 100 milligrams per kilogram (mg/kg) (Nriagu 1988). Chromium most commonly occurs in two valence states: trivalent (Cr[III]) and hexavalent (Cr[VI]).

Cr(III) is the primary form of chromium found in naturally occurring minerals. The Cr(III) mineral chromite is the primary geologic source of chromium. Cr(III) minerals generally have very low solubility at typical groundwater pH conditions, limiting the natural occurrence of dissolved trivalent chromium in groundwater. To a much lesser extent, Cr(VI) also occurs naturally in the environment in groundwater and soil, for example as chromate minerals. Natural occurrence of Cr(VI) in groundwater is discussed in section 4.4.1.

Cr(VI) is the form of chromium typically introduced as a contaminant from industrial activities and is the form of concern as this element is resurfacing in risk

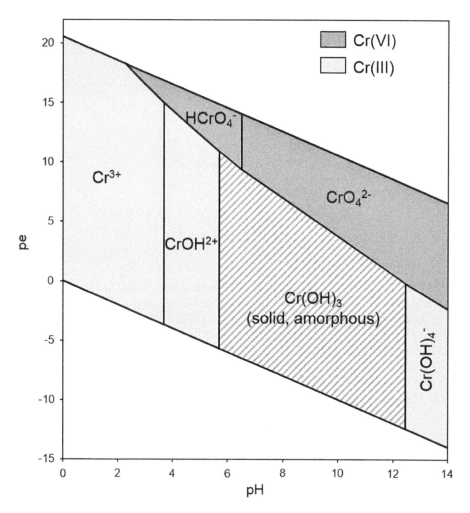

FIGURE 4.1 pe-pH stability diagram for the chromium-water system. This pe-pH stability diagram was calculated assuming 25°C, and a chromium concentration of $1.9 \times 10-6$ molar (99 μg/L). This figure was made using Minteq with the wateq database.

assessment and regulation. Hexavalent chromium tends to exist as an oxyanion/anion complex with negative charge. Figure 4.1 is a stability diagram showing the eH/pH ranges over which the Cr(VI) species and Cr(III) species predominate. The following four oxyanions are the common species of Cr(VI) in groundwater, under various geochemical conditions (based on Palmer and Puls 1994):

- H_2CrO_4 (present under very acidic conditions of pH < 0.9)
- $HCrO_4^-$ (hydrogen chromate): prevalent under acidic conditions, 0.9 < pH < 6.4, when chromium concentrations are <1,000 milligrams per liter [mg/L]

- CrO_4^{2-} (chromate): prevalent under neutral to alkaline conditions, pH > 6.4
- $Cr_2O_7^{2-}$ (dichromate): prevalent under acidic conditions 0.9 < pH < 6.4, becomes significant when chromium concentrations are high (e.g. > 1,000 mg/L)

Chromate salts are generally highly soluble (e.g., Na_2CrO_4 solubility is 876 grams per liter [g/L], K_2CrO_4 solubility is 650 g/L, [Lide 2003–2004]), allowing for high concentrations of Cr(VI) in groundwater as the result of a release to the environment.

4.1.2 Fate and Transport

Although CrO_4^{2-} adsorbs to aquifer minerals, the amount sorption is dependent upon pH, presence of competing anions, and presence and density of sorption sites. The relatively low adsorption at neutral and alkaline pH in the presences of typical groundwater anions and high solubility of the Cr(VI) species allow for the generation of large plumes of Cr(VI) in groundwater in aquifers with little natural attenuation or reducing capacity. The negatively charged chromate ions adsorb to positively charged surfaces, for example, iron oxyhydroxide and aluminum hydroxide minerals. Adsorption generally increases with decreasing pH. The presence of competing anions can inhibit sorption, with the strength of competition in order of HPO_4^{2-}, $H_2PO_4^-$ >> SO_4^{2-} >> Cl^-, NO_3^- (USEPA 1999), although the competition is pH dependent. In a 1999 document, the USEPA created a look up table for estimating the distribution coefficient, K_d, for Cr(VI) based on the data collected in a sorption study conducted by Rai et al. (Rai et al., 1988), as summarized in Table 4.1. This table illustrates the relationship between adsorption, pH, soil iron mineralogy (as a surrogate for amount of sorptive material), and sulfate concentrations.

The conversion of Cr(VI) to Cr(III) is an important process in the natural attenuation and treatment of Cr(VI). Reductants capable of reducing Cr(VI) to Cr(III) will be discussed throughout this chapter. Cr(III) forms hydroxide species through hydrolysis (Figure 4.1). The solubility of Cr(III) is controlled by chromium hydroxide, $Cr(OH)_3$, and co-precipitation with iron oxides and oxyhydroxides ($Cr_xFe_{1-x}[OH]_3$) when iron is present (Sass and Rai 1987; Eary and Rai 1988). In the absence of iron between pH 7 and 10, the neutral species $Cr(OH)_3$ dominates with minimum solubility (Rai et al., 1987). Experimental data indicates that the solubility of Cr(III) incorporated into iron oxyhydroxides is at least an order of magnitude lower than $Cr(OH)_3$ over the pH range of 2 to 6 (Sass and Rai 1987).

Figure 4.2 depicts theoretical solubility curves for both pure amorphous Cr(III) and mixed amorphous iron/Cr(III) hydroxides (Suthersan et al., 2009). The solubility curve for pure amorphous $Cr(OH)_3$ was modeled using a range of published stability constants (Schecher and McAvoy 1998; Rai et al., 2004). The solubility curve for $Cr_{0.25}Fe_{0.75}(OH)_3$ was modeled using a published stability constant for the low pH range (approximately pH less than 6.5) (Sass and Rai 1987). Due to a lack of published data, the $Cr_{0.25}Fe_{0.75}(OH)_3$ solubility curve was extrapolated at higher pH values based on the solubility at lower pH being at least one order of magnitude less soluble than pure amorphous $Cr(OH)_3$. Figure 4.2 illustrates the low solubility of Cr(III) minerals in the near neutral pH range and the decrease in solubility when incorporated in minerals with iron.

TABLE 4.1
Estimated Range of Kd values for Cr(VI) as a function of pH, extractable iron, and soluble sulfate, adapted from Understanding Variation in Partition Coefficient, Kd, Values Table E3 (USEPA 1999)

					K_d (mL/g)					
pH	4.1–5			5.1–6.0			6.1–7.0			>7.1
DCB Extractable Fe (millimoles per gram)	≤0.25	0.26–0.29	≥0.3	≤0.25	0.26–0.29	≥0.3	≤0.25	0.26–0.29	≥0.3	≤0.25	0.26–0.29	≥0.3
Soluble Sulfate (mg/L)												
0–1.9	25–35	400–700	990–1,770	20–34	190–380	390–920	8–22	70–180	80–350	0–7	0–30	1–60
2–18.9	12–15	190–330	460–820	10–15	90–180	180–430	4–10	30–80	40–160	0–3	0–14	1–30
19–189	5–8	90–150	210–380	47–	40–80	80–200	2–5	15–40	20–75	0–2	0–7	0–13
>190	3–4	40–60	100–180	2–3	20–40	40–90	1–2	7–20	8–35	0–1	0–3	0–6

DCB = dithionite-citrate-bicarbonate. This soil extraction is an assessment of the readily reduced form of iron.
Source: USEPA 1999.

Hexavalent Chromium

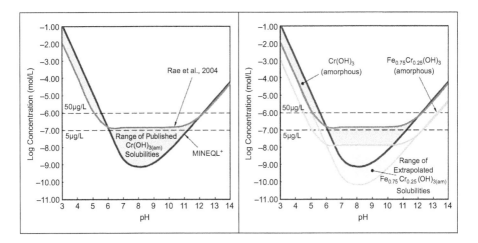

FIGURE 4.2 (Suthersan et al 2009) Solubility of Cr(III). Left panel shows a range of predicted trivalent chromium solubility in equilibrium with amorphous chromium hydroxide. Solubility calculations performed by David Ams using equilibrium constants from Rai et al. (2004) and MINEQL++ (Schecher and McAvoy 1998). (b) Range of predicted trivalent chromium solubility in equilibrium with a mixed amorphous iron-chromium hydroxide superimposed with the amorphous chromium hydroxide solubilities. Solubility calculations for mixed amorphous iron-chromium hydroxide performed by David Ams using equilibrium constants from Sass and Rai (1987) and extrapolating equilibrium constants from Rai et al. (2004) and MINEQL+ (Schecher and McAvoy 1998) (see text for discussion).

4.1.3 Sources of Cr(VI)

Industrial sources of Cr(VI) contamination in groundwater include electroplating, tanning, wood preservation and pulp production, and cooling tower water from industrial operations. Cr(VI) also occurs naturally in groundwater, particularly in alkaline aquifers rich in Cr(III)-bearing minerals where oxidation of Cr(III) from minerals generates dissolved Cr(VI) in groundwater (see section 4.4.1).

4.2 TOXICITY AND RISK ASSESSMENT

Toxicity of chromium is dependent on oxidation state. Cr(VI) is more toxic than Cr(III). Cr(III) toxicity is limited by the ability to enter cells, lack of intracellular accumulation, and limited binding to cell materials (Zhitkovich 2011). Cr(VI) enters cells through general sulfate channels, given its structural similarity to sulfate (Zhitkovich 2011). Exposure to Cr(VI) by inhalation and/or ingestion may produce effects on the respiratory tract, liver, kidney, gastrointestinal and immune systems (ATSDR 2012). It has long been recognized that Cr(VI) is a known human carcinogen by the inhalation route of exposure (USEPA 1998a).

The knowledge around carcinogenicity of Cr(VI) exposure via ingestion has been evolving over the last decade. Epidemiological evidence consists primarily of a case study of a population in China near an alloy plant that smelted chromium, with unclear results. Studies show clear evidence of Cr(VI) carcinogenicity in rats and mice (NTP 2008). However, extrapolation of dose-response data from rats and mice to humans is confounded by the extent to which Cr(VI) may be reduced to Cr(III) within the digestive tract of humans (Zhitkovich 2011). Recent research has sought to use a physiologically based pharmacokinetic models to derive a chronic oral reference dose protective of intestinal cancer in mice and to convert to an equivalent human dose (Thompson et al., 2014).

The USEPA Integrated Risk Information System (IRIS) is currently updating the 1998 risk assessment for Cr(VI). A draft of the risk assessment was originally issued in 2010, but was rescinded and is being revised to take into consideration recent developments in the research on the carcinogenicity via ingestion (USEPA 2016a).

4.3 REGULATORY STATUS

Both the USEPA and the California EPA have been evaluating Cr(VI) in drinking water supplies and considering establishing a Cr(VI)-specific drinking water standard in recent times. As will be discussed in this section, the health-based levels and regulatory levels being considered vary by orders of magnitude. Although there is uncertainty in the understanding of the carcinogenicity via ingestion (see section 4.2), health-based goals and screening levels established in a few U.S. states are in the 10s of parts per trillion (Table 4.2). Evaluation of Cr(VI) at the part per trillion levels has required development of analytical methods with improved sensitivity and evaluation of treatment to low levels for development of regulatory standards. The changing regulatory environment and evaluation of Cr(VI) occurrence and treatment at low levels are the issues surrounding the resurfacing of Cr(VI). A summary of current regulatory standards is provided in Table 4.2.

TABLE 4.2
Summary of Cr(T) and Cr(VI) regulatory standards

Agency	Standard	Concentration
WHO	Drinking Water Guideline Value, Cr(T)	50 µg/L
Italy	Maximum Acceptable Concentration (MAC), Cr(T)	5 µg/L
Canada	Current Cr(T) MAC	50 µg/L
	2015 Proposed Cr(T) MAC Revision	100 µg/L
United States, Federal (USEPA)	Cr(T) MCL	100 µg/L
California, USA	Cr(T) MCL	50 µg/L
	Cr(VI) MCL 2014–2017, no longer in effect	10 µg/L
	Cr(VI) Public Health Goal	0.02 µg/L
North Carolina, USA	Cr(T) EMC 15A NCAC 02L Groundwater Standard	10 µg/L
	Cr(VI) DHHS Health Screening Level	0.07 µg/L

Note: The California Cr(VI) MCL was issued in 2014, followed by a Superior Court of California ruling to delete and replace the MCL in 2017 based on lack of evidence of financial analysis required by the Safe Drinking Water Act. CSWRCB is in the process of proposing a new MCL.

Hexavalent Chromium

4.3.1 U.S. FEDERAL REGULATIONS

Under the Safe Drinking Water Act, USEPA established an MCL for total chromium of 100 µg/L in 1991. Total chromium includes all forms of chromium, including Cr(VI). In 2008, USEPA began a rigorous and comprehensive review of the health effects of Cr(VI), which is still ongoing at the time of publication (IRIS 2017), as described in section 4.2. In 2012, USEPA included Cr(VI) in the Third Unregulated Contaminant Monitoring Rule (UCMR) (Federal Register 2012).

Every six years, the USEPA reviews existing national primary drinking water regulations (NPDWRs) to determine which, if any, need to be revised. A draft of the third Six-Year Review of the NPDWRs was issued for public comment on January 11, 2017 (Federal Register 2017). The draft indicated that Cr(VI) is still undergoing the USEPA IRIS health risk assessment and, consequently, Cr(VI) was not considered for adoption of a new MCL in the third review (Federal Register 2017).

4.3.2 U.S. STATE REGULATIONS

At the state level within the United States, several states, including California, New Jersey, and North Carolina, began development of state regulatory standards around 2008. Since then, California pursued a drinking water standard, while New Jersey and North Carolina are now awaiting the USEPA risk assessment before proceeding further.

4.3.2.1 California

California Department of Health Services (CDHS), which would later become the California Department of Public Health (CDPH), identified chromium for possible MCL review in 1999. The California EPA Office of Environmental Health Hazard Assessment (OEHHA) added Cr(VI) to the unregulated chemicals requiring monitoring in 1999 as well. To gather more data on the toxicology and carcinogenicity of Cr(VI) exposure via ingestion, the California Congressional Delegation, the California EPA, and CDPH nominated Cr(VI) to the National Toxicological Program (NTP) for study (NTP 2008). The NTP studies of rats and mice which showed evidence of carcinogenicity were completed in 2008 (NTP 2008). In 2011, OEHHA deemed there was sufficient evidence of carcinogenicity via ingestion based on the NTP study of rats and mice (OEHHA 2011). OEHHA established a Public Health Goal (PHG) of 0.02 µg/L at that time based on an oral cancer slope factor of 0.5 inverse milligrams per kilogram of body weight per day ([mg/kg bw/day]$^{-1}$) (OEHHA 2011). The OEHHA PHG considered the potential reduction of Cr(VI) to Cr(III) in the digestive tract and concluded it was not a complete conversion (OEHHA 2011). This conclusion was taken into consideration in development of the PHG.

In 2013, the CDHP proposed a new MCL for Cr(VI). In 2014, an MCL of 10 µg/L for Cr(VI) for California went into effect. Following the issuance of the MCL, the California Manufacturers and Technology Association and Solano County Taxpayers Association petitioned the MCL based on a claim that the California State Water Resources Control Board (CSWRCB) failed to consider and determine whether the new standard would be economically feasible (*California Manufacturers and*

Technology Association, a California Corporation, and Solano County Taxpayers Association, a California Corporation, Petitioners v. *State Water Resources Control Board, Respondent*, 2017). On June 1, 2017, the Superior Court of California granted the petition and required the CSWRCB to delete the MCL and set a new one (CRWQCB 2017). The California State Water Resources Control Board did not agree with the conclusion, but decided it would be most expedient to follow the court's order to rescind the MCL and begin the process of adopting a new one (CRWQCB 2017). The proposal of a new MCL is pending.

4.3.2.2 North Carolina

North Carolina has a state standard for groundwater quality for total chromium of 10 µg/L under Environmental Management Law 15A NCAC 02L (NCEQ 2016). The groundwater standard serves to protect groundwater as a public resource. The North Carolina Department of Health and Human Services (DHHS) has a health screening level for Cr(VI) of 0.07 µg/L used for in health risk evaluations for private well water (NCEQ 2016).

4.3.2.3 New Jersey

In the 2009 timeframe, the New Jersey Department of Environmental Protection (NJDEP) was actively assessing the risk of Cr(VI) in soil (Stern 2009) and was considering health-based drinking water goals comparable to California's 0.02 µg/L level (Buchanan 2009). However, NDJEP is currently awaiting the USEPA's risk assessment before proceeding with a recommendation on drinking water standards (NJDEP 2017).

4.3.3 OTHER COUNTRIES

Similar to the United States, the water quality standards internationally vary in magnitude and are subject to the developing understanding of carcinogenicity due to ingestion. Italy adopted a standard for Cr(VI) early, establishing a Maximum Acceptable Concentration (MAC) of 5 µg/L for Cr(VI) in 1999.

The World Health Organization (WHO) also lists a drinking water guideline of 50 µg/L for total chromium (WHO 2017), although this guideline value is designated as "provisional because of uncertainties in the toxicological database" (WHO 2017).

In 2015, Health Canada proposed to increase the Canadian MAC for total chromium from 50 µg/L to 100 µg/L. Public consultation on the proposed guideline ended in fall of 2015. The guideline for total chromium is currently in the queue for publication. Note, the level proposed by Canada is 5,000 times the PHG for California, because of differences in use of the toxicological data. California arrived at 0.02 µg/L PHG by deriving an oral cancer slope factor of 0.5 (mg/kg bw/day)$^{-1}$ based on tumor incidence in male mice in the 2008 NTP study (OEHHA 2011). Health Canada proposed MAC was based on a total daily intake amount of 0.0044 mg/kg bw/day using benchmark dose levels obtained from the 2014 Thompson study incorporating rodent pharmacokinetic modeling and converting to human values using a human pharmacokinetic model.

4.4 OCCURRENCE OF Cr(VI)

4.4.1 NATURALLY OCCURRING (BACKGROUND) Cr(VI) IN GROUNDWATER

While not naturally abundant, Cr(VI) can occur naturally in groundwater under certain mineralogical, geochemical, and hydrogeologic conditions. The source of naturally occurring Cr(VI) in groundwater is Cr(III) in the aquifer minerals.

The average crustal abundance of chromium is 100 mg/kg (Nriagu 1988). However, in certain deposits, chromium is enriched above the continental abundance, including ultramafic rocks and basalts. In ultramafic rocks and serpentinites of ophiolite complexes, chromium concentrations have been observed between 1,000 and 60,000 mg/kg (Oze et al., 2004; Robles-Camacho and Armienta 2000). These deposits occur near convergent plate margins, as in Circum-Pacific and Mediterranean regions (Oze et al., 2007). In addition to ultramafic rocks, chromium is found in basalts, metamorphic rocks, volcanic rocks, shales, sandstones, and granites.

Elevated naturally occurring Cr(VI) concentrations up to 300 µg/L have been detected in groundwaters in aquifers sourced from ultramafic materials, as well as other less mafic granitic, metamorphic, and volcanic rocks. Table 4.3 summarizes measured background values of Cr(VI) in such environments to

TABLE 4.3
Summary of naturally occurring Cr(VI) in groundwater in association with chromium rich deposits

Description and Location	Naturally Occurring Cr(VI) Concentration (µg/L)
Spring waters from ultramafic rocks Cazadero, California, USA (Oze, Fendorf, et al. 2004)	2–12
Groundwater and spring water from ophiolite complex La Spezia Province, Italy (Fantoni, et al. 2002)	5–73
Ultramafic rocks of Sierra de Guanajuato Mexico (Robles Camacho and Armienta 2000)	12
Groundwater from Aromas Red Sands aquifer Santa Cruz County, California, USA (Gonzalez, Ndung'U and Flegal 2006)	4–33
Porewater from phosphorus–amended soils derived from ultramafic rock New Caledonia (Becquer, et al. 2003)	4.6–15.1 *
Groundwater from south west alluvial basins primarily Arizona, USA (Robertson 1991)	0–300
Groundwater in alluvial deposits weathered from mafic rock, Mojave Desert, California, USA (Ball and Izbicki 2004)	Up to 60
Groundwater in alluvial deposits weathered from granitic, metamorphic, volcanic rocks, Mojave Desert, California, USA (Ball and Izbicki 2004)	Up to 36

* Measured as total chromium.

provide the reader with examples of how high naturally occurring concentrations can be encountered.

Cr(VI) in groundwater is generated from the oxidation of the naturally occurring Cr(III) minerals. The most relevant oxidant for Cr(III) minerals are manganese oxide minerals (Oze et al., 2007), although it has also been suspected that anoxic oxidation of Cr(III) in ultramafic minerals may occur, mediated by hydrogen peroxide (Oze et al., 2016). Both Cr(III) and manganese can be found enriched in minerals by the weathering process. For the oxidation reactions with solid manganese to proceed, Cr(III) dissolves into groundwater and sorbs directly to the surface of the manganese oxide mineral (Eary and Rai 1987; Oze et al., 2007). The rate of oxidation is affected in part by surface area of Cr(III) minerals and pH (Oze et al., 2007), which will control the dissolved concentrations of Cr(III) available to react with manganese oxide surfaces. The relationship with pH will depend upon the exact mineral species present, but increased dissolved Cr(III) concentrations and increased rates of oxidation are more likely in alkaline groundwaters.

Ultimately, the concentrations of Cr(VI) in background in groundwater will depend on the factors that determine the rate of oxidation (e.g., the amount of chromium and manganese minerals, pH) and groundwater flow conditions, such as the residence time over which Cr(VI) can accumulate in the groundwater. The geochemical conditions are an additional factor, with aerobic conditions limiting the potential reductive cycling of Cr(VI) once formed.

In some cases, anthropogenic activities other than the source of chromium contamination may add to background. For example, a study of Cr(VI) in groundwater near a dairy in the Mojave Desert indicated that Cr(VI) in groundwater was mobilized from the vadose zone to the water table as a result of irrigation with dairy wastewater (Izbicki 2008). These results indicate the placement and length of screened intervals, that is, within proximity to the water table, which can have profound effects in measurement of Cr(VI) distribution to characterize contamination or anthropogenic background. In another example, elevated Cr(VI) concentrations up to 700 μg/L were detected in porewater from phosphorus-amended soils derived from ultramafic rock in New Caledonia, indicating the potential influence of phosphate-based fertilizers where phosphate displaces Cr(VI) naturally sorbed to aquifer minerals (Becquer et al., 2003).

4.4.2 Cr(VI) in Drinking Water

Data on occurrence of Cr(VI) in Public Water Systems (PWSs) were collected under Unregulated Contaminant Monitoring Rule (UCMR) programs by the USEPA and CSWRCB. Monitoring of PWSs under USEPA UCMR3 was conducted from 2013 to 2015 using a method reporting limit of 0.03 μg/L (USEPA 2016b). Of the more than 62,000 data points collected from drinking water supplies tested, over 75 percent (over 47,000) had detections of Cr(VI) above the Method Reporting Limit (MRL) (USEPA 2016b). The 47,000 detections represented 4,386 public water supplies (USEPA 2016b).

TABLE 4.4
Summary of peak levels of Cr(VI) in drinking water sources detected under the California UCMR (CSWQCB 2014)

Peak Level (μg/L)	Number of Sources	% of Detections
1–5	1,596	66
6–10	496	20
11–20	247	10
21–30	66	3
31–40	17	<1
41–50	5	<1
>50	4	<1
	Total 2,431	

Notes: Data were collected between 2000 and 2012. Sources are active, standby, and pending sources reporting more than a single detection of Cr(VI). Data may include both raw and treated sources, distribution systems, blending reservoirs, and other sampled entities.

Monitoring of PWSs under the California UCMR for Cr(VI) was conducted from 2001 to 2016 using a reporting limit of 1 μg/L, although potentially lower detection limits were allowed in the monitoring rule. The results of the California UCMR are summarized in Table 4.4. More than 2,400 sources reported detections of Cr(VI) greater than 1 μg/L. 12 percent (340) of the sources tested yielded concentrations over the 10 μg/L level that was the California MCL from 2014 to 2017 (see section 4.3.2).

4.5 SITE CHARACTERIZATION

Although standard techniques for site characterization are available for Cr(VI), special considerations apply given the geochemical reactivity of this species. Care must be taken in sampling and preservation to prevent reaction that generates or consumes Cr(VI), as discussed in detail in this section.

4.5.1 INVESTIGATION OF Cr(VI) IN GROUNDWATER

The influence of sampling techniques should be carefully considered when designing investigations for Cr(VI) in groundwater.

Samples that will be analyzed for Cr(VI) can be collected using most traditional methods (e.g., low-flow purge, HydraSleeve™, Snap Sampler, etc.). Passive diffusion bags are not suitable for Cr(VI), because the CrO_4^{2-} molecule does not readily diffuse into the bag.

Understanding mass flux as part of site characterization is key to the design of efficient groundwater remediation systems. High-resolution site characterization can be highly effective for identifying mass transport zones within the aquifer and the extent of contamination and is discussed in greater detail in other references (Suthersan et al., 2016). High-resolution characterization typically includes techniques such as vertical aquifer profiling for permeability and contaminant concentration mapping. Not all of the available techniques are appropriate for Cr(VI), or require awareness of potential bias if they will be used. For example, it has been observed that results from samples collected by grab sampling during drilling activities are depressed compared to samples collected from properly constructed monitoring wells for concentrations in 100 µg/L range and lower. The exact mechanism for this bias is not known but presumably may be related to the reduction of Cr(VI) by reactive soil surfaces, for example, iron, exposed or generated during sampling. Vertical aquifer profiling techniques that allow for purging of the sampling interval prior to sample collection may be more representative but are challenging and not well established for Cr(VI).

4.5.2 ANALYTICAL METHODS

Distinct methods are used to speciate chromium in groundwater and soil, as summarized in Table 4.5. For both soil and water, methods specific to the determination of Cr(VI) and total chromium are available. If needed, trivalent chromium can be determined by the difference between total chromium and Cr(VI).

Care must be taken to properly preserve samples for Cr(VI) in water samples to prevent reactions from altering the results, for example, oxidation of Cr(III) to Cr(VI). USEPA Methods 7196 and Standard Method (SM) 3500 for Cr(VI) in water do not have a preservative and have a short hold time of 24 hours. In 2007, Method 218.6 for Cr(VI) in water was added to the USEPA method update rule (Federal Register 2007). USEPA Method 218.6 uses an ammonium sulfate/ammonium hydroxide buffer to extend the hold time to 28 days. The buffer is titrated to preserve samples at pH 9 to 9.5, the pH required to convert all Cr(VI) to the chromate and dichromate forms appropriate for in chromatography in the analytical method. The ammonia in the buffer reacts with free chlorine present in drinking water samples to form chloramines and prevent the oxidation of Cr(III) to Cr(VI) (Ezebuiro et al., 2012). Despite improved preservation methods, there is still potential for conversion between Cr(III) and Cr(VI) in sample preparation and analysis, for example, in the alkaline extraction of soils in USEPA 3060A. It has been suggested that the Speciated Isotope Dilution Mass Spectrometry, USEPA 6800, can be used to track the interconversion between species, particularly for solids extracted by USEPA 3060A and analyzed by USEPA 7196A. This method uses spikes of isotopically labeled Cr(III) and Cr(VI) to trace potential conversions and apply corrections to the determination of native Cr(VI) and Cr(III) based on the conversions of the isotopic spikes. This method should be used with caution, however, because there is an underlying assumption that oxidation and reduction reactions occur with the spiked isotopes in the same relative amounts as native species, regardless of the relative amount of the spike, which may not be the case.

Total chromium without filtration is the applicable method for analysis of drinking water samples and comparison to total chromium drinking water standards. For total

TABLE 4.5
Summary of analytical methods to speciate chromium in soil and groundwater

Species	Matrix	Analytical Methods	Analysis	Preservative	Hold Time	Typical Reporting Limits[1]
Cr(VI)	Water	USEPA 7196/7196A/ Standard Method 3500	Colorimetric method	None	24 hours	10 µg/L
Cr(VI)	Water	USEPA 218.6/7199 USEPA 218.7	Ion chromatography	Filtration through 0.45 µm filter; ammonium sulfate/ammonium hydroxide buffer to pH 9–9.5	28 days	0.06–0.2 µg/L[4]
Cr(VI)	Soil	Extraction: USEPA 3060a Extract Analysis: USEPA 7196A/7199	Alkaline extraction and standard analyses of extract	None	24 hours	0.04 mg/kg
Cr(VI)/Cr(III)	Water Soil	USEPA 6800	Speciated isotope dilution mass spectrometry	Variable. Isotopic spike as soon as possible after sample collection	Not specified	Variable
Total Cr	Water	USEPA 200.8/ 6020	Inductively coupled plasma—mass spectrometry	Nitric acid to pH <2	6 months	1 µg/L[2]
Total Cr	Water	USEPA 200.7/6010B	Inductively coupled plasma—atomic emission spectrometry	Nitric acid to pH <2	6 months	1 µg/L

(*Continued*)

TABLE 4.5 (Continued)

Species	Matrix	Analytical Methods	Analysis	Preservative	Hold Time	Typical Reporting Limits[1]
Total Cr	Soil	Extraction: USEPA 3050b/ USEPA 3052 Extract Analysis: USEPA 6010B/6020	3050b: strong acid digestion by repeated additions of nitric acid and hydrogen peroxide—does not dissolve silicates 3052: microwave-assisted digestion by nitric acid and hydrofluoric acid—dissolves silicates	None	6 months	0.5 mg/kg
Total Cr	Soil	X-ray fluorescence (XRF)	Handheld XRF can be used to quantify chromium content of soils in the field	None	Not applicable to field method	Varies with instrument. Generally close to crustal abundance (100 mg/kg)[3]

1 Reporting limits obtained by individual laboratories for a specific matrix will vary. Values provided here are based on authors' experience with commercial analytical laboratories.
2 The multi-laboratory method detection limit reported in Method 6020A (USEPA 1998) is generally below 0.1 µg/L for most elements.
3 (USEPA, Innovative Technology Verification Report. Nito XLi 700 Series XRF Analyzer. EPA/540/R-06/003 2006)
4 The multi-laboratory method detection limit reported in Method 218.6 was 0.4 µg/L (Arar et al. 1994). However, lower reporting limits are available from commercial laboratories. Analytical cost may increase as lower reporting limits are specified.

chromium in groundwater samples, filtration with a 45 micron (μm) filter can be used to ensure particulates entrained in samples do not artificially inflate analytical results by contributing more chromium by dissolution in preservation acid, providing a total dissolved chromium result representative of chromium mobile within the aquifer.

4.5.3 Advanced Investigation Techniques

In addition to the standard quantitative analytical methods discussed in section 4.5.1, several advanced techniques are available for deeper evaluation of Cr(VI) occurrence, attenuation, and treatment. These include stable isotope and mineralogic analysis, both of which are discussed in more detail below.

4.5.3.1 Chromium Isotopes

Chromium exists as four stable isotopes: ^{50}Cr, ^{52}Cr, ^{53}Cr, and ^{54}Cr. ^{52}Cr is the most abundant isotope at 83.8 percent, followed by ^{53}Cr at 9.5 percent. Chromium ^{52}Cr/^{53}Cr isotopic analysis is a potential tool for differentiating between chromium from contamination and naturally occurring background or for demonstrating natural attenuation through natural reduction (see section 4.6.4), because some reactions preferentially use the lighter isotope ^{52}Cr and enrich the remaining chromium in the heaver isotope ^{53}Cr in a process referred to as isotopic fractionation. Fractionation of chromium can be an indication that natural reduction is occurring or can be used to differentiate natural from anthropogenic chromium, but data must be interpreted carefully given the potential overlap between the range of fractionation from reduction and the range in native groundwater (Izbicki et al., 2008) and the influences of mixing (Izbicki et al., 2012; Raddatz et al., 2011). Chromium in chromium ores are expected to be composed of primarily ^{52}Cr. For example, δ^{53} Cr, a measure of the ratio of ^{53}Cr to ^{52}Cr normalized to a standard, in basalts from various global locations was measured to be −0.04 to 0.05‰ (Ellis et al., 2002), with negative and near zero values indicating little enrichment of ^{53}Cr compared to the standard. However, δ^{53} values ranging from 0 to 4.6‰ have been reported for native Cr(VI) in groundwater (Izbicki et al., 2012), with the near zero values indicating little enrichment and the high positive values indicating enrichment of the heavy isotope. Sources of chromium used in industrial operations are not expected to be enriched, with values ranging from 0 to 0.36‰ reported to represent sources such as electroplating baths and anthropogenic Cr(VI) in groundwater (Izbicki et al., 2008; Ellis et al., 2002). Lighter isotopes of chromium are preferentially reduced, resulting in a fractionation and increased δ^{53} from reductive processes. δ^{53} values from 0 to 5.7‰ have been measured in chromium plumes thought to have undergone reductive processes (Izbicki et al., 2008; Ellis et al., 2002).

4.5.3.2 Mineralogical Analyses

Given the importance of aquifer solids as a source of naturally occurring Cr(VI) and as the matrix for immobilization of Cr(VI) reduced to Cr(III) for in situ remediation, additional mineralogical analyses may be useful for site characterization or demonstration of treatment and attenuation mechanisms. Table 4.6 summarizes several available mineralogical analyses.

TABLE 4.6
Summary of advanced mineralogical analyses relevant to Cr(VI) groundwater remediation

Method	Description	Uses
Acid volatile sulfide and simultaneously extractable iron (AVS-SEM)	Published methods (Cooper and Morse, 1998); 6 Molar hydrochloric acid (equivalent to 50% HCl) is used to extract metal sulfides	Can be used to evaluate the presence of sulfide minerals and ferrous sulfide minerals as a measure of reductants formed within the aquifer.
Scanning electron microscopy and energy dispersive x-ray spectroscopy (SEM-EDS)	Analysis of soil by electron microscopy.	Allows for qualitative determination of mineralogy and semi-quantitative determination of element association in order to examine morphology of major mineral phases, and map element distribution in the soil at the sub-micron scale. Can be used to evaluate geochemical mechanisms.
Selective Extractions	Extraction by increasingly intense extractants	Can be used to understand Cr(VI) sorbed to soils or available for reactions (Becquer, et al. 2003). Potential applications include assessment of solids-associated Cr(VI) contributing to background and demonstration of treatment mechanism and stability.
X-ray diffraction	X-ray analysis for identification of minerals based on information about the crystalline structure.	Provides bulk mineralogy. May be difficult to detect relevant reductive phases for natural attenuation evaluations.
Micro x-ray fluorescence (μ-XRF) mapping and x-ray adsorption spectroscopy (μ-XAS): x-ray adsorption near edge structure (XANES)	Analysis of soil using synchrotron-based x-rays	μ-XRF to examine element distribution, association, and oxidation state in soil at the sub-mm scale. Provides semi-quantitative determination of the association of chromium with major elements (e.g., iron) in soil. XANES is used to resolve oxidation states of the redox sensitive elements chromium and iron. This advanced technique can potentially be used to demonstrate treatment mechanism. The detection of chromium can be challenging within the background of naturally occurring chromium in soils.

4.6 GROUNDWATER TREATMENT

Chromium cannot be degraded or destroyed. Accordingly, the elimination of Cr(VI) in groundwater relies largely on physical removal and subsequent ex situ treatment or the in situ reduction of Cr(VI) to Cr(III), which sequesters the chromium in the aquifer as a solid precipitate, while also decreasing its solubility and toxicity.

- In situ technologies relying on reduction include chemical and biological processes and can be engineered by the injection of liquid reagents or through the emplacement of solid reactants in fill material, most often as a permeable reactive barrier (PRB).
- Ex situ technologies can also rely on biological and chemical reduction, but include the option of ion exchange that also relies on the sorption properties of the chromate ion to remove it from groundwater.
- Monitored natural attenuation relies largely on the natural abilities of the aquifer to reduce Cr(VI) to Cr(III) and remove Cr(III) from groundwater.

Each of these is discussed further in the sections that follow.

4.6.1 IN SITU REDUCTION

Cr(VI) in groundwater is readily treated by in situ methods. Available in situ methods rely on the reduction of Cr(VI) to Cr(III) and subsequent precipitation and immobilization of Cr(III) as $Cr(OH)_3$ or $Cr_xFe_{1-x}(OH)_3$. In situ reduction can be achieved through chemical reduction or through biologically mediated processes. For injection-based approaches to in situ remediation, the distribution of reagent is essential to establishing In Situ Reactive Zones (IRZs) capable of removing contaminants from groundwater. Primary requirements for establishing effective IRZs are:

1. Achieving sufficient reagent distribution in the subsurface, and
2. Creating a reactive zone of adequate residence time for complete treatment of the contaminant of concern.

For detailed principles of IRZ design and implementation, readers are referred elsewhere (Suthersan et al., 2016). General design and implementation principles are summarized here. Reagent distribution is a function of injection solution volumes, reactivity of the injected reagent, and ability of the reagent to be distributed within the aquifer with the injected solution. Chemical versus biological treatment reagents vary in their reactivity, ability to be distributed in the aquifer, and other properties, which will affect their applicability and selection of the appropriate approach for a given situation.

Other key concepts for design of injection based reduction of Cr(VI) include:

- **A variety of reductants reduce Cr(VI).** Reductants relevant to natural attenuation and engineered treatment of Cr(VI) include organic carbon substrates, ferrous iron, and reduced sulfur.

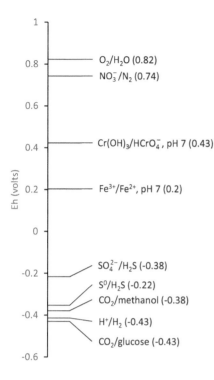

FIGURE 4.3 Redox potentials of common redox couples relevant to Cr(VI) reduction, adapted from Brock Biology of Microorganisms Figure 5.9 (Madigan et al 2003).

- **Cr(VI) reduction occurs within mildly reducing environments.** The redox potential for the Cr(VI)/Cr(III) is between that of nitrate reduction and iron reduction (Figure 4.3), meaning that reductants must be supplied to reduce oxygen and nitrate and to achieve metal reducing conditions. Development of further reducing conditions to stimulate iron and sulfate reduction can also be desirable, given the rapid reduction of Cr(VI) by ferrous iron and sulfide species that form.
- **Cr(VI) reduction is relatively rapid.** Half-lives for reduction of Cr(VI) are relatively short. For example, the half-life for reaction with iron in the presence of aquifer solids ranges from several hours to several days (Bachelor et al., 1998; Eary and Rai 1988; Fendorf and Li 1996). In comparison, half-lives for other common contaminants, such as the chlorinated aliphatics, are over a hundred days (Suthersan et al., 2016). Consequently, Cr(VI) requires relatively shorter residence times that can typically be achieved within the radius of influence (ROI) of injections for various reagents.

4.6.1.1 In Situ Chemical Reduction

Cr(VI) can be reduced abiotically by reduced iron and sulfur, as indicated by the lower redox potential of the oxidation reaction of ferrous iron and sulfur compared to the reduction of Cr(VI) in Figure 4.3. Example chemical reagents containing reduced iron and sulfur are summarized in Table 4.7. In general, chemical reagents generally

Hexavalent Chromium

TABLE 4.7
Summary of example in situ chemical reduction reagents

	Injectability, Longevity and Distribution Considerations	Health and Safety Considerations	Other Considerations	Relative Cost per Reducing Equivalent
Polysulfide, commonly available as calcium polysulfide (CaSx)	Decomposes to thiosulfate, bisulfite and elemental sulfur, limiting distribution • Potential clogging by oxidation of sulfide to elemental sulfur solids • Limit introduction of oxygen into injection stream	Potential generation of hydrogen sulfide at low pH, may require buffering	Residual cation species (e.g., calcium)	$$$
Ferrous sulfate	• Competing reactions with aquifer matrix limits distribution • Potential clogging of injection wells and aquifer in vicinity of injection well by oxidation of ferrous iron to ferric iron precipitates • Limit introduction of oxygen into injection stream. Consider pre-injection of low concentration organic carbon source.	Acidic concentrated ferric sulfate solution requires careful handling and appropriate personal protective equipment.	Residual anion species (sulfate)	$$$$
nZVI	• Distribution very limited by the agglomeration of nZVI particles and incorporation into aquifer solids • Avoid exposure of nZVI to oxygen to limit deactivation	ZVI is reactive and may produce hydrogen.	Lower potential for generating dissolved reduce metals downgradient (iron, manganese, arsenic)	$$$$
Dithionite	• Rapid reaction kinetics a challenge for distribution • Buffering to alkaline pH limits reaction and consumption for greater longevity and distribution	Sodium dithionite is reactive with moisture. Wetting of dry material can cause rapid decomposition and enough heat generation to ignite combustible materials.	Residual sulfate, sulphite and anion (e.g., sodium) Limited availability of reagent for purchase	$$$

have more rapid reaction and consumption kinetics than the soluble organic carbon substrates that are used to support biological reduction. Consequently, reagent distribution for chemical reductants is more limited, and may require more infrastructure at large scales in comparison to organic carbon substrates.

Ferrous Iron

Ferrous iron (Fe^{2+}) reduces Cr(VI) by the following reaction (Eary and Rai 1988):

$$Cr^{6+} + 3Fe^{2+} \rightarrow Cr^{3+} + 3Fe^{3+} \qquad \text{(Reaction 4.1)}$$

The Cr(III) formed will subsequently precipitate with the ferric iron produced (Eary and Rai 1988):

$$xCr^{3+} + (1-x)\,Fe^{3+} + 3H_2O \rightarrow Cr_xFe_{1-x}(OH)_3(s) + 3H^+ \qquad \text{(Reaction 4.2)}$$

where x can vary between 1 and 0. As shown on the solubility diagram in Figure 4.2, the solubility of the resulting $Cr_xFe_{1-x}(OH)_3$ precipitate is very low and lower than $Cr(OH)_3$ that does not contain iron. Fe^{2+} is commonly provided as part of engineered in situ treatment in the form of ferrous sulfate heptahydrate.

Reaction kinetics of Fe^{2+} with Cr(VI) are rapid, with reaction timeframes on the order of minutes to days depending on concentrations of ferrous iron and Cr(VI) (Eary and Rai 1988; Fendorf and Li 1996). For example, 0.5 mg/L Fe^{2+} reacts with a half-life of one day with Cr(VI), equivalent to a pseudo-first order rate coefficient of 0.7 days^{-1}, based on bench work by Fendorf and Li (Fendorf and Li 1996). Fe^{2+} will also react with oxygen and other species in the subsurface. Reaction rates of Fe^{2+} with oxygen are on the order of seconds (Eary and Rai 1988). Injection concentrations must be sufficient to distribute ferrous iron with distance in the aquifer, given the rapid rates of reaction.

Fe^{2+} can also be generated through injection of the strong reducing agent like dithionite, $S_2O_4^{2-}$, in a process that is sometimes referred to as In Situ Redox Manipulation (ISRM). The generation of Fe^{2+} by dithionite is described by the following reaction (Fruchter et al., 2000):

$$S_2O_4^{2-} + 2Fe(III)(s) + 2H_2O \rightarrow 2SO_3^{2-} + 2Fe^{2+} + 4H^+ \qquad \text{(Reaction 4.3)}$$

Dithionite reaction rates are rapid, with first order rate coefficients for dithionite reaction with sediments of 0.13 hr^{-1}, equivalent to a half-life of five hours (Istok et al., 1999). The rapid kinetics may limit the distribution of dithionite within the aquifer in comparison to IRZs biologically generated by organic carbon injections. Heterogeneous distribution and generation of robust, but variable, reaction zones within a ROI of over 20 feet were observed in a high-resolution injection pilot conducted at the Hanford Site in southeastern Washington state in 1995 (Fruchter et al., 2000). Dithionite injections may be followed by an extraction step to manage reaction by-products, such as sulfate, sulfite, and dissolved reduced metals such as manganese and arsenic.

Iron can also be provided in the form of zero valent iron (ZVI). A description of the reaction chemistry for ZVI is provided in section 4.6.1.3 on PRBs. One of the several potential ZVI reduction mechanisms is reduction of Cr(VI) by Fe^{2+} formed by corrosion of ZVI. ZVI can be deployed through injections into the aquifer, amendment in backfill of excavations or PRBs. Injection placement is discussed in further detail here and PRBs are discussed in further detail in section 4.6.1.3. The distribution of ZVI within the subsurface by injection is challenging. ZVI at the 0.1 to 1 millimeter (mm) scale is anticipated to clog injection wells and result in poor distribution. Greater success for ZVI distribution has been achieved with nano-scale ZVI (nZVI) and emulsified ZVI, although achievable ROIs will be more limited than for more soluble reagents. Injection of ZVI slurries can be conducted via direct push injection, potentially coupled with pneumatic fracturing or hydraulic fracturing to create fractures for distribution ROIs more comparable to soluble reagents. Fracture based approaches are most applicable in silts, clays, and weathered or fractured bedrock lithologies.

Reduced Sulfur

Reduced sulfur can be provided in the form of sodium or calcium polysulfide or other sulfide reagents, such as metasulfite or hydrogen sulfite. The suggested reaction mechanism for the reduction of Cr(VI) by calcium polysulfide is (Graham et al., 2006):

$$2CrO_4^{2-} + 3CaS_5 + 10H^+ = 2Cr(OH)_3 + 15S_{solid} + 3Ca^{2+} + H_2O \quad \text{(Reaction 4.4)}$$

For ferrous sulfate and dithionite distribution, the potential for clogging the filter pack of the injection well and immediately surrounding formation can limit the ability to inject and distribute reductants. Sulfide- and ferrous iron-based reagents may oxidize to elemental sulfur and ferric iron precipitates, respectively (Fruchter et al., 2000; USEPA 2000), which can limit injectability and distribution. Clogging by precipitation of oxidation products can be managed by limiting introduction of oxygen into the injection setup by storing reagents at full strength and metering into injection solution near the injection well head. Another strategy to limit oxidation and potential clogging within the aquifer is to anaerobically "prime" the aquifer, through an injection of low concentration organic carbon or other reductant. A combination of ferrous iron with another reductant such as sodium dithionite has been used to limit the immediate oxidation of ferrous iron in the aquifer (Su and Ludwig 2005).

Ferrous Sulfide

Solid-phase ferrous sulfide is also commercially available for use as a chemical reductant. As described earlier, both the Fe^{2+} and sulfide in the ferrous sulfide minerals reduce Cr(VI). Similar to ZVI, the injection of a slurry solid-phase ferrous sulfide is more difficult than soluble substrates. Commercially available materials can be in the micron particle size range. Direct push injection of the slurry can facilitate distribution, potentially coupled with fracturing.

Case Study: In Situ Chemical Reduction at the U.S. Coast Guard Support Center in Elizabeth City, North Carolina

Sodium dithionite was used for in situ chemical reduction of Cr(VI) in groundwater in the source area, an electroplating shop, at the U.S. Coast Guard Support Center in Elizabeth City, North Carolina (Malone et al., 2004). The footprint of remediation under the electroplating shop was approximately 40 feet by 70 feet. A network of 6 horizontal and 34 vertical injection wells on approximately 6 foot spacing was used to inject a 1 molar solution of sodium dithionite into the subsurface. Source area Cr(VI) concentrations rapidly declined from an average of 3.4 mg/L to less than 0.01 to 0.013 mg/L following injections (Malone et al., 2004). Treatment has remained stable through the present.

4.6.1.2 In Situ Biological Reduction

Biologically mediated reduction of Cr(VI) refers to the stimulation of native microorganisms through the delivery of a degradable source of organic carbon. Microorganisms can support the reduction of Cr(VI) by a variety of mechanisms. Figure 4.4 depicts the relevant biogeochemical processes stimulated during in situ biological treatment of Cr(VI). Microbes can directly reduce Cr(VI) through enzymatic reduction coupled to the oxidation of an organic reductant (Beller et al., 2013). Figure 4.3 shows the redox potential for reduction of Cr(VI) is between that of nitrate reduction and iron reduction, while the redox potential for two example organic reductants, methanol and glucose, are lower and therefore favorable for Cr(VI) reduction. Microbes can also indirectly reduce Cr(VI) through the reduction of iron minerals and sulfate in groundwater to generate Fe^{2+} and sulfides (H_2S, HS^-). The Fe^{2+} and sulfides generated in turn react abiotically with Cr(VI), reducing it to trivalent chromium (Cr[III]), according to the following reactions:

$$2CrO_4^{2-} + 3HS^- + 7H^+ \rightarrow 2Cr(OH)_3(s) + 3S_0(s) + 2H_2O \quad \text{(Reaction 4.5)}$$

$$CrO_4^{2-} + 3Fe^{2+} + 8H_2O \rightarrow Cr(OH)_3(s) + 3Fe(OH)_3(s) + 4H^+ \quad \text{(Reaction 4.6)}$$

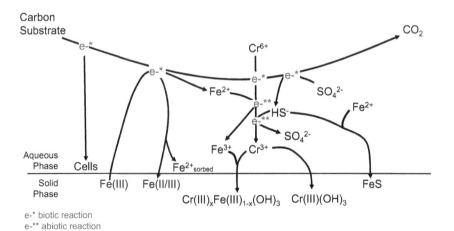

FIGURE 4.4 Biogeochemical oxidation-reduction reactions relevant to biologically mediated Cr(VI) reduction, simplified

In most groundwater systems, Cr(III) would be expected to precipitate with ferric iron to form a $Cr_xFe_{1-x}(OH)_3$ as shown in reaction 4.2 above. $Cr_xFe_{1-x}(OH)_3$ minerals are considerably less soluble and more stable than pure $Cr(OH)_3$ (Eary and Rai 1988; Sass and Rai 1987), as shown in the solubility diagrams in Figure 4.2. Given the relative rates of biological reduction in comparison to abiotic reduction by ferrous iron and sulfide, it is thought that the primary mechanism of Cr(VI) reduction through stimulation with organic carbon is the indirect abiotic mechanism (Weilinga et al., 2001).

A variety of carbon substrates can stimulate the biologically mediated reduction of Cr(VI), including soluble and semi-soluble substrates. A summary of commonly available carbon substrates is provided in Table 4.8.

The distribution of organic carbon for semi-soluble substrates is limited to the ROI of the injections. In comparison, the organic carbon from soluble substrates will travel downgradient of the ROI of the injection, creating a longer IRZ and longer residence time within the IRZ than semi-soluble substrates. For both semi-soluble and soluble organic carbon substrates, Fe^{2+} generated by iron reduction can persist downgradient of the carbon distribution footprint, extending the length of the IRZ. Although longer residence times within a longer IRZ can be created with soluble substrates, particularly in faster groundwater systems, this is rarely a determining factor for substrate selection for Cr(VI)

TABLE 4.8
Summary of example in situ biological reduction reagents

Carbon Source	Chemical Structure	Moles of reducing equivalents (e-) per pound of electron donor	Relative cost per reducing equivalent	Distribution Considerations
Soluble Substrates				
Lactate	$NaC_3H_6O_3$	48	$$$	Soluble organic carbon substrates can be distributed several times further than chemical reductants or semi-soluble organic carbon substrates. Organic carbon substrates and ferrous iron generated can persistent downgradient of injection points, creating relatively long IRZs.
Sucrose, from sources such as molasses	$C_{12}H_{22}O_{11}$	64	$	
Ethanol	C_2H_6O	118	$	
Semi-Soluble Substrates				
Emulsified Vegetable Oil (linoleic acid)	$C_{18}H_{32}O_2$	162	$$	Semi-soluble substrate distribution is limited to the ROI of injection. Ferrous iron generated can persist downgradient of the injection zone, increasing the length of the IRZ.

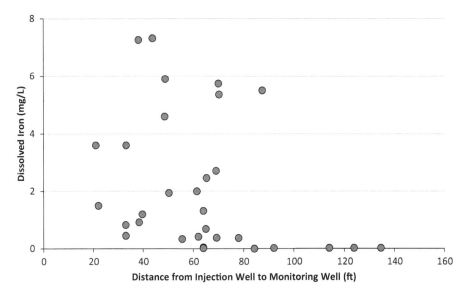

FIGURE 4.5 Maximum dissolved iron concentrations collected from dose response wells versus the distance from the injection well at a site in California where in situ biological reduction is being employed to treat Cr(VI) in groundwater (Sullivan, et al. 2014). Dissolved iron is an indicator of spatial establishment of reducing conditions conducive to Cr(VI) reduction.

reduction given the relatively fast reaction rates and short required residence times for treatment.

As illustrated in Tables 4.7 and 4.8, the costs of biological and chemical reduction reagents vary, with chemical reagents generally being more expensive per reducing equivalent than biological stimulants and with common soluble substrates such as ethanol and molasses being less expensive than semi-soluble substrates, such as emulsified vegetable oil (EVO). However, the unit cost is not the only criteria for substrate selection. The selection of a relatively expensive semi-soluble substrate like EVO versus soluble substrates like molasses or ethanol is typically dependent on scale of the project and a cost benefit balance of:

- Injection frequency: typically lower frequency of injections required with semi-soluble versus soluble substrates.
- Number of injection points: greater ROI, on the order of 75 to 90 feet as shown on Figure 4.5 (Sullivan et al., 2014) and fewer injection wells can be achieved with soluble substrate in comparison with a several times smaller ROI and greater number of injection wells with EVO. The relative ROI and number of injection points may become a determining factor for substrate selection at large scales. Note that the soluble, organic carbon substrates are generally less reactive than chemical reductants, with half-lives of soluble organic substrates in the range of tens of days compared to five days, for example, for dithionite. Therefore, soluble organic substrates can achieve greater distribution with fewer injection locations.

- Injection volumes and time for injection: The distribution and ROI of the semi-soluble substrate EVO is limited by the straining of EVO by the aquifer and requires larger injection volumes and longer injection times than soluble substrates to deliver across the same radius.

Injection concentrations and volumes are typically based on needs for reagent spatial distribution and persistence of the substrate between injections. Injection volumes are based on the volume of the mobile pore space within the target vertical interval and ROI (Payne et al., 2008). For relatively small-scale sites with small (i.e., 10–25 foot) target ROI, injections can be achieved within a few days and injection concentrations will be determined based on the number of electrons needed to achieve chromium reducing conditions and a balance between the rate of washout of the injection solution and degradation rate of the substrate to target persistence of the substrate during washout. For larger-scale injections with a large ROI (e.g., 75 feet), injections can take on the order of weeks and the injection concentration must be sufficiently high to persist as the injection solution travels from the well to the outer reaches of the ROI over the course of many weeks. For these large-scale systems, the amount of substrate needed for distribution will greatly exceed that needed to generate chromium reducing conditions and the redox conditions will vary within the ROI. To deliver sufficient organic carbon at the outer reaches of the ROI to stimulate chromium reducing conditions, much greater concentrations of organic carbon will be delivered to the injection location, generating more strongly reducing conditions, that is, sulfate reducing or even methanogenic conditions, in the vicinity of the injection well.

An additional consideration for the design of in situ biological systems for Cr(VI) reduction is the generation of residual reducing capacity. Within the IRZ for Cr(VI) treatment, several species are formed that have the potential to continue reducing Cr(VI) between active injection periods, after the organic carbon has degraded. These species include microbial biomass (i.e., endogenous decay) and Fe^{2+} and sulfide formed under the reducing conditions of the IRZ, for example, in the form of ferrous sulfide precipitates. The formation of residual reducing capacity can support the reduction of Cr(VI) between active injections for months to years (Murphy et al., 2008), reducing the operations and maintenance activities and costs.

Case Study: In Situ Remediation of Cr(VI) in Groundwater at a Site in the Eastern United States

A site in the eastern United States was impacted with Cr(VI) concentrations in bedrock groundwater greater than 10,000 µg/L. The bedrock is described as alternating layers of reddish-brown, highly fractured siltstones and white to tan sandstones exhibiting variable competency. The permeability and contaminant storage within the bedrock formation is controlled by fractures, bedding planes, joints, and other secondary porosity features. In situ biological reduction of Cr(VI) was implemented in phases across the site, beginning in February 2006 with the installation of three injection wells immediately off-site and

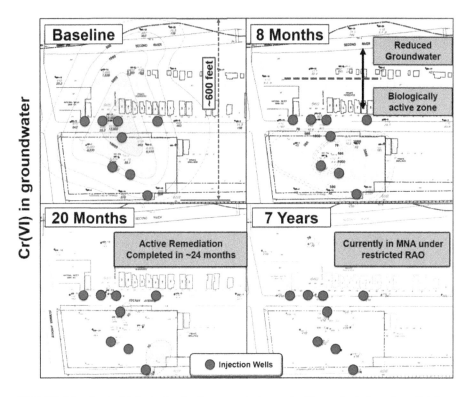

FIGURE 4.6 In situ biological reduction case study results. Hexavalent chromium distribution progress over a 7 year period.

approximately monthly injections of a 0.5 percent by volume molasses solution to stimulate naturally occurring bacteria and establish reducing conditions. Twelve injections were implemented from February 2006 to May 2007. Injection events lasted approximately one week and included approximately 7,000 gallons of 0.5 percent molasses solution into each injection well. A progression of the Cr(VI) plume is shown in Figure 4.6. The difference in the baseline and eight months Cr(VI) distribution is attributed to the injections off-site, with additional treatment by 20 months by on-site injections. After seven years without injections beyond year 1, the Cr(VI) concentration remained below laboratory detection limits.

4.6.1.3 Permeable Reactive Barriers

PRBs are a viable option for treatment of Cr(VI) in groundwater. This section refers to an engineered system in which reactive media are placed within the aquifer such that groundwater is treated as it passes through the media. PRBs are designed to be operated for years with little to no maintenance. PRBs are feasible for groundwater systems with the appropriate hydrogeologic conditions to promote flow through the PRB and to provide adequate residence time for treatment (ITRC 2011). Criteria for evaluation of site-specific conditions and the designs for various configurations of PRBs, the reader is referred to other resources (ITRC 2011; Suthersan et al., 2016).

ZVI is a common media for treatment of Cr(VI). There are several mechanisms by which Cr(VI) may be reduced within a ZVI PRB:

- Direct oxidation of Fe(0) coupled to the reduction of Cr(VI)
- Reduction of Cr(VI) by soluble Fe^{2+} formed by the corrosion of Fe(0)
- Reduction of Cr(VI) by ferrous iron (Fe[II]) bearing solids formed by the corrosion of Fe(0)
- Reduction of Cr(VI) by reduced sulfur minerals. Reduce sulfur minerals may be formed by biological sulfate reduction driven by the production of hydrogen from Fe(0) corrosion.

Studies of the surfaces of iron shavings removed from within constructed PRB cylinders at a site in Willisau, Switzerland, indicated Cr(VI) reduction by both soluble Fe(II) and Fe(II) bearing solids (Flury et al., 2009). The studies also showed the formation of sulfide minerals deeper within the PRB, but Cr(VI) was observed to be reduced before reaching these regions of the PRB (Flury et al., 2009). Studies of surfaces of ZVI granules deployed at another site, described further in the case study below, indicated Cr(VI) reduction was mediated by biologically produced iron sulfides (Wilkin et al., 2005).

Case Study: PRB at the U.S. Coast Guard Support Center in Elizabeth City, North Carolina

One of the early examples of a PRB for the treatment of Cr(VI) is the system installed at the U.S. Coast Guard Support Center in Elizabeth City, North Carolina, in 1996 (Blowes et al., 1999). Cr(VI) was released to groundwater from a chromic acid tank used in an electroplating operation, with concentrations of up to 10 mg/L in groundwater (Blowes et al., 1999). Trichloroethene was also present in groundwater. A granular iron wall 46 meters long, 7.3 meters tall, and 0.6 meters thick (150 foot long, 24 foot tall, and 2 foot thick) was installed approximately perpendicular to groundwater flow to intercept the contaminant plume (Blowes et al., 1999) between the plating shop and the Pasquotank River. Performance monitoring data over the course of the first eight years demonstrated the reduction of up to 5 mg/L of total chromium upgradient of the barrier to less than 5 µg/L within the barrier as demonstrated with a dense network of multilevel samplers (Blowes et al., 1999). A paper published in 2014 reported that the source zone of chromium had been treated and the downgradient concentrations of total chromium remained below 3 µg/L, indicating that the reactive and hydraulic longevity of the PRB lasted 15 years and outlived the mobile chromium plume under the geochemical and hydrogeologic conditions at the site (Wilkin et al., 2014). The source area remedy that reduced Cr(VI) concentrations to less than 0.01 to 0.013 mg/L is discussed in the earlier case study for in situ chemical reduction.

4.6.1.4 Reoxidation of Cr(III) Formed by In Situ Reduction

The permanence of immobilization is critical for the success of in situ metals sequestration strategies. While the reduction of Cr(VI) to Cr(III) is reversible under certain conditions, the kinetics of Cr(VI) reduction are rapid in typical groundwater environments in comparison to the kinetics of Cr(III) oxidation.

Only a few oxidants present in natural systems are known to be capable of oxidizing Cr(III) to Cr(VI), that is, oxygen and manganese oxides. Dissolved oxygen can oxidize Cr(III) to Cr(VI), but the kinetics are very slow at the neutral groundwater pH typical of most aquifer systems, and it has been concluded in the literature that the oxidation of Cr(III) by dissolved oxygen is not a meaningful pathway in typical groundwater systems (Schroeder and Lee 1974; Eary and Rai 1987; Hwang et al., 2002).

Manganese oxides react with Cr(III) at more meaningful rates than dissolved oxygen. Manganese oxides occur in the subsurface primarily as coatings on soil grains. As discussed in section 4.4.1, Cr(III) dissolves into groundwater and sorbs directly to the surface of the manganese oxide mineral to react (Eary and Rai 1987; Oze et al., 2007), and the amount of Cr(III) dissolution and rate of oxidation is affected by surface area of Cr(III) minerals and pH (Oze et al. 2007). Following in situ reduction of Cr(VI), a portion of Cr(III) can be dissolved and oxidize in reaction with manganese oxides. For biological treatment, it has been shown that chromium precipitates with iron as a $Cr_xFe_{1-x}(OH)_3$ mineral species (Beller et al., 2014; Varadharajan et al., 2015; Varadharajan et al., 2017) and chemical treatment by iron reductants would likely form similar species, although the mineral structure may vary with biologically generated ferrous iron. The formation of chromium in minerals with iron limits the amount of aqueous Cr(III) available for adsorption onto manganese oxide surfaces and rates of subsequent oxidation.

Ultimately, the concentrations of Cr(VI) in background in groundwater will depend on the factors that determine the rate of oxidation (e.g., the amount of chromium and manganese minerals, pH) and the residence time over which Cr(VI) can accumulate in the groundwater within the portion of the aquifer where treatment and Cr(III) precipitation occurred. When Cr(III) from in situ treatment forms as precipitates similar or less in solubility and amount to naturally occurring Cr(III), Cr(VI) concentrations from reoxidation are anticipated to be similar to natural background.

4.6.2 Ex Situ Treatment

Cr(VI) is amenable to ex situ treatment, commonly known as the pump and treat strategy. For ex situ treatment, impacted groundwater is captured by extraction wells and treated in an aboveground treatment system. For ex situ treatment, a number of different treatment technologies can address Cr(VI), including chemical treatment, ion exchange, reverse osmosis, bioreactors, and phytostabilization. These technologies are discussed in detail in section 4.8. Pump and treat systems have been widely used to remove mass from groundwater and contain contaminant plumes, but limitations for pump and treat systems to remove residual mass that is strongly sorbed or that has migrated into less mobile pore space in older plumes has also been widely recognized. However, Cr(VI) has the relatively low affinity for sorption under neutral to alkaline groundwater conditions, promoting more successful treatment of extraction approaches for Cr(VI) treatment in comparison to contaminants that sorb to aquifer materials strongly or form separate phases.

4.6.3 Dynamic Groundwater Recirculation

The strategic reinjection of ex situ treated groundwater can also be used to increase the effectiveness of ex situ remedies, particularly for the treatment of large, dilute plumes. The effectiveness of pump, treat, and reinjection strategies can be greatly enhanced by focusing extraction on portions of the aquifer that bear the greatest mass flux and by adaptively changing injection and extraction locations as mass flux changes during remedy implementation, in a strategy that is referred to as dynamic groundwater recirculation (DGR). The DGR strategy promotes treatment of large, dilute plumes, as flushing is promoted through slow advective zones, increasing mass recovery. Additionally, DGR can be combined with in situ treatment approaches by amending the reinjection stream with chemical oxidants or biological substrates. Specific principles for design of DGR remediation systems are discussed in more detail in Remediation Engineering (Suthersan et al., 2016).

4.6.4 Natural Attenuation

In 2011, the USEPA issued a guidance document for the monitored natural attenuation (MNA) of Inorganics (Ford et al., 2007), including a comprehensive chapter on the mechanisms and methods for establishing MNA for Cr(VI) (Kent et al., 2007). An additional guidance document was issued from the USEPA Office of Solid Waste and Emergency Response in 2015 on the use of MNA for inorganics in groundwater at Superfund sites (USEPA 2015). The basic concepts are summarized here.

Reduction of Cr(VI) to Cr(III) by naturally occurring reductants is the primary potential mechanism of natural attenuation. Naturally occurring reductants of Cr(VI) include reduced iron minerals, such as magnetite, sulfide-bearing minerals, and soil organic matter such as humic and fluvic acids (Kent et al., 2007). Aquifer soils can vary widely in reduction capacity, making MNA by this mechanism a viable option at some sites and not at others. Sorption can also be a mechanism for natural attenuation, although the extent of sorption for Cr(VI) is relatively low and is reduced at neutral to alkaline pH and in the presence of competing anions, as discussed in section 4.1.2.

USEPA guidance specifies a tiered approach to the demonstration of MNA for inorganics, summarized as follows.

4.6.4.1 Tier I

The first tier demonstrates that the Cr(VI) plume is static or shrinking. Statistical analyses of plume behavior, for example, linear regression, Sen's Slope Estimate, or Mann-Kendall statistical trend analyses, can be used to quantify trends for this demonstration. In addition, analysis of the plume boundaries and center of mass over time can provide evidence that the plume is stable or shrinking/retreating. This tier is common to establishing MNA for all contaminants and does not require unique analyses for Cr(VI).

4.6.4.2 Tier II

The second tier demonstrates the mechanism and rate of natural attenuation. Rate of attenuation can be determined by interwell comparisons along the plume or intrawell comparisons of trend data over time. Chromium isotopes may also aid in this determination, although the results must be carefully interpreted to sort out the influences of natural background and mixing, as discussed in section 4.5.3. The mechanism of attenuation for Cr(VI) and other inorganics is unique from organics, particularly for the mechanism of reduction that involves solid-phase species. Potential analyses to demonstrate the reductive and sorption mechanism include:

- Identification of mineral phases that can act as reductants or sorbents for Cr(VI) through mineralogical analysis, such as acid volatile sulfide and simultaneously extractable iron (AVS-SEM) and Scanning Electron Microscopy and Energy Dispersive X-ray Spectroscopy (SEM-EDS) (see section 4.5.3 for description of analyses).
- Bench scale demonstration of attenuation, for example, through microcosm batch experiments or column studies.
- Identification of chromium associated with solid-phase reductants, for example, iron and sulfur, through elemental mapping (see section 4.1 for a discussion of XAS). It should be noted that detection of chromium from the contaminant source in mineralogical analysis can be difficult within a background of naturally occurring chromium in the soil matrix.

4.6.4.3 Tier III

The third tier demonstrates the stability of the attenuation mechanism. In this tier, the attenuation capacity of the mechanism identified in Tier II should be compared to the demand, for example, is there enough attenuation capacity to handle the mass of the Cr(VI) plume? Results from bench scale testing and plume mass estimates may aide in this demonstration. If the attenuation capacity is insufficient to treat the mass of the Cr(VI) in the plume, additional source treatment may be needed. Additionally, the potential for changes in geochemical conditions for groundwater passing through areas of chromium immobilized by attenuation should be evaluated, particularly the potential for changes in dissolved oxygen content or pH that could affect stability of Cr(III)-bearing minerals.

4.6.4.4 Tier IV

The fourth tier of evaluation requires the establishment of a monitoring plan to demonstrate MNA is occurring over time and the identification of a contingency plan in case MNA is not adequately addressing the Cr(VI) plume.

4.7 DRINKING WATER TREATMENT

Technologies available for treatment of Cr(VI) in drinking water include: Reduction Coagulation Filtration (RCF), Ion Exchange (IX), and Reverse Osmosis (RO). These well-established technologies, as well as other less well-established technologies are discussed in detail in section 4.8.

For treatment technologies that rely on reduction of Cr(VI) to Cr(III), there is an added layer of complexity due to the common use of chlorine as disinfectant residual in distribution systems. Cr(III) can oxidize to Cr(VI) in the presence of chlorine (Chebeir et al., 2015; Lindsay et al., 2012). As a result, treatment for drinking water will typically require removal of total chromium below the regulatory limit and methods that rely solely on reduction to Cr(III) may be insufficient.

As discussed previously, California was the only state to have adopted a specific MCL for Cr(VI) in 2014, which was deleted in 2017 and in which a new MCL is pending (see section 4.3.2.1). When the California Cr(VI) MCL was in place, Best Available Technologies (BATs) for Cr(VI) treatment were established. At that time, various methods of treatment of Cr(VI) for achieving compliance with the 10 µg/L Cr(VI) standard were published under Title 22 in the California Code of Regulations under BATs, including RCF, IX, and RO. While BATs are reviewed and accepted by California, they have not been proven to work on all water sources for Cr(VI), and are not the only technologies that can be applied in California or elsewhere.

4.7.1 Point-of-Entry and Point-of-Use Treatment

Point-of-entry (POE) treatment is defined as treatment applied to the drinking water entering a house or building for the purpose of reducing contaminants in the drinking water distributed throughout the house or building; point-of-use (POU) treatment is defined as treatment applied to a single tap for the purpose of reducing contaminants in drinking water at that tap. POE/POU treatment is a strategy that may be employed in areas where it is cost-prohibitive to provide centralized treatment systems, such as communities using on-property wells to provide water.

The applicability of the technologies for POE/POU will vary depending on the community for which the systems are intended; however, RCF and Weak Base Anion (WBA) IX are the technologies that are unlikely to be feasible for use as POE/POU devices due to the large quantities of chemicals required. Strong Base Anion (SBA) IX can be feasible in a POE system if operated in single-pass mode so as to avoid the regeneration step (see section 4.7.3.2). It is less likely to be effective for POU systems, as the limited resin life will necessitate very frequent changeouts due to limited space; typically, the space under a sink has insufficient space to allow for an Strong Base Anion Ion Exchange (SBA-IX) system sized to allow for a reasonable changeout frequency. Prepackaged RO units exist for POU treatment and have been successfully deployed for Cr(VI) treatment. RO is less likely to be feasible for POE treatment, as the greater volumes of water treated will require a roughly proportional increase in disposal needs; POE treatment will encompass all household water-using activities, such as showers, washing machines, toilets, etc., and may even include landscape irrigation/filling pools.

POU and POE treatment devices should be appropriately certified. The Safe Drinking Water Act refers first to relevant American National Standards Institute (ANSI) standards to guide appropriate technology selection for contaminants. However, relevant ANSI standards do not all address or cover treatment devices that may be needed to meet the levels that are being considered for regulation, for example, the 10 µg/L level that was previously regulated in California. If California, other states, or the USEPA issue drinking water standards for Cr(VI) that require treatment technologies that are not covered under ANSI standards, those technologies will require approval by the states. Explicit approval will also be required for systems that use POE/POU devices for compliance, but which would not provide such units for every resident/connection.

4.8 Cr(VI) TREATMENT TECHNOLOGIES FOR DRINKING WATER TREATMENT AND EX SITU GROUNDWATER REMEDIATION

The selection of treatment technology for drinking water treatment or ex situ groundwater remediation will depend on several considerations:

- **Influent water quality**: the presence of individual constituents and pH conditions and variability of these constituents over time, strongly influences the effectiveness of some technologies, particularly ion exchange.
- **Operations Intensity/Complexity**
- **Waste management** is often the most critical and costly piece of Cr(VI) treatment. The technologies vary in the type of waste produced, but most have significant quantities of residuals that require disposal. Some waste generated may be hazardous and require special handling.

The available technologies for drinking water treatment and ex situ groundwater remediation are discussed in the following sections and summarized in Table 4.9.

4.8.1 REDUCTION/COAGULATION/FILTRATION WITH FERROUS IRON

Reduction/Coagulation/Filtration (RCF) using Fe(II) as a reductant is a technology that can be used for ex situ groundwater remediation or drinking water treatment and is a California BAT. This method reduces Cr(VI) to Cr(III) (which precipitates), forms larger particles via a coagulation process, and then filters out the particles. A sample process flow diagram is shown in Figure 4.7. Processes that were evaluated for the development of BATs in California used ferrous sulfate as the reductant/coagulant, aeration or chlorine to oxidize excess ferrous (Fe[II]) to ferric (Fe[III]), and media or membrane filtration to remove the flocs (Blute et al., 2013; Chowdhury et al., 2016). Membrane filtration provides more reliable removal in exchange for more troubleshooting when compared to media filtration.

The process is relatively more complex than other technologies considered in this section, except for reductant and regenerant strong base anion ion exchange (SBA-IX, section 4.7.2.2), given the addition of chemicals and several process units required to accomplish the treatment. RCF-Fe(II) is not generally affected by individual constituents in the influent or variations in water quality conditions over time. RCF-Fe(II) generates waste from the backwashing of the filters (and chemical cleaning waste if membrane filtration is used or residual settled solids if coagulation step is used). If local sewers cannot be used to dispose of the liquid waste stream, it may be more cost-effective to dewater the waste (recycling the water to the head of the plant) and dispose of the solid waste generated from the process. This sludge may be hazardous and require appropriate disposal (Blute et al., 2013).

TABLE 4.9
Cr(VI) treatment selection considerations

Treatment Method	Applicability to Ex Situ Remediation	Applicability to Drinking Water Treatment	Influent Water Quality Considerations	Operations Intensity/ Complexity	Waste Quantity/Type/ Difficulty	General/Other
RCF (Fe(II))	Yes	Yes CA BAT	Minimally impacted by influent water quality.	Medium – High: Dependent on configuration.	Medium, backwash: Low quantity, difficulty dependent on disposal method. Potentially hazardous.	Ferrous sulfate can be difficult to obtain. Iron can clog filters quickly. May be too difficult to effectively implement at smaller scales. Can handle water quality fluctuations.
WBA-IX	Yes	Yes CA BAT	Alkalinity and pH drive chemical usage. Uranium influences resin disposal options.	Low – Medium: Initial flush and pH control will require periodic coordination and/or adjustments.	Low – Medium, spent resin: Low quantity, difficulty dependent on presence of uranium. Likely hazardous, TENORM.	Requires pH adjustment to ensure long resin life (100,000+ BVs). If unchecked, uranium can cause waste disposal to be untenable. May require flushing on site before bringing into service.
SBA-IX	Yes	Yes CA BAT	Sulfate drives regeneration/ replacement (R/R) frequency. Co-occurring contaminants can impact R/R frequency as well.	Low – High: Dependent on available disposal options	Medium, spent resin or regenerant (high TDS brine): R/R frequency and disposal method drive quantity and difficulty. Potentially hazardous.	Much lower resin life than WBA. Viable sewer disposal of spent regenerant makes this option attractive. Does not require pH adjustment for performance, but may need it to mitigate corrosion in distribution system.

(*Continued*)

TABLE 4.9 (*Continued*)

Treatment Method	Applicability to Ex Situ Remediation	Applicability to Drinking Water Treatment	Influent Water Quality Considerations	Operations Intensity/ Complexity	Waste Quantity/Type/ Difficulty	General/Other
RO	Not likely selected	Yes CA BAT	Minimally impacted by influent water quality	Low – Medium	High, rejectate (lower TDS brine): High quantity, difficulty dependent on disposal method.	High energy use, water waste makes this untenable in many situations. RO is effective on many different contaminants, making this option viable for water with multiple contaminants. Can handle water quality fluctuations.
Iron-based media	Under development	Under development Not CA BAT	Still under study	Medium: Additional iron removal step required	Low – Medium, backwash and spent media or regenerant: Low quantity, difficulty dependent on disposal method.	Will require bench/pilot-testing to obtain approval for drinking water treatment in California. Iron leaching from media will require additional treatment processes.
Biological	Yes	Not typical Not CA BAT	Presence of alternate electron acceptors (e.g. nitrate, perchlorate) will affect sizing of vessel and reagent demand	Medium – High: Biomass takes time to mature/can be upset	Low – Medium, biomass: Low quantity, difficulty	Will require bench/pilot-testing to obtain approval for drinking water treatment in California. Uses bacteria to reduce chromium.
RF (Sn(II))	Under development	Under development Not CA BAT	Still under study	Low	Low, spent filters – Low quantity, difficulty	Will require bench/pilot-testing to obtain approval for drinking water treatment in California. Untenable if high removal is required.
Phytostabil- ization	Yes	No Not CA BAT	Water quality appropriate for crops	Low	None. Rather, beneficial fodder crop produced	

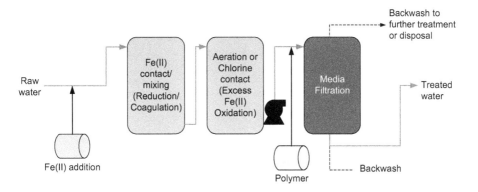

FIGURE 4.7 Example reduction filtration with Fe(II) process flow diagram

In the past, RCF-Fe(II) had a very large footprint compared to the other BATs due to the very long contact times originally used (~ 45 minutes). However, more recent work has reduced the contact time—and therefore the footprint—of the technology (Blute et al., 2013; Chowdhury et al., 2016).

4.8.2 Ion Exchange

Ion exchange (IX) is a technology that can be used for ex situ groundwater remediation or drinking water treatment and is a California BAT for drinking water. This method exchanges Cr(VI) (typically in the anionic form of chromate or dichromate) in the water with another ion (often chloride) on the resin. The BATs evaluated in California included both WBA and SBA resins.

4.8.2.1 Weak Base Anion Resins

A sample process flow diagram for WBA is shown in Figure 4.8. WBA also has a secondary mechanism of Cr(VI) removal in the low pH range (approximately 5.5–6.5). After the IX step, Cr(VI) is converted to Cr(III) and retained on the surface of the resin. The majority of Cr(VI) is removed via this reduction mechanism. pH adjustment provides optimal conditions for this technology.

Individual constituents play a major role in the performance of IX treatment technologies Weak Base Anion Ion Exchange (WBA-IX) and SBA-IX. WBA-IX performs best at pHs below typical groundwater conditions (roughly 5.5–6.0), as this allows for the secondary removal mechanism (conversion of Cr[VI] to Cr[III] on the resin beads) to activate. As a result, WBA-IX systems will first require the pH to be depressed to perform optimally. The pH will then need to be raised after treatment in order to prevent corrosion issues downstream. Accordingly, alkalinity and influent pH will impact the quantity of chemicals required for pH adjustment. Varying influent conditions—such as those experienced at a centralized treatment facility with varying well usages—have an adverse effect on treatment systems that are strongly affected by water quality. WBA-IX may experience premature breakthrough due to water quality instability.

WBA-IX requires little attention during normal operation, but will need to be shut down periodically to exchange resin. If operated properly (pH ~5.5–6), these

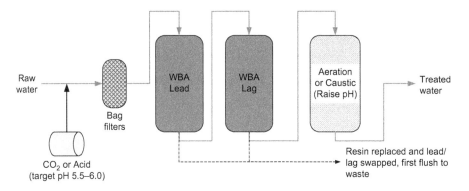

FIGURE 4.8 Example WBA process flow diagram

changeouts will be fairly infrequent—100,000 to 200,000 or more Bed Volumes (BVs).

WBA-IX typically has two waste streams. The first is the initial flush that is used to remove leachates from the virgin resin, such as formaldehyde (Blute et al., 2013). The second is the disposal of the spent resin itself. Due to the long resin life, sufficient Cr(VI) will likely have accumulated on the resin for it to be considered hazardous waste, and will thus require appropriate handling. Additionally, uranium is removed very effectively by WBA-IX and thus accumulates in the resin. This accumulation can give the used resin TENORM status (Technologically Enhanced Naturally Occurring Radioactive Material), which narrows down the locations able to accept the waste. If accumulated in sufficient quantities, the used resin can be designated LLRW (Low Level Radioactive Waste); this designation makes the waste disposal infeasible (Blute et al., 2013).

In general, IX with WBA resins offer the advantage of long operating times in between resin replacements (100,000–200,000 or more BVs).

4.8.2.2 Strong Base Anion Resins

Figures 4.9 and 4.10 show sample regenerable and single-pass configurations for SBA, respectively. pH adjustment is not needed for SBA resins. Raw water is filtered prior to feeding to the SBA resin vessels. In a single-pass system, Figure 4.10, spent vessels are taken off-site for regeneration or disposal. In a regenerable system, Figure 4.9, brine is used to regenerate SBA resins on-site, creating a brine stream that requires further treatment or disposal. Relatively, the regenerable SBA resins are one of the more complex treatment processes of those discussed in this section.

SBA-IX performance is impacted strongly by sulfate, as SBA resins have a higher selectivity for sulfate than other ions, including chromate. Additionally, sulfate tends to be present in concentrations several orders of magnitude higher than chromate. High-sulfate levels can prevent the technology from being feasible due to the high regeneration/replacement (R/R) frequency required (Blute et al., 2013; Chowdhury et al., 2016). While sulfate is the primary driver of SBA feasibility, SBA resins are nonselective, and thus other anionic constituents (e.g., arsenic oxyanions, nitrate, or perchlorate) can control operation, as discussed below. SBA is sensitive to varying influent conditions and may experience premature breakthrough due to water quality instability.

Hexavalent Chromium

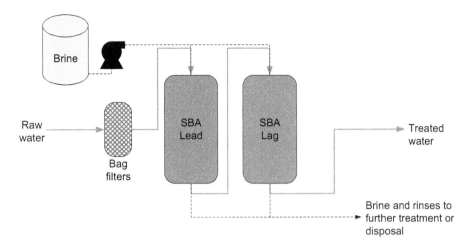

FIGURE 4.9 Example regenerable SBA process flow diagram

SBA-IX requires little attention during normal operation, but may require frequent R/R cycles depending on the influent water quality characteristics (most likely <50,000 BVs, depending highly on sulfate concentrations). The presence of co-occurring contaminants can also lead to chromatographic peaking of less preferentially exchanged ions, such as nitrate.

Chromatographic peaking can lead to concentrations of contaminants multiple times higher than their original concentrations for a short period after breakthrough. Typical SBA-IX configurations mitigate this risk (such as lead/lag or carousel style configurations), but care must be taken to ensure peaking does not cause violations from other contaminants. Finally, SBA-IX may also require anticorrosion treatment due to its nonselective nature; alkalinity removal from the influent will lead to a more aggressive effluent.

SBA-IX waste will consist of high-total dissolved solids (TDS) brine, if regenerated, or spent resin, if the resin is operated as single pass. If there is not a close sewer

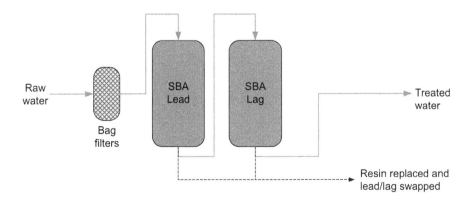

FIGURE 4.10 Example single-pass SBA process flow diagram

capable of handling the spent regenerant (brine), dewatering may be explored to minimize the waste volume. This dewatered sludge/brine (and the spent resin from the single-pass operation) may be hazardous and will thus require appropriate disposal. Uranium accumulation may result in spent resin being designated TENORM, but it is unlikely there will be sufficient accumulation to achieve LLRW designation.

4.8.3 Reverse Osmosis

This method removes Cr(VI) (as well as many other constituents) by forcing water through extremely fine pores, creating a water stream with very low Cr(VI) concentrations and a concentrate stream with more concentrated Cr(VI) levels. A sample process flow diagram is shown in Figure 4.11. As shown on the diagram, acid and scale inhibitors may be needed to prevent scaling and fouling of the RO membranes. This technology was not explored at larger scale as a California BAT specifically for Cr(VI) removal due to high projected costs and water waste generation.

RO is minimally impacted by individual constituents in the influent or by variations in water quality over time.

RO waste will be a lower-TDS brine generated from the high-pressure process in comparison, for example, to SBA-IX waste. The rejectate will contain approximately proportional constituent concentrations (based on the flow of water turned away) that are removed by the RO membranes. This waste will be high quantity, but relatively simple to dispose. RO requires minimal attention during normal operation, but may require additional expertise for troubleshooting issues should they arise. RO generates roughly an order of magnitude more volume of waste than the other technologies considered in this section. Sewers or other on-site infrastructure capable of handling/reducing the waste volume make the process manageable, though facilities have been run successfully without sewer availability.

4.8.4 Bioreactors

This method reduces Cr(VI) via biological reduction. This process is predominantly used in ex situ remediation applications, but has been offered for drinking water by some vendors. Biological treatment is not a California BAT and would require bench testing for approved drinking water use in California, but could be a cost-effective option in some scenarios. Cr(III) formed is removed from water

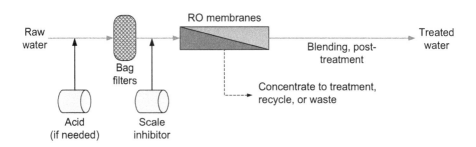

FIGURE 4.11 Example RO process flow diagram

Hexavalent Chromium

either through precipitation as $Cr(OH)_3$ or incorporation into biomass (Williams et. al., 2014; Middleton et al., 2003). Reduction of Cr(VI) in a bioreactor is microbially mediated by bacteria using Cr(VI) as an electron acceptor. Bacteria can use Cr(VI) as an electron acceptor to derive energy for growth (Tebo and Obraztsova 1998). Treated water would likely contain biomass and potentially residual colloidal $Cr(OH)_3$, which may require removal via filtration before discharge to surface water bodies or reinjection into the aquifer. Disinfection would be required for drinking water treatment. In situations not involving drinking water, biological reactors are generally a cost-effective alternative to conventional RO and ion exchange processes.

Bioreactor types for Cr(VI) reduction include fixed film systems, such as submerged packed beds and hollow fiber membranes, in which anoxic conditions appropriate for the development of Cr(VI) reduction can develop.

Submerged Packed Bed Reactors

In a submerged packed bed reactor, biofilms develop on the surface of porous media, such as rocks, stones, or plastic media. An example schematic of a submerged reactor is shown in Figure 4.12. Organic carbon reagent is metered into the influent at quantities to reduce oxygen, Cr(VI), other electron acceptors such as nitrate and perchlorate and to generate biomass. Nutrients, such as nitrogen and phosphorous, are added in quantities required for biomass growth. Within the bioreactor, anoxic conditions develop and the biofilms reduce Cr(VI) in water as the water flows through the media. A second, aerobic reactor may be added if residual organic carbon requires removal. Residual biomass is typically removed using a filter, such as a bag filter. In addition, sand media or additional filtration may be needed to remove colloidal Cr(III). Submerged packed bed bioreactors have been used successfully to achieve greater than 99 percent removal of several mg/L Cr(VI) in groundwater at the pilot and full scale (Williams, et al., 2014).

Potential electron donors include many of the soluble organic compounds discussed in section 4.6.1.2 for in situ biological treatment, such as acetate, glucose, and ethanol. Considerations for selection of organic carbon substrate include cost (see Table 4.8), handling and storage hazards, and yield. Yield is a consideration for ex situ treatment because the mass of solids requiring disposal will be greater for higher yield substrates.

The requirement for electron donors are determined on a site-specific basis, considering the presence and quantity of electron acceptors in the reactor influent, the electron donor, and the electron demand needed for cell growth. The presence of competing electron acceptors, such as nitrate or perchlorate, may increase reagent demand and reactor sizing.

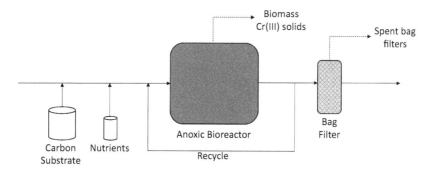

FIGURE 4.12 Example submerged bioreactor process flow diagram

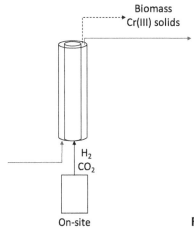

FIGURE 4.13 Example membrane bioreactor process flow diagram

Membrane Bioreactors

In hollow fiber membrane reactors, the hollow fibers serve as a surface for biofilm growth (Figure 4.13). The electron donor used is hydrogen gas, which is provided to the inside of the membrane and diffuses across the membrane into the reaction chamber to feed the bacteria. Hydrogen can be delivered or generated on-site. The bacteria that couple hydrogen oxidation to chromium reduction are autotrophic, generating biomass from inorganic carbon dioxide that is also supplied on the interior of the membranes. The yield of autotrophic bacteria is lower than that of heterotrophs growing on multicarbon organic substrates, and consequently the amount of biomass that is generated and requires handling and disposal is lower with the hydrogen fed membrane bioreactor.

4.8.4.1 Phytostabilization

Phytostablization for Cr(VI) remediation is the use of extracted groundwater for containment or mass removal for large-scale groundwater radiation for the irrigation of crops. Cr(VI) is immobilized by reduction of Cr(VI) to Cr(III) in the root zone of plants (USEPA 1997). Plant roots release inorganic and organic root exudates, creating the appropriate conditions for reduction of Cr(VI). Depending on how water for the plants is managed, a portion of the water treated through the root zone is evaporated by the plants, while a fraction percolates through the root zone to the aquifer. The Cr(III) generated is immobilized in the root zone and incorporated in to soil minerals for the long term. The net increase in soil chromium content due to immobilization is typically insignificant in comparison to the existing chromium in soils. For example, if 100 µg/L of chromium were applied at 570 liters per minute (150 gallons per minute [gpm]) to a 142,000 m^2 (35 acres) field for 15 years and assuming the chromium is immobilized over an aquifer thickness of 1.5 meters (5 feet), the incremental increase in soil chromium content from deposition of 420 kg (930 pounds) of chromium is estimated to be approximately 1 milligram per kilogram (mg/kg). Seasonal fluctuations in plant water demand must be accounted for in remedial designs for phytostablization. Implementation of phytostabilization may result in increased salt concentrations in the water percolating past the root zone due to concentration by evapotranspiration.

Case Study: Ex Situ Remediation of Cr(VI) in Groundwater at the PG&E Hinkley Compressor Station

Several different large-scale phytostablization systems have been operated at the Pacific Gas and Electric Company (PG&E) Hinkley Compressor Station, starting in 1992. At the site, extracted chromium-bearing groundwater is treated by irrigating agricultural lands growing fodder crops. Fodder crops used for phytostabilization at the site have included alfalfa, Bermuda grass, Italian ryegrass, winter oats, winter barley, and winter wheat (Arcadis 2017a). One field approximately 117,000 square meters (29 acres) in area was operated from 1992 to 2001 and treated 3,000 million liters (783 million gallons) of water and an estimated 685 kilograms (1,508 pounds) of chromium (Alisto 2001). Cr(VI) concentrations in groundwater were up to 742 µg/L of Cr(VI), with an average annual concentration of up to 340 µg/L. Lysimeters 1.5 meters (5 feet) beneath the field demonstrated consistent removal over 95 percent during the 10-year operational period. Analysis of plant tissue did not yield detections of Cr(VI). In 2016, there were 1.2 million square meters (306 acres) of agricultural fields in operation which treated 15 million gallons of water at a rate of 2,870 to 5,900 liters per minute (760 to 1,560 gpm) with Cr(VI) concentrations ranging from 2.6 to 210 µg/L (Arcadis 2017b). 685 kilograms (144 pounds) of Cr(VI) were removed from groundwater in 2016 and 992 kilograms (2,186 pounds) of Cr(VI) have been removed from groundwater by phytostabilization from 1992 to 2016 (Arcadis 2017c). A picture of a current drag drip irrigation system used at the site is shown in Figure 4.14.

FIGURE 4.14 Irrigation system used for treatment of Cr(VI) at the PG&E Hinkley Compressor Station in Hinkley, California

4.8.4.2 Iron Media

These methods use iron-based media to adsorb Cr(VI) or exchange Cr(VI) with iron (Blute et al., 2013). Iron-based media technologies have shown promise for ex situ remediation and drinking water treatment. However, this technology is not a California BAT for drinking water treatment and would require bench testing for approved use in California.

Iron-based adsorption/IX operates in a similar fashion as SBA-IX, but requires an additional iron treatment step to remove iron released during the Cr(VI) treatment process. In this process, the oxidation of ZVI is coupled to Cr(VI) reduction and precipitation, with potential reaction mechanisms as discussed in section 4.6.1.3. Various vendors have proprietary iron media on the market, using different particle sizes and processing techniques for a range of available surface area.

Waste streams may consist of backwash and spent regenerant liquid, as well as spent media, depending on the operation of the facility. The regenerant may require special disposal, but the backwash is not anticipated to have special requirements. Some vendors are offering the ability to recycle spent media cost-effectively.

4.8.5 REDUCTION/FILTRATION VIA STANNOUS CHLORIDE (RF-Sn[II])

This method uses stannous chloride as the reductant/coagulant and small-profile filters compared to the comparable BAT (RCF). A sample diagram is shown in Figure 4.15.

RF-Sn(II) is a technology that was previously considered during the research that developed the California BATs (Blute et al., 2013). At the time this technology did not pass screening, as bench testing indicated that the technology did not achieve sufficient removal under test influent concentration and effluent goals (which were 5 µg/L, lower than the California MCL was set from 2014 to 2017). However, the technology can still be applied in situations with lower influent chromium concentrations; for

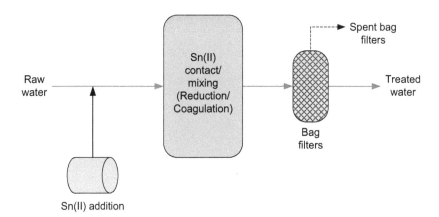

FIGURE 4.15 Example RF-Sn(II) process flow diagram

example, many wells at Coachella Valley Water District have Cr(VI) concentrations between 10–15 μg/L, which could be treated effectively by technologies with lower removal capabilities (RF-Sn[II] can achieve ~ 60 percent removal). This technology is unproven in application in comparison to other options discussed in this chapter.

The reduced complexity and waste compared to RCF make this option attractive, when applicable. RCF-Sn(II) is the simplest of the technologies listed in this section. RF-Sn(II) operation requires minimal attention after optimizing the Sn(II) dose. The bag or cartridge filters used are typically replaced when the differential pressure reaches unacceptable levels, and Sn(II) addition can be flow-paced to accommodate varying schemes if necessary. RCF-Sn(II) waste will consist of spent bag or cartridge filters. There are no anticipated hazardous considerations associated with this method, and the quantity of waste is anticipated to be lower than for other technologies.

4.9 CONCLUSIONS

New studies of Cr(VI) occurrence and consideration of changes to regulatory standards have resurfaced as new developments in toxicology and translation into risk assessment have arisen. Health risk and regulatory standards being established and considered vary by orders of magnitude from health goals in the hundredths of a μg/L range to a deleted standard of 10 μg/L in California to a standard of 100 μg/L under consideration in Canada. The range is due, in part, to variations in the interpretation of the toxicological data. Cr(VI) can naturally occur in drinking water supplies and over three-quarters of U.S. PWSs were found to contain Cr(VI) in the UCMR3. Whether that occurrence will require treatment depends on whether the USEPA and other states set drinking new water standards and the levels that are established. Methods for remediating Cr(VI) in groundwater and treating Cr(VI) in drinking water are well established. However, a greater understanding of the effectiveness and applicability of drinking water treatment to low levels has been an area of active investigation in recent years, providing a strong basis for action if and when standards are established. Similarly, the knowledge developed to date from remediation of Cr(VI) in groundwater can be applied toward treatment of potentially larger groundwater plumes, should standards be lowered.

REFERENCES

Alisto. 2001. Annual Report Evaluation of Corrective Action Activities for Year 2000–2001. Groundwater Remediation Project. Hinkley, California: Pacific Gas and Electric Company. Hinkley Compressor Station.

Arar, E. J., S. E. Long, and J. D. Pfaff. 1994. "Method 218.6. Determination of Dissolved Hexavalent Chromium in Drinking Water, Groundwater, and Industrial Wastewater Effluents by Ion Chromatography, Revision 3.3." Cincinnati, OH: Environmental Monitoring Systems Laboratory, Office of Research and Development, U.S. Environmental Protection Agency.

Arcadis. 2017a. 2016 Annual Agricultural Rate Performance Report. Hinkley, California: Pacific Gas and Electric Company, Hinkley Compressor Station.

Arcadis. 2017b. Fourth Quarter 2016 Agricultural Treatment Units Monitoring Report. Hinkley, California: Pacific Gas and Electric Company, Hinkley Compressor Station.

Arcadis. 2017c. Annual Cleanup Status and Effectiveness Report (January to December 2016). Hinkley, California: Pacific Gas and Electric Company Hinkley Compressor Station.

ATSDR. 2012. Toxicological Profile for Chromium. U.S. Department of Health and Human Services, Public Health Service, Agency for Toxic Substances and Disease Registry.

Bachelor, B., M. Shlautman, H. Inseong, and R. Wang. 1998. Kinetics of Chromium(VI) Reduction by Ferrous Iron. ANRCP-1998-13. Amarillo, TX: Amarillo National Resource Center for Plutonium.

Ball, James A., and J. A. Izibicki. 2004. "Occurrence of Hexavalent Chromium in Ground Water in the Western Mojave Desert, California." Applied Geochemistry 19: 1123–1135.

Becquer, T., C. Quantin, M. Sicot, and J. P. Boudot. 2003. "Chromium Availability in Ultramafic Soils from New Caledonia." *The Science of the Total Environment* 301: 251–261.

Beller, Harry, Li Yang, Charuleka Varadharajan, Ruyang Han, Hsaio Chien Lim, Ulas Karaoz, Sergi Molins, et al. 2014. "Divergent Aquifer Biogeochemical Systems Converge on Similar and Unexpected Cr(VI) Reduction Products." *Environmental Science and Technology 48*:10699–10706.

Beller, Harry R., Ruyang Han, Ulas Karaoz, Hsiao Chien Lim, and Eoin L. Brodie. 2013. "Genomic and Physiological Characterization of the Chromate-Reducing, Aquifer-Derived Firmicute Pelosinus sp. Strain HCF1." *Applied and Environmental Microbiology 79*(1): 63–73.

Blowes, David W., Robert W. Gillham, Carol J. Ptacek, Robert W. Puls, Timothy A. Bennett, Stephanie F. O'Hannesin, Christine J. Hanton-Fong, and Jeffrey G. Bain. 1999. An In Situ Permeable Reactive Barrier for the Treatment of Hexavalent Chromium and Trichloroethylene in Ground Water:Volume 1 Design and Installation. EPA/600/R-99/095a, Cincinnati, Ohio: National Risk Management Research Laboratory, Office of Research and Development, U.S. Environmental Protection Agency.

Blute, Nicole, Xueying Wu, Katie Porter, Greg Imamura, and Michael J. McGuire. 2013. "Hexavalent Chromium Removal Research Supplemental Project Report." Research Managed by City of Glendale, California Department of Water and Power, Report Prepared by Hazen and Sawyer and Arcadis U.S./Malcolm Pirnie.

Buchanan, Gary. 2009. "Regarding PHG Project." Letter from Gary Buchanan, Office of Science New Jersey Department of Environmental Protection, to Michael Baes, President of Environmental Toxicology Branch of Office of Environmental Health Hazard Assessment, California Environmental Protection Agency. Trenton, New Jersey, October 16.

California Manufacturers and Technology Association, a California Corporation, and Solano County Taxpayers Association, a California Corporation, Petitioners v. State Water Resources Control Board, Respondent. 2017. 34-2014-80001850 (Superior Court of California, County of Sacramento, May 31).

Chebeir, Michelle, and Haizhou Liu. 2015. "Kinetics and Mechanisms of Cr (VI) Formation Via the Oxidation of Cr (III) Solid Phases by Chlorine in Drinking Water." *Environmental Science & Technology 50*(2): 701–710.

Chowdhury, Zaid, Steve Bigley, Wil Gonzalez, Katie Porter, Greg Imamura, Chelsea Francis, Nicole Blute, Jacqueline Rhoades, Paul Westerhoff, and Alexandra Bowern. 2016. Compliance Planning and Evaluation of Technologies for Chromium (VI) Removal. Coachella, California: Water Research Foundation.

CRWQCB. 2017. Chromium-6 Drinking Water MCL. June 13. Accessed July 8, 2017. www.waterboards.ca.gov/drinking_water/certlic/drinkingwater/Chromium6.shtml.

CSWQCB. 2014. Chromium-6 in Drinking Water Sources: Sampling Results. July 11. Accessed February 26, 2017. www.waterboards.ca.gov/drinking_water/certlic/drinkingwater/Chromium6sampling.shtml.

Eary, L. E., and Dhanpat Rai. 1988. "Chromate Removal from Aqueous Wastes by Reduction with Ferrous Iron." *Environmental Science and Technology* 22(8): 972–977.

Eary, L. Edmond, and Dhanpat Rai. 1987. "Kinetics of Chromium(III) Oxidation to Chromium(VI) by Reaction with Manganese Dioxide." *Environmental Science and Technology* 21(12): 1187–1193.

Ellis, Andre S., Thomas M. Johnson, and Thomas D. Bullen. 2002. "Chromium Isotopes and the Fate of Hexavalent Chromium in the Environment." *Science* 295(5562): 2060–6026.

Ezebuiro, Prince, Jay Gandhi, Chunlong Zhang, Johnson Mathew, Melvin Ritter, and Marvelyn Humphrey. 2012. "Optimal Sample Preservation and Analysis of Cr(VI) in Drinking Water Samples by High Resolution Ion Chromatography Followed by Post Column Reaction and UV/Vis Detection." *Journal of Analytical Sciences, Methods and Instrumentation 2*: 74–80.

Fantoni, Donatella, Gianpiero Brozzo, Marco Canepa, Francesco Cipolli, Luigi Marini, Giulio Ottonello, and Marino Vetuschi Zuccolini. 2002. "Natural Hexavalent Chromium in Groundwaters Interacting with Ophiolitic Rocks." *Environmental Geology 42*: 871–882.

Federal Register. 2017. Revisions to the Unregulated Contaminant Monitoring Regulation (UCMR 3) for Public Water Systems; Final Rule. Vol 77. No. 85 Part II. January 11. Accessed March 13, 2017. www.federalregister.gov/documents/2017/01/11/2016-31262/nationalprimary-drinking-water-regulations-announcement-of-the-results-of-epas-review-ofexisting.

—. 2012. "Revisions to the Unregulated Contaminant Monitoring Regulation (UCMR 3) for Public Water Systems." 40 CFR Parts 141 and 142. [Docket No. EPA–HQ–OW–2009–0090; FRL–9660–4]. U.S. Environmental Protection Agency, May 2.

—. 2007. "Guidelines Establishing Test Procedures for the Analysis of Pollutants Under the Clean Water Act; National Primary Drinking Water Regulations; and National Secondary Drinking Water Regulations; Analysis and Sampling Procedures; Final Rule." 40 CFR Parts 122, 136, 141, 143, 430, 455, and 465 [EPA–HQ–OW–2003–0070; FRL–8203–8]. U.S. Environmental Protection Agency, March 12.

Fendorf, Scott, and Guangchao Li. 1996. "Kinetics of Chromate Reduction by Ferrous Iron." *Environmental Science and Technology 30*(5): 1614–1617.

Flury, Bettina, Jakob Frommer, Urs Eggenberger, Urs Mader, Maarten Nachtegaal, and Ruben Kretzschmar. 2009. "Assessment of Long-Term Performance and Chromate Reduction Mechanisms in a Field Scale Permeable Reactive Barrier." *Environmental Science and Technology 43*(17): 6786–6792.

Ford, Robert G., Richard T. Wilkin, and Robert W., Eds. Puls. 2007. Monitored Natural Attenuation of Inorganic Contaminants in Ground Water. Cincinnati, Ohio: National Risk Management Research Laboratory, Office of Research and Development, U.S. Environmental Protection Agency.

Fruchter, J. S., C. R. Cole, M. D. Williams, J. E. Vermeul, J. E. Amonette, J. E. Szecsody, J. D. Istock, and M. D. Humphrey. 2000. "Creation of a Subsurface Permeable Treatment Zone of Aqueous Chromate Contamination Using In Situ Redox Manipulation." *Ground Water Monitoring and Remediation*, Spring: 66–77.

Gonzalez, A. R., K. Ndung'U, and A. R. Flegal. 2006. "Natural Occurrence of Hexavalent Chromium in the Aromas Red Sands Aquifer, California." *Environmental Science and Technology 39*(16): 5505–5511.

Graham, Margaret C., John G. Farmer, Peter Anderson, Edward Paterson, Stephen Hillier, David G. Lumsdon, and Richard J. F. Bewley. 2006. "Calcium Polysulfide Remediation of Hexavalent Chromium Contamination from Chromite Ore Processing Residue." *Science of the Total Environment 364*: 32–44.

Hwang, Inseong, Bill Batchelor, Mark A. Schlautman, and Renjin Wang. 2002. "Effects of Ferrous Iron and Molecular Oxygen on Chromium(VI) Redox Kinetics in the Presence of Aquifer Solids." *Journal of Hazardous Materials*, 143–159.

IRIS. 2017. Chromium (VI). Accessed April 27, 2017. https://cfpub.epa.gov/ncea/iris2/chemicalLanding.cfm?substance_nmbr=144#fragment-2.

Istok, J.D., J. E. Amonette, C. R. Cole, J. S. Fructer, M. D. Humphrey, J. E. Szecsody, S. S. Teel, V. R. Vermeul, M. D. Williams, and S. B. Yabusaki. 1999. "In Situ Redox Manipulation by Dithionite Injection: Intermediate-Scale Laboratory Experiments." *Groundwater 37*(6): 884–889.

ITRC. 2011. Permeable Reactive Barrier: Technology Update. Washington, D.C.: Interstate Technology and Regulatory Council.

Izbicki, J. A., James W. Ball, Thomas D. Bullen, and Stephen J. Sutley. 2008. "Chromium, Chromium Isotopes and Selected Trace Elements, Western Mojave Desert, USA." *Applied Geochemistry 23*: 1325–1352.

Izbicki, J. A., Thomas D. Bullen, Peter Martin, and Brian Schroth. 2012. "Delta Chromium-53/52 Isotopic Composition of Native and Contaminated Groundwater, Mojave Desert, USA." *Applied Geochemistry 27*: 841–853.

Izbicki, John. 2008. "Chromium Concentrations, Chromium Isotopes, and Nitrate in the Unsaturated Zone and at the Water-Table Interface, El Mirage, California." Letter from Peter Martin to California Regional Water Quality Control Board. San Diego, CA: United States Department of the Interior, United States Geological Survey, December 24.

Kent, Douglas B., Robert W. Puls, and Robert G. Ford. 2007. "Chromium." In Monitored Natural Attenuation of Inorganic Contaminants in Groundwater. Volume 2: Assessment for Non-Radionuclides Including Arsenic, Cadmium, Chromium, Copper, Lead, Nickel, Nitrate, Perchlorate, and Selenium, by Robert G. Ford, Richard T. Wilkin and Robert W. Puls, 43–55. Cincinnati, Ohio: National Risk Management Research Laboratory, Office of Research and Development, U.S. Environmental Protection Agency.

Lide, David R. (Ed.) 2003–2004. *Handbook of Chemistry and Physics*. Boca Raton, FL: CRC Press.

Lindsay, Dana R., Kevin J. Farley, and Richard F. Carbonaro. 2012. "Oxidation of Cr III to Cr VI During Chlorination of Drinking Water." *Journal of Environmental Monitoring* 0789–1797.

Malone, Donald R., J. P. Messier, Frank Blaha, and Fred Payne. 2004. "In Situ Immobilization of Hexavalent Chromium in Groundwater with Ferrous Iron: A Case Study." Remediation of Chlorinated and Recalcitrant Compounds 2004, Proceedings for the 4th International Conference. Monterey, California: Battelle.

Middleton, Sarah S., Rizlan Bencheikh Latmani, Mason R. Mackey, Mark H. Ellisman, Bradley M. Tebo, and Craig S. Criddle. 2003. "Cometabolism of Cr(VI) by Shewenella oneidensis MR-1 Produces Cell-Associated Reduced Chromium and Inhibits Growth." *Biotechnology and Bioengineering 83*(6): 627–637.

Murphy, Richard, Jessica Ely, Jeff Gillow, Margaret Gentile, and Jim Harrington. 2008. "Development of Long-Term Residual Treatment Capacity During In Situ Cr(VI) Reduction." Sixth International Conference on Remediation of Chlorinated and Recalcitrant Compounds. Monterey, California: Battelle.

NCEQ. 2016. Final Report on the Study of Standards and Health Screening Levels for Hexavalent Chromium and Vanadium. Raleigh, North Carolina: North Carolina Department of Environmental Quality.

NJDEP. 2017. Status of Hexavalent Chromium Standards—April 2014. January 14. Accessed July 8, 2017. www.state.nj.us/dep/dsr/chromium/status-2014.html.

Nriagu, Jerome O. 1988. "Production and Uses of Chromium." In Chromium in the Natural and Human Environments, by Jerome O Nriagu and Evert, Eds. Neiboer, 81–103. New York: John Wiley & Sons.

NTP. 2008. NTP Technical Report on the Toxicology and Carcinogenesis Studies of Sodium Dichromate Dihydrate (Cas No. 7789-12-0) in F344/N Rats and B6C3F1 Mice (Drinking Water Studies): NTP TR 546.

OEHHA. 2011. "Public Health Goal for Hexavalent Chromium (CrVI) in Drinking Water." Office of Environmental Health Hazard Assessment, California Environmental Protection Agency, July.
Oze, Christopher, Norman H. Sleep, Robert G. Coleman, and Scott Fendorf. 2016. "Anoxic oxidation of chromium." *The Geological Society of America 44*(7): 543–546.
Oze, Christopher, Dennis K Bird, and Scott Fendorf. 2007. "Genesis of Hexavalent Chromium from Natural Sources in Soil and Groundwater." *Proceedings of the National Academy of Sciences 104*(16): 6544–6549.
Oze, Christopher, Scott Fendorf, Dennis K Bird, and Robert G Coleman. 2004. "Chromium Geochemistry of Serpentine Soils." *International Geology Review 46*: 97–126.
Palmer, Carl D., and Robert W. Puls. 1994. Natural Attenuation of Hexavalent Chromium in Groundwater and Soils. EPA/540/5-94/505. Ada, Oklahoma: Office of Research and Development, Office of Solid Waste and Emergency Response, United States Environmental Protection Agency.
Payne, Fred C., Joseph A. Quinnan, and Scott T. Potter. 2008. *Remediation Hydraulics*. Boca Raton, FL: CRC Press.
Raddatz, Amanda L., Thomas M. Johnson, and Travis L. Mcling. 2011. "Cr Stable Isotopes in Snake River Plain Aquifer Groundwater: Evidence of Natural Reduction of Dissolved Cr(VI)." *Environmental Science and Technology 45*(2): 502–507.
Rai, D., D. A. Moore, N. J. Hess, L. Rao, and S. B. Clark. 2004. "Chromium(III) hydroxide solubility in the aqueous Na+–OH—H2PO2-4–_PO3-4–H2O system: A thermodynamic model." *Journal of Solution Chemistry 33*: 1213–1242.
Rai, D., J. M. Zachara, L. E Eary, C.C. Ainsworth, J.E. Amonette, C.E. Cowan, R.W. Szelmeczka, C.T. Resch, R.L. Schmidt, D.C. Girvin and S.C. Smith. 1988. "Chromium Reactions in Geological Materials." EPRI-EA-5741. Electric Power Research Institute, Palo Alto, California.
Rai, Dhanpat, Bruce M. Sass and Dean A. Moore. 1987. "Chromium(III) Hydrolysis Constants and solubility of Chromium(III) Hydroxide. *Inorganic Chemistry* 26:345-349.
Rai, Dhanpat, Bruce M. Sass, and Dean A. Moore. 1986. "Chromium(III) Hydrolysis Constants of Chromium(III) Hydroxide." *Inorganic Chemistry 26*: 345–349.
Robertson, Frederick N. 1991. Geochemistry of Ground Water in Alluvial Basins of Arizona and Adjacent Parts of Nevada New Mexico, and California. U.S. Geological Survey Professional Paper 1406-C, Washington D.C.: United States Government Printing Office.
Robles-Camacho, J., and M. A. Armienta. 2000. "Natural Chromium Contamination of Groundwater at León Valley, Mexico." *Journal of Geochemical Exploration* 167–181.
Sass, Bruce, and Dhanpat Rai. 1987. "Solubility of Amorphous Chromium (III)-Iron(III) Hydroxide Solid Solutions." *Inorganic Chemistry 26*: 2228–2232.
Schecher, W. D., and D. C. McAvoy. 1998. MINEQL+, 4.5 ed. Hallowell, Maine: Environmental Research Software.
Schroeder, David C., and G. Fred Lee. 1974. "Potential Transformations of Chromium in Natural Waters." *Water, Air, and Soil Pollution* 4: 355–365.
Stern, Alan H. 2009. Derivation of an Ingestion-Based Soil Remediation Criterion for Cr^{6+} Based on the NTP Chronic Bioassay Data for Sodium Dichromate Dihydrate. Trenton, New Jersey: Office of Science, New Jersey Department of Environmental Protection.
Su, Chunming, and Ralph D. Ludwig. 2005. "Treatment of Hexavalent Chromium in Chromite Ore Processing Waste Using a Mixed Reductant Solution Of Ferrous Sulfate and Sodium Dithionite." *Environmental Science and Technology 39*(16): 6208–6216.
Sullivan, Kevin, Margaret Gentile, Jay Erickson, Frank Lenzo, and Scott Seyfried. 2014. "Large-Scale Implementation of Recirculation Systems for In-Situ Treatment of Hexavalent Chromium in Groundwater in Hinkley, California." Conference on Remediation of Chlorinated and Recalcitrant Compounds. Monterey, California: Battelle.

Suthersan, Suthan S., John Horst, Matthew Schnobrich, Nicklaus Welty, and Jeff McDonough. 2016. Remediation Engineering: Design Concepts, Second Edition. Boca Raton, FL: CRC Press.

Suthersan, Suthan, John Horst, and David Ams. 2009. "In Situ Metals Precipitation: Meeting the Standards." *Ground Water Monitoring & Remediation 29*(3): 44–50.

Tebo, Bradley M., and Anna Ya Obraztsova. 1998. "Sulfate-reducing Bacterium Grows with Cr(VI), U(VI), Mn(IV), and Fe(III) as Electron Acceptors." *FEMS Microbiology Letters 162*:193–198.

Thompson, C. M., C. R. Kirman, D. M. Proctor, L. C. Haws, M. Suh, S. M. Hays, J. G. Hixon, and M. A. Harris. 2014. "A Chronic Oral Reference Dose for Hexavalent Chromium-induced Intestinal Cancer." *Journal of Applied Toxicology 34*(5): 525–536.

USEPA. 2016a. https://cfpub.epa.gov/ncea/iris2/chemicalLanding.cfm?substance_nmbr=144. June 4. https://cfpub.epa.gov/ncea/iris2/chemicalLanding.cfm?substance_nmbr=144.

—. 2016b. Third Uncontaminated Monitoring Rule (UCMR 3): Data Summary. EPA 815-S-16-004. Washington, D.C.: Office of Water, United State Environmental Protection Agency.

—. 2015. Use of Monitored Natural Attenuation for Inorganic Contaminants in Groundwater at Superfund Sites. Directive 9283.1-36, U.S. Environmental Protection Agency, Office of Solid Waste and Emergency Response.

—. 2006. "Innovative Technology Verification Report. Nito XLi 700 Series XRF Analyzer. EPA/540/R-06/003." Prepared by Tetra Tech EM Inc. National Exposure Research Laboratory, Office of Research and Development, U.S. Environmental Protection Agency, February.

—. 2000. In Situ Treatment of Soil and Groundwater Contaminated with Chromium. EPA/625/R-00/005. U.S. Environmental Protection Agency.

—. 1999. Understanding Variation in Partition Coefficient, Kd, Values. Volume II: Review of Geochemistry and Available Kd Values for Cadmium, Cesium, Chromium, Lead, Plutonium, Radon, Strontium, Thoriu, Tritium (3H), and Uranium. Office of Air and Radiation. EPA 402-R-99-004B.

—. 1998a. Integrated Risk Information System Chemical Assessment Summary. Chromium (VI); CASRN 18540-29-9. U.S. Environmental Protection Agency National Center for Environmental Assessment.

—. 1998b. "Method 6020A Inductively Coupled Plasma-Mass Spectroscopy, Revision 1." United States Environmental Protection Agency.

—. 1997. Recent Developments for In Situ Treatment of Metal Contaminated Sites. United States Environmental Protection Agency, Technology Innovation Office. EPA 542-R-97-004.

Varadharajan, Charuleka, Harry R. Beller, Markus Bill, Eoin L. Brodie, Mark E. Conrad, Ruyang Han, Courtney Irwin, Joern T. Larsen, Hsiao-Chien Lim, Sergi Molins, Carl I. Steefel, April Van Hise, Li Yang, and Peter S. Nico. 2017. "Re-oxidation of Chromium (III) Products Formed under Different Biogeochemical Regimes." *Environmental Science and Technology 51*(9): 4918–4927.

Varadharajan, Charuleka, Ruyang Han, Harry R Beller, Li Yang, Matthew A Marcus, Marc Michel, and Peter S Nico. 2015. "Characterization of Chromium Bioremediation Products in Flow-Through Column Sediments Using Micro–X-ray Fluorescence and X-ray Absorption Spectroscopy." *Journal of Environmental Quality* 729–738.

Weilinga, Bruce, Midori M. Mizuba, Colleen M. Hansel, and Scott Fendorf. 2001. "Iron Promoted Reduction of Chromate by Dissimilatory Iron-Reducing Bacteria." *Environmental Science and Technology 35*(3): 522–527.

WHO. 2017. Guidelines for Drinking-Water Quality. World Health Organization.

Wilkin, Richard T., Steven D. Acree, Randall R. Ross, Robert W. Puls, Tony R. Lee, and Leilani L. Woods. 2014. "Fifteen-year Assessment of a Permeable Reactive Barrier for

Treatment of Chromate and Trichloroethylene in Groundwater." *Science of the Total Environment*, 186–194.

Wilkin, Richard T., Chunming Su, Robert G. Ford, and Cynthia J. Paul. 2005. "Chromium-Removal Process during Groundwater Remediation by a Zerovalent Iron Permeable Reactive Barrier." *Environmental Science and Technology 39*(12): 4599–4605.

Williams, Peter J., Elsabe Botes, Maleke M. Maleke, Abidemi Ojo, Mary F. DeFlaun, Jim Howell, Robert Borch, Robert Jordan, and Esta VanHeerden. 2014. "Effective Bioreduction of Hexavalent Chromium–Contaminated Water in Fixed-Film Bioreactors." Water SA, 549–554.

Zhitkovich, Anatoly. 2011. "Chromium in Drinking Water: Sources, Metabolism, and Cancer Risks." *Chemical Research in Toxicology 24*: 1617–1629.

Acronyms

µg/L	microgram per liter
µm	micron
µ-XAS	micro x-ray adsorption spectroscopy
µ-XRF	micro x-ray fluorescence
ANSI	American National Standards Institute
AVS-SEM	acid volatile sulfide and simultaneously extractable iron
BAT	Best Available Technology
BV	bed volume
CDHS	California Department of Health Services
CDPH	California Department of Public Health
Cl^-	chloride
Cr(III)	trivalent chromium
$Cr_xFe_{1-x}(OH)_3$	chromium-iron oxyhydroxide
CrO_4^{2-}	chromate
$Cr_2O_7^-$	dichromate
$Cr(OH)_3$	chromium hydroxide
Cr(T)	total chromium
Cr(VI)	hexavalent chromium
CSWRCB	California State Water Resources Control Board
DCB	dithionite-citrate-bicarbonate
DGR	dynamic groundwater recirculation
DHHS	Department of Health and Human Services
EVO	emulsified vegetable oil
Fe(II)	ferrous iron
$FeCr_2O_4$	chromite
ft	feet
g/L	grams per liter
gpm	gallons per minute
$HCrO_4^-$	hydrogen chromate
HPO_4^{2-}	hydrogen phosphate
$H_2PO_4^-$	dihydrogen phosphate
IRIS	Integrated Risk Information System
IRZ	in situ reactive zones
ISRM	in situ redox manipulation
IX	ion exchange
K_d	distribution coefficient
km	kilometer
LLRW	low-level radioactive waste
NO_3^-	nitrate
m	meter
MCL	maximum contaminant level
mg/kg	milligrams per kilogram

mg/L	milligrams per liter
mm	millimeter
MNA	Monitored Natural Attenuation
MRL	Method Reporting Limit
NCEQ	North Carolina Department of Environmental Quality
NJDEP	New Jersey Department of Environmental Protection
NPDWR	National Primary Drinking Water Regulations
NTP	National Toxicological Program
nZVI	nano-scale zero valent iron
OEHHA	Office of Environmental Health Hazard Assessment
PHG	public health goal
POE	point-of-entry
POU	point-of-use
PRB	permeable reactive barriers
PWS	public water system
RCF	reduction coagulation filtration
RF	reduction and filtration
RO	reverse osmosis
ROI	radius of influence
R/R	regeneration/replacement
SBA-IX	strong base anion ion exchange
SEM-EDS	scanning electron microscopy and energy dispersive X-ray spectroscopy
SM	standard method
Sn(II)	stannous tin
SO_4^{2-}	sulfate
TDS	total dissolved solids
TENORM	technologically enhanced naturally occurring radioactive material
UCMR	Unregulated Contaminant Monitoring Rule
USA	United States of America
USEPA	United States Environmental Protection Agency
WBA-IX	weak base anion ion exchange
WHO	World Health Organization
XANES	X-ray adsorption near edge structure
XRF	X-ray fluorescence
ZVI	Zero Valent Iron

5 1,2,3-Trichloropropane

Margaret Gentile
(Arcadis U.S. Inc., Margaret.Gentile@arcadis.com)

Shandra Justicia-Leon
(Arcadis U.S. Inc., Shandra.Justicia-Leon@arcadis.com)

Sarah Page
(City of Ann Arbor, SPage@a2gov.org)

5.1 BASIC INFORMATION

1,2,3-Trichloropropane (TCP) is an anthropogenic chlorinated alkane (Figure 5.1), used industrially and also present in industrial chemicals as a by-product of production and in the formulation of agricultural chemicals. TCP is an emerging contaminant; because risk evaluations have been completed only relatively recently, occurrence has become better understood in the last 5 to 15 years by monitoring under the Unregulated Contaminant Monitoring Rules (UCMRs), and new regulations are on the horizon. In the United States, attention became focused on TCP, particularly in California, when it was discovered at the Burbank Operable Unit, a southern California Superfund hazardous waste site. TCP has been identified as a candidate contaminant for regulation at the federal level in the United States and screening levels in the hundreds of nanograms per liter (ng/L) level have been established. However, it is unclear whether a federal Maximum Contaminant Limit (MCL) is imminent and what the level of a new federal MCL might be. Greater activity is taking place at the state level. In particular, a very low 5 ng/L state MCL was established in 2017 for the state of California in the United States.

Industrial sources of TCP in groundwater include locations of manufacturing and mechanical operations. TCP has been used directly as a cleaning and maintenance solvent, paint and varnish remover, and degreasing agent (ASTDR 1992). TCP is used as an intermediate in the production of other chemicals, including polysulfone and epichlorohydrin (ASTDR 1992; USEPA 2009).

TCP is also a contaminant associated with agriculture. TCP was historically formulated with dichloropropenes in the manufacture of soil fumigants for nematodes, including D-D (NTP 2016) and Telone (WHO 2003). Agriculture-related sources of TCP in groundwater include point sources from agricultural chemical manufacturing and storage/distribution and nonpoint sources from the use of fumigants in farming.

Data on occurrence of TCP in Public Water Systems (PWSs) was collected under UCMR programs by the United States Environmental Protection Agency (USEPA)

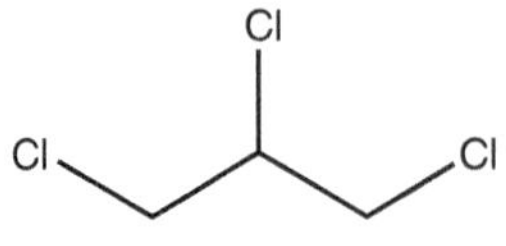

FIGURE 5.1 1,2,3-Trichloropropane chemical structure.

and California State Water Resources Control Board (CSWRCB). Monitoring of PWSs under USEPA UCMR3 was conducted from 2013 to 2015 using a method reporting limit (MRL) of 0.03 micrograms per liter (μg/L, 30 ng/L) and reference level of 0.0004 μg/L (0.4 ng/L), representing a 10^{-6} cancer risk (USEPA 2016a). Of the more than 36,000 data points collected from drinking water supplies tested, only approximately 253 (0.7 percent) had detections above the MRL and reference level (USEPA 2016a). The 36,000 data points represented 4,905 public water supplies and 65 of these PWSs had detections above the MRL and reference level (1.3 percent) (USEPA 2016a). TCP detections in PWSs in UCMR3 were limited to 14 states, as shown on Figure 5.2. The UCMR3 data suggest that occurrences in water supplies are low, but the analysis was conducted with an MRL of 30 ng/L, which is six times higher than the new California MCL of 5 ng/L and 36 times higher than the federal screening level for tap water based on cancer of 0.74 ng/L, as shown on Figure 5.2. The occurrence of TCP in drinking water at these lower levels was not evaluated in the federal UCMR, but was evaluated in the California UCMR, discussed next.

Monitoring of PWSs under the California UCMR was conducted from 2001 to 2016 using a detection limit of 0.005 μg/L (5 ng/L), which is the current notification level and new MCL for TCP in California (CSWRCB 2017a). The results of the California UCMR are summarized in Table 5.1. A greater number of detections above the 5 ng/L detection limit were observed in the CSWRCB dataset, 891, than in the USEPA

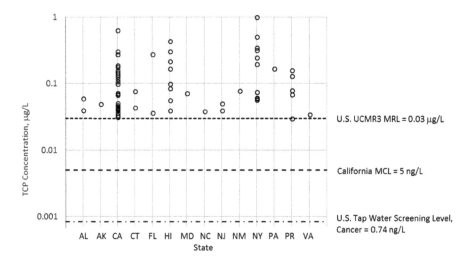

FIGURE 5.2 TCP detections by state in the US UCMR3. Data presented are maximum values from each PWS within each state with detections.

TABLE 5.1
Summary of occurrence of TCP in drinking water sources detected under the California UCMR

Average Concentration of TCP μg/L (ng/L)	Number of Wells Impacted at Each Concentration	Percentage of Wells Impacted at Each Concentration
0.15 (150)	20	2
0.07 (70)	60	7
0.035 (35)	104	12
0.015 (15)	173	19
0.007 (7)	245	27
0.005 (5)	289	32
Total	**891**	

Source: CSWRCB 2016.

database, potentially due to the lower detection limit and longer period of monitoring (CSWRCB 2016a). Almost 80 percent of the detections were in the range from greater than 5 ng/L to 35 ng/L, generally lower than the USEPA UCMR3 MRL of 30 ng/L. The great majority of detections in the California database were located in the Central Valley, a region where there is intense agricultural activity (CSWRCB 2017b,c).

Physical and chemical properties for TCP are provided in Appendix C and summarized in comparison with the common chlorinated alkene, trichloroethene (TCE), and the common chlorinated alkane 1,1,1-trichloroethane (TCA) in Table 5.2. TCP has a specific gravity 1.38 g/mL, greater than water, and high octanol water partition coefficient, indicating potential for TCP, if sourced in a pure form, to sink and form dense nonaqueous phase liquids (DNAPL). The comparison with TCE and TCA demonstrates the relatively low volatility of TCP based on low vapor pressure and low Henry's constant, indicating limited partitioning to the vapor phase. Solubility and sorption properties are comparable to TCE and TCA, indicating comparable fate and transport in groundwater.

5.2 TOXICITY AND RISK ASSESSMENT

Human health toxicity is evaluated from two perspectives: potential cancer effects and noncancer effects. The primary routes of human exposure to TCP include ingestion and inhalation. In the past, dermal exposure may have occurred when products such as paint and varnish removers contained TCP (NTP 2016). No human toxicological or epidemiological studies are available for TCP (USEPA 2009; NTP 2016). Studies on animals, particularly studies on rats conducted by the United States Department of Health and Human Services National Toxicology Program (NTP 1993), inform the current understanding of noncancer and cancer effects, as summarized in the reference document *Toxicological Review of 1,2,3-Trichloropropane* (USEPA 2009).

Based on animal studies, it has been concluded that long-term oral exposure can lead to damage to liver and kidneys (NTP 1993). Key toxicological reference values

TABLE 5.2
Comparison of relative physical and chemical properties of TCP with other volatile organic compounds

Physical or Chemical Property	Benzene[1]	TCE[1]	TCA[1]	1,4-D[1]	TCP
Molecular formula	C_6H_6	C_2HCl_3	$C_2H_3Cl_3$	$C_4H_8O_2$	$C_3H_5Cl_3$
Molecular weight	78.1	131.4	133.4	88.1	147.4
Specific gravity(referenced to water @ 1 g/mL)	0.86 (Lower than water)	1.465 (Higher than water)	1.339 (Higher than water)	1.033 (Comparable to water)	1.38 (Higher than water)
Solubility(g/mL)	1.8 (Moderate)	1.1 (Moderate)	0.91 (High)	Miscible	1.75 (2) (Moderate)
Vapor pressure (mm Hg)	95.2 (Moderate)	72.6 (Moderate)	124 (High)	38.1 (Low)	3.1 (3) (Very Low)
Henry's constant (atm × m³/mol)	5.5×10^{-3} (Moderate)	9.1×10^{-3} (Moderate)	1.6×10^{-2} (High)	4.8×10^{-6} (Very Low)	3.4×10^{-4} (4) (Low)
Sorption (log K_{oc})	1.92 (Moderate)	1.81 (Moderate)	2.18 (Moderate)	1.23 (Moderate)	1.99 (4) (Moderate)
Octanol-water partition coefficient (log K_{ow})	2.13 (High)	2.29 (High)	2.49 (High)	−0.27 (Very Low)	1.98 to 2.27 (5) (High)

Notes and References: Full list of values for TCP can be found in Appendix A. Properties provided at 25°C unless noted otherwise.
1. Properties listed in Suthersan & Payne 2005.
2. Riddick, Bunger, & Sakano 1986.
3. MacKay, Bobra, Chan, & Shiu 1982.
4. Lyman, Reehl, & Rosenblatt 1982.
5. EPA 1988.
6. Qualitative comparison relative to other volatile organic compounds.

Abbreviations: atm × m³/mol = atmosphere meter cubed per mol, g/mL = grams per milliliter, mm Hg = millimeters of mercury, TCE = trichloroethene, TCA = 1,1,1-trichloroethane, 1,4-D = 1,4-dioxane

TABLE 5.3
Summary of toxicological reference values for chronic oral and inhalation noncarcinogenic exposures to TCP

Parameter	Chronic Oral Exposure	Chronic Inhalation Exposure
Toxicity reference value	Reference dose (RfD)	Reference concentration (RfC)
Value	0.004 mg/kg-d	0.0003 mg/m^3
POD	BMD_{10}: 3.8 mg/kg-d	BMC_{10}: 1.6 ppm
	$BMDL_{10}$: 1.6 mg/kg-d	$BMCL_{10}$: 0.84 ppm
	$BMDL_{ADJ}$: 1.1 mg/kg-d	$BMCL_{ADJ}$: 0.90 mg/m^3
		$BMCL_{HEC}$: 0.90 mg/m^3
POD type	BMD	BMC
Critical effect	Increased absolute liver weight in male rats	Peribronchial lymphoid hyperplasia in male rats
Key study	2-year bioassay in rats (NTP 1993)	13-week inhalation study in rats (Johannsen et al. 1988)
Uncertainty factor	300 ($UF_A = 10$, $UF_H = 10$, $UF_D = 3$)	3,000 ($UF_A = 3$, $UF_H = 10$, $UF_S = 10$, $UF_D = 3$)
Overall confidence	Medium-to-high	Low-to-medium

Abbreviations: BMD = benchmark dose, BMC = benchmark concentration, POD = point of departure, UF_A = interspecies extrapolation factor, UF_H = intraspecies variability factor, UF_S = subchronic to chronic exposure duration extrapolation factor, UF_D = toxicological database uncertainty factor, ppm = part per million.
Source: IRIS 2009.

are summarized in Table 5.3. A Reference Dose (RfD) for chronic oral exposure is an estimate (with uncertainty spanning perhaps an order of magnitude) of a daily oral exposure to the human population (including sensitive subgroups) that is likely to be without an appreciable risk of deleterious effects. An RfD of 0.004 milligrams per kilogram per day (mg/kg-d) has been established for TCP, with an overall confidence of medium-to-high (IRIS 2009). A Reference Concentration (RfC) for chronic inhalation exposure is an estimate (with uncertainty spanning perhaps an order of magnitude) of a continuous inhalation exposure to the human population (including sensitive subgroups) that is likely to be without an appreciable risk of deleterious effects during a lifetime. An RfC of 0.0003 milligrams per meter cubed (mg/m^3) has been established for chronic inhalation exposure by the USEPA (IRIS 2009).

With regards to carcinogenicity, TCP is "reasonably anticipated to be a human carcinogen based on sufficient evidence of carcinogenicity from studies in experimental animals" (NTP 2016). Sites of cancer in rats include the oral cavity, forestomach, pancreas, kidney, preputial gland, clitoral gland, mammary gland, and Zymbal's gland (NTP 1993). The oral slope factor for use in risk assessment established by the USEPA in 2009 in the Integrated Risk Information System (IRIS) is summarized in Table 5.4. The USEPA recommended an upper-bound estimate on human extra cancer risk from continuous lifetime oral exposure to TCP is 28 (mg/kg-d)$^{-1}$ (IRIS 2009). The carcinogenicity evaluation in the *IRIS Toxicological Review*

TABLE 5.4
Summary of cancer slope factors and inhalation unit risks for TCP

Parameter	Oral Exposures	
Weight of evidence classification	Likely to be carcinogenic to humans	
Toxicity Reference	Overall cancer slope factor (CSF)	
Value	28 (mg/kg-d)$^{-1}$	
POD (mg/kg/d)	Alimentary system, total squamous neoplasms- incidental	BMD_{10} = 0.0032 mg/kg-d $BMDL_{10}$ = 0.00065 mg/kg-d Slope Factor = 150 (mg/kg-d)$^{-1}$
	Alimentary system, total squamous neoplasms- fatal	BMD_{10} = 0.0095 mg/kg-d $BMDL_{10}$ = 0.0039 mg/kg-d Slope Factor = 26 (mg/kg-d)$^{-1}$
	Liver: adenoma or carcinoma	BMD_{10} = 0.3 mg/kg-d $BMDL_{10}$ = 0.14 mg/kg-d Slope Factor = 0.73 (mg/kg-d)$^{-1}$
	Harderian gland adenoma	BMD_{10} = 0.42 mg/kg-d $BMDL_{10}$ = 0.2 mg/kg-d Slope Factor = 0.50 (mg/kg-d)$^{-1}$
	Uterus: adenoma or carcinoma	BMD_{10} = 0.42 mg/kg-d $BMDL_{10}$ = 0.21 mg/kg Slope Factor = 0.47 (mg/kg-d)$^{-1}$
POD type	BMD	
Critical effect	alimentary system squamous cell neoplasms[1]; liver hepatocellular adenomas or carcinomas, Harderian gland adenomas, uterine/cervix adenomas or carcinomas in female rats	
Key study	(NTP 1993)	

1. Squamous papillomas or squamous cell carcinomas of the pharynx/palate, tongue, or forestomach
Source: IRIS 2009.

included tumor sites formed in the forestomach of rats in the analysis. External peer reviewers of the *IRIS Toxicological Review* recommended removing the forestomach tumor data given that humans do not have a forestomach. The USEPA responded that they consider the forestomach tumors to be relevant to humans given the comparability of human epithelial tissues in the esophagus to the rodent forestomach. However, oral slope factors excluding the forestomach tumor data were derived and provided in the final toxicological review in 2009 (USEPA 2009).

It should be noted that risk assessments conducted by others have yielded very different conclusions from the *IRIS Toxicological Review*. In particular, a risk assessment by Tardiff and Carson questioned the substantial premature mortality in the NTP 2003 study (Tardiff and Carson 2010). Tardiff and Carson recommended use of a nonlinear dose response model and exclusion of data from the Harderian gland, Zymbal gland, and forestomach based on lack of a human homologue (Tardiff and Carson 2010). They determined a Drinking Water Equivalent Level (DWEL) of 200 to 280 µg/L for cancer with those assumptions (Tardiff and Carson 2010).

The differences in interpretation of the toxicological studies has resulted in orders of magnitude differences in regulatory standards, for example, in Hawaii versus California, as discussed in more detail in section 5.3.

5.3 REGULATORY STATUS

There is currently no federal MCL in the United States for TCP. In the United States, states that have either set a regulatory standard or guidance value or are actively developing regulations include Hawaii, Alaska, New Jersey, Florida, Minnesota, and California. A summary of the regulatory standard establishment and development is listed in Table 5.5 for the USEPA and these states, as detailed further in this section. A map is provided in Figure 5.3.

5.3.1 U.S. Federal Regulations

In 2011, the USEPA listed screening levels for tap water of 0.00084 µg/L (0.84 ng/L) for noncancer and 0.00075 µg/L (0.75 ng/L) for cancer, assuming a target risk of 10^{-6} (USEPA 2016b), based on IRIS 2009. These levels are actually below the detection limits of available analytical methods (see section 5.4).

TABLE 5.5
Summary of TCP regulatory standards and guidance levels for the United States of America and several U.S. states

Agency	Standard	Concentration
United States, Federal (USEPA)	Tap water screening level ingestion-non-cancer	0.00084 µg/L (0.84 ng/L)
	Tap water screening level ingestion-cancer	0.00075 µg/L (0.75 ng/L)
	Carcinogenic MCL	Not established
Alaska, USA	Groundwater human health cleanup level	0.0075 µg/L (7.5 ng/L)
California, USA	PHG	0.0007 µg/L (0.7 ng/L)
	MCL	0.005 µg/L (5 ng/L)
Florida, USA	HAL	40 µg/L
Hawaii, USA	MCL	0.6 µg/L (600 ng/L)
Minnesota, USA	Non-cancer HRL	7 µg/L (7,000 ng/L)
	Cancer HRL	0.003 µg/L (3 ng/L)
Missouri, USA	HAL, drinking water supply and groundwater	40 µg/L
New Jersey, USA	Recommended health-based MCL (NJDEP 2016)	0.0005 µg/L (0.5 ng/L)
	Recommended MCL (NJDEP 2016)	0.03 µg/L (30 ng/L)
New York, USA	POC MCL	5 µg/L

FIGURE 5.3 Map of United States TCP standards

Under the Safe Drinking Water Act amendments, the USEPA issues a new list of unregulated contaminants to be monitored by PWSs every five years. TCP was listed for Assessment Monitoring in the UCMR 3 issued on May 2, 2012 (Federal Register 2012). The purpose of the UCMR is to collect occurrence data for contaminants without health-based standards.

Subsequently, The USEPA listed TCP on the Final Contaminant Candidate List (CCL) 4 that was issued on February 4, 2015 (Appendix A). The USEPA must determine whether or not to regulate five of the contaminants on CCL4 in the Regulatory Determination 4. TCP may or may not be selected and have a determination made based on inclusion in CCL4.

5.3.2 U.S. State Regulations

This section presents a summary of the regulatory status of TCP in several states with recent regulatory developments. Regulatory standards and guidance levels vary drastically at the state level, spanning four orders of magnitude, depending on how states are incorporating the IRIS 2009 chemical assessment. Since the completion of the IRIS chemical assessment in 2009, several states have incorporated the risk assessment into the establishment of new guidance levels or standards (Alaska, California, and Minnesota) or are considering regulatory standards (New Jersey) at the ng/L level. Other states have either not incorporated the IRIS assessments into their standards (Missouri and Florida) or have interpreted the NTP studies that form the basis of the IRIS assessment less conservatively (Hawaii), resulting in levels in the µg/L range. Further details on the regulatory standards and guidance values for each state and their basis are provided in this section.

Hawaii: Hawaii was the first to adopt an MCL for TCP in 2005 of 0.6 µg/L (HDOH 2013). The early adoption of this MCL was supported by a review of health effects by Dr. Robert Tardiff commissioned by the state in 1992 and updated in 2001, contained in unpublished reports (Hooker et al., 2012). The Tardiff evaluation was based on the same 1993 NTP study on mice and rats as the USEPA IRIS 2009 risk assessment, but with different interpretation.

Notably, the Tardiff study did not consider tumor development in certain organs specific to rodents. Based on the evaluation, an MCL of 0.6 μg/L (600 ng/L) was recommended.

California: In 1999, the CSWRCB established a 0.005 μg/L (5 ng/L) drinking water notification level for TCP after the discovery of TCP in the Burbank Operable Unit, a southern California Superfund site. In 2001, California added TCP to the state UCMR (CSWRCB 2017a). Based on the UCMR detections, the CSWRCB requested a Public Health Goal (PHG) from the Office of Environmental Health Hazard Assessment (OEHHA), and OEHHA established the PHG at 0.0007 μg/L (0.7 ng/L) in 2009. PHGs are based on health effects, do not consider technological or economic feasibility, and are not enforceable. OEHHA based the 0.7 ng/L PHG on an oral slope factor of 25 $(mg/kg\text{-}d)^{-1}$ based on tumors in the forestomach of female mice in the 1993 NTP study and used a linear dose response model with a lifetime risk of 10^{-6} (OEHAA 2009).

Subsequently, the CSWRCB evaluated the technical and economic feasibility and protection of public health considerations. The evaluation of technical feasibility assumed Granular Activated Carbon (GAC) treatment as the Best Available Technology (BAT). In July 2016, the CSWRCB recommended adoption of a 5 ng/L MCL for TCP (CSWRCB 2016a).

New Jersey: In 2009, the New Jersey Drinking Water Quality Institute (NJDWQI), recommended a health-based MCL of 0.0013 μg/L (1.3 ng/L) and an MCL of 0.03 μg/L (30 ng/L) based on the Practical Quantitation Limit (PQL) of 0.03 μg/L determined by the Testing Subcommittee (NJDWQI 2009). In September 2015, the Commissioner from the New Jersey Department of Environmental Protection (NJDEP) asked the NJDWQI to reevaluate the 2009 recommendations (NJDWQI 2016). In this review, the Health Effects Subcommittee considered current USEPA risk assessment guidance and decreased the health-based MCL recommendation to 0.5 ng/L and the Testing Subcommittee verified the 0.03 μg/L PQL. The recommended health-based MCL of 0.5 ng/L was based on a cancer potency factor of 26 $(mg/kg\text{-}day)^{-1}$ and a risk of 10^{-6} (NJDWQI 2016). The 0.5 ng/L health-based MCL is comparable to the California OEHAA PHG of 0.7 ng/L.

The Treatment Subcommittee concluded "that TCP can be reliably and feasibly removed by carefully designed GAC below the PQL of 0.03 μg/L" (NJDWQI 2016). The NJDWQI recommended that NJDEP propose and adopt an MCL of 0.03 μg/L (30 ng/L) for drinking water based on this assessment of health effects, testing, and treatment. Note that the PQL determined by the NJDWQI is 30 ng/L using USEPA method 524.2 (NJDWQI 2009), over an order of magnitude higher than the MDL of 5 ng/L achievable with Sanitation and Radiation Laboratories (SRL) Method 524M-TCP (Purge and Trap Gas Chromatography/Mass Spectrometry [GC/MS]) and SRL 525M-TCP developed by California Department of Public Health's SRL, now the Drinking Water and Radiation Laboratories (DWRL), see section 5.4. This discrepancy in quantitation limits factored into the difference in recommended MCL values.

Alaska: Although Alaska has not specified a drinking water standard for TCP, Alaska Department of Environmental Conservation (ADEC) regulations specify a groundwater human health cleanup level of 0.0075 µg/L (7.5 ng/L) for TCP (ADEC 2016a). ADEC has also set cleanup levels for TCP in soil of 0.089 milligrams per kilogram (mg/kg) for the sites in the Arctic zone, 0.0066 mg/kg for sites with under 40 inches of annual precipitation, and 0.054 mg/kg for sites with over 40 inches of annual precipitation (ADEC 2016a). The Alaska cleanup levels were calculated using the ADEC Procedures for Calculating Cleanup Levels adopted by 18AAC 75 in 2016 (ADEC 2016b), which referenced the IRIS slope factors and reference dose for TCP (see section 5.3).

Florida: The Florida Department of Health Environmental Health (FDHEH) has a Health Advisory Level (HAL) of 40 µg/L for TCP (FDHEH 2016). The HAL for Florida does not appear to be based on IRIS 2009.

Minnesota: In 2013, Minnesota set Health Risk Limits (HRLs) for TCP. The noncancer HRL is 7 µg/L for acute exposure of one day, chronic exposures of 1 to 30 days and chronic exposure for greater than 10 percent of a lifetime (Minnesota Department of Health [MDH], 2015). The cancer HRL for 0 to 70 years of age is 0.003 µg/L (MDH 2015). Minnesota used the IRIS 2009 reference doses and slope factors to develop the HRLs.

Missouri: The Missouri Department of Natural Resources (MDNR) lists a HAL for drinking water and groundwater of 40 µg/L for TCP (MDNR 2014). The HAL for Missouri does not appear to be based on IRIS 2009.

New York: The New York State Department of Health (NYSDH) lists TCP as a principal organic contaminant (POC) under Title 10, Part 5 (NYSDH 2011). NYSDH requires routine monitoring for POCs. A generic MCL of 5 µg/L for POCs has been established.

5.3.3 INTERNATIONAL GUIDANCE

TCP was selected as a priority by the International Programme on Chemical Safety, a cooperative program of the World Health Organization (WHO). Under this program, a summary of the hazards and dose response information from exposure for TCP was prepared in 2003 (WHO 2003). Within the international community, the attention around emerging contaminants tends to focus on contaminants that are persistent and bioaccumulate and are listed as Persistent Organic Pollutants (POP) under the Stockholm Convention, through the European Union Registration, Evaluation, Authorization, and Restriction of Chemicals (REACH) regulation, and more recently through evaluations of persistent, mobile, toxic (PMT) and very persistent very mobile (vPvM) criteria. TCP does not tend to bioaccumulate, is not a POP under the Stockholm Convention, and has not been included in the current Drinking Water Guidelines, Fourth Edition Incorporating the First Addendum by WHO (WHO 2017). TCP is on the REACH candidate list (Appendix B). In a preliminary list of substances that could be considered to meet PMT or vPvM criteria by the Norwegian Geotechnical Institute (NGI), a private foundation, prepared for the German Environment Agency (Umweltbundesamt—UBA), TCP was listed as potentially persistent/very persistent, very mobile, and toxic (Appendix D; NGI 2018).

5.4 SITE CHARACTERIZATION

TCP is not particularly challenging to analyze and characterize at environmental sites. However, there are considerations relevant to characterization of potentially large TCP plumes and nuances of the analytical methods that must be understood when analyzing to the ng/L levels relevant to emerging regulations and guidance levels. The details of these issues are presented in this section.

5.4.1 INVESTIGATION

The physical and chemical properties of TCP and its resulting behavior in the subsurface make it amenable to traditional environmental site investigation approaches such as the advancement of soil borings to collect soil samples and the installation of groundwater monitoring wells to collect groundwater samples. Groundwater sample collection for TCP analysis can be conducted via numerous methods including low-flow purge, HydraSleeve™, Snap Sampler, and passive diffusion bags.

As the scale of groundwater impacts increases, the importance of characterizing mass flux as part of site characterization becomes key to the design of efficient groundwater remediation systems. The potential for large scale TCP plumes is suggested by the non-point agricultural sources and occurrence in drinking water wells, indicating the potential need for investigation of mass flux as part of site characterization. High-resolution site characterization can be highly effective for identifying mass transport zones within the aquifer and the extent of contamination, and is discussed in greater detail in other references (Suthersan et al. 2016). High resolution characterization typically includes techniques such as vertical aquifer profiling for permeability and contaminant concentration mapping. Routine methods for vertical aquifer profiling used for other common chlorinated organics, such as TCE, would be readily applicable and useful for TCP.

5.4.2 ANALYTICAL METHODS

TCP is amenable to standard volatile organic compound (VOC) analysis, such as USEPA Methods 504.1, 551.1, and 8260 or 524.2 (Table 5.6). Although these methods are capable of detection of TCP in the 10s of ng/L range, typical reporting limits can be in the 100 ng/L to µg/L range, limiting the use of these methods for characterization at the ng/L range that is becoming of interest in several states within the U.S. (see section 5.3.2). For standard VOC methods USEPA 524.2 and 8260, TCP retention times for some gas chromatography columns can extend beyond a typical run time and/or TCP may not be included in the standard analyte list. Therefore, particular attention is required when using these methods for TCP to determine if the methods are sufficient for the concentration range anticipated and that TCP is included in the calibration and reporting.

To establish analytical methods with more sensitive detection limits, California's Department of Public Health's SRL, now the DWRL, developed several analytical methods for detecting and/or monitoring TCP in environmental media with analytical reporting limits that matched the notification level of

TABLE 5.6
Summary of analytical methods for TCP

Analytical Methods	Developed by	Analysis	Quantitation	Method Detection Limit	Recommended Use
USEPA 504.1	USEPA	Microextraction and gas chromatography and electron capture detector; specific to TCP, EDB and DBCP	Standard calibration curve	0.02 µg/L (20 ng/L)[1]	Greater than 100 ng/L
USEPA 551.1	USEPA	Liquid-liquid extraction and gas chromatography with electron-capture detection; specific to chlorination disinfection by-products and halogenated pesticides/herbicides	Standard calibration curve	0.008 µg/L (8 ng/L);	Greater than 100 ng/L
USEPA 8260	USEPA	Gas chromatography/ mass spectrometry	Standard calibration curve	0.09–0.32 µg/L (90–320 ng/L)[1]	Greater than 500 ng/L
USEPA 524.2	USEPA	Purge and trap gas chromatography and mass spectrometry	Standard calibration curve	0.03–0.32 µg/L (30–320 ng/L)[1,2]	Greater than 500 ng/L
SRL 524M-TCP	California SRL DWRL	Purge and trap gas chromatography and mass spectrometry, with quadrupole MS operated in SIM mode or ion trap in SIS mode	Isotopic dilution using TCP-D5	0.005 µg/L (5 ng/L)	Down to 5 ng/L
SRL 525M-TCP	California SRL DWRL	Liquid-liquid extraction and gas chromatography and mass spectrometry, with quadrupole MS operated in SIM mode or ion trap in SIS mode	Isotopic dilution using TCP-D5	0.005 µg/L (5 ng/L)	Down to 5 ng/L

[1] Note that while MDLs in the 10s of ng/L range are achievable, reporting limits can be one to two orders of magnitude higher.
[2] MDL for USEPA 524.2 depends on column used.

Abbreviations: EDB = 1,2,-dibromoethane, DBCP = 1,2-dibromo-3chloropropane, SIM = selected ion monitoring, SIS = selected ion storage

5 ng/L established by California in 1999. SRL Method 524M-TCP (Purge and Trap GC/MS) and SRL 525M-TCP (Liquid-Liquid Extraction GC/MS), developed by the DWRL and summarized in Table 5.6, can determine TCP in water at concentrations below the quantifiable ranges of USEPA Methods 504.1, 524.2 551.1, and 8260, and have calibration ranges for TCP from 5 to 500 ng/L (Okamoto et al. 2002a; Okamoto et al. 2002b). The California State Environmental Laboratory Accreditation Program (ELAP) has issued a list of nation-wide laboratories that are certified to run the two SRL methods (CRWQCB 2016b).

Method SRL 524M-TCP is very similar to USEPA Methods 524.2 and 8260, except that this single analyte method is designed to quantitate TCP at concentrations as low as 5 ng/L. SRL 524M-TCP analysis is performed using purge and trap and GC/MS to identify TCP by matching the retention time and fragment ions from the sample with those of the reference standard. TCP-D5 is used as the internal standard. The retention time for both TCP and TCP-D5 is approximately 20 minutes. The applicable matrices for method SRL 524M-TCP include drinking water, groundwater, surface water, and, potentially, wastewaters.

Method SRL 525M-TCP shares similar detection limits and applicable matrices as SRL 524M-TCP; however, instead of purge and trap, this method employs an overnight liquid-liquid extraction of TCP in methylene chloride prior to analysis by GC/MS. The retention time for TCP via method SRL 525M-TCP is between 7 and 8 minutes. Both methods SRL 524M-TCP and SRL 525M-TCP were designed for trace-level analysis of TCP. It is advisable to use USEPA Methods 504.1 or 551.1 for TCP concentrations equal or greater than 100 ng/L, and USEPA Method 524.2 (8260) for TCP concentrations equal or greater than 500 ng/L.

5.4.3 Advanced Investigation Techniques

The development of advanced analytical methods for interrogating the fate of TCP in the environment is just beginning, and there are several techniques that are becoming available or that may be applicable.

In section 5.5.1.2, the biological degradation of TCP is discussed, in particular the reductive dechlorination of TCP by select species of the *Dehalogenimonas* genus and cometabolic aerobic degradation. Molecular biology tools (MBTs) for assessing the presence of *Dehalogenimonas* spp. in environmental samples have been recently developed and are commercially available (e.g., Microbial Insights' CENSUS® quantitative polymerase chain reaction [qPCR] or QuantArray®, SiREM's Gene-Trac® qPCR, Pace Energy Services' qPCR). The developed MBTs are proving to be useful for the targeted detection and quantification of *Dehalogenimonas* strains as TCP degradation progresses during the research and development stages of this technology. Additional functional genes may also prove useful targets for evaluating the activity of TCP biodegradation. For example, the soluble methane monooxygenase (sMMO) could be a useful target for evaluation of the aerobic cometabolism of TCP in the presence of methane as a primary substrate.

5.5 GROUNDWATER REMEDIATION TECHNOLOGIES

Although Hawaii's drinking water standard has been established for years, actual treatment of drinking water in Hawaii is not common given the relatively high value of the standard in comparison to other standards being considered and to the occurrence in drinking water. The first ng/L drinking water standard was just established in California and is being considered in other states. Researchers have actively been testing potential remedial technologies such as biodegradation, hydrolysis, chemical oxidation and chemical reduction by zero-valent metals in the lab for years. Initial data indicates some technologies may be able to achieve ng/L levels from low initial concentrations, but it remains to be seen whether it is feasible to remediate groundwater to these levels across a range of conditions and technologies.

5.5.1 IN SITU TREATMENT

Processes by which TCP may be remediated in groundwater in situ have been studied at the bench scale. There are a number of different potential degradation processes, including hydrolysis, biodegradation and chemical reduction and oxidation. However, field application of these technologies is limited, with application of several technologies limited to a few pilot studies as summarized in this section. In situ treatment is likely viable to remove mass in source zones associated with point releases, although there are potential limitations in the case where DNAPL has formed.

5.5.1.1 In Situ Hydrolysis

TCP degrades by hydrolysis. Studies by Pagan et al. indicate that base-mediated hydrolysis through the intermediate 2,3-dichloro-1-propene followed by hydrolysis to give 2-chloro-2-propen-1-ol is the dominant hydrolysis pathway (Pagan et al. 1998), Figure 5.4. The breakdown pathways of 2-chloro-2-propen-1-ol have not been well evaluated in the literature and the toxicity of this compound is not documented

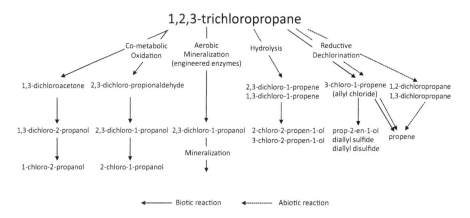

FIGURE 5.4 Likely degradation pathways and intermediates for TCP hypothesized (Sarathy et al. 2010, Tratnyek 2010) and determined experimentally (Bosma and Janssen 1998, Yan et al. 2009, Pagan et al. 1998 Schmitt, et al. 2017).

1,2,3-Trichloropropane

by the Agency for Toxic Substances and Disease Registry (ASTDR) or IRIS. Under neutral conditions, the hydrolysis yields 1,3-dichloro-propene and then 3-chloro-2-propene-1-ol, which is readily degraded to acetaldehyde, a typical metabolic intermediate (Wan 2011; McCornack et al., 2011; Gao et al., 2010).

Under typical ambient groundwater temperature and pH conditions, rates of hydrolysis are too slow for meaningful degradation, as shown on Figure 5.5. Pagan et al. measured a half-life for hydrolysis of 680 years at 25°C and pH 7 (Pagan et al. 1998). Typical groundwater temperatures are close to the mean air temperature above land surface. While the average annual global temperature is around 15°C, temperatures vary with latitude, for example in the United States average groundwater temperatures range from 5 to 25°C. Typical groundwater pH conditions range from 6 to 8.5, although acidic (near pH 6) and corrosive (near pH 8.5) groundwater

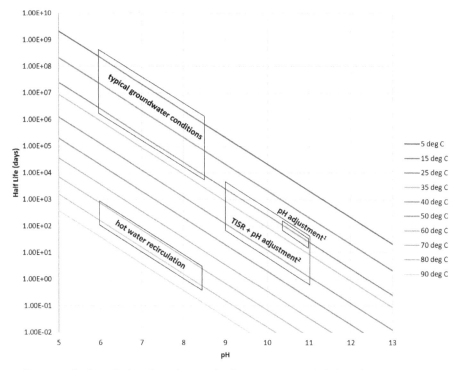

[1] Most appropriate for small subset of groundwaters with ambient temperatures at the high end of the range.
[2] Achieveable temperature depends upon initial groundwater temperature and other factors, more appropriate for higher ambient temperature groundwater.

FIGURE 5.5 TCP hydrolysis half-lives as a function of pH and temperature, modified from diagrams presented by Tratnyek P. G., et al., 2010 and Sarathy, et al., 2010. Half-life estimates based on the Arrhenius expression $k_{base} = A\, e^{-Ea/(RT)}$ for the base-mediated hydrolysis of TCP where A is the Arrhenius pre-exponential factor given by $Log_{10}A = 13.3 \pm 0.53$ (A in $M^{-1}\,s^{-1}$), Ea = 95.9 ± 3.5 kJ/mol is the activation energy (Pagan et al., 1998), R is the gas constant and T is temperature. Rate dependence on pH was calculated by the expression $k = k_{base}*[OH^-]$, where $[OH^-]$ is the hydroxide concentration, accounting for the temperature dependence of the dissociation of water to form hydroxide.

is less common than near neutral (near pH 7) groundwater. Slow ambient hydrolysis rates under typical groundwater conditions indicate this process is not a meaningful pathway for natural attenuation. However, hydrolysis rates increase to meaningful ranges at increased temperature and pH. Pagan et al. estimated Arrhenius parameters for the base-mediated hydrolysis of TCP as $\text{Log}_{10}A = 13.3 \pm 0.53$ where the pre-exponential factor, A, is units of per molar per second (M^{-1} s^{-1}) and activation energy E_a is 95.9 ± 3.5 kilojoules per mole kJ/mol (Pagan et al. 1998). Figure 5.5 presents estimated hydrolysis half-lives as a function of pH and temperature.

The isotherms on Figure 5.5 illustrate that rates may increase meaningfully with pH. For instance, at 25°C, the half-life may decrease from several hundred thousand days at pH 7 to less than 100 days above pH 10. This decrease in half-life suggests that degradation can be stimulated through addition of alkalinity to increase pH. However, increasing hydrolysis rates by increasing pH is most viable for higher ambient groundwater temperatures on the very high end of the typical range, e.g. 25°C, as shown on Figure 5.5, rather than more typical groundwater temperature of 15°C. In addition, titration of groundwater pH within an aquifer system is challenging, given the buffering capacity of aquifer soils and the on-going flux of groundwater with ambient pH and acidity through a potential target treatment zone. pH adjustment may also produce other undesirable side effects, such as dissolution of metals from aquifer minerals, that must be managed. In addition to testing of buffering capacity of the aquifer, evaluation of pH adjustment to enhance hydrolysis as a potential remedial strategy should include careful consideration of the timescales over which pH adjustment would be needed to obtain complete degradation, the amount of base that would be needed to maintain the target pH over those timescales, and the method for sustained delivery.

As illustrated on the figure, as temperature increases at a neutral pH of 7, half-lives decrease from several hundred thousand days at 25°C to less than 100 days at 70°C and to several days at 90°C. Traditional heating technologies, such as thermal conductive heating or electrical resistive heating, are typically optimized for mass recovery and are designed to achieve temperatures at or above the boiling point of water. These approaches are very energy intensive given the heat capacity of the subsurface. To increase the temperatures to less than the boiling point for enhanced hydrolysis rates, a much simpler and less resource- intensive approach can be taken, such as hot water recirculation. Cost-efficient, lower energy requirement in-line heaters can be coupled with groundwater recirculation to increase aquifer temperatures, as shown on Figure 5.5. This approach was used to increase aquifer temperatures to enhance the hydrolysis rates for TCA at a site in North Carolina (Suthersan et al. 2012). In this case study, a 108 kW electric water heater was used to heat recirculated groundwater up to 77°C. Following the heating, concentrations of TCA decreased from 8 milligrams per liter (mg/L) to less than 1 μg/L, indicating enhanced decomposition of TCA.

Another proprietary method of low temperature in situ thermal remediation under development is thermal in situ sustainable remediation (TISR™). In TISR™, solar collectors and a closed loop heating system are used to increase subsurface temperatures by thermal conduction and advection. Solar collectors can increase temperatures of effluent water to 90 to 140°C for use in a borehole heat exchanger. Achievable temperatures within the aquifer will depend upon starting temperature and other properties affecting heat transfer. Aquifer temperatures of 35 to 40°C are theoretically achievable, but lower temperatures would be achievable for lower starting temperatures at more northern latitudes. For TCP, this temperature increase would enhance hydrolysis rates to a meaningful range for an alkaline aquifer (e.g., pH 9) or could be used in conjunction with base addition to raise pH for sites with more neutral ambient pH.

5.5.1.2 In Situ Biological Treatment

Biological treatment for TCP is in a nascent state: laboratory testing indicates it may be a viable option, but field testing is limited. Biodegradation of TCP has been documented for aerobic environments by direct and cometabolic mechanisms and for anaerobic environments under controlled laboratory conditions.

5.5.1.2.1 Aerobic Cometabolic Degradation

Aerobically, methanotrophic and ammonia oxidizing bacteria have been shown capable of co-metabolizing TCP via monooxygenase enzymes with wide-range substrate specificities. Bosma and Janssen (1998) described the co-metabolic oxidation of TCP (Bosma and Janssen 1998), as well as other lesser chlorinated chloropropanes, in resting cell suspensions of *Methylosinus trichosporium* OB3b, an sMMO-expressing methanotroph. The transformation of TCP by *M. trichosporium* was tested with mg/L levels of TCP and followed first-order kinetics. Studies showed there was transformation capacity for TCP with turnover-dependent inactivation of the sMMO (Bosma and Janssen 1998), resulting in slower transformation of TCP than other lesser chlorinated propanes. Biological oxidation of TCP yielded a combination of 2-chloro-1-propanol, 1,3-dichloro-2-propanol, 2,3-dichloro-1-propanol, and chloride, (Bosma and Janssen 1998). The latter steps in the transformation pathway to the mono chlorinated alcohols and beyond are not well understood and are potentially reductive mechanisms (Bosma and Janssen 1998). A more complete understanding of the longevity of these products and their toxicity is needed as laboratory studies transition to the field scale. Cell suspensions of the ammonia-oxidizing bacterium *Nitrosomas europaea* were also capable of catalyzing the ammonia monooxygenase-mediated, co-metabolic oxidation of TCP (Vannelli et al. 1990). In a more recent study, propane oxidizing bacteria were observed to co-metabolically oxidize TCP (Wang and Chu 2017).

Although co-metabolic transformation of TCP has been demonstrated in the lab, it is uncertain if this approach will work in the field. The transformation kinetics associated with the co-metabolic oxidation of TCP are over an order of magnitude slower than those associated with co-metabolic conversion of TCE (Bosma and Janssen 1998) and are comparable to TCA.

Kinetics at µg/L and ng/L concentration ranges are less understood than at the mg/L level. To employ in the field, the primary substrate, e.g. methane, and oxygen would have to be distributed throughout the volume of aquifer where biological activity is targeted.

5.5.1.2.2 Aerobic Mineralization

Aerobic mineralization (hydrolytic dehalogenation) of TCP has also been documented as a result of the engineering of haloalkane dehalogenase enzymes known to catalyze the hydrolytic removal of chlorine from compounds structurally similar to TCP. Bosma et al. (2002) described the directed evolution of the haloalkane dehalogenase (DhaA) from *Rhodococcus* sp. m15-3, which hydrolyses carbon-halogen bonds in a wide range of haloalkanes, and its subsequent expression in the 2,3-dichloro-1-propanol-utilizing bacterium *Agrobacterium radiobacter* (Bosma et al. 2002). The active site of the evolved (mutant) DhaA was smaller in size, resulting in increased preference for smaller molecules such as TCP and, thus, increased TCP-dehalogenase activity. This modified DhaA catalyzed the transformation of TCP to 2,3-dichloro-1-propanol, which was then easily mineralized by the host organism *A. radiobacter*. Similarly, Monincová et al. (2007) identified haloalkane dehalogenase LinB from *Sphingobium japonicum* UT26 as a potential target for protein engineering to develop a new biocatalyst for the hydrolytic dehalogenation of TCP.

Although far from ready for implementation as an in situ approach, these studies suggest that directed evolution (engineering) of natural biocatalysts can become a powerful tool for the design and development of in situ biodegradation approaches for TCP, as well as other emerging contaminants (Parales et al. 2002). In situ implementation of such an approach would entail the insertion of the engineered TCP-dehalogenase enzyme into a microbial host capable of its expression, and the subsequent introduction of the modified microorganism into the contaminated subsurface (i.e., bioaugmentation). While bioaugmentation approaches are generally accepted in the United States, the introduction of genetically modified organisms into the subsurface might meet public resistance. Use of the genetically modified organism may be more readily acceptable if used in an ex situ treatment system with post-treatment disinfection.

5.5.1.2.3 Anaerobic Dechlorination

The anaerobic dehalogenation of TCP has been observed in anaerobic microcosms prepared with sediment from Red Cedar Creek in Okemos, Michigan, in anaerobic slurries prepared with material from a slow-moving stream located near Nieuwersluis, The Netherlands, and in an anaerobic bioreactor inoculated with Saale River sediment (Germany) (Peijnenburg et al. 1998) (Loffler, et al. 1997; Hauck and Hegemann 2000). More recently, reductive dechlorination of TCP has been confirmed in cultures of select species of the *Dehalogenimonas* genus, the closest cultured relatives of *Dehalococcoides*, although sharing only 90 percent 16S rRNA gene sequence similarity. The TCP-dechlorinating *Dehalogenimonas* spp. utilize hydrogen as electron donor and produce allyl chloride (3-chloro-1-propene) via dehaloelimination when grown with TCP as electron acceptor (Yan et al. 2009; Bowman et al. 2013; Mukherjee et al. 2014), as shown in Figure 5.4. In practice, the hydrogen utilized by *Dehalogenimonas* could be produced within a microbial community by other organisms that ferment commonly applied organic carbon substrates, such as sugars or vegetable oil. Allyl chloride is a transient intermediate that reacts abiotically to form allyl alcohol, diallyl sulfide, or diallyl disulphide, and potentially other sulfide-bearing compounds in the presence of cysteine or sulfide (Yan et al. 2009). Allyl alcohol is toxic, but readily biodegradable, while the sulfide-bearing compounds are metabolites (Yan et al. 2009). Results from field testing of bioaugmentation approaches show production of propene (Schmitt et al. 2017), suggesting an alternative biological reduction pathway for allyl chloride to propene, as shown on Figure 5.4. The toxicity of propene is not documented by the ASTDR or IRIS.

Dehalogenimonas spp. have been identified in the commercial bioaugmentation consortium KB-1® plus, available from SiREM (Schmitt et al. 2017). Laboratory testing of the KB-1® consortium has demonstrated TCP degradation ability over TCP concentrations ranging from 10 to over 10,000 µg/L (Schmitt et al. 2017).

Albeit limited, recent field experiences have demonstrated the potential for in situ biodegradation to remediate TCP contamination. In one case study, HRC™, a slow-release carbon donor, was injected into a TCP-impacted aquifer via direct push technology (Schmitt 2016). The injections resulted in two to four order-of-magnitude decreases in TCP concentrations within approximately three years from the injection event. Conversely, in a second case study, biostimulation with lactate and emulsified vegetable oil did not produce significant reductions in dissolved-phase TCP concentrations (Schmitt 2016). Bioaugmentation with the commercial bacterial consortium KB-1® plus with lactate and emulsified vegetable oil was implemented following the ineffective biostimulation attempt. TCP degradation was observed from initial concentration of approximately 1 µg/L to below the laboratory reporting limit of 0.005 µg/L (5 ng/L), accompanied by the generation of propene and a transient increase in the gene copies of *Dehalogenimonas* (Schmitt et al. 2017). These results from these couple of field studies suggest potential viability of in situ anaerobic biological treatment for TCP to low levels.

5.5.1.3 In Situ Chemical Reduction

TCP is amenable to in situ chemical reduction. The postulated reductive dechlorination pathways for TCP include hydrogenolysis and β-elimination as shown in Figure 5.4 (Tratnyek et al. 2008).

The two most well-tested potential chemical reductants for TCP are the metals iron and zinc. Testing of iron minerals including Fe(III) and Fe(II/III) oxides (goethite and magnetite) indicate very slow or little degradation. In contrast, faster reduction with pseudo-first order kinetics has been observed with zero-valent iron (ZVI) as well as zero-valent zinc (ZVZ). Surface normalized and mass normalized first-order rate coefficients for ZVI and ZVZ with common chlorinated contaminants and TCP from various studies are presented in Figure 5.6. Early batch studies conducted in deionized water with reagent-grade ZVZ indicated faster TCP degradation rates by ZVZ than by ZVI (Sarathy et al. 2010; Hui et al. 2015), with surface normalized reaction rates (k_{sa}) of TCP measured 1 to 2.5 orders of magnitude greater by ZVZ (10^{-2} to 10^{-3} liters per meter squared per hour [L/m²/hr]) than by ZVI (10^{-5} to 10^{-6} L/m²/hr) (Sarathy et al. 2010). Figure 5.6 depicts the normalized reaction coefficients for TCP with ZVI as open diamonds in comparison with the reaction coefficients for TCP reaction with ZVZ in batch experiments by Sarathy et al. (2010) in closed diamonds. Based on these results, more recent research has focused on ZVZ, evaluating industrial-grade ZVZ with groundwater in batch, column, and pilot scale systems (Salter et al. 2010; Salter-Blanc and Tratnyek 2011; Salter-Blanc et al. 2012).

Table 5.7 provides information on two industrial-grade zinc products: Zn 1210 powder, which is micron sized, and Zn 64 dust, which is nanometer sized. The reaction rates of these industrial products are comparable to reagent grade in DI water

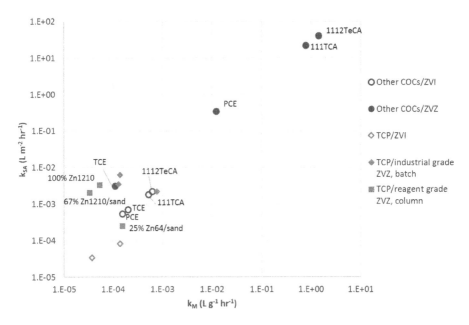

FIGURE 5.6 Mass normalized rate constants (k_M) and surface-area normalized rate constants (k_{SA}) for the reduction of TCP and other chlorinated compounds. Figure adapted from Sarathy et al. 2010 and Salter-Blanc et al. 2012. TCP rates from Sarathy et al. 2010 and Salter-Blanc et al. 2012 and Focht 1994, with values from Sarathy et al 2010 referenced from Arnold and Roberts 1998 and Arnold et al. 1998.

with and without oxygen (Salter et al. 2010; Salter-Blanc and Tratnyek 2011). Reaction rates are slower in groundwater at alkaline pH due to the formation of a film on the zinc that passivates the surface (Salter-Blanc and Tratnyek 2011). However, surface normalized reaction rates derived from long-term column testing of mixtures of Zn 64 and Zn 1210 with groundwater of 3.3×10^{-3} to 2.5×10^{-4} L/m²/hr (shown as squares in Figure 5.6) have been reported, which are comparable to surface normalized reaction rates of chlorinated alkanes and chlorinated alkenes with ZVI (shown as open gray circles on Figure 5.6) and sufficient for further consideration for in situ remediation and potentially drinking water treatment (Salter-Blanc et al. 2012).

TABLE 5.7
ZVZ properties

Type	Mesh	Particle Size	Specific Surface Area (m²/g)	Bulk Density (g/cm³)
Zinc dust 64	Through 325	<40 nm	0.620	2.60
Zinc powder 1210	20 to 60	<850 µm	0.016	2.34

Abbreviations: nm = nanometer, m²/g = meter squared per gram, g/cm³ = grams per cubic centimetre
Source: Adapted from Salter-Blanc et al 2012.

In situ remediation approaches using zero-valent metals require the placement of reactive materials in the subsurface to contact groundwater with adequate residence time to degrade the contaminant to cleanup goals. Common approaches for placement of reactive materials include construction of permeable reactive barriers (PRBs) within a trench, in situ mixing, and injection-based placement, which may also be considered a form of PRB. Trench installation and injection-based placement are discussed here.

For PRBs within a trench, aquifer materials are removed from the trench and backfilled with reactive media and fill to create the PRB. For general design principles on PRBs, readers are directed to other references (ITRC 2011; Suthersan et al. 2016). To achieve adequate residence time, the design of a PRB needs to consider groundwater velocity, contaminant flux, reaction kinetics, and cleanup goals, among other factors, to design a PRB of adequate width to provide the needed residence time for complete reaction. Kinetics from the literature were summarized earlier and can also be determined on a site-specific basis. Table 5.8 provides estimates for the residence times that would be required to reduce concentrations of TCP with ZVZ, and also provides residence times with ZVI for comparison. The estimates provided in Table 5.8 assume common design parameters of 30 percent total porosity and 30 percent metal content in fill (with the other 70 percent being structural fill, e.g., sand).

Although field experience with injection of ZVZ is limited, we can draw on experience with ZVI to ascertain the feasibility of distributing ZVZ in the subsurface via injection-based approaches. Similar to ZVI, ZVZ at the 0.1 to 1 millimeter (mm) scale (e.g., Zn 1210) is anticipated to clog injection wells and result in poor distribution. Other approaches to injection of ZVZ slurries include delivery via direct push

TABLE 5.8
PRB residence times for TCP concentration reduction

		ZVZ	ZVI
Surface area/mass[1]	m^2/g	0.62	1.68
Surface area per volume	m^2/L	247	673
k_{SA}[1]	$L/m^2/hr$	3.50E-03	8.10E-05
k	Hr^{-1}	0.86	0.05
Half-life	hr	0.80	12.71
Residence time to achieve reduction of:			
3 orders of magnitude	day	0.3	5.3
4 orders of magnitude	day	0.4	7.0
5 orders of magnitude	day	0.6	8.8
[1]Material and Reference		Zinc dust 64 Salter-Blanc et al. 2012	Fe, Hepure Technologies Sarathy et al. 2010

Notes: Residence times estimated assuming total porosity of 30% and metal composition of 30% by mass.
Abbreviations: m^2/L = meters squared per liter, hr^{-1} = per hour, hr = hour.

injection, potentially coupled with pneumatic fracturing or hydraulic fracturing to create fractures for distribution. Distribution of microscale particles can still be limited with direct push and fracturing approaches, and greater success for ZVI distribution has been achieved with nano-scale ZVI and emulsified ZVI, indicating that injection of Zn 64 may be more successful than Zn 1210.

In one case study, ZVZ injections using Zn 1210 were conducted in a pilot test for TCP treatment at a field site in southern California (Suchomel et al. 2015). Injections were conducted via direct push with fracturing. Following observations of surfacing of ZVZ with pneumatic fracturing, injections were conducted with hydraulic fracturing and guar gum was added (Suchomel et al. 2015). Observation of distribution of ZVZ to 10-foot dose response wells in the pilot test varied, with limited distribution of zinc observed in one location (Suchomel et al. 2015). These results illustrate the anticipated challenge of controlling and ensuring relatively homogeneous distribution by fractured placement of particulate reagents and should be avoided unless other placement techniques and remedy options are not available. Regardless, TCP concentrations were reduced by over an order of magnitude to the 10 ng/L range in the location where zinc distribution was observed in the case study (Suchomel et al. 2015).

5.5.1.4 In Situ Chemical Oxidation

TCP is amenable to in situ chemical oxidation (ISCO). For an overview of chemical oxidation, reagents, and design consideration the reader is referred to other resources (Suthersan et al. 2016), and specifics for TCP are presented here. Bench study results show that oxidation of TCP is negligible with mild/specific oxidants like permanganate, while stronger oxidants, especially involving the nonspecific hydroxyl and sulfate radicals, oxidize TCP (Tratnyek et al. 2010). A summary of effectiveness of various chemical oxidants based on bench testing results reported in the literature are provided in Table 5.9. The limited number of bench-scale studies available indicate that TCP reacts with catalyzed hydrogen peroxide, ozone, activated persulfate and permanganate (Tratnyek et al. 2010; Suchomel 2013; Hui et al. 2015; Hunter 1996).

Persulfate is a particularly promising oxidant. Persulfate can be activated by several different methods. Activators tested for TCP oxidation with persulfate have included heat, alkalinity, and iron. Preliminary studies indicate that persulfate oxidizes TCP rapidly when activated by heat. Half-lives of four hours under anoxic conditions and 1.8 hours under oxic conditions were measured and conversion to carbon dioxide and chloride were observed (Tratnyek et al. 2010). Alkaline activation has also been shown reactive toward TCP (Suchomel 2013). It should be noted that at elevated temperature and pH under heat and alkaline activation, the rates of hydrolysis increase and may account for observed degradation. TCP was oxidized by iron-activated persulfate in one bench study, but was not in a different bench study (Suchomel 2013).

As with any ISCO application, the selection of chemical oxidant will depend on factors such as the required residence time and distribution specifications. Given the variation of efficacy and oxidant demand with varying site conditions, bench-scale studies are always recommended to determine site-specific applicability, reagent dosing, and reaction kinetics for design.

TABLE 5.9
Summary of oxidants for chemical oxidation of TCP

Oxidant	Resultant Reactive Species[4]	Results of Bench Testing for TCP	Observed Intermediates and End Products
Permanganate	MnO_4^-	Negligible reactivity, very slow kinetics[1]	Not applicable
Persulfate			
Heat activation	$SO_4^{\cdot-}$	Reactive, favorable kinetics when formation of sulfate radical is efficient[1]	Complete mineralization to CO_2 and chloride indicated
Alkaline activation	$SO_4^{\cdot-}$	Reactive[2]	Not presented
Iron activation	$SO_4^{\cdot-}$	Uncertain reactivity[2]	Not presented
Catalyzed hydrogen peroxide	$OH^\cdot, O_2^{\cdot-}, HO_2^\cdot, HO_2^-$	Reactive, but kinetics slow and unfavorable[3]	1,3-dichloroacetone, chloroacetic acid, formic acid
Ozone	$O_3, OH^\cdot$	Reactive[2]	Not presented

Sources:
1. Tratnyek et al. 2010.
2. Suchomel 2013.
3. Hunter 1996.
4. Huling & Pivetz 2006.

Although bench-scale ISCO testing indicates ISCO is a viable option for in situ treatment of TCP, field-scale demonstrations are limited, and it has not been determined if this technology is capable of treatment to ng/L levels. Field-pilot testing is recommended prior to full-scale implementation.

5.5.2 Ex Situ Treatment

TCP is amenable to ex situ treatment, commonly known as the pump and treat strategy. For ex situ treatment, impacted groundwater is captured by extraction wells and treated in an aboveground treatment system. A few different treatment technologies can address TCP for ex situ treatment, with GAC being the best established. These technologies are discussed in detail in section 5.7. Pump and treat systems have been widely used to remove mass from groundwater and contain contaminant plumes, but limitations for pump and treat systems to remove residual mass that is strongly sorbed or that has migrated into less-mobile pore space in older plumes has also been widely recognized. TCP sorbs moderately to aquifer materials, similar to TCE (Table 5.2) and can also form DNAPL. This behavior indicates that pump and treat remedies are not likely effective for source zones of TCP and is more likely applicable to the dilute portions of groundwater plumes. It is unclear whether remediation of large, dilute TCP plumes will be required. Point of use treatment in those situations may be more viable.

5.6 WATER TREATMENT

TCP has not traditionally been a specific target for removal from wastewater or drinking water. Data on the potential removal of TCP in wastewater treatment are limited. A study of data collected from nine water treatment plants in Hawaii indicated removal of 57 to 96 percent of TCP in seven of the nine plants (Hooker et al. 2012) while data from two of the plants indicated a 2 to 5 percent increase in concentrations. Average influent concentrations at the seven plants demonstrating removal ranged from 61 to 2,400 ng/L (Hooker et al. 2012). The removal at these plants resulted in average effluent concentrations ranging from 20 to 110 ng/L (Hooker et al. 2012), well above the California MCL of 5 ng/L but well below the Hawaii MCL of 600 ng/L. Potential removal of TCP in wastewater treatment could occur through sorption to solids in the removal of suspended solids, volatilization in aerated activated sludge, or biodegradation in the aerobic activated sludge or anaerobic sludge digestion. However, these processes will likely be limited by the moderate sorption and low volatility of TCP (Table 5.2) and relatively slow biodegradation kinetics (section 5.5.1.2).

Drinking water treatment unit processes typically focus on removal of suspended solids, bacteria, algae, viruses, fungi, and minerals such as iron and manganese. Of conventional water treatment processes, TCP could potentially be removed through solids separation, although sorption to organic materials is only moderate. Of the advanced unit processes that may be used at drinking water treatment facilities, GAC, and Advanced Oxidation Processes (AOPs) can treat TCP and are discussed in section 5.7. Other advanced processes, such as aeration towers, are not typically completely effective for TCP removal and would need to be supplemented for drinking water treatment (Ozekin 2016).

5.7 TCP TREATMENT TECHNOLOGIES FOR DRINKING WATER TREATMENT AND EX SITU GROUNDWATER REMEDIATION

Several treatment technologies are available for TCP, based on the physical and chemical properties and reactivity of the compound. GAC is currently the BAT for TCP and is discussed in depth here. Other sorptive methods, such as resins, have not yet been developed for TCP. Relatively less is understood about chemical destruction via AOPs, but testing indicates promise as discussed here. The volatility of TCP is low in comparison to chlorinated alkanes and alkenes, and as such, removal via air stripping is not likely to be effective for complete treatment. Other processes, such as the use of zero-valent metals or biological treatment technologies are potentially viable as indicated by basic science demonstrating the potential, but have not been proven out for TCP.

5.7.1 GRANULAR ACTIVATED CARBON

GAC is the BAT for treatment of TCP. In fact, the Water Research Foundation's State of the Science evaluation of TCP concludes that GAC is "the only viable technology for TCP removal" (Ozekin 2016).

GAC is made from various types of carbon-based materials (e.g., coal, wood, or coconut shells) that are pyrolytically carbonized and then activated to generate a crystalline, carbonaceous material with a network of internal pores and a high surface-to-volume ratio. GAC treatment is usually in the form of a fixed bed reactor in which the contaminated water is brought into contact with the GAC. Organic molecules can accumulate on the GAC surface and in the pores of the GAC (sorption) until all of the surface area is covered by organic compounds and the GAC sorption capacity is exhausted. At this point, breakthrough of the organic compounds will be observed in the GAC and the GAC must be replaced.

The capacity of GAC to absorb organic molecules depends on the type of carbon and the chemical properties of the organic compounds (e.g., size, hydrophobicity). Additionally, any organic compounds in the GAC contactor influent water will compete for sorption sites in the GAC, and the lower the concentration of the target contaminant in comparison with other organics, the lower the efficiency of removal of the target contaminant. Increasing the amount of time that the influent is in contact with the GAC (i.e., empty bed contact time [EBCT]) is one method to increase removal of a low-concentration target contaminant. Increasing EBCT can significantly increase treatment costs through increasing treatment system footprint, number of contactors, and GAC use. Site-specific testing is recommended to get the most accurate estimate of removal capacity (Chowdhury et al. 2013).

GAC contactors have been implemented for removal of TCP for drinking water treatment in many locations in the United States, including Hawaii, California, and New York (Chowdhury et al. 2016; Hooker et al. 2012). A summary of several case studies from California can be found in Table 5.10.

When TCP is compared to other organic compounds that are also likely present in the GAC influent (e.g., other contaminants and organic matter), TCP is often present at lower concentrations, creating an inherent challenge for treatment, because the other organic compounds present will compete with TCP for sorption sites and speed up breakthrough. For example, an increase in organic matter from 0.3 to 1.3 mg/L in the influent for rapid small-scale column tests (RSSCTs) was demonstrated to decrease the bed volumes (BV) before TCP breakthrough by 35 to 77 percent, depending on the GAC type (Figure 5.7; Mital 2013). This challenge is compounded by the very low regulatory standards for TCP (i.e., treatment to 5 ng/L under the California drinking water MCL). A potential low-level ng/L standard is much lower than the standards for many other co-occurring contaminants that tend to be in the µg/L range.

Additionally, the removal of TCP by GAC contactors has been demonstrated to have a mass transfer zone that is two to five times larger than those of other VOCs that are typically treated with GAC (CH2M Hill 2005), as illustrated in Figure 5.8. A longer mass transfer zone results in TCP breakthrough before the adsorption capacity of the GAC in the lower part of the bed is exhausted and increases the carbon usage rates (CH2M Hill 2005).

The sorption of TCP to GAC has been shown to range from 6.7 to 75 micrograms TCP per gram of GAC before breakthrough of 5 ng/L (Mital 2013), with higher utilization achieved at higher influent concentrations of TCP. Because of the site-specific nature of water quality and composition and concentration of organic

TABLE 5.10
Summary of case studies from California for TCP drinking water treatment

Utility	Contaminants of Concern – Average Influent	Treatment Processes	GAC Design Parameters	Notes
Valley County Water District, Baldwin, California[1]	TCP 34 ng/L TCE 70 µg/L PCE 210 µg/L CT 0.8 µg/L 1,2-DCA 0.5 µg/L 1,1-DCE 18 µg/L cDCE 9 µg/L	Air stripping GAC Ion exchange UV/hydrogen peroxide advance oxidation	EBCT = 6.1 – 8.7 min Flow = 780 gpm per vessel GAC = virgin coconut shell Changeout = after 34,000 – 46,000 BV	• Carbon changeout initiated when TCP is detected at 75% of bed depth • Air stripping at 55:1 air:water ratio removed 40–50% of TCP
City of Glendale, California[2]	TCP 17 ng/L TCE 208 µg/L PCE 26.5 µg/L CT 2.4 µg/L	Packed tower aeration GAC	EBCT = 8.6 min Flow = 625 gpm per vessel GAC = virgin bituminous coal	—
City of Shafter, California[3]	TCP 138 ng/L	GAC	Flow = 700 gpm per vessel GAC = virgin bituminous coal Changeout = 12 months	—
City of Burbank, California[4]	TCP 100 ng/L TCE – PCE –	Air stripping GAC	—	—

Sources:
1. (Sosher and O'Keefe 2007, Stetson Engineers, Inc. 2011)
2. (Vecchio 2001)
3. (Stallworth 2014)
4. (Santos 2015)

Abbreviations: CT = carbon tetrachloride, 1,2-DCA = 1,2-dichloroethane, 1,1-DCE = 1,1-dichloroethylene, cDCE = cis-dichloroethylene, min = minutes, gpm = gallons per minute, – = not available or not applicable

1,2,3-Trichloropropane

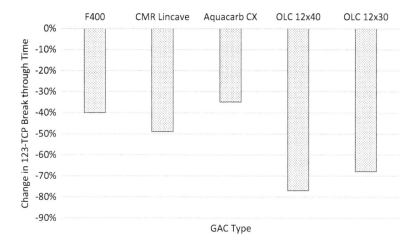

FIGURE 5.7 Change in TCP breakthrough time between Rapid Small Scale Column Tests (RSSCT) with 0.3 mg/L TOC and 1.3 mg/L TOC for various GAC types (adapted from Mital 2013).

compounds, RSSCTs or pilot tests are recommended to quantify the site-specific conditions that will impact TCP adsorption and to most appropriately size the GAC treatment system.

In comparison to other common VOC contaminants, GAC has a sorption efficiency for TCP that is greater than for some compounds (e.g., 1,1-dichloroethane). In one pilot study, 30 percent breakthrough was observed prior to 24,000 bed volumes for a range of VOCs using various GAC types, while TCP, perchloroethene (PCE),

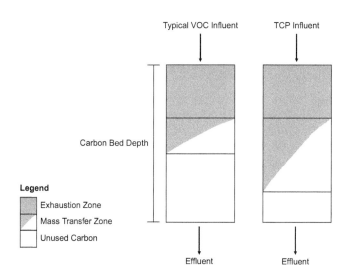

FIGURE 5.8 Difference in mass transfer zone between a typical VOC influent and a TCP-containing influent.

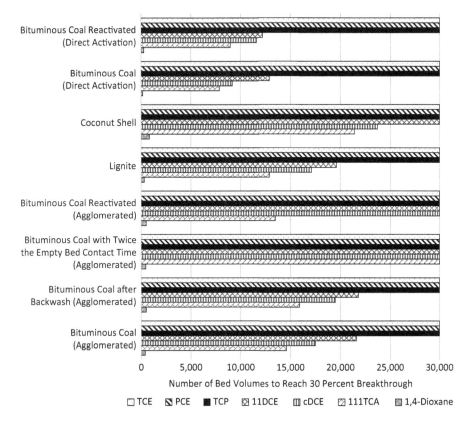

FIGURE 5.9 Bed volumes before 30 percent breakthrough for VOCs and various GACs (adapted from Fotta 2012).

and TCE did not break through (Figure 5.9; Fotta 2012; background TOC 0.3 mg/L). Similar results were observed for RSSCT experiments conducted with the same GAC types and influent (Fotta 2012).

Similar results were observed in a separate pilot study conducted for a Water Research Foundation Project (Chowdhury et al. 2016). The pilot columns were virgin, direct-activated bituminous GAC, with VOC influent concentrations as summarized in Table 5.11. Breakthrough of TCP was observed after 1,1-DCA, 1,2-dichloroethane (1,2-DCA), and carbon tetrachloride (CT) but before TCE and PCE (Figure 5.10; Chowdhury et al. 2016). In this pilot, 11 different GAC types were tested, including various coal, lignite, and coconut shell-based GACs. Breakthrough of TCP was only observed for coal-based GACs, suggesting that lignite and coconut shell-based GACs are better alternatives for TCP removal (Chowdhury et al. 2016).

The study by Chowdhury et al. (2016) also assessed a variety of influent and regulatory scenarios to determine what parameter is most likely to control GAC replacement frequency. Under the scenarios tested, meeting a regulatory standard for the sum of total carcinogenic VOCs of less than 2 µg/L was the primary driver of GAC replacement, with TCP removal being the secondary driver. It is worth noting that

TABLE 5.11
Compounds in TCP GAC pilot influent

Compound	Average Influent Concentration (µg/L)
1,1-dichloroethane (1,1-DCA)	2.35
1,2-dichloroethane (1,2-DCA)	0.81
carbon tetrachloride (CT)	0.75
1,2,3-trichloropropane (TCP)	0.73
trichloroethene (TCE)	3.73
tetrachloroethene (PCE)	3.29

Source: Chowdhury et al. 2016.

the regulatory limit used to assess these scenarios was 30 ng/L for TCP (Chowdhury et al. 2016), which is considerably higher than the 5 ng/L MCL in California.

5.7.2 ADVANCED OXIDATION PROCESSES

AOPs can destroy most organic pollutants in water. In the AOP, highly reactive radicals are produced and oxidize organics. For general information on commercially

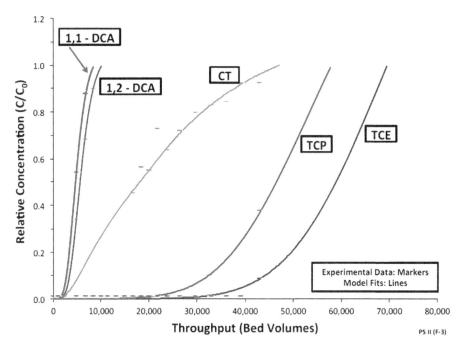

FIGURE 5.10 VOC breakthrough from pilot-scale adsorber. Source: Chowdhury et al. 2016. Reprinted with permission. © Water Research Foundation.

available AOP technologies and selection considerations, see the AOP discussion in Chapter 2, section 2.9.1 for 1,4-dioxane.

The use of an ozone and hydrogen peroxide-based advanced oxidation process has been proposed for removal of TCP from highly contaminated waters but has not been widely applied. This technique has been demonstrated to successfully reduce TCP influent concentrations of 950 ng/L to less than 5 ng/L using a dose of 53 mg/L of ozone (Dombeck and Borg 2015). The relatively high cost of this technique (i.e., the system was estimated to cost three times what a typical GAC system would cost [CH2M Hill 2005]) makes it less competitive with GAC for wide application.

5.7.3 AIR STRIPPING

Air stripping is not likely to be suitable as a primary treatment technology for TCP, given the relatively low volatility of TCP. Table 5.2 presents a comparison of Henry's law constants as a measure of volatility and viability of air stripping for TCP and other common volatile contaminants. The Henry's constant for TCP is 1-2 orders of magnitude lower than that of more readily strippable compounds like PCE and BTEX (i.e., the gasoline constituents benzene, toluene, ethylbenzene, xylenes). The lower the value the less efficient air stripping will be. For influent concentrations from 100 to 1,000 ng/L, a removal efficiency of 95 to 99.5 percent would be needed to achieve the proposed California MCL of 5 ng/L. However, performance of air strippers at ambient temperatures indicate the removal efficiency is 50 percent or less. An optimistic estimate of the air-to-water ratio required to achieve the desired removal of a given compound can be described by Equation 5.1.

$$A/W = (C_o - C_f)/(H \times C_f) \tag{5.1}$$

where:
  A/W = volumetric air-to-water ratio
  C_o = initial contaminant concentration
  C_f = final concentration
  H = Henry's law constant (unitless)

The air-to-water ratio needed to achieve 90 percent removal of TCP (unitless Henry's law constant of 0.014) is 620 compared to 23 for TCE (unitless Henry's law constant of 0.389). And when looking at 95 or 99.5 percent removal that may be needed to achieve ng/L level cleanup goals, the air-to-water ratio is 1,300 to 13,700.

However, air stripping as part of a drinking water treatment train can provide partial removal of TCP. In the Lante Plant in the Valley County Water District in Baldwin Park, California, 45 to 49 percent of TCP was removed by air stripping on average, with influent concentrations in the 25 to 60 ng/L range (Stetson Engineers 2011). In addition, air stripping can be used to remove more volatile compounds, such as TCE and PCE, prior to treatment with GAC, reducing the overall loading on the GAC system.

5.7.4 OTHER PROCESSES

Reverse osmosis can remove 0 to 85 percent of TCP depending on the membrane type (Fronk et al. 1990). If reverse osmosis is employed to achieve other treatment goals, the removal of TCP may be an extra benefit provided. However, reverse osmosis is an expensive technique that only provides partial removal of TCP and is not recommended if TCP removal is the primary treatment goal.

ZVZ has been also suggested for use in drinking water treatment (Salter-Blanc and Tratnyek 2011), but has not been fully developed and is not considered a BAT. In addition, given the ability of microorganisms to aerobically and anaerobically degrade TCP, biofilters are a potentially viable treatment option worthy of exploration.

5.8 CONCLUSIONS

The information presented in this chapter illustrates the wide range of approaches being taken by different states to apply toxicological studies on rats to form the understanding of occurrence and risk of TCP in the environment to human health. Although using the same toxicological studies, the U.S. federal IRIS program and states like Hawaii and California have used very different assumptions in extrapolating the data to assess human health risk. The resulting screening levels and drinking water standards vary by orders of magnitude, with extremely low ng/L levels established as the drinking water standard in the state of California. In addition, analytical methods and reporting limits are playing a role in the understanding of occurrence in the environment and regulatory levels proposed. In the U.S. UCMR3, an analytical method with an MDL of 30 ng/L was used and indicated that 1 percent of water supplies contain TCP above that level, but what that data mean depends on the risk assumptions and potential regulatory levels to which the data are compared. In comparison to the Hawaii MCL of 0.6 µg/L, less than a third of a tenth of a percent of the water supplies would exceed the drinking water standard. In the case of the California MCL of 5 ng/L, it is unknown how many water supplies in the United States would exceed given the elevated reporting limit used in national occurrence studies in comparison to the standard. Regardless, it is clear that the water supplies in about one-quarter of U.S. states are most vulnerable to high concentrations. While regulatory activity has been mainly on the state level, it is uncertain whether and when additional state or federal standards may be adopted.

Because the regulation of low levels where occurrence increases is brand new, there is not a lot of experience to date with implementing treatment in water supplies or for groundwater remediation. For drinking water treatment, GAC is the current BAT available for implementation, but other options for resins or biological treatment may arise as exploration to meet new standards begins. For groundwater remediation, a number of in situ and ex situ options appear viable, but whether remediation to low levels will be required, for example, for large-scale nonpoint source pollution, or will be achievable remains to be seen.

REFERENCES

ADEC. (2016a). *18 AAC 75 Oil and Other Hazardous Substanced Pollution Control. Articles 3 and 9 only*. Alaska Department of Environmental Conservation.

ADEC. (2016b). *Procedures for Calculating Cleanup Levels*. Alaska Department of Environmental Conservation, Division of Spill Prevention and Response Contaminated Sites Program.

ASTDR. (1992). *Toxicological Profile for 1,2,3-trichloropropane*. Agency for Toxic Substance and Disease Registry.

Bosma, T., Damborsky, J., Stucki, G., & Janssen, D. B. (2002). Biodegradation of 1,2,3-trichloropropane through directed evolution. *Applied and Environmental Microbiology*, 3582–3587.

Bosma, T., & and Janssen, D. (1998). Conversion of chlorinated propanes by *Methylosinus trichosporium* OB3b. *Applied Microbiology Biotechnology*, 50:105–112.

Bowman, K. S., Nobre, M. F., da Costa, M. S., Rainey, F. A., & Moe, W. M. (2013). *Dehalogenimonas alkenigegnens* sp. nov., a chlorinated-alkane-dehalogenating bacterium isolated from groundwater. *International Journal of Systematic and Evolutionary Microbiology*, 63:1492–1498.

CH2M Hill. (2005). *Interim Guidance for Investigating Potential 1,2,3-trichloropropane Sources in San Gabriel Valley Area 3*. EPA contract 68-W-98-225.

Chowdhury, Z., Porter, K. L., Francis, C., Cornwell, D., Brown, R., & Knappe, D. U. (2016). *Survey of Existing VOC Treatment Installations*. Denver, CO: Water Research Foundation.

Chowdhury, Z. K., Summers, R. S., Westerhoff, G. P., Leto, B. J., Nowack, K. O., Corwin, C. J., & Passantino, L. B. (2013). *Activated Carbon: Solutions for Improving Water Quality*. American Water Works Association.

CSWRCB. (2017a). *1,2,3-trichloropropane*. Retrieved from California Environmental Protection Agency. State Water Resources Control Board (CRWQCB): www.waterboards.ca.gov/drinking_water/certlic/drinkingwater/123TCP.shtml

CSWRCB. (2017b). *1,2,3-TCP concentrations above 150 ppt*. Retrieved from California Environmental Protection Agency, California State Water Resources Control Board: www.waterboards.ca.gov/drinking_water/certlic/drinkingwater/documents/123-tcp/123tcp_map_150ppt.pdf

CSWRCB. (2017c). *1,2,3-TCP concentrations above 5 ppt*. Retrieved from California Environmental Protection Agency, California State Water Resources Control Board: www.waterboards.ca.gov/drinking_water/certlic/drinkingwater/documents/123-tcp/123tcp_map_5ppt.pdf

CSWRCB. (2016a). *1,2,3-Trichloropropane Maximum Contaminant Level Development Process*. Retrieved from California Environmental Protection Agency. State Water Resources Control Board. Division of Drinking Water. www.waterboards.ca.gov/drinking_water/certlic/drinkingwater/documents/123-tcp/tcp_mcl_presentation.pdf

CRWQCB. (2016b). *California Environmental Protection Agency State Water Resources Control Board. 1,2,3-Trichloropropane*. Retrieved from California State Environmental Laboratory Accreditation Program. Laboratories Certified for 1,2,3-Trichloropropane: www.waterboards.ca.gov/drinking_water/certlic/drinkingwater/documents/123-tcp/123tcp_lablist.pdf

Dombeck, G., & Borg, C. (2015). Multi-contaminant treatment for 1,2,3-trichloropropane destruction using the HiPOx reactor. *Proceedings of the 2005 National Groundwater Association Conference on MTBE and Perchlorate: Assessment, Remediation, and Public Policy*.

FDHEH. (2016). *Maximum Contaminant Levels and Health Advisory Levels*. Florida Department of Health and Environmental Health.

Federal Register. (2012). *Revisions to the Unregulated Contaminant Monitoring Regulation (UCMR 3) for Public Water Systems; Final Rule. Vol. 77. No. 85 Part II*. U.S. Environmental Protection Agency 40CFR Parts 141.142.

Focht, R. M. (1994). *Bench-Scale Treatability Testing to Evaluate the Applicability of Metallic Iron for Above-Ground Remediation of 1,2,3-Trichloropropane Contaminated Groundwater. Master's Thesis*. Waterloo, Ontario, Canada: University of Waterloo.

Fotta, M. E. (2012). *Effect of Granular-Activated Carbon Type of Adsorber Performance and Scale-Up Approaches for Volatile Organic Compound Removal. Master's Thesis*. North Carolina State University.

Fronk, C. A., Lykins, B. W., & Carswell, J. K. (1990). Membranes for removing organics from drinking water. *Proceedings of 1990 American Filtration Society Annual Meeting*. Washington, D.C.

Gao, J., Ellis, L. M., & Wackett, L. P. (2010). The University of Minnesota biocatalysis/biodegradation database: Improving public access. *Nucleic Acids Research, 38*:D488-D491.

Hauck, R., & W. Hegemann. (2000). Anaerobic degradation of 1,2,-dichloropropane in batch and continuous culture. *Water Science and Technology, 41*(12): 7–13.

HDOH. (2013). *Amendment and Compilation of Hawaii Administrative Rules*. Hawaii Department of Health.

Hooker, E. P., Fulcher, K. G., & Gibb, H. J. (2012). *Report to the Hawaii Department of Health, Safe Drinking Water Branch, Regarding the Human Health Risks of 1,2,3-Trichloropropane in Tap Water*. Arlington, VA: Tetra Tech.

Hui, L., Zhan-tao, H., Chun-xiao, M., & Jian-ye, G. (2015). Comparison of 1,2,3-trichloropropane reduction and oxidation by nanoscale zero-valent iron, zinc, and activated persulfate. *Journal of Groundwater Science and Engineering, 3*(2):156–163.

Huling, S. G., & Pivetz, B. E. (2006). *In-Situ Chemical Oxidation. EPA 600-R-06-072*. Cincinnati, OH: U.S. Environmental Protection Agency Office of Research and Development National Risk Management Laboratory.

Hunter, F. (1996). Fenton's treatment of 1,2,3-trichloropropane: chemical reaction byproducts, pathway, and kinetics. *Proceedings of the Sixth International Symposium Chemical Oxidation: Technologies for the Nineties* (pp. 50–71). Nashville, TN: Technomic Publishing.

IRIS. (2009). *Integrated Risk Information System (IRIS) Chemical Assessment Summary: 1,2,3-Trichloropropane*. U.S. Environmental Protection Agency. National Center for Environmental Assessment.

ITRC. (2011). *Permeable Reactive Barrier: Technology Update*. Interstate Technology and Regulatory Council.

Loffler, F., Champine, J. E., Ritalahti, K. M, Sprague, S. J., & J. M. Tiedje. (1997). Complete reductive dechlorination of 1,2-dichloropropane by anaerobic bacteria. *Applied and Environmental Microbiology, 63*(7):2870–2875.

McCornack, M. A., McLeish, R., & Kang, J. (2011). *3-Chloroacrylic Acid Family Pathway*. Retrieved from Biocatalysis/Biodegradation Database: http://eawag-bbd.ethz.ch/caa/caa_map.html

MDH. (2015). *Human Health-Based Water Guidance Table*. Retrieved from Minnesota Department of Health. www.health.state.mn.us/divs/eh/risk/guidance/gw/table.html

MDNR. (2014). *Rules of Department of Natural Resources, Division 20—Clean Water Commission, Chapter 7—Water Quality*. Missouri Department of Natural Resources.

Mital, J. (2013). *Granular Activated Carbon Treatment of 1,2,3-Trichloropropane, Master's Thesis*. University of California, Davis.

Monincova, M., Prokop, Z., Vevodova, J., Nagata, Y., & Damborsky, J. (2007). Weak activity of haloalkane dehalogenase linb with 1,2,-trichloropropane revealed by X-ray crystallography and microcalorimetry. *Applied and Environmental Microbiology, 73*(6):2005–2008.

Mukherjee, K., Bowman, K. S., Rainey, F. A., Siddaramappa, S., Challacombe, J. F., & Moe, W. M. (2014). *Dehalogenimonas lykanthorporepellens* BL-DC-9T simultaneously transcribes many rdhA genes during organohalide respiration with 1,2-DCA, 1,2-DCP, and 1,2,3-TCP as electron acceptor. *FEMS Microbiology Letters*, 354:111–1128.

NGI. 2018. *Technical Note: Preliminary Assessment of Substances Registered under REACH That Could Fulfil the Proposed PMT/vPvM Criteria*. Document no. 20160526-TN-01, Trondheim, Norway: Norwegian Geotechnical Institute.

NJDWQI. (2009). *Maximum Contaminant Level Recommendations for Hazardous Contaminants in Drinking Water*. New Jersey Drinking Water Quality Institute.

NJDWQI. (2016). *Maximum Contaminant Level Recommendations for 1,2,3-trichloropropane in Drinking Water: Basis and Background*. New Jersey Drinking Water Quality Institute.

NTP. (2016). 1,2,3-Trichloropropane. In National Toxicology Program, *14th Report on Carcinogens*. Research Triangle Park, NC: U.S. Department of Health and Human Services.

NTP. (1993). *Toxicology and Carcinogenesis. Studies of 1,2,3,-trichloropropane (CAS No. 96-18-4) in F344/N Rats and B6C3F1 Mice (Gavage Studies)*. Research Triangle Park, NC: National Toxicology Program, U.S. Department of Health and Human Services, Public Health Service, National Institutes of Health.

NYSDH. (2011). *Part 5, Subpart 5-1 Public Water Systems—Tables. New York State Department of Health (NYSDH)*. Retrieved April 16, 2017, from www.health.ny.gov/regulations/nycrr/title_10/part_5/subpart_5-1_tables.htm

OEHAA. (2009). *Public Health Goal for 1,2,3-Trichloropropane in Drinking Water*. Pesticide and Environmental Toxicology Branch, Office of Environmental Health Hazard Assessment, California Environmental Protection Agency.

Okamoto, H. S., Dhoot, J., & Perera, S. K. (2002a). *Determination of 1,2,3-Trichloropropane in Drinking Water by Continuous Liquid-Liquid Extraction and Gas Chromatography/Mass Spectrometry*. Berkeley, CA: California Department of Health Services, Division of Drinking Water Management, Sanitation and Radiation Laboratories Branch.

Okamoto, H. S., Steeber, W. R., Remoy, J., Hill, P., & Perera, S. K. (2002b). *Determination of 1,2,3-Trichloropropane in Drinking Water by Purge and Trap Gas Chromatography/Mass Spectrometry*. Berkeley, CA: California Department of Health Services, Division of Drinking Water Management, Sanitation and Radiation Laboratories Branch.

Ozekin, K. (2016). *1,2,3-Trichloropropane State of the Science*. Water Research Foundation.

Pagan, M., Cooper, W., & Joens, J. A. (1998). Kinetic studies of the homogeneous abiotic reactions of several chlorinated aliphatic compounds in aqueous solution. *Applied Geochemistry, 13*(6):779–785.

Parales, R. E., Bruce, N. C., Schmid, A., & Wackett, L. P. (2002). Biodegradation, biotransformation, and biocatalysis (B3). *Applied and Environmental Microbiology, 68*(10):4699–4709.

Peijnenburg, W., Eriksson, L., de Groot, A., Sjostrom, M., & Verboom, H. (1998). The kinetics of reductive dehalogenation of a set of halogenated aliphatic hydrocarbons in anaerobic sediment slurries. *Environmental Science and Pollution Research, 5*(1)12–16.

Salter, A. J., Tratnyek, P. G., & Johnson, R. L. (2010). Degradation of 1,2,3-trichloropropane by zero-valent zinc: Laboratory assessment for field application. *Proceedings of the 7th International Conference on Remediation of Chlorinated and Recalcitrant Compounds, May 2010, Monterey California*. Columbus, OH: Battelle.

Salter-Blanc, A. J., Suchomel, E. J., Fortuna, J. H., Nurmi, J. T., Walker, C., Krug, T., O'Hara, S. Ruiz, N. Morley, T., & P. Tratnyek. (2012). Evaluation of zero-valent zinc for treatment of 1,2,3-trichloropropane-contaminated groundwater: Laboratory and field assessment. *Groundwater Monitoring and Remediation, 4*:42–52.

Salter-Blanc, A. J., & Tratnyek, P. (2011). Effects of solution chemistry on the dechlorination of 1,2,3-trichloropropane by zero-valent zinc. *Environmental Science and Technology*, 45:4073–4079.
Santos, M. (2015). *System 1910179—City of Burbank 2013 Sanitary Survey.* Hollywood District, Division of Drinking Water, State Water Quality Resources Control Board.
Sarathy, V., Salter, A. J., Nurmi, J. T., Johnson, G. O., Johnson, R. L., & Tratnyek, P. (2010). Degradation of 1,2,3-trichloropropane (TCP): Hydrolysis, elimination, and reduction by iron and zinc. *Environmental Science and Technology*, 44(2):787–793.
Schmitt, M., Varadhan, S., Dworatzek, S., Webb, J., & E. Suchomel. (2017). Optimization and validation of enhanced biological reduction of 1,2,3-trichloropropane in groundwater. *Remediation. 28*:17–25.
Schmitt, M. (2016). 1,2,3-trichloropropane in situ remediation technologies. *Northwest Environmental Conference.* Portland, OR.
Sosher, A., & O'Keefe, J. (2007). *Engineering Report for Consideration of an Amended Water Supply Permit for the Valley County Water District Serving the City of Baldwin Park, and Portions of the Cities of West Covina and Irwindale, Los Angeles County.* Drinking Water Field Operations Branch, California Department of Public Health, Metropolitan District.
Stallworth, S. (2014). *City of Shafter Sanitary Survey Report 1510019.* Southern California Branch, Drinking Water Field Operations, California Department of Public Health.
Stetson Engineers. (2011). *Valley County Water District 2010 Annual Technical Performance Report for Lante Plant.* Baldwin Park, CA.
Suchomel, E., McHugh, D., & Butler, H. (2015). *Pilot Study Report Zero Valent Zinc Injection for 1,2,3-Trichloropropane 22/23 Area Groundwater Camp Peddleton, CA.* Prepared for Department of the Navy, Naval Facilities Engineering Command Southwest.
Suchomel, E. (2013). 1,2,3-trichloropropane occurrence, fate, transport and remediation. *New Webinar Series on Problematic Contaminants.* Midwest Geosciences Group.
Suthersan, S. S., Horst, J., Schnobrich, M., Welty, N., & McDonough, J. (2016). *Remediation Engineering: Design Concepts, Second Edition.* Boca Raton, FL: CRC Press.
Suthersan, S., Horst, J., Klemmer, M., & Malone, D. (2012). Temperature-activated auto-decomposition temperature-activated auto-decomposition remediation solution. *Ground Water Monitoring and Remediation*, 34–40.
Tardiff, R. G., & Carson, M. L. (2010). Derivation of a reference dose and drinking water equivalent level for 1,2,3-trichloropropane. *Food and Chemical Toxicology*, 48:1488–1510.
Tratnyek, P. G., Sarathy, V., Salter, A. J., Nurmi, J. T., Johnson, G. O., DeVoe, T., & Lee, P. (2010). *Prospects for Remediation of 1,2,3-Trichloropropane by Natural and Engineered Abiotic Degradation Reactions.* Strategic Environmental Research and Development Program.
Tratnyek, P., Sarathy, V., and J. Fortuna. 2008. Fate and remediation of 1,2,3-trichloropropane. *Proceedings of the Sixth International Conference on Remediation of Chlorinated and Recalcitrant Compounds (Monterey, CA).* Columbus, OH: Battelle.
USEPA. (2016a). *Third Uncontaminated Monitoring Rule (UCMR 3): Data Summary. EPA 815-S-16-004.* Washington, D.C.: Office of Water, U.S. Environmental Protection Agency.
USEPA. (2016b, May). *Regional Screening Level Resident Tapwater Table (TR=10E-06, HQ=1).* Retrieved May 11, 2017, from www.epa.gov/sites/production/files/2016-06/documents/restap_sl_table_run_may2016.pdf
USEPA. (2009). *Toxicological Review of 1,2,3-Trichloropropane.* Integrated Risk Information System, U.S. Environmental Protection Agency.
Vannelli, T., Logan, M., Arciero, D. M., & Hooper, A. B. (1990). Degradation of halogenated aliphatic compounds by the ammonia-oxidizing bacterium *Nitrosomonas Europaea. Applied and Environmental Microbiology*, 56(4):1169–1171.

Vecchio, V. M. (2001). *System No. 1910043—Amended Domestic Water Supply Permit No. 04-15-00PA-000*. Los Angeles, CA: Department of Health Services, Drinking Water Field Operations Branch.

Wang, B., & Chu, K-H. (2017). Cometabolic biodegradation of 1,2,3-trichloropropane by propane-oxidizing bacteria. *Chemosphere 168*:1494–1497.

Wan, Y. (2011). *1,3-Dichloropropene Family Pathway Map*. Retrieved from Biocatalysis/Biodegradation Database: http://eawag-bbd.ethz.ch/cpr/cpr_map.html

WHO. (2017). *Guidelines for Drinking-Water Quality*. World Health Organization.

WHO. (2003). *Concise International Chemical Assessment Document 56: 1,2,3-Trichloropropane*. Geneva: World Health Organization.

Yan, J., Rash, B. A., Rainey, F. A., & Moe, W. M. (2009). Isolation of novel bacteria within the cloroflexi capable of reductive dechlorination of 1,2,3-trichloropropane. *Environmental Microbiology, 11*(40):833–843.

Acronyms

1,2-DCA	1,2-dichloroethane
1,1-DCE	1,1-dichloroethene
1,4-D	1,4-dioxane
A	Arrhenius equation pre-exponential factor
ADEC	Alaska Department of Environmental Conservation
AOP	Advanced Oxidation Process
ASTDR	Agency for Toxic Substance and Disease Registry
atm*m³/mol	atmosphere meter cubed per mol
A/W	volumetric air-to-water ratio
BAT	best available technology
BMC	benchmark concentration
BMD	benchmark dose
BTEX	benzene, toluene, ethylbenzene, xylenes
BV	bed volume
CCL	contaminant candidate list
cDCE	cis-dichloroethene
C_o	initial contaminant concentration
C_f	final contaminant concentration
CSF	cancer slope factor
CSWRCB	California State Water Resources Control Board
CT	carbon tetrachloride
DhaA	dehalogenase
DBCP	1,2-dibromo-3chloropropane
DNAPL	dense non-aqueous phase liquid
DWEL	drinking water equivalent level
DWRL	Drinking Water Radiation Laboratories
E_a	Arrhenius equation activation energy
EBCT	empty bed contact time
EDB	1,2,-dibromoethane
ELAP	Environmental Laboratory Accreditation Program
FDHEH	Florida Department of Health Environmental Health
g/cm³	grams per cubic centimeter
GAC	granular activated carbon
GC/MS	gas chromatography/mass spectrometry
gpm	gallons per minute
H	Henry's constant
HAL	health advisory level
HRL	heath risk limit
IRIS	Integrated Risk Information System
ISCO	in situ chemical oxidation
k_M	mass normalized reaction rate
k_{SA}	surface normalized reaction rate

kJ/mol	kilojoules per mole
L/m²/hr	liters per meter squared per hour
MBT	molecular biological tools
MCL	maximum concentration level
m²/g	meter squared per gram
m²/L	meters squared per liter
MDH	Minnesota Department of Health
MDNR	Missouri Department of Natural Resources
mg/kg	milligrams per kilogram
mg/kg-d	milligrams per kilogram per day
mg/L	milligram per liter
mg/m³	milligrams per meter cubed
min	minutes
mm	millimeter
mm Hg	millimeter of mercury
$M^{-1}s^{-1}$	per molar per second
MRL	method reporting limit
nm	nanometer
NGI	Norwegian Geotechnical Institute
ng/L	nanograms per liter
NJDEP	New Jersey Department of Environmental Protection
NJDWQI	New Jersey Drinking Water Quality Institute
NTP	National Toxicology Program
NYSDH	New York State Department of Health
OEHHA	Office of Environmental Health Hazard Assessment
PCE	perchloroethene
POP	persistent organic pollutants
PHG	public health goal
PMT	persistent, mobile, toxic
ppm	part per million
ppt	part per trillion
POC	principal organic contaminant
POD	point of departure
PQL	practical quantification limit
PRB	permeable reactive barrier
PWS	public water system
R	gas constant
REACH	registration, evaluation, authorization, and restriction of chemicals
RfC	reference concentration
RfD	reference dose
qPCR	quantitative Polymerase Chain Reaction
RSSCT	rapid small-scale column tests
SIM	selected ion monitoring
SIS	selected ion storage
sMMO	soluble methane monooxygenase
SRL	Sanitation and Radiation Laboratories

TCA	trichloroethane
TCE	trichloroethene
TCP	1,2,3-trichloropropane
TISR	thermal in situ sustainable remediation
TOC	total organic carbon
UBA	German Environment Agency (Umweltbundesamt)
UCMR	unregulated contaminant monitoring rules
UF_A	interspecies extrapolation factor
UF_H	intraspecies variability factor
UF_S	subchronic to chronic exposure duration extrapolation factor
UF_D	toxicological database uncertainty factor
USEPA	United States Environmental Protection Agency
vPvM	very persistent very mobile
VOC	volatile organic compound
WHO	World Health Organization
ZVI	zero-valent iron
ZVZ	zero-valent zinc
µg/L	micrograms per liter

6 Considerations for Future Contaminants of Emerging Concern

Caitlin H. Bell
(Arcadis U.S., Inc., caitlin.bell@arcadis.com)

John Horst
(Arcadis U.S., Inc., john.horst@arcadis.com)

Paul Anderson
(Arcadis U.S., Inc., paul.anderson@arcadis.com)

Norman Forsberg
(Arcadis U.S., Inc., norman.forsberg@arcadis.com)

6.1 INTRODUCTION

While the previous chapters have focused on four emerging contaminants relevant to groundwater quality that were receiving both public and regulatory attention at the time of this publication, they represent a tiny fraction of the ever-evolving list of contaminants of emerging concern (CECs). With this in mind, this chapter discusses key concepts relevant to how emerging contaminants are classified, the challenges posed in emerging contaminant management, and ideas for navigating future emerging contaminants. These topics have broad relevance to CECs, be those compounds that are currently identified as emerging contaminants or compounds that may be identified as such in the future. While we have focused the sections that follow mainly on chemical compounds, the concepts can also be extended to biological and material CECs. Developing a collective understanding around these topics can help stakeholders be better prepared to address emerging contaminants as they come onto the public stage.

6.2 CATEGORIZING FUTURE EMERGING CONTAMINANTS

As the list of emerging contaminants evolves, it is natural that we seek ways to categorize them for purposes of making comparisons. Comparing new chemicals to existing categories allows us to identify similarities and differences, make initial assumptions about environmental fate and potential risks, and compare the effectiveness of management approaches. This section discusses various approaches to categorizing emerging contaminants and points out benefits and challenges. It is meant to serve as a starting point for the reader when faced with future emerging contaminants and when looking to glean lessons learned from the past.

One approach to categorizing emerging contaminants is by intended purpose or use. Examples include categories such as pharmaceuticals, pesticides, nanomaterials, and personal care products. However, these groupings are very broad and include compounds with a variety of characteristics, environmental occurrence, and potential for exposure. This approach can make it difficult to find a particular contaminant, or may see it end up in multiple categories.

With this in mind it is sometimes possible to create a separate category for a class of compounds based on their structural similarity. Per- and polyfluoroalkyl substances (PFASs) are an example of such a category. They represent a large and diverse group of chemicals with multiple uses and a range of chemical and toxicological characteristics.

Another approach to categorizing emerging contaminants is by their potential adverse effects. A well-known example is the category of endocrine disrupting compounds (EDCs). EDCs are grouped together because they are believed to have the potential to cause adverse effects by interacting with or affecting hormonal systems in humans or animals. EDCs fall into many different use categories (e.g., pesticides, pharmaceuticals, plastics) and can have other potential human health effects in addition to endocrine disruption (e.g., developmental and carcinogenic effects). Because emerging contaminants can have adverse effects on more than one toxicological endpoint, such overlaps between "effect" categories can also create confusion.

How emerging contaminants are classified may depend on the intent of a certain publication, the relevant audience, or how the remainder of the discussion is framed. For example, it may be useful to categorize emerging contaminants by their toxicological effects if a discussion regarding human health risks is what follows. While it is not the intent of this writing to identify the "best" method for classifying emerging contaminants, it seems appropriate to acknowledge that the most common approach to classifying emerging contaminants seems to be by their intended purpose or use with noted exceptions where separate categories are created for groups of compounds with particular relevance, but many uses (e.g., PFASs).

This type of classification is presented in Table 6.1. In some cases, the compounds (or groups of compounds) that fall into these categories tend to exhibit similar chemical/physical characteristics and similar potential adverse effects (e.g., disinfection by-products). This helps streamline discussions around toxicology, fate in the environment, and treatment options. In many other cases, compounds within a category can have diverse chemical/physical characteristics and potential adverse effects. Pharmaceuticals, personal care products, and pesticides are examples of three groups that have a wide variety of physical/chemical characteristics and potential adverse effects. This approach to categorization avoids "branding" a particular compound based on assumptions regarding potential adverse effects and toxicity, when there is a chance that the assumptions may prove to be incorrect.

Considerations for Future Contaminants 357

TABLE 6.1
Example classes of emerging contaminants

Class of Emerging Contaminant	Example Compounds	Typical Environmental Presence	Example Potential Adverse Effects	Characterization/ Analysis Considerations	Treatment Considerations
PFASs	Perfluorooctane sulfonate (PFOS), Perfluorooctanoic acid (PFOA), 6:2 Fluorotelomer Alcohol (6:2 FTOH)	Groundwater, drinking water, waste water, sediment, landfills	Developmental and reproductive effects, hepatoxicity	USEPA Methods (and modifications) available for water and soil; other methods available to provide additional characterization details	Focus on granular activated carbon and treatment trains; large research efforts on in situ remediation and destructive treatment
Pharmaceuticals	Erythromycin (antibiotic), Ibuprofen (anti-inflammatory), Sertraline (antidepressant)	Domestic wastewater, surface water bodies	Pharmacological effect if dose is high enough	Analytical methods focused on quantifying active ingredients	May not be needed when below the effective dose; comprehensive treatment options are limited
Personal Care Products	Triclosan (anti-microbial), Benzophenone (sunscreen), Acetophenone (fragrance)	Domestic wastewater, surface water bodies	Varies widely	Analytical methods vary on a case-by-case basis	Varies widely and many ingredients are not the focus of development
Metals	Strontium, Vanadium, Germanium	Drinking water, groundwater	Heavy metals can damage the brain, lungs, kidney, liver, blood composition, bones, and other important organs	Dissolved versus total metals; speciation may affect toxicity	Metals cannot be destroyed; treatment technologies may include adsorption, precipitation, or reverse osmosis
Flame Retardants	Polybrominated diphenyl ethers (PBDEs), Polybrominated biphenyls (PBBs), Brominated aliphatic cyclics	Household dust, groundwater and surface water from landfill discharge	Endocrine-, reproductive- and behavioral effects	Analytical methods available for individual compounds and groups of compounds	Photodegradation, thermal incineration, and biodegradation are viable options; but degradation products may also be harmful (e.g., dioxins)

(*Continued*)

TABLE 6.1 (*Continued*)

Class of Emerging Contaminant	Example Compounds	Typical Environmental Presence	Example Potential Adverse Effects	Characterization/ Analysis Considerations	Treatment Considerations
Disinfection By-products	Bromochloroacetic acid, Tribromoacetic acid, N-nitrosodimethylamine (NDMA)	Drinking water	May be associated with increased risk of bladder cancer	USEPA methods available	Focus has been on limiting generation, rather than treatment after formation
Cyanotoxins	Microcystin-LR, Cylinderspermopsin	Surface water bodies, drinking water	Headache, vomiting, diarrhea, among others	Field screening kits widely used, quantitative methods also available, special handling considerations apply	Conventional water/ wastewater treatment processes effectively remove cyanotoxins/cells
Pesticides	α-Hexachlorocyclohexane, Permethrin, Tribufos	Soil, groundwater	Many have endocrine disrupting qualities	Well-developed analytical methods for a small list of pesticides and components	Many instances of disperse soil distribution require excavation; chemical and biological destruction in water
Nanomaterials	Carbon nanotubes, Titanium dioxide, Quantum dots, Dendrimers, Nanosilver	Domestic wastewater, groundwater and surface water from landfill discharge	Respiratory diseases, neurodegenerative diseases, and oxidative stress	Current methods focus on particle size; methods related to shape, surface area, and reactivity are needed	Partial removal through conventional water/ wastewater treatment processes; innovative methods subject of current research
Microplastics	Microbeads, Secondary microplastics from physical break down of larger polyethylene, polypropylene, or polystyrene products	Surface water bodies (particularly the ocean), drinking water, domestic wastewater	Unknown	Quantification currently focuses on number of plastic fragments present and/or particles size	Depending on size, conventional water/ wastewater treatment processes can be effective; molecules sorbed to plastics may require treatment

6.3 THE CHALLENGES POSED IN EMERGING CONTAMINANT MANAGEMENT

Many times, the state of the science is not well developed when awareness of an emerging contaminant is rising, and management decisions need to be made. For example, the analytical methods used to quantify an emerging contaminant in environmental media may still be developing. Likewise, the understanding of a chemical's risk to human health and the environment takes time to develop. However, many times there is a need for regulatory action to be taken—with a trickledown effect on various stakeholders (e.g., responsible parties, the public, regulatory bodies, advocacy groups). These stakeholders may be faced with the need to develop and implement management strategies ahead of the data. This is where a decision to implement a proactive or reactive management strategy may come into play (see section 6.4 for more detail). The next few sections discuss the challenges that may be faced by various stakeholders for current and future emerging contaminants. While they are similar to some of the concepts in Chapter 1, the discussion here focuses on the very early stages of developing an understanding of emerging contaminants.

6.3.1 Challenges Associated with Release to the Environment

One of the first issues explored with any emerging contaminant is how it enters the environment, so the source(s) can be addressed. Many times, emerging contaminants are associated with both manufacturing and products in the public marketplace. Entry points to the environment are not just industrial facilities, but also locations where products are used and disposed. Consequently, emerging contaminants are exposure considerations for consumers as well as industrial or commercial entities that use them. For example, dermal contact with emerging contaminants for consumers may occur through clothing or personal care products, or inhalation may occur from building materials at home or in the workplace.

Ultimately, there are many pathways for release of chemicals (including future emerging contaminants) into the environment, and many of these pathways are interconnected. The following are some of the routes of entry to the environment that can affect drinking water supplies, with the interrelation of many of these shown in Figure 6.1:

- Industrial wastewater discharges to groundwater, surface water, and combined sewer
- Public wastewater treatment plant discharges to surface water
- Industrial release to soil, groundwater, or surface water
- Sustainable reuses (rain water reuse, gray water)
- Landscape irrigation from groundwater sources or public water supply
- Crop irrigation from groundwater or surface water sources, or public water supply
- Human-made recreational waters
- Constructed wetlands
- Aquaculture
- Dust control

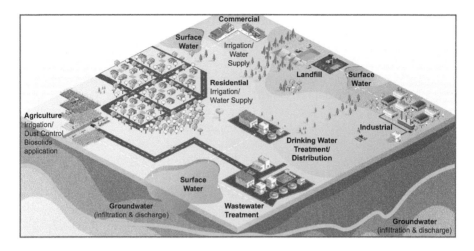

FIGURE 6.1 Example routes of entry to the environment for emerging contaminants: Visual depiction of example routes of entry into the environment for emerging contaminants. The focus here is on discharges/intakes from groundwater and surface water for both industrial and residential/commercial sources (Adapted from Raghav et al., 2013)

- Use of recycled water to augment surface water reservoirs
- Biosolids application to farmland
- Landfill discharge to groundwater and surface water

The variety and interconnectedness of these release pathways can result in a wide distribution of a chemical in the environment over a brief period of time. This is challenging for future emerging contaminants, because awareness of potential human health risks may come after the compound has made its way into the environment via these (and other) release mechanisms. Likewise, it may take time to develop reliable methods for quantifying an emerging contaminant. This can result in a delay in understanding its occurrence in environmental media.

6.3.2 Challenges Associated with Assessing Toxicological Risk

Emerging contaminants pose a unique challenge to the hazard, toxicity, and risk characterization process, for both new substances coming into the marketplace as well as existing compounds recently detected in the environment. In the latter case, human and ecological exposure may already be occurring with unknown consequences. Thus, the critical challenge is to rapidly generate the appropriate type and quality of toxicity data with which to understand quickly if existing exposures pose an unacceptable risk. For this reason, the testing approach should be objective and expedient such that potential risks can be quickly determined and appropriately managed.

New compounds being evaluated for the marketplace (e.g., industrial chemical, pharmaceutical, cosmetic, etc.) are generally considered for potential hazards, toxicity, and risks to determine the most promising to synthesize and bring to market. The goal of these evaluations is to identify the compounds that pose the best performance for their intended application and pose the least likelihood of unintended adverse effects to people and the environment, as demonstrated by compliance with applicable regulatory requirements. Thus, the toxicological evaluation of emerging contaminants should leverage the same accepted testing strategies used to bring new chemicals to market.

In the United States, regulatory agencies such as the Food and Drug Administration (FDA) and the United States Environmental Protection Agency (USEPA) are charged with evaluating the hazards, toxicity, and potential risks of new chemicals. The FDA is responsible for registration of pharmaceuticals, food additives, and cosmetics, while the USEPA regulates pesticides and industrial chemicals. Similar regulatory agencies exist in Europe (e.g., the European Chemicals Agency [ECHA], that oversee the Registration, Evaluation, Authorisation, and Restriction of Chemicals program), Japan, Australia, and others. All of these entities have developed guidelines for evaluating the safety of new chemical compounds. Many of these agencies commonly rely on national or harmonized approaches provided by the Organization for Economic Cooperation and Development (OECD), International Council for Harmonisation of Technical Requirements for Pharmaceuticals for Human Use (ICH), and International Cooperation on Harmonisation of Technical Requirements for Registration of Veterinary Medicinal Products (VICH), and others. These same guidelines and the approaches described therein can be readily applied to evaluate the toxicity and potential risk of future emerging contaminants that may have limited or no information on toxicity.

In the absence of any empirical toxicity testing information, a prudent approach is to initially rely on computational or predictive approaches (i.e., in silico methods) to predict an emerging contaminant's toxicological profile. This process consists of using (Quantitative) Structure Activity Relationships [(Q)SAR], analog, and read-across methods to predict important metrics of toxicity from the physical characteristics of the structure of chemicals, specifically evaluating the new chemical for structural alerts that have been associated with toxic outcomes in other tested chemicals. Publicly available tools for supporting this effort include USEPA's Toxicity Estimation Software Tool (TEST) (USEPA, 2016), ECOSAR (USEPA, 2011), EPI Suite™ (USEPA, 2012), OncoLogic™ (USEPA, 2010), and Analog Identification Methodology (AIM) (USEPA, 2013); FDA's Thresholds of Toxicological Concern (TTC) (FDA, 2018b); OECD's QSAR Toolbox (OECD, 2009); the Istituto di Ricerche Farmacologiche Mario Negri Milano's Virtual models for property Evaluation of chemical within a Global Architecture (VEGA, 2018); and the National Library of Medicine's ChemIDplus (National Library of Medicine, 2018).

Beyond (Q)SARs, in vitro tests represent a second set of screening methods that are routinely used to evaluate the potential toxicity of chemicals. Commonly used in vitro tests include those associated with identification of chemicals that are genotoxic to somatic cells (mutagenic and clastogenic); irritating or corrosive to skin; skin sensitizers; inducers of serious eye damage; and agonists or antagonists of the estrogen receptor. Additional tests are also available to evaluate the physical-chemical properties of emerging contaminants, such as partition coefficient and hydrolysis as a function of pH. Further details about these standard guideline methods and approaches for designing testing strategies are provided by the OECD (OECD, 2018), USEPA (USEPA, 2018), FDA (FDA, 2018a), ICH (ICH, 2018), and ECHA (ECHA, 2017). Additionally, a great deal of effort is currently being invested by regulatory agencies to develop and validate high-throughput in vitro screening tests that allow the potential of a chemical to disrupt biological pathways associated with toxicity to be evaluated (Tox21, 2018). Furthermore, new "omics"-based assays (i.e., genomics, transcriptomics, proteomics, metabolomics, etc.) are also being developed and validated to evaluate the activity of chemicals via different metabolic processes and mechanisms of action that can lead to toxicity in humans (Klassen, 2013). While omics methods are relatively new, their utility continues to expand as the number of tests (each corresponding to a unique metabolic process or mechanism of action) increases and activity observed in the tests is linked to adverse outcome pathways in organisms and/or populations. Thus, in vitro tools with which to evaluate hazards, toxicity, and risks of emerging contaminants are currently available and are forecasted to become broader in coverage in the near future.

If the results of the QSAR and in vitro tests suggest that an emerging contaminant has sufficient toxicity to pose a risk to human health at concentrations typical of those found in the environment, then a specific sequence of in vivo tests using whole animals can be undertaken. These are the classic animal tests used in traditional toxicology and are required for safety testing by regulatory agencies (USEPA, FDA, ECHA, etc.). The tests can range in duration from days to years (e.g., acute to subchronic to chronic), can evaluate numerous endpoints (e.g., mortality, growth, reproduction, development, behavior), and can investigate exposure through a variety of routes (e.g., ingestion, dermal, inhalation). Approaches to developing testing strategies have been described by several authoritative groups (FDA, ECHA, ICH, VICH, etc.) and in Casarett and Doull's *Toxicology* (Klassen, 2013).

In summary, in the absence of any information regarding the toxicity of an emerging contaminant, a reasonable progression of toxicological investigation would follow a series of steps. (1) In silico (Q)SAR, analog, and read-across evaluations to quickly predict the potential hazard of the new chemical. (2) Perform a battery of in vitro tests sequenced based on findings from the in silico evaluation. (3) Consider developing provisional no effect levels based on in silico evaluations, in vitro evaluations, and Thresholds of Toxicological Concern (Cramer et al., 1976; Kroes et al., 2004; Kroes et al., 2005) or other similar banding approaches. (4) Perform in vivo testing based on findings in steps 1 through 3. (5) Refine no effect levels on a rolling basis as additional information from in vivo studies becomes available.

In most cases the classic in vivo tests take longer and cost substantially more than either in vitro tests or in silico evaluations. However, the in vivo tests allow toxicokinetics (i.e., absorption, distribution, metabolism, and excretion) and toxicodynamics to be integrated into the same study. In vivo tests cover a multitude of endpoints (acute, subacute, subchronic, chronic, developmental, reproduction, neurological, etc.) and lead to greater understanding of the potential toxicity of an emerging contaminant in a whole organism. Ideally, before progressing to in vivo testing, information gathered using in silico and in vitro tests would be used to develop conservative estimates of toxicity of a future emerging contaminant. Those estimates could be used to develop provisional no effect levels in relevant environmental media. Assuming analytical methods are available to estimate environmental concentrations of an emerging contaminant, or environmental concentrations of the emerging contaminant can be predicted using other methods, the environmental concentrations could be compared to the provisional no effect levels. The need for further in vivo toxicity testing could then be evaluated based on a comparison of the environmental concentration of the emerging contaminant to its provisional no effect levels. If the environmental concentrations are well below the provisional no effect levels, the extent of further toxicity testing (e.g., in vivo studies) could be distilled to focus on a limited number of key confirmation studies. If environmental concentrations are similar to or exceed the allowable concentrations, then additional toxicological investigations are likely warranted.

6.3.3 CHALLENGES ASSOCIATED WITH REGULATION

Many of the challenges associated with the regulatory climate around emerging contaminants are related to the challenges with toxicology. For example, in the U.S., the USEPA will issue the final toxicological values through the Integrated Risk Information System (IRIS) and may develop health advisory levels (HALs) or screening levels for various exposure pathways (e.g., tap water regional screening levels [RSLs]), but may not develop drinking water maximum contaminant levels (MCLs) right away. As detailed more in Chapter 1, the process for developing federal drinking water MCLs can be lengthy and the USEPA may decide not to undertake that effort for all emerging contaminants. Consequently, determining whether to regulate and develop standards falls to state agencies.

In the absence of an MCL, many states choose to adopt HALs or RSLs. Other states will use the IRIS values in conjunction with the human health risk assessment approach that the state has adopted to generate their drinking water and groundwater guidance and/or promulgated standards. Different states use differing assumptions in human health risk assessment. A common example of this is assuming a 1-in-10,000 excess lifetime cancer risk versus a 1-in-1,000,000 risk when calculating a drinking water standard. Because states take different approaches to developing guidance/regulations, this can result in guidance values and/or promulgated standards that vary by orders of magnitude from state to state. The variation in guidance values and/or promulgated standards is exacerbated by toxicological data that come into the public domain from various sources, which may or may not be reliable or interpreted correctly. Achieving greater collaboration between academics, industry, and regulatory agencies in this regard could improve the consistency and concurrence in the regulatory values.

As mentioned previously when considered as part of the emerging contaminant life cycle (Chapter 1), regulations may be developed while the toxicological understanding of a compound is still being refined. Consequently, toxicological values contained in IRIS may change after regulations are developed and could affect the human health risk estimated by the USEPA or various state agencies. If these updates imply a greater risk to human health, then many states will often move to quickly adopt lower regulatory levels to be more protective, a position that is typically readily accepted by the public (e.g., 1,4-dioxane in Michigan [Chapter 2], or the PFOS/PFOA HAL [Chapter 1]). However, in instances where updated toxicology data imply a lower risk to human health, it can take to change regulatory values given the greater challenge to communicate and gain acceptance for what may be viewed as a less protective regulatory level.

In the end, this regulatory situation poses a challenge for potentially responsible parties, particularly when they have potential environmental liability in multiple states. These challenges are due to different guidance values and/or promulgated standards in different geographical areas, as well as changing values over time. The approach for managing emerging contaminants on a property located in a state where there is no regulation may be different from the approach employed in a state with a regulation. Likewise, some states generate guidance values while others are generating promulgated standards; each with various levels of application. The evolution of regulatory guidance/standards (e.g., as toxicological benchmarks are updated) should be included in a management strategy (see more in section 6.4), particularly when the regulatory climate around an emerging contaminant is fluctuating and different regulatory agencies have unique environmental standards.

6.3.4 Challenges Associated with Characterization and Analysis

The challenges associated with characterization and analysis of future emerging contaminants are those of any new compound. The chemical and physical characteristics of a compound will dictate the analytical method that will produce the most accurate results, as well as the fate and transport of the compound in the environment. For example, individual volatile compounds are generally easily separated from a mixture via gas chromatography while other compounds require use of liquid chromatography to achieve separation and quantification.

Most commercial analytical laboratories use the USEPA-approved suite of methods, as shown by their various lab certifications. Therefore, it is most useful to begin the search for an appropriate analytical method for a future emerging contaminant among this approved list. In some cases, the compound can be quantified using one or more of the established methods, but in many cases, a different starting point will be needed. Once a starting point is established, it is important that the quality assurance/quality control (QA/QC) steps associated with a given method are also appropriate for the new compound.

1,4-Dioxane serves as an example where the properties of an emerging contaminant need to be compared to those typically quantified by the accepted analytical methods and how challenges can be overcome by laboratories who are willing to do so. As described in more detail in Chapter 2, the physical/chemical properties of 1,4-dioxane introduce a potential bias when quantified via USEPA Method 8270 for semi-volatile organic compounds (SVOCs). In the case of USEPA Method 8270, the compounds identified as analytical surrogates and internal standards associated with the QA/QC procedures do not have the same physical/chemical properties as 1,4-dioxane. Therefore, the normal QA/QC procedure that would account for low recoverability of other SVOCs in the sample will not account for the low recoverability of 1,4-dioxane. This challenge can be mitigated by use of a special analytical surrogate/internal standard specifically chosen to account for the low recoverability of 1,4-dioxane. Unfortunately, not all commercial laboratories are set up to include this additional QA/QC step.

In the case where the existing acceptable analytical methods will not quantify the new compound, it may be necessary to develop new methods. New method development commonly occurs in commercial, government, or university laboratories.

Ideally, these newly developed methods would be incorporated into the USEPA's collection of methods, but there would certainly be in an interim period without a USEPA method. For methods that are not yet accepted by USEPA, the QA/QC procedure may be called into question. Depending on the data quality objectives for a given site, it may be necessary to verify that the QA/QC procedure meets the needs of intended use of the data. Examples of QA/QC items to review include the use of analytical surrogates, use of internal standards, development of appropriate calibration curves, use of matrix spike and matrix spike duplicates, etc.

In some cases, it may be necessary to develop new analytical methods to quantify groups of compounds that fall within the emerging contaminant class. An example of this are PFASs. As detailed more in Chapter 3, PFASs comprise an extensive list of individual compounds, many of which are not quantified by the current analytical techniques. However, in the subsurface, many of these compounds will be biotransformed to a handful of end products that are quantifiable (and regulated). If the starting concentration of compounds that could result in these regulated end products is not known, it can make development of management techniques difficult. In the case of PFASs, the total oxidizable precursor assay (Chapter 3) was developed to quantify precursors that could be biotransformed to the regulated end products. Quantifying precursors allows for a more complete understanding of the mass of PFASs present that could need to be addressed compared to only what is visible to standard analytical techniques, supporting the development of a more complete management strategy.

As with laboratory analysis of a new compound, a similar set of questions needs to be asked when it comes to characterization efforts to understand the nature of, extent of, and fate and transport of a future emerging contaminant. If the new compound is expected to be colocated with other constituents (e.g., 1,4-dioxane with chlorinated solvents or methyl tertiary butyl ether with petroleum hydrocarbons), the first challenge is determining whether the physical/chemical characteristics of the new compound(s) will be similar to the other constituents they occur with. If so, then the existing characterization approach may be appropriate to apply to the new compound(s). If not, then adjustments to the characterization approach should be considered. This may include adjusting sampling programs to focus more on one media over another (e.g., groundwater over soil), adding investigation locations that may be downgradient or cross gradient to those already in place, or collecting additional samples in the vicinity of discharge locations (e.g., sediment in a surface water body).

As investigation practices have evolved for traditional contaminants, advances have been made in investigation technologies and strategies for building conceptual site models that inform remedial design. A challenge facing practitioners is applying these advanced technologies and strategies to future emerging contaminants. For example, Smart Characterization™ (Curry et al., 2016) is based on collection of high-resolution data to define key aspects of the conceptual site model that contribute to mass flux. A mobile laboratory is often used to quickly quantify chemical concentrations and make near-real-time decisions about investigation locations—typically resulting in an optimized investigation footprint and overall cost benefit. To apply these tools to emerging contaminants, the analytical capabilities of the mobile laboratory and the properties of an emerging contaminant need to be considered. Similar to the earlier discussion on method development for brick-and-mortar laboratories, mobile laboratories may not be readily available with the equipment needed to accurately quantify the target compound(s). In some cases, it may be appropriate to use a mobile laboratory with analytical capabilities that can provide semi-quantitative results. An example of this is the use of a mobile laboratory that operates a full-scan volatile organic compound analysis like USEPA Method 8260 for investigation of chlorinated solvents and 1,4-dioxane. While this analytical method may underestimate the actual concentration of 1,4-dioxane in the sample (see Chapter 2), it can serve as a useful screening tool for understanding the presence of 1,4-dioxane in the subsurface. In these cases, it is important to conduct side-by-side brick-and-mortar laboratory analyses to confirm the values quantified by the mobile laboratory and understand the biases. With these considerations, Smart Characterization™ can become a powerful tool for future emerging contaminants, particularly when investigation areas may only be a portion of an existing investigation/monitoring network.

With the above in mind, in most cases where there are gaps in analytical capability and accuracy, the challenge is taken up by multiple entities—ranging from commercial to academic. Depending on the sense of urgency around a particular emerging contaminant, multiple methods may be developed by different entities. These various methods may have different intentions, such as field screening, evaluating toxicological effects, accurate quantification of a single compound, or quantification of a group of compounds simultaneously. These methods all become tools in the toolbox for environmental site characterization.

6.3.5 Challenges Associated with Treatment

The topic of emerging contaminant treatment is still developing for the emerging contaminants already in the limelight, let alone those of the future. As identified in the emerging contaminant life cycle (Chapter 1), application of existing treatment technologies and development of new ones generally comes later in the process. This is because the treatment endpoints are nebulous until the toxicology is better understood and regulations have been put in place, motivating treatment.

As is the common theme running through this chapter, a worthwhile first step in treatment technology development is to compare the future emerging contaminant in question to those better known and identify if established treatment technologies can be applied to the new compound(s). This may be particularly relevant when determining if conventional wastewater treatment or conventional drinking water treatment unit processes will be effective at removing the future emerging contaminant.

Typically, more aggressive treatment technologies have been initially adopted to address emerging contaminants in environmental media. Examples include the use of chemical oxidation processes or incineration. These treatment technologies may be readily applied early on because they are widely accepted and effective at contaminant destruction. However, they can be expensive and less sustainable than technologies under development. The early use of aggressive treatment technologies means they typically gain more regulatory and public acceptance, which can later pose challenges when more innovative, sustainable, and affordable technologies come onto the market.

In the vein of innovation and technology development, it is common for physical treatment techniques to be developed before innovative chemical treatment techniques. Likewise, it is common that both physical and chemical treatment techniques be developed prior to innovative biological treatment techniques. Academia plays a large role in the development of these new treatment technologies. Biological treatment approaches can be very attractive, because they are generally more sustainable and can be more cost effective than physical or chemical treatment techniques. However, it typically takes longer to understand how the microbial community interacts with a new chemical, identify relevant biodegradation mechanisms, and articulate appropriate implementation strategies. Therefore, when the biological treatment technologies become available there may be challenges with gaining regulatory and public acceptance over use of the more established techniques. While challenging, there is certainly precedent for it, so it should not be a deterrent. Lessons learned from application of various treatment technologies for existing and other emerging contaminants is expected to shorten the overall technology development and acceptance timeline for future emerging contaminants.

6.4 THE FUTURE OF EMERGING CONTAMINANTS

What does the future of emerging contaminants look like? Our society will certainly continue the ever-increasing pace of technological development. This in turn will result in the continued development of new compounds that may, under certain circumstances, have an adverse effect on human health and the environment. Unfortunately, the pace for development does not always allow us to fully understand the potential adverse effects prior to use. Programs like the European Union's Registration, Evaluation, Authorisation, and Restriction of Chemicals regulation or the United States' Toxic Substances Control Act and associated Frank R. Lautenberg Chemical Safety for the 21st Century Act are trying to better characterize these potential human health effects and control substances in the marketplace. Even with these efforts, we can expect to have an ever-increasing list of emerging contaminants to address.

As has been the case with emerging contaminants in the past, the importance of public perception as a driver for the next compound or class of compounds to receive heightened attention and focus cannot be understated. The fact that a new compound has been classified as an "emerging contaminant" means that is has been detected in the environment (or has been predicted to be in the environment), that such detection is new or recent, and that we likely have limited information on the potential risks associated with exposure to this compound. The combination of these three characteristics often creates fear in the public and also makes it likely that the situation will be reported in the press. Attention in the press can reinforce and heighten the fear in the public. This fear can lead to an emerging contaminant receiving more attention and higher priority than might be warranted based on available information about environmental concentration and toxicity.

The development of a management strategy is therefore key to help address and minimize the uncertainty and risk associated with emerging contaminants. Management strategies may be reactive, acting when and only if outside drivers such as regulatory standards demand; or they can be proactive, taking actions as awareness develops in advance of regulatory standards. Different stakeholders may opt for different management approaches depending on their individual situation, such as cost management priorities, real or perceived human health or ecological exposure, reputational risks, and litigation potential. A comparison of reactive and proactive management approaches, examples of when they are appropriate, and the benefits and drawbacks of each are described in Table 6.2. Some stakeholders may find benefit in assessing their portfolio early on in the decision process to identify potential, or actual, risks associated with a particular emerging contaminant. Likewise, while some stakeholders may adopt a particular management strategy for all emerging contaminants, others may find it beneficial to adopt differing management strategies for each emerging contaminant relevant to their portfolio.

Regardless of the approach, proactive management strategies benefit all stakeholders and minimize long-term risk and costs. Taking a proactive approach may be particularly useful when a portfolio crossing multiple geographies must be addressed. The specific elements of a proactive management strategy may differ between stakeholders, but some common considerations are outlined below:

- **Vulnerability:** For stakeholders with the potential for emerging contaminant entry to the environment at their facilities, the first step can include assessing that potential and using that to prioritize actions. This requires an examination of the facility processes with quantitative analysis of the material handling for a given operation, with emphasis on the primary potential sources of various emerging contaminants. This book provides one reference on the potential sources and facilities where the individual compounds covered in each chapter might occur. For example, facilities that use Class B firefighting foams may evaluate storage conditions and the frequency and manner of use to assess the vulnerability for an environmental release of PFASs.
- **Exposure:** The other half of the vulnerability equation relates to the potential risk of exposure to human and ecological receptors. Approaching this may include a simple assessment of pathways or a more conservative screening approach that combines conservative estimates of toxicity with predicted environmental concentrations for a range of use scenarios. Such assessments can be used by stakeholders to prioritize action.
- **Regulation:** A management strategy will also need to consider the basis for varying interpretations of allowable concentrations of an emerging contaminant in the environment and how to match treatment actions and goals between facilities and unique regulatory settings. A manufacturer could select the most conservative of all standards and have all facilities comply, regardless of whether or not all locations have a local regulation. Or a manufacturer can have each facility comply with the standard specific to its location. Whether to adopt one of these approaches or develop a different one will depend on the manufacturer's interpretation of the protectiveness of the various environmental standards and the compliance costs and liabilities associated with each.

  To further manage changing regulatory requirements, regulatory review should be incorporated into the management strategy, which could include a regular review of state regulations to determine if new ones have been

TABLE 6.2
Comparison of emerging contaminant management strategies

Management Strategy	Description	Relevant Situations	Benefits/Drawbacks
Reactive	Action in response to regulatory pressures, known human health or ecological risks, litigation, or reputational risk due to public concern	– No federal/state/local regulations or requirements are in place for a given chemical – No known human or ecological exposure – Near-term spend minimization is a priority – Flexible resources are available to address emergency situations	– Avoids unnecessary effort and spend – Increased chance of upsets to long-term planning – Resources (e.g., money/staff) will need to be allocated quickly if/when action is required – Potential damage to relationships with stakeholders – Limited data may also cause challenges in decision making (e.g., not having enough data to perform useful analyses)
Proactive	Developing a strategy for a portfolio ahead of risk drivers; this may include quantification of risk, implementation of readiness measures, and collection of relevant data	– Particularly helpful when a portfolio of sites need to be managed holistically – Sites are in different geographical locations (e.g., different states) where regulatory pressures differ – Reputation is a significant driver and public concern is not desirable – Liability removal is a near-term priority – Minimizing lifetime cost is a priority	– Resources (e.g., money/staff) can be appropriately allocated to areas of highest risk – Data can be collected ahead of regulatory/health/reputational pressures, allowing for quick responses to inquiries or allegations – Long-term data sets can be built to allow for effective decision making and potential selection of more economical alternatives (e.g., monitored natural attenuation) – Technologies can be employed to treat current and emerging contaminants concurrently

implemented and/or setting up specific notifications of new regulations in states of interest. Multinational stakeholders should also consider regulatory review in all countries in which the emerging contaminant in question is either manufactured or used.
- **Characterization:** Before implementing any characterization effort, it is important to identify the data quality objectives and whether the current state of the science will allow those objectives to be met. For many emerging contaminants, it is possible that currently available analytical techniques preclude meeting such objectives. This situation suggests characterization efforts in the near-term may not produce environmental concentration data that can be used for decision-making. For such emerging contaminants stakeholders should consider developing conservative estimates of environmental concentrations based on fate and transport modeling. The modeling would rely on the physio-chemicals properties of an emerging contaminant as well as its expected production and use.
- **Treatment:** An understanding of the treatment options for emerging contaminants can be incorporated into the management strategy. For many emerging contaminants these may be still evolving. Understanding what treatment processes exist, which ones may already be in place, and where gaps exist can be used for planning. For emerging contaminants that lack technologies to treat water to acceptable levels, planning for alternative water supplies may be needed or investment in technology development may be warranted. In cases where technologies exist, cost planning for future retrofits or construction can be included in financial forecasting.

While each future emerging contaminant will have its own set of challenges and may require specialized approaches, the concepts laid out in this chapter are meant to help stakeholders successfully chart and navigate their own path. Consideration of these concepts before emerging contaminants "emerge" in the public eye can help bring together stakeholders that may have been conflicted on other issues in the past (e.g., regulatory agencies and potentially responsible parties) given the common goals of protecting public health and the environment and providing a balanced assessment of potential risk. Developing effective management strategies also offers the opportunity for enhanced collaboration between stakeholders like water supply providers, academia, and other industries. Working together, stakeholders can find innovative, effective, and practical solutions to these challenges, such as combining point source treatment, public water supply treatment, and point of entry/point of use treatment to ensure protectiveness. Together, we can shorten, and perhaps even eliminate, portions of the life cycle associated with future emerging contaminants.

REFERENCES

Cramer, G. M., R. A. Ford, and R. L. Hall. 1976. "Estimation of Toxic Hazard: A Decision Tree Approach." *Food and Cosmetics Toxicology* 255–276.

Curry, Patrick, Nicklaus Welty, Jesse Wright, Dave Favero, and Joseph Quinnan. 2016. "Smart Characterization: An Integrated Approach for Evaluating a Comples 1,4-Dioxane Site." *Remediation* 29–45.

ECHA. 2017. "Guidance on Information Requirements and Chemical Safety Assessment, Chapter R.7a: Endpoint Specific Guidance (version 6.0)." https://echa.europa.eu/documents/10162/13632/information_requirements_r7a_en.pdf.

FDA. 2018a. "Guidance, Compliance & Regulatory Information: Pharmacology/Toxicology." www.fda.gov/Drugs/GuidanceComplianceRegulatoryInformation/Guidances/ucm065014.htm.

———. 2018b. M7(R1) Assessment and Control of DNA Reactive (Mutagenic) Impurities in Pharmaceuticals to Limit Potential Carcinogenic Risk. U.S. Department of Health and Human Services. Silver Spring, Maryland. March.

ICH. 2018. "Safety Guidelines." www.ich.org/products/guidelines/safety/article/safety-guidelines.html.

Klassen, Curtis D, ed. 2013. *Casarett and Doull's Toxicology: The Basic Science of Poisons*, Eighth Edition. New York, NY: McGraw-Hill Education.

Kroes, R., J Kleiner, and A Renwick. 2005. "The Threshold of Toxicological Concern Concept in Risk Assessment." *Toxicological Sciences* 226–230.

Kroes, R., A. G. Renwick, M. Cheeseman, J. Kleiner, I. Mangelsdorf, A. Piersma, B. Schilter, et al. 2004. "Structure-based Thresholds of Toxicological Concern (TTC): Guidance for Application to Substances Present at Low Levels in the Diet." *Food and Chemical Toxicology* 65–83.

National Library of Medicine. 2018. "ChemIDplus: A TOXNET Database." Bethesda, MD. https://chem.nlm.nih.gov/chemidplus/.

OECD. 2018. "OECD Guidelines for the Testing of Chemicals." www.oecd.org/chemicalsafety/testing/oecdguidelinesforthetestingofchemicals.htm.

OECD. 2009. "Guidance Document for Using the OECD (Q)SAR Application Toolbox to Develop Chemical Categories According to the OECD Guidance on Grouping Chemicals." www.oecd.org/publications/the-guidance-document-for-using-the-oecd-q-sar-application-toolbox-to-develop-chemical-categories-according-to-the-oecd-guidance-9789264221482-en.htm.

Raghav, Madhumitha, Susanna Eden, Katharine Mitchell, and Becky Witte. 2013. *Contaminants of Emerging Concern in Water*. Tuscon, AZ: Water Resources Reseach Center, College of Agriculture and Life Sciences, University of Arizona.

Tox21. 2018. *Goals and Objectives*. June 5. Accessed June 2018. https://tox21.gov/page/overview/goals-and-objectives.

USEPA. 2018. "Final Test Guidelines for Pesticides and Toxic Substances." www.epa.gov/test-guidelines-pesticides-and-toxic-substances/final-test-guidelines-pesticides-and-toxic.

USEPA. 2016. *User's Guide for T.E.S.T. (version 4.2) (Toxicity Estimation Software Tool): A Program to Estimate Toxicity from Molecular Structure*. Cincinnati, OH: National Risk Management Research Laboratory.

USEPA. 2013. "Analog Identification Methodoloy (AIM) User's Manual (version 1.01)." www.epa.gov/tsca-screening-tools/analog-identification-methodology-aim-tool.

USEPA. 2012. *EPI Suit: Estimation Program Interface*. Washington, D.C.: Office of Pollution Prevention Toxics and Syracuse Research Corporation.

USEPA. 2011. *The ECOSAR (ECOlogical Structure Activity Relationship) Class Program for Microsoft Windows (version 1.10)*. Washington, D.C.: Office of Pollution Prevention Toxics: Risk Assessment Division.

USEPA. 2010. "OncoLogic, An Expert System for Prediction of the Carcinogenic Potential of Chemicals: User's Manual (version 7.0)." www.epa.gov/sites/production/files/2015-09/documents/usermanual.pdf.

VEGA. 2018. *Virtual Models for Property Evaluation of Chemical with a Global Architecture (VEGA)*. Milan, Italy: Instituto di Ricerche Farmacologiche Mario Negri Milano. www.vegahub.eu/.

Acronyms

AIM	analog identification methodology
CEC	contaminant of emerging concern
ECHA	European Chemicals Agency
EDCs	endocrine disrupting compounds
FDA	Food and Drug Administration
HAL	health advisory level
ICH	International Council for Harmonisation of Technical Requirements for Pharmaceuticals for Human Use
IRIS	Integrated Risk Information System
MCL	maximum contaminant level
NDMA	n-nitrosodimethylamine
OECD	Organization for Economic Cooperation and Development
PBDEs	polybrominated diphenyl ethers
PBBs	polybrominated biphenyls
PFASs	per- and polyfluoroalkyl substances
QA/QC	quality assurance/quality control
(Q)SAR	(quantitative) structure activity relationships
RSL	regional screening level
SVOC	semi-volatile organic compound
TEST	USEPA's Toxicity Estimation Software Tool
TTC	FDA's Thresholds of Toxicological Concern
USEPA	United States Environmental Protection Agency
VICH	International Cooperation on Harmonisation of Technical Requirements for Registration of Veterinary Medicinal Products

Appendix A: USEPA Candidate Contaminant List

Contaminant	CAS#	Addition Date (CCL#/Year)	Removal Date (Process/Year)
Acanthamoeba	–	CCL1/1998	RD1/2003
Adenoviruses	–	CCL1/1998	
Aeromonas hydrophila	–	CCL1/1998	CCL3/2009
Caliciviruses	–	CCL1/1998	
Campylobacter jejuni	–	CCL3/2009	
Coxsackieviruses	–	CCL1/1998	CCL3/2009
Cyanobacteria (blue-green algae), other freshwater algae, and their toxins	–	CCL1/1998	CCL3/2009
Echoviruses	–	CCL1/1998	CCL3/2009
Enterovirus	–	CCL3/2009	
Escherichia coli (O157)	–	CCL3/2009	
Helicobacter pylori	–	CCL1/1998	
Hepatitis A virus	–	CCL3/2009	
Legionella pneumophila	–	CCL3/2009	
Microsporidia (Enterocytozoon & Septata)	–	CCL1/1998	CCL3/2009
Mycobacterium avium-intracellulare	–	CCL1/1998	CCL3/2009
Mycobacterium avium	–	CCL3/2009	
Naegleria fowleri	–	CCL3/2009	
Salmonella enterica	–	CCL3/2009	
Shigella sonnei	–	CCL3/2009	
1,1,2,2-Tetrachloroethane	79-34-5	CCL1/1998	RD2/2008
1,1,1,2-Tetrachloroethane	630-20-6	CCL3/2009	
1,2,4-Trimethylbenzene	95-63-6	CCL1/1998	CCL3/2009
1,2,3-Trichloropropane	96-18-4	CCL3/2009	
1,1-Dichloroethane	75-34-3	CCL1/1998	
1,1-Dichloropropene	563-58-6	CCL1/1998	CCL3/2009
1,2-Diphenylhydrazine	122-66-7	CCL1/1998	CCL3/2009
1,3-Butadiene	106-99-0	CCL3/2009	
1,3-Dichloropropane	142-28-9	CCL1/1998	CCL3/2009
1,3-Dichloropropene	542-75-6	RD2/2008	CCL3/2009
1,3-Dinitrobenzene	99-65-0	CCL3/2009	RD3/2016
1,4-Dioxane	123-91-1	CCL3/2009	
2,4,6-Trichlorophenol	88-06-2	CCL1/1998	CCL3/2009
2,2-Dichloropropane	594-20-7	CCL1/1998	CCL3/2009
2,4-Dichlorophenol	120-83-2	CCL1/1998	CCL3/2009
2,4-Dinitrophenol	51-28-5	CCL1/1998	CCL3/2009

(Continued)

Contaminant	CAS#	Addition Date (CCL#/Year)	Removal Date (Process/Year)
2,4-Dinitrotoluene	121-14-2	CCL1/1998	RD2/2008
2,6-Dinitrotoluene	606-20-2	CCL1/1998	RD2/2008
1-Butanol	71-36-3	CCL3/2009	
2-Methoxyethanol	109-86-4	CCL3/2009	
2-Methyl-phenol (o-cresol)	95-48-7	CCL1/1998	CCL3/2009
2-Propen-1-ol	107-18-6	CCL3/2009	
3-Hydroxycarbofuran	16655-82-6	CCL3/2009	
17α-Estradiol	57-91-0	CCL3/2009	
4,4′-Methylenedianiline	101-77-9	CCL3/2009	
Acephate	30560-19-1	CCL3/2009	
Acetaldehyde	75-07-0	CCL3/2009	
Acetamide	60-35-5	CCL3/2009	
Acetochlor	34256-82-1	CCL1/1998	
Acetochlor ethanesulfonic acid	187022-11-3	CCL3/2009	
Acetochlor oxanilic acid	184992-44-4	CCL3/2009	
Acrolein	107-02-8	CCL3/2009	
Alachlor ethanesulfonic acid	142363-53-9	CCL3/2009	
Alachlor ethanesulfonic acid & other acetanilide pesticide degradation products	–	CCL1/1998	CCL3/2009
Alachlor oxanilic acid	171262-17-2	CCL3/2009	
Aldrin	309-00-2	CCL1/1998	RD1/2003
α-Hexachlorocyclohexane	319-84-6	CCL3/2009	
Aluminum	7429-90-5	CCL1/1998	CCL3/2009
Aniline	62-53-3	CCL3/2009	
Bensulide	741-58-2	CCL3/2009	
Benzyl chloride	100-44-7	CCL3/2009	
Boron	7440-42-8	CCL1/1998	RD2/2008
Bromobenzene	108-86-1	CCL1/1998	CCL3/2009
Butylated hydroxyanisole	25013-16-5	CCL3/2009	
Captan	133-06-2	CCL3/2009	
Chlorate	14866-68-3	CCL3/2009	
Chloromethane (methyl chloride)	74-87-3	CCL3/2009	
Clethodim	110429-62-4	CCL3/2009	
Cobalt	7440-48-4	CCL3/2009	
Cumene hydroperoxide	80-15-9	CCL3/2009	
Cyanotoxins	–	CCL3/2009	
Dimethyl tetrachloroterephthalate mono-acid degradate	887-54-7	CCL1/1998	RD2/2008
Dimethyl tetrachloroterephthalate di-acid degradate	2136-79-0	CCL1/1998	RD2/2008
DDE (dichlorodiphenyldichloroethylene)	72-55-9	CCL1/1998	RD2/2008
Diazinon	333-41-5	CCL1/1998	CCL3/2009
Dicrotophos	141-66-2	CCL3/2009	
Dieldrin	60-57-1	CCL1/1998	RD1/2003
Dimethipin	55290-64-7	CCL3/2009	
Dimethoate	60-51-5	CCL3/2009	RD3/2016
Disulfoton	298-04-4	CCL1/1998	CCL4/2016
Diuron	330-54-1	CCL1/1998	

USEPA Candidate Contaminant List

Contaminant	CAS#	Addition Date (CCL#/Year)	Removal Date (Process/Year)
EPTC (S-ethyl-N, N-dipropylthiocarbamate)	759-94-4	CCL1/1998	RD2/2008
Equilenin	517-09-9	CCL3/2009	
Equilin	474-86-2	CCL3/2009	
Erythromycin	114-07-8	CCL3/2009	
17β-Estradiol	50-28-2	CCL3/2009	
Estriol	50-27-1	CCL3/2009	
Estrone	53-16-7	CCL3/2009	
Ethinyl estradiol (17α ethynyl estradiol)	57-63-6	CCL3/2009	
Ethoprop	13194-48-4	CCL3/2009	
Ethylene glycol	107-21-1	CCL3/2009	
Ethylene oxide	75-21-8	CCL3/2009	
Ethylene thiourea	96-45-7	CCL3/2009	
Fenamiphos	22224-92-6	CCL3/2009	CCL4/2016
Fonofos	944-22-9	CCL1/1998	RD2/2008
Formaldehyde	50-00-0	CCL3/2009	
Germanium	7440-56-4	CCL3/2009	
Halon 1011 (bromochloromethane)	74-97-5	CCL3/2009	
HCFC-22 (chlorodifluoromethane or difluoromonochloromethane)	75-45-6	CCL3/2009	
Hexane	110-54-3	CCL3/2009	
Hexachlorobutadiene	87-68-3	CCL1/1998	RD1/2003
Hydrazine	302-01-2	CCL3/2009	
p-Isopropyltoluene (p-cymene)	99-87-6	CCL1/1998	CCL3/2009
Linuron	330-55-2	CCL1/1998	CCL3/2009
Manganese	7439-96-5	CCL1/1998	RD1/2003
Mestranol	72-33-3	CCL3/2009	
Methamidophos	10265-92-6	CCL3/2009	
Methanol	67-56-1	CCL3/2009	
Methyl bromide (bromomethane)	74-83-9	CCL1/1998	
Methyl-tert-butyl ether	1634-04-4	CCL1/1998	
Metolachlor	51218-45-2	CCL1/1998	
Metolachlor ethanesulfonic acid	171118-09-5	CCL3/2009	
Metolachlor oxanilic acid	152019-73-3	CCL3/2009	
Metribuzin	21087-64-9	CCL1/1998	RD1/2003
Molinate	2212-67-1	CCL1/1998	CCL4/2016
Molybdenum	7439-98-7	CCL3/2009	
Naphthalene	91-20-3	CCL1/1998	RD1/2003
Nitrobenzene	98-95-3	CCL1/1998	
Nitroglycerin	55-63-0	CCL3/2009	
N-Methyl-2-pyrrolidone	872-50-4	CCL3/2009	
N-Nitrosodiethylamine	55-18-5	CCL3/2009	
N-Nitrosodimethylamine	62-75-9	CCL3/2009	
N-Nitrosodipropylamine	621-64-7	CCL3/2009	
N-Nitrosodiphenylamine	86-30-6	CCL3/2009	
N-Nitrosopyrrolidine	930-55-2	CCL3/2009	
Nonylphenol	25154-52-3	CCL4/2016	
Norethindrone (19-norethisterone)	68-22-4	CCL3/2009	

(*Continued*)

Contaminant	CAS#	Addition Date (CCL#/Year)	Removal Date (Process/Year)
n-Propylbenzene	103-65-1	CCL3/2009	
Organotins	–	CCL1/1998	CCL3/2009
o-Toluidine	95-53-4	CCL3/2009	
Oxirane, methyl-	75-56-9	CCL3/2009	
Oxydemeton-methyl	301-12-2	CCL3/2009	
Oxyfluorfen	42874-03-3	CCL3/2009	
Perchlorate	14797-73-0	CCL1/1998	CCL4/2016
Perfluorooctane sulfonic acid (PFOS)	1763-23-1	CCL3/2009	
Perfluorooctanoic acid (PFOA)	335-67-1	CCL3/2009	
Permethrin	52645-53-1	CCL3/2009	
Profenofos	41198-08-7	CCL3/2009	
Prometon	1610-18-0	CCL1/1998	CCL3/2009
Quinoline	91-22-5	CCL3/2009	
RDX (hexahydro-1,3,5-trinitro-1,3,5-triazine)	121-82-4	CCL1/1998	
sec-Butylbenzene	135-98-8	CCL3/2009	
Sodium	7440-23-5	CCL1/1998	RD1/2003
Strontium	7440-24-6	CCL3/2009	RD3/2016
Sulfate	14808-79-8	CCL1/1998	RD1/2003
Tebuconazole	107534-96-3	CCL3/2009	
Tebufenozide	112410-23-8	CCL3/2009	
Tellurium	13494-80-9	CCL3/2009	
Terbacil	5902-51-2	CCL1/1998	RD2/2008
Terbufos	13071-79-9	CCL1/1998	RD3/2016
Terbufos sulfone	56070-16-7	CCL3/2009	RD3/2016
Thiodicarb	59669-26-0	CCL3/2009	
Thiophanate-methyl	23564-05-8	CCL3/2009	
Toluene diisocyanate	26471-62-5	CCL3/2009	
Triazines & degradation products of triazines		CCL1/1998	CCL3/2009
Tribufos	78-48-8	CCL3/2009	
Triethylamine	121-44-8	CCL3/2009	
Triphenyltin hydroxide	76-87-9	CCL3/2009	
Urethane	51-79-6	CCL3/2009	
Vanadium	7440-62-2	CCL1/1998	
Vinclozolin	50471-44-8	CCL3/2009	
Ziram	137-30-4	CCL3/2009	

Acronyms
CAS- Chemical Abstract Service
CCL- Candidate Contaminant List
RD- Regulatory Determination

Appendix B: REACH Candidate List

Name	EC#	CAS#	Date of inclusion
1,6,7,8,9,14,15,16,17,17,18,18-Dodecachloropentacyclo[12.2.1.16,9.02,13.05,10]octadeca-7,15-diene ("Dechlorane Plus"™)	-	-	15/01/2018
Benz[a]anthracene	200-280-6	56-55-3, 1718-53-2	15/01/2018
Cadmium carbonate	208-168-9	513-78-0	15/01/2018
Cadmium hydroxide	244-168-5	21041-95-2	15/01/2018
Cadmium nitrate	233-710-6	10022-68-1, 10325-94-7	15/01/2018
Chrysene	205-923-4	218-01-9, 1719-03-5	15/01/2018
Reaction products of 1,3,4-thiadiazolidine-2,5-dithione, formaldehyde and 4-heptylphenol, branched and linear (RP-HP)	-	-	15/01/2018
Perfluorohexane-1-sulphonic acid and its salts (PFHxS)	-	-	07/07/2017
4,4'-isopropylidenediphenol (BPA)	201-245-8	80-05-7	12/01/2017
4-heptylphenol, branched and linear	-	-	12/01/2017
Nonadecafluorodecanoic acid (PFDA) and its sodium and ammonium salts	-	-	12/01/2017
Nonadecafluorodecanoic acid	206-400-3	335-76-2	12/01/2017
Decanoic acid, nonadecafluoro-, sodium salt	-	3830-45-3	12/01/2017
Ammonium nonadecafluorodecanoate	221-470-5	3108-42-7	12/01/2017
p-(1,1-dimethylpropyl)phenol	201-280-9	80-46-6	12/01/2017
Benzo[def]chrysene (Benzo[a]pyrene)	200-028-5	50-32-8	20/06/2016
1,3-propanesultone	214-317-9	1120-71-4	17/12/2015
2,4-di-tert-butyl-6-(5-chlorobenzotriazol-2-yl) phenol (UV-327)	223-383-8	3864-99-1	17/12/2015
2-(2H-benzotriazol-2-yl)-4-(tert-butyl)-6-(sec-butyl)phenol (UV-350)	253-037-1	36437-37-3	17/12/2015
Nitrobenzene	202-716-0	98-95-3	17/12/2015
Perfluorononan-1-oic-acid and its sodium and ammonium salts	-	-	17/12/2015
Ammonium salts of perfluorononan-1-oic-acid	-	-, 4149-60-4	17/12/2015
Perfluorononan-1-oic-acid	206-801-3	375-95-1	17/12/2015
Sodium salts of perfluorononan-1-oic-acid	-	-, 21049-39-8	17/12/2015

(*Continued*)

Name	EC#	CAS#	Date of inclusion
1,2-benzenedicarboxylic acid, di-C6-10-alkyl esters or mixed decyl and hexyl and octyl diesters	-	-	15/06/2015
1,2-Benzenedicarboxylic acid, mixed decyl and hexyl and octyl diesters	272-013-1	68648-93-1	15/06/2015
1,2-Benzenedicarboxylic acid, di-C6-10-alkyl esters	271-094-0	68515-51-5	15/06/2015
5-sec-butyl-2-(2,4-dimethylcyclohex-3-en-1-yl)-5-methyl-1,3-dioxane [1], 5-sec-butyl-2-(4,6-dimethylcyclohex-3-en-1-yl)-5-methyl-1,3-dioxane [2]	-	-	15/06/2015
5-sec-butyl-2-(4,6-dimethylcyclohex-3-en-1-yl)-5-methyl-1,3-dioxane	-	-	15/06/2015
5-sec-butyl-2-(2,4-dimethylcyclohex-3-en-1-yl)-5-methyl-1,3-dioxane	-	-	15/06/2015
2-(2H-benzotriazol-2-yl)-4,6-ditertpentylphenol (UV-328)	247-384-8	25973-55-1	17/12/2014
2-benzotriazol-2-yl-4,6-di-tert-butylphenol (UV-320)	223-346-6	3846-71-7	17/12/2014
2-ethylhexyl 10-ethyl-4,4-dioctyl-7-oxo-8-oxa-3,5-dithia-4-stannatetradecanoate (DOTE)	239-622-4	15571-58-1	17/12/2014
Cadmium fluoride	232-222-0	7790-79-6	17/12/2014
Cadmium sulphate	233-331-6	10124-36-4, 31119-53-6	17/12/2014
Reaction mass of 2-ethylhexyl 10-ethyl-4,4-dioctyl-7-oxo-8-oxa-3,5-dithia-4-stannatetradecanoate and 2-ethylhexyl 10-ethyl-4-[[2-[(2-ethylhexyl)oxy]-2-oxoethyl]thio]-4-octyl-7-oxo-8-oxa-3,5-dithia-4-stannatetradecanoate (reaction mass of DOTE and MOTE)	-	-	17/12/2014
1,2-Benzenedicarboxylic acid, dihexyl ester, branched and linear	271-093-5	68515-50-4	16/06/2014
Cadmium chloride	233-296-7	10108-64-2	16/06/2014
Sodium perborate, perboric acid, sodium salt	-	-	16/06/2014
Sodium perborate	239-172-9	15120-21-5	16/06/2014
Perboric acid, sodium salt	234-390-0	11138-47-9	16/06/2014
Sodium peroxometaborate	231-556-4	7632-04-4	16/06/2014
Cadmium sulphide	215-147-8	1306-23-6	16/12/2013
Dihexyl phthalate	201-559-5	84-75-3	16/12/2013
Disodium 3,3'-[[1,1'-biphenyl]-4,4'-diylbis(azo)]bis(4-aminonaphthalene-1-sulphonate) (C.I. Direct Red 28)	209-358-4	573-58-0	16/12/2013

Name	EC#	CAS#	Date of inclusion
Disodium 4-amino-3-[[4'-[(2,4-diaminophenyl)azo][1,1'-biphenyl]-4-yl]azo]-5-hydroxy-6-(phenylazo)naphthalene-2,7-disulphonate (C.I. Direct Black 38)	217-710-3	1937-37-7	16/12/2013
Imidazolidine-2-thione (2-imidazoline-2-thiol)	202-506-9	96-45-7	16/12/2013
Lead di(acetate)	206-104-4	301-04-2	16/12/2013
Trixylyl phosphate	246-677-8	25155-23-1	16/12/2013
4-Nonylphenol, branched and linear, ethoxylated	-	-	20/06/2013
Ammonium pentadecafluorooctanoate (APFO)	223-320-4	3825-26-1	20/06/2013
Cadmium	231-152-8	7440-43-9	20/06/2013
Cadmium oxide	215-146-2	1306-19-0	20/06/2013
Dipentyl phthalate (DPP)	205-017-9	131-18-0	20/06/2013
Pentadecafluorooctanoic acid (PFOA)	206-397-9	335-67-1	20/06/2013
1,2-Benzenedicarboxylic acid, dipentyl ester, branched and linear	284-032-2	84777-06-0	19/12/2012
1,2-diethoxyethane	211-076-1	629-14-1	19/12/2012
1-bromopropane (n-propyl bromide)	203-445-0	106-94-5	19/12/2012
3-ethyl-2-methyl-2-(3-methylbutyl)-1,3-oxazolidine	421-150-7	143860-04-2	19/12/2012
4,4'-methylenedi-o-toluidine	212-658-8	838-88-0	19/12/2012
4,4'-oxydianiline and its salts	-	-	19/12/2012
4,4'-oxydianiline	202-977-0	101-80-4	19/12/2012
4-(1,1,3,3-tetramethylbutyl)phenol, ethoxylated	-	-	19/12/2012
4-aminoazobenzene	200-453-6	60-09-3	19/12/2012
4-methyl-m-phenylenediamine (toluene-2,4-diamine)	202-453-1	95-80-7	19/12/2012
4-Nonylphenol, branched and linear	-	-	19/12/2012
6-methoxy-m-toluidine (p-cresidine)	204-419-1	120-71-8	19/12/2012
[Phthalato(2-)]dioxotrilead	273-688-5	69011-06-9	19/12/2012
Acetic acid, lead salt, basic	257-175-3	51404-69-4	19/12/2012
Biphenyl-4-ylamine	202-177-1	92-67-1	19/12/2012
Bis(pentabromophenyl) ether (decabromodiphenyl ether) (DecaBDE)	214-604-9	1163-19-5	19/12/2012
Cyclohexane-1,2-dicarboxylic anhydride	-	-	19/12/2012
Cyclohexane-1,2-dicarboxylic anhydride	201-604-9	85-42-7	19/12/2012
trans-cyclohexane-1,2-dicarboxylic anhydride	238-009-9	14166-21-3	19/12/2012
cis-cyclohexane-1,2-dicarboxylic anhydride	236-086-3	13149-00-3	19/12/2012
Diazene-1,2-dicarboxamide (C,C'-azodi(formamide)) (ADCA)	204-650-8	123-77-3	19/12/2012
Dibutyltin dichloride (DBTC)	211-670-0	683-18-1	19/12/2012
Diethyl sulphate	200-589-6	64-67-5	19/12/2012
Diisopentyl phthalate	210-088-4	605-50-5	19/12/2012
Dimethyl sulphate	201-058-1	77-78-1	19/12/2012

(Continued)

Name	EC#	CAS#	Date of inclusion
Dinoseb (6-sec-butyl-2,4-dinitrophenol)	201-861-7	88-85-7	19/12/2012
Dioxobis(stearato)trilead	235-702-8	12578-12-0	19/12/2012
Fatty acids, C16-18, lead salts	292-966-7	91031-62-8	19/12/2012
Furan	203-727-3	110-00-9	19/12/2012
Henicosafluoroundecanoic acid	218-165-4	2058-94-8	19/12/2012
Heptacosafluorotetradecanoic acid	206-803-4	376-06-7	19/12/2012
Hexahydromethylphthalic anhydride	-	-	19/12/2012
Hexahydromethylphthalic anhydride	247-094-1	25550-51-0	19/12/2012
Hexahydro-4-methylphthalic anhydride	243-072-0	19438-60-9	19/12/2012
Hexahydro-3-methylphthalic anhydride	260-566-1	57110-29-9	19/12/2012
Hexahydro-1-methylphthalic anhydride	256-356-4	48122-14-1	19/12/2012
Lead bis(tetrafluoroborate)	237-486-0	13814-96-5	19/12/2012
Lead cyanamidate	244-073-9	20837-86-9	19/12/2012
Lead dinitrate	233-245-9	10099-74-8	19/12/2012
Lead monoxide (lead oxide)	215-267-0	1317-36-8	19/12/2012
Lead oxide sulfate	234-853-7	12036-76-9	19/12/2012
Lead titanium trioxide	235-038-9	12060-00-3	19/12/2012
Lead titanium zirconium oxide	235-727-4	12626-81-2	19/12/2012
Methoxyacetic acid	210-894-6	625-45-6	19/12/2012
Methyloxirane (Propylene oxide)	200-879-2	75-56-9	19/12/2012
N,N-dimethylformamide	200-679-5	68-12-2	19/12/2012
N-methylacetamide	201-182-6	79-16-3	19/12/2012
N-pentyl-isopentylphthalate	-	776297-69-9	19/12/2012
o-aminoazotoluene	202-591-2	97-56-3	19/12/2012
o-toluidine	202-429-0	95-53-4	19/12/2012
Orange lead (lead tetroxide)	215-235-6	1314-41-6	19/12/2012
Pentacosafluorotridecanoic acid	276-745-2	72629-94-8	19/12/2012
Pentalead tetraoxide sulphate	235-067-7	12065-90-6	19/12/2012
Pyrochlore, antimony lead yellow	232-382-1	8012-00-8	19/12/2012
Silicic acid (H2Si2O5), barium salt (1:1), lead-doped	272-271-5	68784-75-8	19/12/2012
Silicic acid, lead salt	234-363-3	11120-22-2	19/12/2012
Sulfurous acid, lead salt, dibasic	263-467-1	62229-08-7	19/12/2012
Tetraethyllead	201-075-4	78-00-2	19/12/2012
Tetralead trioxide sulphate	235-380-9	12202-17-4	19/12/2012
Tricosafluorododecanoic acid	206-203-2	307-55-1	19/12/2012
Trilead bis(carbonate) dihydroxide	215-290-6	1319-46-6	19/12/2012
Trilead dioxide phosphonate	235-252-2	12141-20-7	19/12/2012
1,2-bis(2-methoxyethoxy)ethane (TEGDME,triglyme)	203-977-3	112-49-2	18/06/2012
1,2-dimethoxyethane,ethylene glycol dimethyl ether (EGDME)	203-794-9	110-71-4	18/06/2012
1,3,5-Tris(oxiran-2-ylmethyl)-1,3,5-triazinane-2,4,6-trione (TGIC)	219-514-3	2451-62-9	18/06/2012

Name	EC#	CAS#	Date of inclusion
1,3,5-tris[(2S and 2R)-2,3-epoxypropyl]-1,3,5-triazine-2,4,6-(1H,3H,5H)-trione (β-TGIC)	423-400-0	59653-74-6	18/06/2012
4,4'-bis(dimethylamino)-4"-(methylamino) trityl alcohol	209-218-2	561-41-1	18/06/2012
4,4'-bis(dimethylamino)benzophenone (Michler's ketone)	202-027-5	90-94-8	18/06/2012
[4-[4,4'-bis(dimethylamino) benzhydrylidene] cyclohexa-2,5-dien-1-ylidene] dimethylammonium chloride (C.I. Basic Violet 3)	208-953-6	548-62-9	18/06/2012
[4-[[4-anilino-1-naphthyl][4-(dimethylamino) phenyl]methylene]cyclohexa-2,5-dien-1-ylidene] dimethylammonium chloride (C.I. Basic Blue 26)	219-943-6	2580-56-5	18/06/2012
Diboron trioxide	215-125-8	1303-86-2	18/06/2012
Formamide	200-842-0	75-12-7	18/06/2012
Lead(II) bis(methanesulfonate)	401-750-5	17570-76-2	18/06/2012
N,N,N',N'-tetramethyl-4,4'-methylenedianiline (Michler's base)	202-959-2	101-61-1	18/06/2012
α,α-Bis[4-(dimethylamino)phenyl]-4 (phenylamino)naphthalene-1-methanol (C.I. Solvent Blue 4)	229-851-8	6786-83-0	18/06/2012
1,2-dichloroethane	203-458-1	107-06-2	19/12/2011
2,2'-dichloro-4,4'-methylenedianiline	202-918-9	101-14-4	19/12/2011
2-Methoxyaniline, o-Anisidine	201-963-1	90-04-0	19/12/2011
4-(1,1,3,3-tetramethylbutyl)phenol	205-426-2	140-66-9	19/12/2011
Aluminosilicate Refractory Ceramic Fibres	-	-	19/12/2011
Arsenic acid	231-901-9	7778-39-4	19/12/2011
Bis(2-methoxyethyl) ether	203-924-4	111-96-6	19/12/2011
Bis(2-methoxyethyl) phthalate	204-212-6	117-82-8	19/12/2011
Calcium arsenate	231-904-5	7778-44-1	19/12/2011
Dichromium tris(chromate)	246-356-2	24613-89-6	19/12/2011
Formaldehyde, oligomeric reaction products with aniline	500-036-1	25214-70-4	19/12/2011
Lead diazide, Lead azide	236-542-1	13424-46-9	19/12/2011
Lead dipicrate	229-335-2	6477-64-1	19/12/2011
Lead styphnate	239-290-0	15245-44-0	19/12/2011
N,N-dimethylacetamide	204-826-4	127-19-5	19/12/2011
Pentazinc chromate octahydroxide	256-418-0	49663-84-5	19/12/2011
Phenolphthalein	201-004-7	77-09-8	19/12/2011
Potassium hydroxyoctaoxodizincatedichromate	234-329-8	11103-86-9	19/12/2011
Trilead diarsenate	222-979-5	3687-31-8	19/12/2011
Zirconia Aluminosilicate Refractory Ceramic Fibres	-	-	19/12/2011

(*Continued*)

Name	EC#	CAS#	Date of inclusion
1,2,3-trichloropropane	202-486-1	96-18-4	20/06/2011
1,2-Benzenedicarboxylic acid, di-C6-8-branched alkyl esters, C7-rich	276-158-1	71888-89-6	20/06/2011
1,2-Benzenedicarboxylic acid, di-C7-11-branched and linear alkyl esters	271-084-6	68515-42-4	20/06/2011
1-Methyl-2-pyrrolidone (NMP)	212-828-1	872-50-4	20/06/2011
2-ethoxyethyl acetate	203-839-2	111-15-9	20/06/2011
Hydrazine	206-114-9	302-01-2, 7803-57-8	20/06/2011
Strontium chromate	232-142-6	7789-06-2	20/06/2011
2-ethoxyethanol	203-804-1	110-80-5	15/12/2010
2-methoxyethanol	203-713-7	109-86-4	15/12/2010
Acids generated from chromium trioxide and their oligomers	-	-	15/12/2010
Dichromic acid	236-881-5	7738-94-5	15/12/2010
Oligomers of chromic acid and dichromic acid	-	-	15/12/2010
Chromic acid	231-801-5	13530-68-2	15/12/2010
Chromium trioxide	215-607-8	1333-82-0	15/12/2010
Cobalt(II) carbonate	208-169-4	513-79-1	15/12/2010
Cobalt(II) diacetate	200-755-8	71-48-7	15/12/2010
Cobalt(II) dinitrate	233-402-1	10141-05-6	15/12/2010
Cobalt(II) sulphate	233-334-2	10124-43-3	15/12/2010
Ammonium dichromate	232-143-1	7789-09-5	18/06/2010
Boric acid	-	-	18/06/2010
Boric acid, crude natural	234-343-4	11113-50-1	18/06/2010
Boric acid	233-139-2	10043-35-3	18/06/2010
Disodium tetraborate, anhydrous	215-540-4	12179-04-3, 1303-96-4, 1330-43-4	18/06/2010
Potassium chromate	232-140-5	7789-00-6	18/06/2010
Potassium dichromate	231-906-6	7778-50-9	18/06/2010
Sodium chromate	231-889-5	7775-11-3	18/06/2010
Tetraboron disodium heptaoxide, hydrate	235-541-3	12267-73-1	18/06/2010
Trichloroethylene	201-167-4	79-01-6	18/06/2010
Acrylamide	201-173-7	79-06-1	30/03/2010
2,4-dinitrotoluene	204-450-0	121-14-2	13/01/2010
Anthracene oil	292-602-7	90640-80-5	13/01/2010
Anthracene oil, anthracene paste	292-603-2	90640-81-6	13/01/2010
Anthracene oil, anthracene paste, anthracene fraction	295-275-9	91995-15-2	13/01/2010
Anthracene oil, anthracene paste, distn. lights	295-278-5	91995-17-4	13/01/2010
Anthracene oil, anthracene-low	292-604-8	90640-82-7	13/01/2010
Diisobutyl phthalate	201-553-2	84-69-5	13/01/2010
Lead chromate	231-846-0	7758-97-6	13/01/2010

REACH Candidate List

Name	EC#	CAS#	Date of inclusion
Lead chromate molybdate sulphate red (C.I. Pigment Red 104)	235-759-9	12656-85-8	13/01/2010
Lead sulfochromate yellow (C.I. Pigment Yellow 34)	215-693-7	1344-37-2	13/01/2010
Pitch, coal tar, high-temp.	266-028-2	65996-93-2	13/01/2010
Tris(2-chloroethyl) phosphate	204-118-5	115-96-8	13/01/2010
4,4'- Diaminodiphenylmethane (MDA)	202-974-4	101-77-9	28/10/2008
5-tert-butyl-2,4,6-trinitro-m-xylene (Musk xylene)	201-329-4	81-15-2	28/10/2008
Alkanes, C10-13, chloro (Short Chain Chlorinated Paraffins)	287-476-5	85535-84-8	28/10/2008
Anthracene	204-371-1	120-12-7	28/10/2008
Benzyl butyl phthalate (BBP)	201-622-7	85-68-7	28/10/2008
Bis (2-ethylhexyl)phthalate (DEHP)	204-211-0	117-81-7	28/10/2008
Bis(tributyltin) oxide (TBTO)	200-268-0	56-35-9	28/10/2008
Cobalt dichloride	231-589-4	7646-79-9	28/10/2008
Diarsenic pentaoxide	215-116-9	1303-28-2	28/10/2008
Diarsenic trioxide	215-481-4	1327-53-3	28/10/2008
Dibutyl phthalate (DBP)	201-557-4	84-74-2	28/10/2008
Hexabromocyclododecane (HBCDD)	-	-	28/10/2008
Hexabromocyclododecane	247-148-4	25637-99-4	28/10/2008
1,2,5,6,9,10-hexabromocyclodecane	221-695-9	3194-55-6	28/10/2008
alpha-hexabromocyclododecane	-	134237-50-6	28/10/2008
beta-hexabromocyclododecane	-	134237-51-7	28/10/2008
gamma-hexabromocyclododecane	-	134237-52-8	28/10/2008
Lead hydrogen arsenate	232-064-2	7784-40-9	28/10/2008
Sodium dichromate	234-190-3	10588-01-9, 7789-12-0	28/10/2008
Triethyl arsenate	427-700-2	15606-95-8	28/10/2008

Source: European Chemical Agency website https://echa.europa.eu/candidate-list-table, retrieved June 1, 2018
EC#: European Community Number
CAS#: Chemical Abstract Service Number

Appendix C: Emerging Contaminants and Their Physical and Chemical Properties

Chemical Name(s)	CAS Registry Number	Molecular Formula	MW	ρ	K_H	V_p	S_W	log K_{ow}	log K_{oc}	MP	BP
Featured Chapters											
1,4-Dioxane	123-91-1	$C_4H_8O_2$	88.11	1.0	4.80×10^{-6} (see note 1)	38.1	Miscible	−0.27	0.54–1.46	12	101
Hexavalent chromium	18540-29-9	Cr^{6+}	51.99	7.2	Not applicable	Not applicable*	Chromate salts are highly soluble (e.g., Na_2CrO_4: 876,000 K_2CrO_4: 650,000)	not hydrophobic	sorption varies based on soil mineralogy and water chemistry	1,907	2,672
1,2,3-trichloropropane; TCP	96-18-4	$C_3H_5Cl_3$	147.43	1.4	3.43×10^{-4}	3.1	1,750	1.98–2.27	1.70–1.99	−15	157
Perfluoroalkyl Substances (PFASs)											
				Perfluoroalkyl Carboxylates / Perfluoroalkyl Carboxylic Acids (PFCAs)							
Perfluorobutanoic Acid; PFBA	375-22-4	$F(CF_2)_3COOH$	214.04	1.65	–	9.83	Miscible	2.82 (p)	1.88	−17.5	121
Perfluoropentanoic Acid; PFPeA	2706-90-3	$F(CF_2)_4COOH$	264.05	1.7	3.31×10^{-10} (p)	7.95	112,600 (p)	3.43 (p)	1.37	–	124.4
Perfluorohexanoic Acid; PFHxA	307-24-4	$F(CF_2)_5COOH$	314.05	1.72	2.50×10^{-10} (p)	3.44 (p)	21,700 (p)	4.06 (p)	1.91	14	143
Perfluoroheptanoic Acid; PFHpA	375-85-9	$F(CF_2)_6COOH$	364.06	1.79	2.21×10^{-10} (p)	1.19 (p)	4,200 (p)	4.67 (p)	2.19	30	175
Perfluorooctanoic Acid; PFOA	335-67-1	$F(CF_2)_7COOH$	414.07	1.8	0.04–0.0910^{-10}	0.02–9.77	3,400–9,500	5.3 (p)	1.31–2.35	37–60	188–192
Perfluorononanoic Acid; PFNA	375-95-1	$F(CF_2)_8COOH$	464.08	1.75	1.24×10^{-09} (p)	9.77×10^{-03}	9,500	5.92 (p)	2.39	59–66	218
Perfluorodecanoic Acid; PFDA	335-76-2	$F(CF_2)_9COOH$	514.09	1.76	3.58×10^{-10} (p)	1.50×10^{-05}	9,500	6.5 (p)	2.76	77–88	218
Perfluoroundecanoic Acid; PFUnA	2058-94-8	$F(CF_2)_{10}COOH$	564.09	1.76	–	7.52×10^{-04}	4 (p)	7.15 (p)	3.3	83–101	160–230
Perfluorododecanoic Acid; PFDoA	307-55-1	$F(CF_2)_{11}COOH$	614.1	1.77	–	7.52×10^{-05}	0.7 (p)	7.77 (p)	–	107–109	245
Perfluorotridecanoic Acid; PFTrdA	72629-94-8	$F(CF_2)_{12}COOH$	664.11	1.77	–	2.26×10^{-3} (p)	0.2 (p)	8.25 (p)	–	–	–
Perfluorotetradecanoic Acid; PFTeDA	376-06-7	$F(CF_2)_{13}COOH$	714.12	1.78	–	7.52×10^{-4} (p)	0.03 (p)	8.9 (p)	–	–	276
Perfluoropentadecanoic Acid; PFPeDA	141074-63-7	$F(CF_2)_{14}COOH$	764.12	–	–	–	–	–	–	–	–
Pentadecafluorooctanoic Acid Ammonium Salt (Ammonium Pentadecafluorooctanoate); APFO	3825-26-1	$C_8H_4NF_{15}NO_2$	445.11	–	–	7.52×10^{-05}	14,200	–	–	157–165	–

Emerging Contaminant Properties

Chemical Name(s)	CAS Registry Number	Molecular Formula	MW	ρ	K_H	V_p	S_W	log K_{ow}	log K_{oc}	MP	BP
Perfluoroalkyl Sulfonates / Perfluoroalkyl Sulfonic Acids (PFSAs)											
Perfluorobutane Sulfonate; PFBS	375-73-5	F(CF$_2$)$_4$SO$_3$H	300.1	1.81	3.19×10^{-10} (p)	4.74	46,200–56,600	3.9 (p)	1	76–84	211
Perfluorohexane Sulfonate; PFHxS	432-50-8	F(CF$_2$)$_6$SO$_3$H	400.11	—	1.95×10^{-10} (p)	4.43×10^{-01}	2,300 (p)	5.17 (p)	1.78	—	—
Perfluoroheptane Sulfonate; PFHpS	357-92-8	F(CF$_2$)$_7$SO$_3$H	450.12	—	—	—	—	—	—	—	—
Perfluorooctane Sulfonate; PFOS	1763-23-1	F(CF$_2$)$_8$SO$_3$H	500.13	—	<2×10^{-16} to 3×10^{-10}	5.04×10^{-02}	520–570	6.43 (p)	2.5–3.1	54	260
Perfluorodecane Sulfonate; PFDS	333-77-3	F(CF$_2$)$_{10}$SO$_3$H	600.14	—	—	5.34×10^{-03}	2 (p)	7.66 (p)	3.53	—	—
Perfluoroalkyl Phosphonic Acids (PFPAs)											
Perfluorobutyl Phosphonic Acid; PFBPA	52299-24-8	F(CF$_2$)$_4$P(O)(OH)$_2$	350.02	—	—	1.35×10^{-03}	1,4,259,100 (p)	2.19 (p)	—	—	—
Perfluorohexyl Phosphonic Acid; PFHxPA	40143-76-8	F(CF$_2$)$_6$P(O)(OH)$_2$	400.03	—	—	3.01×10^{-04}	515,300 (p)	3.48 (p)	—	—	—
Perfluorooctyl Phosphonic Acid; PFOPA	40143-78-0	F(CF$_2$)$_8$P(O)(OH)$_2$	500.05	—	—	7.52×10^{-05}	24,500 (p)	4.73 (p)	—	—	—
Perfluorodecyl Phosphonic Acid; PFDPA	52299-26-0	F(CF$_2$)$_{10}$P(O)(OH)$_2$	600.06	—	—	1.50×10^{-06}	500 (p)	5.98 (p)	—	—	—
Perfluorooctane Sulfonamide and Derivatives											
Perfluorooctane Sulfonamide; PFOSA	754-91-6	F(CF$_2$)$_8$SO$_2$NH$_2$	499.14	—	1.93×10^{-10} (p)	0.024	—	—	2.5–2.62	154–155	—
N-Methyl-Perfluorooctane Sulfonamidoethanol; FOSE	10116-92-4	F(CF$_2$)$_8$SO$_2$NH(CH$_2$)$_2$OH	543.19	—	—	0 (p)	0.9 (p)	5.78 (p)	—	—	—
N-Methyl-Perfluorooctane Sulfonamide; N-MeFOSA	31506-32-8	F(CF$_2$)$_8$SO$_2$NHCH$_3$	513.17	—	—	2.26×10^{-3} (p)	0.2 (p)	6.07 (p)	3.14	—	—
N-Ethyl-Perfluorooctane Sulfonamide; N-EtFOSA	4151-50-2	F(CF$_2$)$_8$SO$_2$NHCH$_2$CH$_3$	527.2	—	—	9.02×10^{-04} (p)	0.1 (p)	6.71 (p)	3.23	—	—
N-Methyl-Perfluorooctane Sulfonamidoethanol; N-Me-FOSE	24448-09-7	F(CF$_2$)$_8$SO$_2$N(CH$_3$)(CH$_2$)$_2$OH	557.22	—	—	3.01×10^{-06}	0.3 (p)	6 (p)	—	—	—
N-Ethyl-Perfluorooctane Sulfonamidoethanol; N-EtFOSE	1691-99-2	F(CF$_2$)$_8$SO$_2$N(CH$_2$CH$_3$)(CH$_2$)$_2$OH	571.25	—	—	1.50×10^{-05}	0.1 (p)	6.52 (p)	—	55–60	—

(Continued)

Chemical Name(s)	CAS Registry Number	Molecular Formula	MW	ρ	K_H	V_p	S_w	log K_{ow}	log K_{oc}	MP	BP
Fluorotelomer Sulfonic Acids											
1H, 1H, 2H, 2H-Perfluorobutanesulfonic Acid; H4-PFBS (2:2 FTS)	149246-63-9	$F(CF_2)_2CH_2CH_2SO_3H$	228.13	–	–	–	–	–	–	–	–
1H, 1H, 2H, 2H-Perfluorohexanesulfonic Acid; H4-PFHxS (4:2 FTS)	757124-72-4	$F(CF_2)_4CH_2CH_2SO_3H$	328.15	–	–	2.48×10^{-03}	27,900 (p)	3.21 (p)	–	–	–
1H, 1H, 2H, 2H-Perfluorooctanesulfonic Acid; H4-PFOS (6:2 FTS)	276619-97-2	$F(CF_2)_6CH_2CH_2SO_3H$	428.17	–	–	8.27×10^{-04}	1,300 (p)	4.44 (p)	–	–	–
1H, 1H, 2H, 2H-Perfluorodecanesulfonic Acid; H4-PFDeS (8:2 FTS)	39108-34-4	$F(CF_2)_8CH_2CH_2SO_3H$	528.18	–	–	7.52×10^{-05}	60 (p)	5.66 (p)	0.01	–	–
1H, 1H, 2H, 2H-Perfluoroundecanesulfonic Acid; H4-PFUdS (10:2 FTS)	120226-60-0	$F(CF_2)_{10}CH_2CH_2SO_3H$	628.2	–	–	7.52×10^{-06}	2 (p)	6.91 (p)	–	–	–
1H, 1H, 2H, 2H-Perfluorotetradecanesulfonic Acid; H4-PFTeS (12:2 FTS)	149246-64-0	$F(CF_2)_{12}CH_2CH_2SO_3H$	728.21	–	–	7.52×10^{-06}	0.2 (p)	7.94 (p)	–	–	–
Fluorotelomer Alcohols											
Perfluoromethylethanol 2; 2:2 FTOH	54949-74-5	$F(CF_2)_2CH_2CH_2OH$	164.08	–	–	–	–	–	–	–	–
Perfluoroethylethanol 4:2; 4:2 FTOH	2043-47-2	$F(CF_2)_4CH_2CH_2OH$	264.09	–	–	1.61	980	3.3	0.93	–	–
Perfluorobutylethanol 6:2; 6:2 FTOH	647-42-7	$F(CF_2)_6CH_2CH_2OH$	364.11	–	5.73×10^{-03}	1.37×10^{-01}	20	4.54	2.43	-33	172
Perfluorocyclethanol 8:2; 8:2 FTOH	865-86-1	$F(CF_2)_8CH_2CH_2OH$	464.12	–	5.04×10^{-03}	2.99×10^{-02}	0.1	5.58	3.84	45	114
Perfluorodecylethanol 10:2; 10:2 FTOH	678-39-8	$F(CF_2)_{10}CH_2CH_2OH$	564.14	–	7.78×10^{-03}	1.50×10^{-03}	0.01	6.63	6.2	–	–
Perfluorododecylethanol 12:2; 12:2 FTOH	39239-77-5	$F(CF_2)_{12}CH_2CH_2OH$	664.15	–	–	–	–	–	–	–	–
Polyfluorinated Alkyl Phosphates (PAPs)											
Monoester (monoPAP)											
4:2 Fluorotelomerphosphatemonoester; 4:2 monoPAP	150065-76-2	$F(CF_2)_4CH_2CH_2OP(O)(OH)_2$	344.07	–	–	0 (p)	11,900 (p)	1.99 (p)	–	–	–
6:2 Fluorotelomerphosphatemonoester; 6:2 monoPap	57678-01-0	$F(CF_2)_6CH_2CH_2OP(O)(OH)_2$	444.09	–	–	0 (p)	2,600 (p)	3.39 (p)	–	–	–
8:2 Fluorotelomerphosphatemonoester; 8:2 monoPAP	57678-03-2	$F(CF_2)_8CH_2CHOP(O)(OH)_2$	544.1	–	–	0 (p)	160 (p)	4.67 (p)	–	–	–
10:2 Fluorotelomerphosphatemonoester; 10:2 monoPAP	57678-05-4	$F(CF_2)_{10}CH_2CH_2OP(O)(OH)_2$	644.12	–	–	0 (p)	10 (p)	5.92 (p)	–	–	–
12:2 Fluorotelomerphosphatemonoester; 12:2 monoPAP	57678-07-6	$F(CF_2)_{12}CH_2CH_2OP(O)(OH)_2$	744.13	–	–	0 (p)	0.3 (p)	7.21 (p)	–	–	–

Emerging Contaminant Properties 389

Chemical Name(s)	CAS Registry Number	Molecular Formula	MW	ρ	K_H	V_p	S_w	log K_{ow}	log K_{oc}	MP	BP
4:2 Fluorotelomerphosphatediester; 4:2 diPAP	135098-69-0	F(CF$_2$)$_4$CH$_2$CH$_2$OP(OH)OCH$_2$CH$_2$	590.15	—	—	0 (p)	0.4 (p)	6.16 (p)	—	—	—
6:2 Fluorotelomerphosphatediester; 6:2 diPAP	57677-95-9	F(CF$_2$)$_6$CH$_2$CH$_2$OP(OH)OCH$_2$CH$_2$	790.18	—	—	0 (p)	8.00×10^{-4} (p)	8.41 (p)	—	—	—
8:2 Fluorotelomerphosphatediester; 8:2 diPAP	678-41-1	F(CF$_2$)$_8$CH$_2$CH$_2$OP(OH)OCH$_2$CH$_2$	990.21	—	—	0 (p)	5.00×10^{-7} (p)	10.93 (p)	—	—	—
10:2 Fluorotelomerphosphatediester; 10:2 diPAP	1895-26-7	F(CF$_2$)$_{10}$CH$_2$CH$_2$OP(OH)OCH$_2$CH$_2$	1,190.24	—	—	0 (p)	2.00×10^{-9} (p)	12.88 (p)	—	—	—
12:2 Fluorotelomerphosphatediester; 12:2 diPAP	57677-99-3	F(CF$_2$)$_{12}$CH$_2$CH$_2$OP(OH)OCH$_2$CH$_2$	1,390.27	—	—	0 (p)	3.00×10^{-12} (p)	15.15 (p)	—	—	—
Metals					*Diester (diPAP)*						
Cobalt	7440-48-4	Co	58.93	8.9	Not applicable	Not applicable*	cobalt (II) carbonate 1.4 cobalt (II) chloride 562,000	not hydrophobic	sorption dependent on soil mineralogy and organic matter	1,495	2,927
Strontium	7440-24-6	Sr	87.62	2.5	Not applicable	Not applicable*	strontium carbonate 3 strontium chloride 538,000 strontium sulfate 135	not hydrophobic	sorption dependent on cation exchange capacity of soil, pH, presence of calcium	752	1,390
Vanadium	7440-62-2	V	50.94	6.0	Not applicable	Not applicable*	vanadium (V) oxide 700	not hydrophobic	sorption varies with soil mineralogy and pH	1,910	3,407
Molybdenum	7439-98-7	Mo	95.94	10.2	Not applicable	Not applicable*	molybdenum (IV) dioxide 64,700 molybdenum (VI) trioxide 1,400	not hydrophobic	sorption varies with soil mineralogy, pH and water chemistry	2,622	4,639
Manganese	7439-96-5	Mn	54.94	7.3	Not applicable	Not applicable*	manganese (II) chloride 773,000 manganese (II) carbonate 0.8	not hydrophobic	Sorption varies with mineralogy	1,246	2,061
Germanium	7440-56-4	Ge	72.59	5.3	Not applicable	Not applicable*	germanium (IV) dioxide 4,470	not hydrophobic	sorption varies with soil mineralogy and pH	937	2,700

(Continued)

Chemical Name(s)	CAS Registry Number	Molecular Formula	MW	ρ	K_H	V_p	S_w	log K_{ow}	log K_{oc}	MP	BP
Flame Retardants											
Polybrominated diphenyl ethers; PBDEs											
Heptabromodiphenyl Ethers; HeptaBDE (e.g., 2,2',3,4,4',5',6-heptabromodiphenyl ether)	189084-67-1	$C_{12}H_3Br_7O$	722.48	2.6	4.96×10^{-4} (p)	1.99×10^{-4} (p)	0.15 (p)	7.50 (p)	Not available	219 (p)	450 (p)
Hexabromodiphenyl Ethers; HexaBDE (e.g., 2,2',4,4',5,5'-Hexabromodiphenyl ether)	68631-49-2	$C_{12}H_4Br_6O$	643.62	2.5 (p)	2.56×10^{-6}	3.80×10^{-8}	9.00×10^{-4}	7.90	Not available	183	471 (p)
Pentabromodiphenyl Ethers; PentaBDE (e.g., 2,2',4,4',5-Pentabromodiphenyl ether)	60348-60-9	$C_{12}H_5Br_5O$	564.69	2.3 (p)	2.00×10^{-5} (p)	3.50×10^{-7}	0.013	6.84	4.34	−5	434 (p)
Tetrabromodiphenyl Ethers; TetraBDE (e.g., 2,2',4,4'-Tetrabromodiphenyl ether)	5436-43-1	$C_{12}H_6Br_4O$	485.79	2.2 (p)	6.23×10^{-4} (p)	0.11	100,000 (p)	6.81	4.27 (p)	83	395 (p)
Polybrominated biphenyls; PBBs											
Hexabromobiphenyl; HexaBB (e.g., 2,2',4,4',5,5'-Hexabromobiphenyl)	36355-01-8	$C_{12}H_4Br_6$	627.58	2.5 (p)	4.30×10^{-4} (p)	5.20×10^{-8}	0.011	6.39	3.33–3.87 (p)	72	484 (p)
Chlorinated organophosphates											
Tris(1,3-dichloroisopropyl) phosphate; TDCPP	13674-87-8	$C_9H_{15}C_{l6}O_4P$	430.89	1.5	2.61×10^{-9} (p)	2.86×10^{-7} (p)	7	3.65	3.04–3.97 (p)	27	315
Tris(2-chloroethyl) phosphate; TCEP	115-96-8	$C_6H_{12}Cl_3O_4P$	285.49	1.4	3.29×10^{-6} (p)	0.061	7,000	1.44	2.48–2.59 (p)	−55	330
Pharmaceuticals/EDCs/Personal Care Products (includes other pesticides)											
Ibuprofen; (RS)-2-(4-(2-methylpropyl)phenyl) propanoic acid	15687-27-1	$C_{13}H_{18}O_2$	206.28	1.0	8.20×10^{-5} (p)	4.74×10^{-5}	21	3.97	2.60–3.53 (p)	77	364
Di-2-ethylhexyl phthalate; bis(2-ethylhexyl) phthalate; DHEP	117-81-7	$C_{24}H_{38}O_4$	390.57	1.0	3.58×10^{-10} (p)	1.42×10^{-7}	0.27	7.60	4–6	−50	460
2,3,7,8-Tetrachlorodibenzo-para-dioxin; TCDD	1746-01-6	$C_{12}H_4Cl_4O_2$	321.97	1.8	3.94×10^{-4} (p)	1.50×10^{-9}	2.00×10^{-4}	6.80	7.39	305	417 (p)
Triclosan	3380-34-5	$C_{12}H_7Cl_3O_2$	289.54	1.5	2.10×10^{-8} (p)	4.60×10^{-6} (p)	10	4.76	3.38–4.20	56	120
Erythromycin	114-07-8	$C_{37}H_{67}NO_{13}$	733.93	1.2	1.27×10^{-11} (p)	2.12×10^{-25} (p)	2,000	3.06	2.76 (p)	191	818 (p)
Bisphenol A; BPA; 4,4'-(propane-2,2-diyl)diphenol	80-05-7	$C_{15}H_{16}O_2$	228.29	1.2	4.00×10^{-11} (p)	4.00×10^{-8}	120	3.32	2.06–3.59	155	200
Titanium dioxide	13463-67-7	TiO_2	79.87	4.2	Not applicable	Not applicable	Insoluble	not hydrophobic	1.38 (p)	1,855	2,500
Dichlorodiphenyltrichloroethane; p,p'-DDT; DDT	50-29-3	$C_{14}H_9Cl_5$	354.49	1.0	8.30×10^{-6}	1.60×10^{-7}	Insoluble	6.91	5.05–6.26	109	230 (decomposes)
Permethrin	52645-53-1	$C_{21}H_{20}Cl_2O_3$	391.29	1.2	2.40×10^{-8} (p)	5.18×10^{-7}	0.0011	6.50	4.02–4.93	34	200

Emerging Contaminant Properties

Chemical Name(s)	CAS Registry Number	Molecular Formula	MW	ρ	K_H	V_p	S_w	log K_{ow}	log K_{oc}	MP	BP
Tributos	78-48-8	$C_{12}H_{27}OPS_3$	314.54	1.1	2.94×10^{-7}	5.30×10^{-6}	0.7	5.70	3.69–4.10	39.7 (p)	210
Bromomethane; methyl bromide	74-83-9	CH_3Br	94.94	4.0	7.34×10^{-3}	1,620	15,200	1.19	0.95–1.92	−94	4
Other											
Sulfolane	126-33-0	$C_4H_8O_2S$	120.16	1.3	4.80×10^{-6} (p)	6.20×10^{-3}	Miscible	−0.77	0.60–1.54	27	285
Sulfur hexafluoride	2551-62-4	SF_6	146.05	0.007	4.52	1.78×10^4	31	1.68	1.46–1.93 (p)	−51 (sublimes)	Not applicable
1,1-Dichloroethane; 1,1-DCA	75-34-3	$C_2H_4Cl_2$	98.96	1.2	5.62×10^{-3}	287	5,040	1.79	0.96–1.48	−97	57
1,3-Butadiene	106-99-0	C_4H_6	54.09	0.6	7.36×10^{-2}	2,110	735	1.99	2.08	−109	−5
Bis(2-chloroethyl)ether; BCEE	111-44-4	$C_4H_8Cl_2O$	143.01	1.2	1.21×10^{-4} (p)	1.55	17,000	1.29	1.15	−50	139
Benzidine; 1,1'-biphenyl-4,4'-diamine	92-87-5	$C_{12}H_{12}N_2$	184.24	1.3	7.81×10^{-7} (p)	1.09×10^{-5} (p)	322	1.34	1.60–5.95	128	400
1-Butanol	71-36-3	$C_4H_{10}O$	74.12	0.8	8.81×10^{-6}	6.7	63,200	0.88	0.51	−89	118
2-Propen-1-ol; allyl alcohol	107-18-6	C_3H_6O	58.08	0.9	4.99×10^{-6}	25.4	Miscible	0.17	0.51	−129	97
1,1-Dimethylhydrazine; UDMH	57-14-7	$C_2H_8N_2$	60.10	0.8	1.30×10^{-5}	163	Miscible	−0.77 (p)	1.30 (p)	−58	63
N-nitrosodimethylamine; NDMA	62-75-9	$C_2H_6N_2O$	74.08	1.0	1.08×10^{-6}	2.7	Miscible	−0.57	1.83–2.07	−39 (p)	152
Cyclotrimethylenetrinitramine; RDX; cyclonite; 1,3,5-trinitroperhydro-1,3,5-triazine	121-82-4	$C_3H_6N_6O_6$	222.26	1.8	2.10×10^{-11} (p)	5.40×10^{-5}	59.7	0.87	1.62–2.22	205	280
Perchlorate (examples below)											
Ammonium perchlorate	7790-98-9	NH_4ClO_4	117.49	2.0	Not available	Very low	200,000	−5.84	1.98 (p)	130 (decomposes)	Not applicable
Sodium perchlorate	7601-89-0	$NaClO_4$	122.44	2.5	Not available	Very low	2,100,000	−7.18 (p)	Not available	482 (decomposes)	Not applicable
Potassium perchlorate	7778-74-7	$KClO_4$	138.55	2.5	Not available	Very low	15,000	−7.18 (p)	Not available	400 (decomposes)	Not applicable
Perchloric acid	7601-90-3	$HClO_4$	100.47	1.8	7.34×10^{-39} (p)	6.8	Miscible	−4.63 (p)	1.69 (p)	−64	203
Chlorodifluoromethane; HCFC-22; R-22; difluoromonochloromethane	75-45-6	$CHClF_2$	86.47	1.2	4.06×10^{-2}	7,250	2,770	1.08	0.93–1.54 (p)	−157	−41

(Continued)

Chemical Name(s)	CAS Registry Number	Molecular Formula	MW	ρ	K_H	V_p	S_w	log K_{ow}	log K_{oc}	MP	BP
Cyanotoxins											
Microcystin-LR	101043-37-2	$C_{49}H_{74}N_{10}O_{12}$	995.17	1.3	Not applicable	Not applicable	Highly	Not applicable	Not applicable	Not applicable	Not applicable
Disinfection by-product											
Bromochloroacetic acid	5589-96-8	$C_2H_2BrClO_2$	173.39	1.9	7.07×10^{-9}	0.14 (p)	250,000 (p)	0.61 (p)	0.28 (p)	Not available	215
Bromodichloroacetic acid	71133-14-7	$C_2HBrCl_2O_2$	207.84	2.3 (p)	7.26×10^{-9} (p)	0.17 (p)	4,900 (p)	1.55 (p)	0.44 (p)	51 (p)	219 (p)
Chlorodibromoacetic acid	5278-95-5	$C_2HBr_2ClO_2$	252.29	2.7 (p)	2.56×10^{-9} (p)	0.022 (p)	2,400 (p)	1.99 (p)	0.44 (p)	68 (p)	263 (p)
Tribromoacetic acid	75-96-7	CBr_3CO_2H	296.74	3.1 (p)	3.34×10^{-9}	2.80×10^{-4} (p)	200,000	1.98 (p)	0.44–0.72 (p)	132	245

Units in Table:
MW as g/mole
ρ is density as g/mL; typically at 20–25°
K_H as atm m3 mol-1
V_p as mmHg (see Note 2)
S_w is solubility in water as mg/L (see Note 3)
log K_{ow} is unitless
log K_{oc} is dimensionless or as L/kg
MP is melting point as °C
BP is boiling point as °C (see Note 4)

Primary Sources:
PubChem: Open Chemistry Database (https://pubchem.ncbi.nlm.nih.gov/)
ChemSpider (http://www.chemspider.com/)
USEPA Chemistry Dashboard (https://comptox.epa.gov/dashboard/)
PFAS values are from peer reviewed literature as summarized by Concawe (2016) Environmental fate and effects of poly and perfluoroalkyl substances (PFAS). Concawe Soil and Groundwater Taskforce, Brussels, Belgium. Prepared by ARCADIS: T. Pancras, G. Schrauwen, T. Held, K. Baker, I. Ross, H. Slenders, Report # 8/16. Accessed March 15, 2017. Concawe (www.concawe.eu) is a division of the European Petroleum Refiners Association based in Brussels, Belgium.

Notes:
*Not applicable under environmentally relevant conditions
Some values are approximate, based on the range of available literature values
1 Several experimentally derived values from other sources were also reviewed and were found to be within a factor of 1.5 of the value reported by (Mohr 2010).
2 typically at 25 °C and atmospheric pressure
3 typically at 25 °C
4 typically at sea level

Abbreviations: °C– degrees Celsius, atm m3 mol-1– atmosphere cubic meter per mole, CAS – Chemical Abstract Service, g/mL – grams per milliliter, g/mole– grams per mole, K_H– Henry's constant, K_{ow}– octanol:water partition coefficient, L/kg- liters per kilogram, mmHg– millimeters of mercury, V_p – vapor pressure (p)– predicted values are provided when experimentally derived values could not be readily located. K_{oc}– organic carbon:water partition coefficient.

Appendix D: NGI Preliminary List of Substances That Could Be Considered to Meet the PMT or vPvM Criteria

Name	EC#	CAS#
Dapsone	201-248-4	80-08-0
Dinoseb	201-861-7	88-85-7
Hexamethyldisiloxane	203-492-7	107-46-0
Melamine	203-615-4	108-78-1
Phenol	203-632-7	108-95-2
1,4,5,6,7,7-heachloro-8,9,10-trinorborn-5-ene-2,3-dicarboxylic anhydride	204-077-3	115-27-5
2,4,6-tribromophenol	204-278-6	118-79-6
4-nitrotoluene-2-sulphonic acid	204-445-3	121-03-9
tetrachloroethylene	204-825-9	127-18-4
Diuron	206-354-4	330-54-1
Cyanamide	206-992-3	420-04-2
4,4'-bis(dimethylamino)-4"-(methylamino)trityl alcohol	209-218-2	561-41-1
(Z)-N-9-octadecenylpropane-1,3-diamine	230-528-9	7173-62-8
Disodium 6-hydroxy-5-[(2-methoxy-4-sulphonato-m-tolyl)azo]naphthalene-2-sulphonate	247-368-0	25956-17-6
[[(phosphonomethyl)imino]bis[hexam ethylenenitrilobis(methylene)]]tetrakis phosphonic acid	252-156-6	34690-00-1
1,3,5-triazine-2,4,6(1H,3H,5H)-trione, compound with 1,3,5-triazine-2,4,6-triamine (1:1)	253-575-7	37640-57-6
Tetraethylammonium heptadecafluorooctanesulphonate	260-375-3	56773-42-3
Amines, N-C12-18-alkyltrimethylenedi-	268-957-9	68155-37-3
Quaternary ammonium compounds, C20-22-alkyltrimethyl, chlorides	271-756-9	68607-24-9
Ethanol, 2,2'-iminobis-, N-C12-18-alkyl derivs.	276-014-8	71786-60-2
4-mesyl-2-nitrotoluene	430-550-0	1671-49-4
sodium (4,6-dimethoxypyrimidin-2-yl)carbamoyl-[[3-(2,2,2-trifluoroethoxy)-2-pyridyl]sulfonyl]azanide	688-332-8	199119-58-9
ammonium 2,3,3,3-tetrafluoro-2-(heptafluoropropoxy)propanoate	700-242-3	62037-80-3
2,2'-dimethyl-2,2'-azodipropiononitrile	201-132-3	78-67-1
6-phenyl-1,3,5-triazine-2,4-diyldiamine	202-095-6	91-76-9
1,2-dichlorobenzene	202-425-9	95-50-1
Calcium cyanamide	205-861-8	156-62-7

(Continued)

Name	EC#	CAS#
2,2,3,3,5,5,6,6-octafluoro-4-(trifluoromethyl)morpholine	206-841-1	382-28-5
6-methyl-1,3,5-triazine-2,4-diyldiamine	208-796-3	542-02-9
1-nitroguanidine	209-143-5	556-88-7
Bis(2-dimethylaminoethyl)(methyl)amine	221-201-1	3030-47-5
N,N,N',N'-tetramethyl-2,2'-oxybis(ethylamine)	221-220-5	3033-62-3
4,4'-bis[4-[bis(2-hydroxyethyl)amino]-6-anilino-1,3,5-triazin-2-yl]amino]stilbene-2,2'-disulphonic acid	224-548-7	4404-43-7
2,2'-[(3,3'-dichloro[1,1'-biphenyl]-4,4'-diyl)bis(azo)]bis[N-(2-methylphenyl)-3-oxobutyramide]	226-789-3	5468-75-7
2,2'-[(3,3'-dichloro[1,1'-biphenyl]-4,4'-diyl)bis(azo)]bis[N-(4-chloro-2,5-dimethoxyphenyl)-3-oxobutyramide]	226-939-8	5567-15-7
Disodium 4,4'-diamino-9,9',10,10'-tetrahydro-9,9',10,10'-tetraoxo[1,1'-bianthracene]-3,3'-disulphonate	227-877-4	6022-22-6
[[(phosphonomethyl)imino]bis[(ethyle nenitrilo)bis(methylene)]] tetrakisphosp honic acid, sodium salt	244-751-4	22042-96-2
2,2'-[(2,2',5,5'-tetrachloro[1,1'-biphenyl]-4,4'-diyl)bis(azo)]bis[N-(2,4-dimethylphenyl)-3-oxobutyramide]	244-776-0	22094-93-5
5-amino-2,4,6-triiodoisophthalic acid	252-575-4	35453-19-1
1,3,5-triazine-2,4,6-triamine phosphate	255-449-7	41583-09-9
Pentasodium pentahydrogen [[(phosphonatomethyl)imino]bis[ethan e-2,1-diylnitrilobis(methylene)]]tetrakisphos phonate	263-212-4	61792-09-4
Butanamide, 2,2'-[(3,3'-dichloro[1,1'-biphenyl]-4,4'-diyl) bis(azo)]bis[3-oxo-, N,N'-bis(o-anisyl and 2,4-xylyl) derivs.	271-878-2	68610-86-6
Disodium oxybis[methylbenzenesulphonate]	277-242-0	73037-34-0
2-[(2-[2-(dimethylamino)ethoxy]ethyl)methyla mino]ethanol	406-080-7	83016-70-0
sodium 4,4-dimethyl-2,5-dioxoimidazolidin-1-ide	476-160-4	-
rac-5-Amino-N-(2,3-dihydroxypropyl)-2,4,6-triiodoisophthalamic acid	601-093-6	111453-32-8
Benzene, 1,1'-oxybis-, tetrapropylene derivs., sulfonated, sodium salts	601-601-6	119345-04-9
bis(nonafluorobutyl)phosphinic acid	700-183-3	52299-25-9
1,6-Bis[2,2-dimethyl-3-(N-morpholino)-propylideneamino]-hexane	700-570-7	1217271-49-2
Tetra potassium 5,5'-[ethane-1,2-diylbis[thio-1,3,4-thiadiazole-5,2-diyldiazene-2,1-diyl(5-amino-3-tert-buyl-1H-pyrazole-4,1-diyl)]}diisophthalate	700-616-6	849608-59-9
Diguanidinium carbonate	209-813-7	593-85-1
5,5'-(1H-isoindole-1,3(2H)-diylidene)dibarbituric acid	253-256-2	36888-99-0
4-amino-N-(1,1-dimethylethyl)-4,5-dihydro-3-(1-methylethyl)-5-oxo-1H-1,2,4-triazole-1-carboxamide	603-373-3	129909-90-6
1,2-dibromo-3-chloropropane	202-479-3	96-12-8
Propargite	219-006-1	2312-35-8
Tricarbonyl(methylcyclopentadienyl) manganese	235-166-5	12108-13-3
2,2'-(octadec-9-enylimino)bisethanol	246-807-3	25307-17-9
Alcohols, C8-14, γ-ω-perfluoro	269-927-8	68391-08-2

Preliminary List of PMT/vPvM Substances

Name	EC#	CAS#
N-[9-(dichloromethylidene)-1,2,3,4-tetrahydro-1,4-methanonaphthalen-5-yl]-3-(difluoromethyl)-1-methyl-1H-pyrazole-4-carboxamide	691-719-4	1072957-71-1
Trichloroethylene	201-167-4	79-01-6
2,6-xylidine	201-758-7	87-62-7
Toluene-2-sulphonamide	201-808-8	88-19-7
1,2-dichloroethane	203-458-1	107-06-2
4-aminophenol	204-616-2	123-30-8
1,4-dioxane	204-661-8	123-91-1
2,4,7,9-tetramethyldec-5-yne-4,7-diol	204-809-1	126-86-3
2-piperazin-1-ylethylamine	205-411-0	140-31-8
1,2,4-triazole	206-022-9	288-88-0
2-morpholinoethanol	210-734-5	622-40-2
Amantadine	212-201-2	768-94-5
Ametryn	212-634-7	834-12-8
Decyldimethylamine	214-302-7	1120-24-7
4-hydroxy-2,2,6,6-tetramethylpiperidinoxyl	218-760-9	2226-96-2
3-aminomethyl-3,5,5-trimethylcyclohexylamine	220-666-8	2855-13-2
Sulisobenzone	223-772-2	4065-45-6
Calcium 3-hydroxy-4-[(4-methyl-2-sulphonatophenyl)azo]-2-naphthoate	226-109-5	5281-04-9
2-phosphonobutane-1,2,4-tricarboxylic acid	253-733-5	37971-36-1
Amines, C16-18-alkyldimethyl	269-915-2	68390-97-6
Amines, C12-18-alkyldimethyl	269-923-6	68391-04-8
Amines, C12-14-alkyldimethyl	283-464-9	84649-84-3
rac-[(2R,4R)-2-(2,4-dichlorophenyl)-2-(1H-1,2,4-triazol-1-ylmethyl)-1,3-dioxolan-4-YL]methyl methanesulfonate monohydrochloride	445-550-6	155661-07-7
Hexadecyldimethylamine	203-997-2	112-69-6
Dimethyl(tetradecyl)amine	204-002-4	112-75-4
Anthracene	204-371-1	120-12-7
Dimantine	204-694-8	124-28-7
2,2'-azobis[2-methylbutyronitrile]	236-740-8	13472-08-7
Amines, (C16-18 and C18-unsatd.alkyl)dimethyl	268-220-1	68037-96-7
Amines, C12-16-alkyldimethyl	270-414-6	68439-70-3
Benzyltrimethylammonium chloride	200-300-3	56-93-9
Chloroform	200-663-8	67-66-3
Dimethyl sulfoxide	200-664-3	67-68-5
Iodomethane	200-819-5	74-88-4
1,1-dichloroethylene	200-864-0	75-35-4
Chlorotrimethylsilane	200-900-5	75-77-4
tert-butyl hydroperoxide	200-915-7	75-91-2
Triethyl phosphate	201-114-5	78-40-0
1,2-dichloropropane	201-152-2	78-87-5
2-nitropropane	201-209-1	79-46-9
4,4'-isopropylidenediphenol	201-245-8	80-05-7

(*Continued*)

Name	EC#	CAS#
4,4'-sulphonyldiphenol	201-250-5	80-09-1
4,4'-oxydi(benzenesulphonohydrazide)	201-286-1	80-51-3
1,3-dimethylimidazolidin-2-one	201-304-8	80-73-9
4,4'-diaminostilbene-2,2'-disulphonic acid	201-325-2	81-11-8
cyclohexane-1,2-dicarboxylic anydride	201-604-9	85-42-7
4-aminotoluene-3-sulphonic acid	201-831-3	88-44-8
2-nitrotoluene	201-853-3	88-72-2
1-chloro-2-nitrobenzene	201-854-9	88-73-3
2-nitrophenol	201-857-5	88-75-5
Naphthalene	202-049-5	91-20-3
benzotriazole	202-394-1	95-14-7
o-phenylenediamine	202-430-6	95-54-5
1,2,3-trichloropropane	202-486-1	96-18-4
1,3-di-o-tolylguanidine	202-577-6	97-39-2
Tetramethylthiuram monosulphide	202-605-7	97-74-5
4-tert-butylphenol	202-679-0	98-54-4
2-phenylpropene	202-705-0	98-83-9
Nitrobenzene	202-716-0	98-95-3
4-isopropylaniline	202-797-2	99-88-7
4-nitrotoluene	202-808-0	99-99-0
1-chloro-4-nitrobenzene	202-809-6	100-00-5
4-vinylpyridine	202-852-0	100-43-6
N-methylaniline	202-870-9	100-61-8
Methenamine	202-905-8	100-97-0
4,4'-oxydianiline	202-977-0	101-80-4
N-ethylaniline	203-135-5	103-69-5
1,4-dichlorobenzene	203-400-5	106-46-7
4-chloroaniline	203-401-0	106-47-8
4-chlorophenol	203-402-6	106-48-9
Dibutyl hydrogen phosphate	203-509-8	107-66-4
Diisopropyl ether	203-560-6	108-20-3
3-chloroaniline	203-581-0	108-42-9
Mesitylene	203-604-4	108-67-8
1-methylpiperazine	203-639-5	109-01-3
2-chloropyridine	203-646-3	109-09-1
Bis(2-methoxyethyl) ether	203-924-4	111-96-6
Dimethyl ether	204-065-8	115-10-6
1,2,3,4-tetrahydronaphthalene	204-340-2	119-64-2
Isoquinoline	204-341-8	119-65-3
2,4-dichlorophenol	204-429-6	120-83-2
Tetrahydrothiophene 1,1-dioxide	204-783-1	126-33-0
1,8-naphthylenediamine	207-529-8	479-27-6
2-(1,3-dihyrod-3-oxo-2H-indol-2-ylidene)-1,2-dihydro-3H-indol-3-one	207-586-9	482-89-3
1,3-dichlorobenzene	208-792-1	541-73-1
3,3'-sulphonyldianiline	209-967-5	599-61-1

Preliminary List of PMT/vPvM Substances

Name	EC#	CAS#
4-(α,α-dimethylbenzyl)phenol	209-968-0	599-64-4
Triphenylphosphine	210-036-0	603-35-0
1,3-dichloro-4-nitrobenzene	210-248-3	611-06-3
4-chloro-o-xylene	210-438-6	615-60-1
1,1,1,3,3,3-hexafluoropropan-2-ol	213-059-4	920-66-1
Dimethoxydimethylsilane	214-189-4	1112-39-6
Nitrotoluene	215-311-9	1321-12-6
Aminoiminomethanesulphinic acid	217-157-8	1758-73-2
4,4'-methylenebis(cyclohexylamine)	217-168-8	1761-71-3
Butan-2-one O,O',O"-(vinylsilylidyne)trioxime	218-747-8	2224-33-1
O,O-dimethylphosphorochloridothioate	219-754-9	2524-03-0
N-N-dimethylacrylamide	220-237-5	284-95-7
Cyclohexylidenebis[tert-butyl]peroxide	221-111-2	3006-86-8
N-,4-dimethylpentyl)-N'-phenylbenzene-1,4-diamine	221-374-3	3018-01-4
N,N'-bis(1,4-dimethylpentyl)-p-phenylenediamine	221-375-9	3081-14-9
Triclosan	222-182-2	3380-34-5
Tetraoctyltin	222-733-7	3590-84-9
N,N-diethylhydroxylamine	223-055-4	3710-84-7
p-(2,3-epoxypropoxy)-N,N-bis(2,3-epoxypropyl)aniline	225-716-2	5026-74-4
Sodium-hydrogen 4-amino-5-hydroxynaphthalene-2,7-disulphonate	226-736-4	5460-09-3
4-methylthiosemicarbazide	229-563-2	6610-29-3
1-(m-chlorophenyl)piperazine	229-654-7	6640-24-0
1,8-diazabicyclo[5.4.0]undec-7-ene	229-713-7	6674-22-2
2,2'-dimethyl-4,4'-methylenebis(cyclohexylamine)	229-962-1	6864-37-5
Tetrahydro-1,3-dimethyl-1H-pyrimidin-2-one	230-625-6	7226-23-5
Disodium 4,4'-diaminostilbene-2,2'-disulphonate	230-847-3	7336-20-1
3,4-diaminobenzenesulphonic acid	231-274-1	7474-78-4
1,1'-(methylenedi-p-phenylene)bismaleimide	237-163-4	13676-54-5
Salbutamol	242-424-0	18559-94-9
Butan-2-one O,O',O"-(methylsilylidyne)trioxime	245-366-4	22984-54-9
Dinitrotoluene	246-836-1	25321-14-6
7-aminonaphthalene-1,3,5-trisulphonic acid	248-394-5	27310-25-4
O-(4-bromo-2-chlorophenyl) O-ethyl S-propyl phosphorothioate	255-255-2	41198-08-7
4-amino-2,5-dimethoxy-N-methylbenzenesulphonamide	256-435-3	49701-24-8
Sodium 3-(allyloxy)-2-hydroxypropanesulphonate	258-004-5	52556-42-0
1,3-dihydro-4(or 5)-methyl-2H-benzimidazole-2-thione	258-904-8	53988-10-6
Barium bis[5-chloro-4-ethyl-2-[(2-hydroxy-1-naphthyl)azo] benzenesulphonate]	267-122-6	67801-01-8
Pyridine, alkyl derivs	269-929-9	68391-11-7
m-(2,3-epoxypropoxy)-N,N-bis(2,3-epoxypropyl)aniline	275-662-9	71604-74-5
Distillates (coal tar), benzole fraction	283-482-7	84650-02-2
4-[[(2-aminophenyl)methyl]amino]cyclohexy l acetate	298-842-9	93839-71-5
2-methyl-1-(4-methylthiophenyl)-2-morpholinopropan-1-one	400-600-6	71868-10-5
4,6-bis(octylthiomethyl)-o-cresol	402-860-6	110553-27-0
4,4',4"-(ethan-1,1,1-triyl)triphenol	405-800-7	27955-94-8

(Continued)

Name	EC#	CAS#
2,5-bis-isocyanatomethyl-bicyclo[2.2.1]heptane	411-280-2	74091-64-8
2-amino-5-methylthiazole	423-800-5	7305-71-7
2-fluoro-6-trifluoromethylpyridine	428-100-3	94239-04-0
4-(1-Phenylethyl)-benzene-1,3-diol	480-070-0	85-27-8
Tin, dioctylbis(2,4-pentanedionato-κO2,κO4)-	483-270-6	-
4,4'-Isopropylidenediphenol, oligomeric reaction products with 1-chloro-2,3-epoxypropane	500-033-5	25068-38-6
[2-butyl-4-chloro-1-({4-[2-(2H-1,2,3,4-tetrazol-5-yl)phenyl] phenyl}methyl)-1H-imidazol-5-yl]methanol	601-329-8	114798-26-4
N,N-dimethyl-1-adamantanamine	609-335-2	3717-40-6
sodium 2-amino-5-methylbenzenesulfonate	611-210-2	54914-95-3
No International Union of Pure and Applied Chemistry name	614-637-2	68603-75-8
(2E)-N,6,6-trimethyl-N-(1-naphthylmethyl) hept-2-en-4-yn-1-amine	618-706-8	91161-71-6
4,4'-methylenebis(N-sec-butylcyclohexamine	679-514-8	154279-60-4
Norethisterone	200-681-6	68-22-4
3a,4,7,7a-tetrahydro-4,7-methanoindene	201-052-9	77-73-6
α,α-dimethylbenzyl hydroperoxide	201-254-7	80-15-9
Bis(α,α-dimethylbenzyl) peroxide	201-279-3	80-43-3
Carbazole	201-696-0	86-74-8
α,α,α-trifluoro-o-toluidine	201-806-7	88-17-5
Benzothiazole	202-396-2	95-16-9
1-chloro-2,4-dinitrobenzene		97-00-7
4-nitro-o-anisidine	202-588-6	97-52-9
1,3-diphenylguanidine	203-002-1	102-06-7
1,3-diphenyl-2-thiourea	203-004-2	102-08-9
1-bromo-3-chloropropane	203-697-1	109-70-6
2,2'-iminodi(ethylamine)	203-865-4	111-40-0
Benzophenone	204-337-6	119-61-9
6-methoxy-m-toluidine	204-419-1	120-71-8
2-chloro-4-nitrotoluene	204-501-7	121-86-8
Ethisterone	207-096-5	434-03-7
1,5-naphthylenediamine	218-817-8	2243-62-1
1-(4-methyl-2-nitrophenylazo)-2-naphthol	219-372-2	2425-85-6
2-(2H-benzotriazol-2-yl)-p-cresol	219-470-5	2440-22-4
2,4,6,8-tetramethyl-2,4,6,8-tetravinylcyclotetrasiloxane	219-863-1	2554-06-5
Dibutyltin bis(2-ethylhexanoate)	220-481-2	2781-10-4
4'-chloro-2',5'-dimethoxyacetoacetanilide	224-638-6	4433-79-8
4-chloro-2,5-dimethoxyaniline	228-782-0	6358-64-1
Tris[2-chloro-1-(chloromethyl)ethyl] phosphate	237-159-2	13674-87-8
Diisopropylbenzene	246-835-6	25321-09-9
Diaminotoluene	246-910-3	25376-45-8
Hexahydromethylphthalic anhydride	247-094-1	25550-51-0
Dibenzyltoluene	248-097-0	26898-17-9
4-[[4-(aminocarbonyl)phenyl]azo]-3-hydroxy-N-(2-methoxyphenyl)naphthalene-2-carboxamide	248-097-0	26898-17-9

Preliminary List of PMT/vPvM Substances

Name	EC#	CAS#
Climbazole	253-775-4	38083-17-9
Resin acids and Rosin acids, esters with ethylene glycol	270-986-7	68512-65-2
Phenol, isopropylated, phosphate (3:1)	273-066-3	68937-41-7
1,4-Benzenediamine, N,N'-mixed Ph and tolyl derivs.	273-227-8	68953-84-4
Benzyl(3-hydroxyphenacyl)methylammonium chloride	276-017-4	71786-67-9
Barium bis[6-chloro-4-[(2-hydroxy-1-naphthyl)azo]toluene-3-sulphonate]	277-553-1	73612-34-7
Phenyl bis(2,4,6-trimethylbenzoyl)-phosphine oxide	423-340-5	162881-26-7
4'-Bromomethylbiphenyl-2-carbonitrile	425-280-5	114772-54-2
(2S)-3-methyl-2-(N-{[2'-(1H-1,2,3,4-tetrazol-5-yl)-[1,1-biphenyl]-4-yl]methyl}pentanamido)butanoic acid	604-045-2	137862-53-4
3-(4-Methyl-1H-imidazol-1-yl)-5-(trifluoromethyl)aniline	688-269-6	641571-11-1

Source: NGI. 2018.

Technical Note: Preliminary Assessment of Substances Registered under REACH that could Fulfil the Proposed PMT/vPvM Criteria. Document Number 20160526-TN-01, Trondheim, Norway: Norwegian Geotechnical Institute.

Appendix E.1: Summary of PFAS Environmental Standards: Soil

Region	Country	State/Jurisdiction	Compound	Type	Units	Value(s)	Additional PFASs Included in Value	Link
Australasia	Australia	Federal (Airports)	6:2 FTS	HISLs	mg/kg	60(i); 900(i)	N/A	1
Australasia	Australia	Federal (Airports)	PFOA	HISLs	mg/kg	3.73(o); 16(i); 240(i)	PFOA and 8:2 FtS	1
Australasia	Australia	Federal (Airports)	PFOS	HISLs	mg/kg	0.373 to 4.71(o); 6(i); 90(i)	N/A	1
Australasia	Australia	N/A	PFOA	CRC CARE Health Screening Levels	mg/kg	220(res); 900(res); 6,400(i); 450(o)	N/A	2
Australasia	Australia	N/A	PFOS	CRC CARE Health Screening Levels	mg/kg	22(res); 90(res); 640(i); 45(o)	PFOS and PFHxS	2
Australasia	Australia	N/A	6:2 FTS	DoD HH	mg/kg	60(res)	N/A	3
Australasia	Australia	N/A	PFOA	DoD HH	mg/kg	16(res)	N/A	3
Australasia	Australia	N/A	PFOS	DoD HH	mg/kg	6(res)	PFOS and PFHxS	3
Australasia	Australia	N/A	PFOA	DEE	mg/kg	29(res); 81(i); 1(o)	N/A	4
Australasia	Australia	N/A	PFOS	DEE	mg/kg	0.01(res); 0.13(i); 0.19(i); 60(i); 0.01(o); 6.6(o); 32(o)	N/A	4
Australasia	Australia/New Zealand	N/A	PFOA	HEPA	mg/kg	0.1(res); 20(res); 50(i); 10(o); 10(o)	N/A	5
Australasia	Australia/New Zealand	N/A	PFOS	HEPA	mg/kg	0.009(res); 2(res); 20(i); 1(o); 1.9(o); 4.6(o); 8.2(o)	PFOS and PFHxS	5
Australasia	Australia/New Zealand	N/A	PFOS	HEPA	mg/kg	0.01(res); 0.14(i); 1(o)	N/A	5
Australasia	Australia	Victoria	PFOA	Various	mg/kg	0.1(res); 20(res); 100(i)	N/A	6
Australasia	Australia	Victoria	PFOS	Various	mg/kg	0.009(res); 2(res); 20(i)	PFOS and PFHxS	6
Australasia	Australia	Western Australia	PFOS	Department of Regulation - HH	mg/kg	4(res); 100(i)	N/A	7

PFAS Standards: Soil

Region	Country	State/Jurisdiction	Compound	Type	Units	Value(s)	Additional PFASs Included in Value	Link
Europe	Denmark	N/A	PFOS	Health-Based Value	mg/kg	0.4	PFOS, PFOA, PFOSA, PFBS, PFBA, PFPeA, PFHxA, PFHpA, PFNA, PFDA, PFHxS, 6:2 FtS	8
Europe	Germany	Baden-Wurttemberg	Sum of PFASs	Landfilling at Landfill Class DKII	µg/kg	<20,000 (plus leachate requirements)	PFOS, PFOA, 6:2 FtS, PFNA, PFDA, PFHpS, PFHpA, PFHxS, PFHxA, PFPeS, PFPeA, PFBS, PFBA	10
Europe	Germany	Baden-Wurttemberg	Sum of PFASs	Landfilling at Landfill Class DKIII	µg/kg	≥20,000 to <50,000 (plus leachate requirements)	PFOS, PFOA, 6:2 FtS, PFNA, PFDA, PFHpS, PFHpA, PFHxS, PFHxA, PFPeS, PFPeA, PFBS, PFBA	10
Europe	Germany	Baden-Wurttemberg	Sum of PFASs	Landfill Disposal Not Allowed	µg/kg	≥50,000 (plus leachate requirements)	PFOS, PFOA, 6:2 FtS, PFNA, PFDA, PFHpS, PFHpA, PFHxS, PFHxA, PFPeS, PFPeA, PFBS, PFBA	10
Europe	Italy	N/A	PFOA	Risk-Based Value	mg/kg	5(i)	N/A	11
Europe	Netherlands	Gemeente Haarlemmermeer and Provincie Noord-Holland	PFOA	Contamination Determination	µg/kg	0.1 to >674	N/A	12, 13

(Continued)

Region	Country	State/Jurisdiction	Compound	Type	Units	Value(s)	Additional PFASs Included in Value	Link
Europe	Netherlands	Gemeente Haarlemmermeer and Provincie Noord-Holland	PFOS	Contamination Determination	µg/kg	0.1 to >8	N/A	12, 13
Europe	Netherlands	Gemeente Haarlemmermeer and Gemeente Aalsmeer	PFOA	Based on Land Use	µg/kg	0.1 to 674	N/A	12, 97
Europe	Netherlands	Gemeente Haarlemmermeer and Gemeente Aalsmeer	PFOS	Based on Land Use	µg/kg	0.1 to 8	N/A	12, 97
Europe	Netherlands	Zuid-Holland-Zuid	PFOA	Based on Soil Type	µg/kg	2.5 (Zone A); 10 (Zone B)	N/A	98
Europe	Netherlands	N/A	PFOA	Risk-Based Value – Ecological	µg/kg	7 to 50,000	N/A	14
Europe	Netherlands	N/A	PFOA	Risk-Based Value – Human	µg/kg	0.6 to 4,200	N/A	14
Europe	Netherlands	N/A	PFOS	Generic Boundaries	µg/kg	0.1 to 6,600	N/A	15
Europe	Netherlands	N/A	PFOS	Risk-Based Value – Ecological	µg/kg	7.1 to 16,000	N/A	15
Europe	Netherlands	N/A	PFOS	Risk-Based Value – Human	µg/kg	0.78 to 16,000	N/A	15
Europe	Norway	N/A	PFOS	Risk-Based Value	mg/kg	0.002; 0.1	N/A	16
Europe	Sweden	N/A	PFOS	Based on Class of Soil	mg/kg	0.003; 0.02	N/A	17
N. Am	Canada	N/A	PFOA	Soil Screening Values/ Federal Quality Guidelines	mg/kg	0.85(res); 1.28(i); 12.1(i)	N/A	18
N. Am	Canada	N/A	PFOS	Soil Screening Values/ Federal Quality Guidelines	mg/kg	0.01(res); 2.1(res); 0.14(i); 0.21(i); 0.21(i); 3.2(i); 30.5(i); 0.01(o)	N/A	18, 19
N. Am	USA	USEPA	PFBS	Risk- Based Screening Level	mg/kg	1,260(res); 1.82(i); 1,640(i); 4,670(i); 0.13(o)	N/A	20

PFAS Standards: Soil

Region	Country	State/Jurisdiction	Compound	Type	Units	Value(s)	Additional PFASs Included in Value	Link
N. Am	USA	USEPA	PFOA	Risk- Based Screening Level	mg/kg	1.26(res); 1.82(i); 16.4(i); 46.7(i); 0.000172(o)	N/A	20
N. Am	USA	USEPA	PFOS	Risk- Based Screening Level	mg/kg	1.26(res); 1.82(i); 16.4(i); 46.7(i); 0.000378(o)	N/A	20
N. Am	USA	Alaska	PFOA	Risk-Based Value – Based on Class of Soil	mg/kg	0.0017 to 2.2	N/A	21
N. Am	USA	Alaska	PFOS	Risk-Based Value – Based on Class of Soil	mg/kg	0.003 to 2.2	N/A	21
N. Am	USA	Delaware	PFBS	Screening Levels - Hazardous Substance Cleanup Act	mg/kg	130	N/A	22
N. Am	USA	Delaware	PFOA	Screening Levels - Hazardous Substance Cleanup Act	mg/kg	16	N/A	22
N. Am	USA	Delaware	PFOS	Screening Levels - Hazardous Substance Cleanup Act	mg/kg	6	N/A	22
N. Am	USA	Indiana	PFBS	Screening Levels	mg/kg	1,800(res); 16,000(i); 34,000(o)	N/A	23
N. Am	USA	Iowa	PFOA	Statewide Brownfield Standards for Soil	mg/kg	1.2	N/A	24
N. Am	USA	Iowa	PFOS	Statewide Brownfield Standards for Soil	mg/kg	1.8	N/A	24
N. Am	USA	Maine	PFOA	Remedial Action Guideline - Based on Activity	mg/kg	0.8 to 6.2	N/A	25
N. Am	USA	Maine	PFOS	Remedial Action Guideline - Based on Activity	mg/kg	11 to 82	N/A	25
N. Am	USA	Michigan	PFOA	Risk-Based Criteria	mg/kg	2.1(res); 25(i); 0.059(o)	N/A	26

(Continued)

Region	Country	State/Jurisdiction	Compound	Type	Units	Value(s)	Additional PFASs Included in Value	Link
N. Am	USA	Michigan	PFOS	Risk-Based Criteria	mg/kg	2.1(res); 25(i); 0.0014(o)	N/A	26
N. Am	USA	Minnesota	PFBA	Soil Reference Value	mg/kg	63(res); 150(i)	N/A	27
N. Am	USA	Minnesota	PFBS	Soil Reference Value	mg/kg	30(res); 450(i)	N/A	27
N. Am	USA	Minnesota	PFOA	Soil Reference Value	mg/kg	0.33(res); 0.33(i)	N/A	27
N. Am	USA	Minnesota	PFOS	Soil Reference Value	mg/kg	1.7(res); 21(i)	N/A	27
N. Am	USA	Nevada	PFBS	Basic Comparison Levels	mg/kg	125(res); 125(i)	N/A	28
N. Am	USA	Nevada	PFOA	Basic Comparison Levels	mg/kg	1.56(res); 26(i)	N/A	28
N. Am	USA	Nevada	PFOS	Basic Comparison Levels	mg/kg	1.56(res); 26(i)	N/A	28
N. Am	USA	New Hampshire	PFOA	Risk-Based Soil Concentration	mg/kg	0.5(res); 4.3(i)	N/A	29
N. Am	USA	New Hampshire	PFOS	Risk-Based Soil Concentration	mg/kg	0.5(res); 4.3(i)	N/A	30
N. Am	USA	North Carolina	PFBS	Preliminary Soil Remediation Goal	mg/kg	250(res); 3,300(i)	N/A	31
N. Am	USA	North Carolina	PFOA	Preliminary Soil Remediation Goal	mg/kg	0.017(o)	N/A	31
N. Am	USA	Texas	PFBA	Risk-Based Criteria (res)	mg/kg	0.098 to 1,000,000	N/A	32
N. Am	USA	Texas	PFBA	Risk-Based Criteria (i)	mg/kg	0.29 to 1,000,000	N/A	32
N. Am	USA	Texas	PFBS	Risk-Based Criteria (res)	mg/kg	0.053 to 1,000,000	N/A	32
N. Am	USA	Texas	PFBS	Risk-Based Criteria (i)	mg/kg	0.16 to 1,000,000	N/A	32
N. Am	USA	Texas	PFPeA	Risk-Based Criteria (res)	mg/kg	0.00016 to 1.4	N/A	32
N. Am	USA	Texas	PFPeA	Risk-Based Criteria (i)	mg/kg	0.00048 to 7.8	N/A	32
N. Am	USA	Texas	PFHxA	Risk-Based Criteria (res)	mg/kg	0.00024 to 1.4	N/A	32
N. Am	USA	Texas	PFHxA	Risk-Based Criteria (i)	mg/kg	0.00072 to 7.8	N/A	32
N. Am	USA	Texas	PFHxS	Risk-Based Criteria (res)	mg/kg	0.001 to 6,300	N/A	32
N. Am	USA	Texas	PFHxS	Risk-Based Criteria (i)	mg/kg	0.003 to 8,800	N/A	32
N. Am	USA	Texas	PFHpA	Risk-Based Criteria (res)	mg/kg	0.0023 to 8.2	N/A	32
N. Am	USA	Texas	PFHpA	Risk-Based Criteria (i)	mg/kg	0.0068 to 47	N/A	32
N. Am	USA	Texas	PFOA	Risk-Based Criteria (res)	mg/kg	0.0015 to 910	N/A	32

PFAS Standards: Soil 407

Region	Country	State/Jurisdiction	Compound	Type	Units	Value(s)	Additional PFASs Included in Value	Link
N. Am	USA	Texas	PFOA	Risk-Based Criteria (i)	mg/kg	0.0044 to 1,300	N/A	32
N. Am	USA	Texas	PFOS	Risk-Based Criteria (res)	mg/kg	0.025 to 160,000	N/A	32
N. Am	USA	Texas	PFOS	Risk-Based Criteria (i)	mg/kg	0.075 to 230,000	N/A	32
N. Am	USA	Texas	PFOSA	Risk-Based Criteria (res)	mg/kg	0.011 to 92	N/A	32
N. Am	USA	Texas	PFOSA	Risk-Based Criteria (i)	mg/kg	0.015 to 280	N/A	32
N. Am	USA	Texas	PFNA	Risk-Based Criteria (res)	mg/kg	0.0015 to 6,800	N/A	32
N. Am	USA	Texas	PFNA	Risk-Based Criteria (i)	mg/kg	0.0046 to 9,500	N/A	32
N. Am	USA	Texas	PFDA	Risk-Based Criteria (res)	mg/kg	0.011 to 78,000	N/A	32
N. Am	USA	Texas	PFDA	Risk-Based Criteria (i)	mg/kg	0.033 to 110,000	N/A	32
N. Am	USA	Texas	PFTrDA	Risk-Based Criteria (res)	mg/kg	0.03 to 6.1	N/A	32
N. Am	USA	Texas	PFTrDA	Risk-Based Criteria (i)	mg/kg	0.091 to 25	N/A	32
N. Am	USA	Texas	PFTeDA	Risk-Based Criteria (res)	mg/kg	0.056 to 11	N/A	32
N. Am	USA	Texas	PFTeDA	Risk-Based Criteria (i)	mg/kg	0.17 to 34	N/A	32
N. Am	USA	Texas	PFDS	Risk-Based Criteria (res)	mg/kg	0.02 to 4.3	N/A	32
N. Am	USA	Texas	PFDS	Risk-Based Criteria (i)	mg/kg	0.06 to 25	N/A	32
N. Am	USA	Texas	PFUA	Risk-Based Criteria (res)	mg/kg	0.0092 to 4.3	N/A	32
N. Am	USA	Texas	PFUA	Risk-Based Criteria (i)	mg/kg	0.027 to 25	N/A	32
N. Am	USA	Texas	PFDoA	Risk-Based Criteria (res)	mg/kg	0.017 to 64,000	N/A	32
N. Am	USA	Texas	PFDoA	Risk-Based Criteria (i)	mg/kg	0.05 to 89,000	N/A	32
N. Am	USA	Vermont	PFOA	Soil Screening Value - Resident Soil	mg/kg	0.3	N/A	33
N. Am	USA	West Virginia	PFOA	Residential Soil	mg/kg	240	N/A	34
N. Am	USA	Wisconsin	PFBS	Residual Contaminant Level	mg/kg	1,260(res); 16,400(i)	N/A	35
N. Am	USA	Wisconsin	PFOA	Residual Contaminant Level	mg/kg	1.26(res); 16.4(i)	N/A	35
N. Am	USA	Wisconsin	PFOS	Residual Contaminant Level	mg/kg	1.26(res); 16.4(i)	N/A	35

Appendix E.2: Summary of PFAS Environmental Standards: Groundwater

PFAS Standards: Groundwater 409

Region	Country	State or Jurisdiction (if applicable)	Compound	Type	Units	Value	Additional PFASs Included in Value	Link
Australasia	Australia	N/A	PFOA	Contact and Maintenance Values	ng/L	50,000	N/A	2
Australasia	Australia	N/A	PFOS	Contact and Maintenance Values	ng/L	5,000	PFOS and PFHxS	2
Australasia	Australia	New South Wales	Sum of PFASs	Trigger Points 1, 2, and 3	ng/L	1=10,000; 2=100; 3=50	All PFAS analytes measured	36
Australasia	Australia	Western Australia	PFOS	Non-Potable and Recreational Uses	ng/L	5,000	N/A	7
Europe	Denmark	N/A	Sum of PFASs	Health-Based Value	ng/L	100	PFOS, PFOA, PFOSA, PFBS, PFBA, PFPeA, PFHxA, PFHpA, PFNA, PFDA, PFHxS, 6:2 FtS	8
Europe	Germany	Baden-Wurttemberg	6:2 FtS	Provisional Threshold Value	µg/L	0.3	N/A	37
Europe	Germany	Baden-Wurttemberg	PFBA	Provisional Threshold Value	µg/L	7	N/A	37
Europe	Germany	Baden-Wurttemberg	PFBS	Provisional Threshold Value	µg/L	3	N/A	37
Europe	Germany	Baden-Wurttemberg	PFPeA	Provisional Threshold Value	µg/L	3	N/A	37
Europe	Germany	Baden-Wurttemberg	PFPeS	Provisional Threshold Value	µg/L	1	N/A	37
Europe	Germany	Baden-Wurttemberg	PFHxA	Provisional Threshold Value	µg/L	1	N/A	37
Europe	Germany	Baden-Wurttemberg	PFHxS	Provisional Threshold Value	µg/L	0.3	N/A	37
Europe	Germany	Baden-Wurttemberg	PFHpA	Provisional Threshold Value	µg/L	0.3	N/A	37
Europe	Germany	Baden-Wurttemberg	PFHpS	Provisional Threshold Value	µg/L	0.3	N/A	37
Europe	Germany	Baden-Wurttemberg	PFOA	Provisional Threshold Value	µg/L	0.3	N/A	37
Europe	Germany	Baden-Wurttemberg	PFOS, alone	Provisional Threshold Value	µg/L	0.23	N/A	37
Europe	Germany	Baden-Wurttemberg	PFOS, if other PFASs present	Provisional Threshold Value	µg/L	0.3	N/A	37
Europe	Germany	Baden-Wurttemberg	PFNA	Provisional Threshold Value	µg/L	0.3	N/A	37

(*Continued*)

Region	Country	State or Jurisdiction (if applicable)	Compound	Type	Units	Value	Additional PFASs Included in Value	Link
Europe	Germany	Baden-Wurttemberg	PFDA	Provisional Threshold Value	µg/L	0.3	N/A	37
Europe	Germany	Baden-Wurttemberg	PFASs listed above	Provisional Threshold Value - Additional Rule	Sum of quotients	$\sum$ (Measured Concentration / Provisional Threshold Value) ≤ 1	PFASs listed above	37
Europe	Germany	Baden-Wurttemberg	PFASs other than those listed above	Provisional Threshold Value	µg/L	1 (each)	N/A	37
Europe	Germany	Bavaria	6:2 FtS	Provisional Threshold Value	ng/L	100	N/A	38
Europe	Germany	Bavaria	PFBA	Provisional Threshold Value	ng/L	10,000	N/A	38
Europe	Germany	Bavaria	PFBS	Provisional Threshold Value	ng/L	6,000	N/A	38
Europe	Germany	Bavaria	PFPeA	Provisional Threshold Value	ng/L	3,000	N/A	38
Europe	Germany	Bavaria	PFHxA	Provisional Threshold Value	ng/L	6,000	N/A	38
Europe	Germany	Bavaria	PFHxS	Provisional Threshold Value	ng/L	100	N/A	38
Europe	Germany	Bavaria	PFHpA	Provisional Threshold Value	ng/L	300	N/A	38
Europe	Germany	Bavaria	PFHpS	Provisional Threshold Value	ng/L	300	N/A	38
Europe	Germany	Bavaria	PFOA	Provisional Threshold Value	ng/L	100	N/A	38
Europe	Germany	Bavaria	PFOS	Provisional Threshold Value	ng/L	100	N/A	38
Europe	Germany	Bavaria	PFOSA	Provisional Threshold Value	ng/L	100	N/A	38
Europe	Germany	Bavaria	PFNA	Provisional Threshold Value	ng/L	60	N/A	38
Europe	Germany	Bavaria	PFDA	Provisional Threshold Value	ng/L	100	N/A	38
Europe	Germany	Bavaria	PFNA, PFOS, PFOA, PFHxS, PFHxA, PFBS, PFBA	Provisional Threshold Value	Sum of quotients	$\sum$ (Measured Concentration / Provisional Threshold Value) ≤ 1	See left	38

PFAS Standards: Groundwater 411

Region	Country	State or Jurisdiction (if applicable)	Compound	Type	Units	Value	Additional PFASs Included in Value	Link
Europe	Netherlands	Gemeente Haarlemmermeer	PFOA	Contamination Determination	µg/L	0.01 to >0.39	N/A	12
Europe	Netherlands	Gemeente Haarlemmermeer	PFOS	Contamination Determination	µg/L	0.01 to >4.7	N/A	12, 13
Europe	Netherlands	N/A	PFOA	Risk-Based Value – Ecological	µg/L	30 to 7,000	N/A	99
Europe	Netherlands	N/A	PFOA	Risk-Based Value – Human	µg/L	0.0875 to 607	N/A	99
Europe	Netherlands	N/A	PFOS	Generic Boundaries	µg/L	0.00023 to 4.7	N/A	15
Europe	Netherlands	N/A	PFOS	Risk-Based Value – Ecological	µg/L	0.33 to 930	N/A	15
Europe	Netherlands	N/A	PFOS	Risk-Based Value – Human	µg/L	0.038 to 740	N/A	15
Europe	Switzerland	N/A	6:2 FtS	Provisional Threshold Value	µg/L	7	N/A	39
Europe	Switzerland	N/A	PFBA	Provisional Threshold Value	µg/L	700	N/A	39
Europe	Switzerland	N/A	PFBS	Provisional Threshold Value	µg/L	700	N/A	39
Europe	Switzerland	N/A	PFPeA	Provisional Threshold Value	µg/L	100	N/A	39
Europe	Switzerland	N/A	PFHxA	Provisional Threshold Value	µg/L	40	N/A	39
Europe	Switzerland	N/A	PFHxS	Provisional Threshold Value	µg/L	0.7	N/A	39
Europe	Switzerland	N/A	PFHpA	Provisional Threshold Value	µg/L	9	N/A	39
Europe	Switzerland	N/A	PFOA	Provisional Threshold Value	µg/L	4	N/A	39
Europe	Switzerland	N/A	PFOS	Provisional Threshold Value	µg/L	0.7	N/A	39
Europe	UK	N/A	PFOS	Maintenance Concentration	µg/L	1	N/A	40
N. Am	Canada	NA	PFOS	Federal Groundwater Quality Guideline	ng/L	6,800	N/A	19
N. Am	USA	Alaska	PFOA	Risk-Based Level	ng/L	400	N/A	21
N. Am	USA	Alaska	PFOS	Risk-Based Level	ng/L	400	N/A	21
N. Am	USA	Colorado	PFOA and PFOS	Site-Specific Groundwater Standard	ng/L	70	PFOS and PFOA	41

(*Continued*)

Region	Country	State or Jurisdiction (if applicable)	Compound	Type	Units	Value	Additional PFASs Included in Value	Link
N. Am	USA	Connecticut	Sum of PFASs	Groundwater Action Level	ng/L	70	PFOA, PFOS, PFNA, PFHxS, PFHpA	42
N. Am	USA	Delaware	PFBS	Screening Levels - Hazardous Substance Cleanup Act	ng/L	40,000	N/A	22
N. Am	USA	Delaware	PFOA and PFOS	Screening Levels - Hazardous Substance Cleanup Act	ng/L	70	PFOA and PFOS	22
N. Am	USA	Iowa	PFOA	State Brownfield Standards	ng/L	70 (protected)	N/A	24
N. Am	USA	Iowa	PFOS	State Brownfield Standards	ng/L	70 (protected); 1,000 (non-protected)	N/A	24
N. Am	USA	Maine	PFOA	Remedial Action Guideline	ng/L	130(res); 220(i)	N/A	25
N. Am	USA	Maine	PFOS	Remedial Action Guideline	ng/L	560(res); 5,300(i)	N/A	25
N. Am	USA	Michigan	PFOA	Surface Water Interface Criteria	ng/L	12,000	N/A	26
N. Am	USA	Michigan	PFOS	Surface Water Interface Criteria	ng/L	12	N/A	26
N. Am	USA	New Hampshire	PFOA and PFOS	Ambient Groundwater Quality Standard	ng/L	70	PFOA and PFOS	43
N. Am	USA	New Jersey	PFNA	Risk-Based Level	ng/L	10	N/A	44
N. Am	USA	North Carolina	PFOA	Groundwater Protection Standard	ng/L	2,000	N/A	31, 45
N. Am	USA	Rhode Island	PFOA and PFOS	Groundwater Quality Standard	ng/L	70	PFOA and PFOS	46
N. Am	USA	Texas	PFBA	Risk-Based Criteria (res)	mg/L	0.071 to 850,000	N/A	32
N. Am	USA	Texas	PFBA	Risk-Based Criteria (i)	mg/L	0.000021 to 1,000,000	N/A	32
N. Am	USA	Texas	PFBS	Risk-Based Criteria (res)	mg/L	0.034 to 540,000	N/A	32
N. Am	USA	Texas	PFBS	Risk-Based Criteria (i)	mg/L	0.000010 to 760,000	N/A	32

PFAS Standards: Groundwater

Region	Country	State or Jurisdiction (if applicable)	Compound	Type	Units	Value	Additional PFASs Included in Value	Link
N. Am	USA	Texas	PFPeA	Risk-Based Criteria (res)	ng/L	93 to 9,300	N/A	32
N. Am	USA	Texas	PFPeA	Risk-Based Criteria (i)	ng/L	0.028 to 28,000	N/A	32
N. Am	USA	Texas	PFHxA	Risk-Based Criteria (res)	ng/L	93 to 9,300	N/A	32
N. Am	USA	Texas	PFHxA	Risk-Based Criteria (i)	ng/L	0.028 to 28,000	N/A	32
N. Am	USA	Texas	PFHxS	Risk-Based Criteria (res)	ng/L	79 to 290,000,000	N/A	32
N. Am	USA	Texas	PFHxS	Risk-Based Criteria (i)	ng/L	0.028 to 400,000,000	N/A	32
N. Am	USA	Texas	PFHpA	Risk-Based Criteria (res)	ng/L	560 to 56,000	N/A	32
N. Am	USA	Texas	PFHpA	Risk-Based Criteria (i)	ng/L	0.17 to 170,000	N/A	32
N. Am	USA	Texas	PFOA	Risk-Based Criteria (res)	ng/L	16 to 90,000,000	N/A	32
N. Am	USA	Texas	PFOA	Risk-Based Criteria (i)	ng/L	0.088 to 130,000,000	N/A	32
N. Am	USA	Texas	PFOS	Risk-Based Criteria (res)	ng/L	340 to 1,800,000,000	N/A	32
N. Am	USA	Texas	PFOS	Risk-Based Criteria (i)	ng/L	0.17 to 2,600,000,000	N/A	32
N. Am	USA	Texas	PFOSA	Risk-Based Criteria (res)	ng/L	0.0000096 to 29,000	N/A	32
N. Am	USA	Texas	PFOSA	Risk-Based Criteria (i)	ng/L	0.088 to 88,000	N/A	32
N. Am	USA	Texas	PFNA	Risk-Based Criteria (res)	ng/L	120 to 650,000,000	N/A	32
N. Am	USA	Texas	PFNA	Risk-Based Criteria (i)	ng/L	0.088 to 120,000,000	N/A	32
N. Am	USA	Texas	PFDA	Risk-Based Criteria (res)	ng/L	230 to 1,300,000,000	N/A	32

(*Continued*)

Region	Country	State or Jurisdiction (if applicable)	Compound	Type	Units	Value	Additional PFASs Included in Value	Link
N. Am	USA	Texas	PFDA	Risk-Based Criteria (i)	ng/L	0.11 to 1,800,000,000	N/A	32
N. Am	USA	Texas	PFTrDA	Risk-Based Criteria (res)	ng/L	290 to 29,000	N/A	32
N. Am	USA	Texas	PFTrDA	Risk-Based Criteria (i)	ng/L	0.088 to 88,000	N/A	32
N. Am	USA	Texas	PFTeDA	Risk-Based Criteria (res)	ng/L	290 to 29,000	N/A	32
N. Am	USA	Texas	PFTeDA	Risk-Based Criteria (i)	ng/L	0.088 to 88,000	N/A	32
N. Am	USA	Texas	PFDS	Risk-Based Criteria (res)	ng/L	290 to 29,000	N/A	32
N. Am	USA	Texas	PFDS	Risk-Based Criteria (i)	ng/L	0.088 to 88,000	N/A	32
N. Am	USA	Texas	PFUA	Risk-Based Criteria (res)	ng/L	290 to 29,000	N/A	32
N. Am	USA	Texas	PFUA	Risk-Based Criteria (i)	ng/L	0.088 to 88,000	N/A	32
N. Am	USA	Texas	PFDoA	Risk-Based Criteria (res)	ng/L	100 to 550,000,000	N/A	32
N. Am	USA	Texas	PFDoA	Risk-Based Criteria (i)	ng/L	0.088 to 780,000,000	N/A	32
N. Am	USA	Vermont	PFOA and PFOS	Enforcement Standard	ng/L	20	PFOA and PFOS	47
N. Am	USA	Vermont	PFOA	Preventive Action Level	ng/L	10	N/A	47
N. Am	USA	Vermont	PFOS	Preventive Action Level	ng/L	10	N/A	47

Appendix E.3: Summary of PFAS Environmental Standards: Surface Water

Region	Country	State or Jurisdiction (if applicable)	Compound	Type	Units	Value	Additional PFASs Included in Value	Link
Australasia	Australia	Federal (Airports)	6:2 FTS	HISLs	ng/L	6.5	N/A	1
Australasia	Australia	Federal (Airports)	PFOA	HISLs and EISLs	ng/L	300(o); 2,900,000(o)	PFOA and 8:2 FtS	1
Australasia	Australia	Federal (Airports)	PFOS	HISLs and EISLs	ng/L	0.65(o); 6,660(o)	N/A	1
Australasia	Australia	N/A	PFOA	CRC CARE HSLs and ESLs	ng/L	210 to 1,824,000	N/A	2
Australasia	Australia	N/A	PFOS	CRC CARE HSLs (Fresh/Marine Water)	ng/L	21 to 616	PFOS and PFHxS	2
Australasia	Australia	N/A	PFOS	CRC CARE ESLs (Fresh Water)	ng/L	0.23 to 31,000	N/A	2
Australasia	Australia	N/A	PFOS	CRC CARE Marine Water	ng/L	6.6 to 60	N/A	2
Australasia	Australia	N/A	6:2 FTS	DoD HH	ng/L	50,000	N/A	3
Australasia	Australia	N/A	PFOA	DoD HH	ng/L	50,000	N/A	3
Australasia	Australia	N/A	PFOS	DoD HH	ng/L	5,000	PFOS and PFHxS	3
Australasia	Australia	N/A	PFOA	DEE (Fresh Water)	µg/L	19 to 1,824	N/A	4
Australasia	Australia	N/A	PFOA	DEE (Marine Water)	µg/L	3,000 to 22,000	N/A	4
Australasia	Australia	N/A	PFOS	DEE (Fresh Water)	ng/L	0.23 to 31,000	N/A	4
Australasia	Australia	N/A	PFOS	DEE (Marine Water)	ng/L	290 to 130,000	N/A	4
Australasia	Australia	N/A	PFOA	Dept of Health (Rec Water)	ng/L	5,600	N/A	48
Australasia	Australia	N/A	PFOS	Dept of Health (Rec Water)	ng/L	700,000	PFOS and PFHxS	48
Australasia	Australia	N/A	PFOA	EnHealth (Rec Water)	ng/L	50,000	N/A	49
Australasia	Australia	N/A	PFOS	EnHealth (Rec Water)	ng/L	5,000	PFOS and PFHxS	49
Australasia	Australia/New Zealand	N/A	PFOA	HEPA (Fresh/Marine Water)	ng/L	19,000 to 1,824,000	N/A	5

PFAS Standards: Surface Water   417

Region	Country	State or Jurisdiction (if applicable)	Compound	Type	Units	Value	Additional PFASs Included in Value	Link
Australasia	Australia/New Zealand	N/A	PFOA	HEPA (Rec Water)	ng/L	5,600	N/A	5
Australasia	Australia/New Zealand	N/A	PFOS	HEPA (Fresh/Marine Water)	ng/L	0.23 to 31,000	N/A	5
Australasia	Australia/New Zealand	N/A	PFOS	HEPA (Rec Water)	ng/L	700	PFOS and PFHxS	5
Australasia	Australia	New South Wales	Sum of PFASs	Trigger Points 1, 2, and 3	ng/L	1=10,000; 2=100; 3=50	All PFAS analytes measured	36
Australasia	Australia	Victoria	PFOA	Fresh/Rec Water	ng/L	5,600 to 1,824,000	N/A	6
Australasia	Australia	Victoria	PFOS	Recreational Water	ng/L	700	PFOS and PFHxS	6
Australasia	Australia	Victoria	PFOS	Fresh Water	ng/L	0.23 to 31,000	N/A	6
Australasia	Australia	Western Australia	PFOA	Ecological-Based Values	ng/L	19,000 to 1,824,000	N/A	7
Australasia	Australia	Western Australia	PFOS	Ecological-Based Values	ng/L	0.23 to 31,000	N/A	7
Australasia	Australia	Western Australia	PFOS	Non-Potable and Recreational Uses	ng/L	5,000	N/A	7
Asia	Japan	N/A	PFOS	Ecological-Based Value	ng/L	23,000	N/A	95
Europe	Austria	N/A	PFOS	Inland and Other Surface Waters	ng/L	0.13 to 36,000	N/A	50
Europe	Belgium	N/A	PFOS	Inland and Other Surface Waters	ng/L	0.13 to 36,000	N/A	51
Europe	Croatia	N/A	PFOS	Inland and Other Surface Waters	ng/L	0.13 to 36,000	N/A	52
Europe	Czech Republic	N/A	PFOS	Inland and Other Surface Waters	ng/L	0.13 to 36,000	N/A	53

(*Continued*)

Region	Country	State or Jurisdiction (if applicable)	Compound	Type	Units	Value	Additional PFASs Included in Value	Link
Europe	Denmark	N/A	PFOS	Inland and Other Surface Waters	ng/L	0.13 to 36,000	N/A	9
Europe	France	N/A	PFOS	Inland and Other Surface Waters	ng/L	0.13 to 36,000	N/A	54
Europe	Germany	N/A	PFOS	Inland and Other Surface Waters	ng/L	0.13 to 36,000	N/A	55
Europe	Germany	Bavaria	6:2 FTS	PNEC Aquatic	µg/L	870	N/A	38
Europe	Germany	Bavaria	PFBA	PNEC Aquatic	µg/L	1,260	N/A	38
Europe	Germany	Bavaria	PFBS	PNEC Aquatic	µg/L	3,700	N/A	38
Europe	Germany	Bavaria	PFPeA	PNEC Aquatic	µg/L	320	N/A	38
Europe	Germany	Bavaria	PFHxA	PNEC Aquatic	µg/L	1,000	N/A	38
Europe	Germany	Bavaria	PFHxS	PNEC Aquatic	µg/L	250	N/A	38
Europe	Greece	N/A	PFOS	Inland and Other Surface Waters	ng/L	0.13 to 36,000	N/A	56
Europe	Hungary	N/A	PFOS	Inland and Other Surface Waters	ng/L	0.13 to 36,000	N/A	57
Europe	Italy	N/A	PFOS	Inland and Other Surface Waters	ng/L	0.13 to 36,000	N/A	58
Europe	Netherlands	N/A	PFOA	Fresh and Salt Surface Waters	µg/L	0.048 to 2,800	N/A	59
Europe	Netherlands	N/A	PFOS	Inland and Other Surface Waters	ng/L	0.13 to 36,000	N/A	60
Europe	Poland	N/A	PFOS	Inland and Other Surface Waters	ng/L	0.13 to 36,000	N/A	61
Europe	Portugal	N/A	PFOS	Inland and Other Surface Waters	ng/L	0.13 to 36,000	N/A	62
Europe	Romania	N/A	PFOS	Inland and Other Surface Waters	ng/L	0.13 to 36,000	N/A	63
Europe	Slovak Republic	N/A	PFOS	Inland and Other Surface Waters	ng/L	0.13 to 36,000	N/A	64
Europe	Spain	N/A	PFOS	Inland and Other Surface Waters	ng/L	0.13 to 36,000	N/A	65
Europe	Turkey	N/A	PFOS	Inland and Other Surface Waters	ng/L	0.13 to 36,000	N/A	66
N. Am	Canada	N/A	PFOS	Federal Environmental Quality Guideline	ng/L	6,800	N/A	19

Appendix E.4: Summary of PFAS Environmental Standards: Drinking Water

Region	Country	State or Jurisdiction (if applicable)	Compound	Type	Units	Value	Additional PFASs Included in Value	Link
Australia	Australia	Federal (Airports)	6:2 FtS	Drinking water	ng/L	5,000	N/A	1
Australia	Australia	Federal (Airports)	PFOA	Drinking water	ng/L	400	PFOA and 8:2 FtS	1
Australia	Australia	Federal (Airports)	PFOS	Drinking water	ng/L	200	N/A	1
Australasia	Australia	N/A	PFOA	CRC CARE Health Screening Levels	ng/L	5,000	N/A	2
Australasia	Australia	N/A	PFOS	CRC CARE Health Screening Levels	ng/L	500	PFOS and PFHxS	2
Australasia	Australia	N/A	6:2 FtS	DoD HH	ng/L	5,000	N/A	3
Australasia	Australia	N/A	PFOA	DoD HH	ng/L	5,000	N/A	3
Australasia	Australia	N/A	PFOS	DoD HH	ng/L	500	PFOS and PFHxS	3
Australasia	Australia	N/A	PFOA	Dept of Health Drinking Water	ng/L	560	N/A	48
Australasia	Australia	N/A	PFOS	Dept of Health Drinking Water	ng/L	70	PFOS and PFHxS	48
Australasia	Australia	N/A	PFOA	EnHealth Drinking Water	ng/L	5,000	N/A	49
Australasia	Australia	N/A	PFOS	EnHealth Drinking Water	ng/L	500	PFOS and PFHxS	49
Australasia	Australia/New Zealand	N/A	PFOA	HEPA	ng/L	560	N/A	5
Australasia	Australia/New Zealand	N/A	PFOS	HEPA	ng/L	70	PFOS and PFHxS	5
Australia	Australia	Victoria	PFOA	Drinking water	ng/L	560	N/A	6
Australia	Australia	Victoria	PFOS	Drinking water	ng/L	70	PFOS and PFHxS	6
Australia	Australia	Western Australia	PFOS	Drinking water	ng/L	500	N/A	7
Europe	Denmark	N/A	Sum of PFASs	Health Based Value	ng/L	100	PFOS, PFOA, PFOSA, PFBS, PFBA, PFPeA, PFHxA, PFHpA, PFNA, PFDA, PFHxS, 6:2 FtS	8, 9

PFAS Standards: Drinking Water 421

Region	Country	State or Jurisdiction (if applicable)	Compound	Type	Units	Value	Additional PFASs Included in Value	Link
Europe	European Union	N/A	Sum of PFASs	Maximum Tolerable Level	ng/L	500	All PFASs (not specified)	67
Europe	European Union	N/A	6:2 FtS	Maximum Tolerable Level	ng/L	100	N/A	67
Europe	European Union	N/A	PFBA	Maximum Tolerable Level	ng/L	100	N/A	67
Europe	European Union	N/A	PFBS	Maximum Tolerable Level	ng/L	100	N/A	67
Europe	European Union	N/A	PFPeA	Maximum Tolerable Level	ng/L	100	N/A	67
Europe	European Union	N/A	PFHxA	Maximum Tolerable Level	ng/L	100	N/A	67
Europe	European Union	N/A	PFHxS	Maximum Tolerable Level	ng/L	100	N/A	67
Europe	European Union	N/A	PFHpA	Maximum Tolerable Level	ng/L	100	N/A	67
Europe	European Union	N/A	PFOA	Maximum Tolerable Level	ng/L	100	N/A	67
Europe	European Union	N/A	PFOS	Maximum Tolerable Level	ng/L	100	N/A	67
Europe	European Union	N/A	PFNA	Maximum Tolerable Level	ng/L	100	N/A	67
Europe	European Union	N/A	PFDA	Maximum Tolerable Level	ng/L	100	N/A	67
Europe	Germany	N/A	6:2 FtS	Health Precautionary Value	ng/L	100	N/A	68
Europe	Germany	N/A	PFBA	Health-Related Indication Value	ng/L	10,000	N/A	68
Europe	Germany	N/A	PFBS	Health-Related Indication Value	ng/L	6,000	N/A	68
Europe	Germany	N/A	PFPeA	Health Precautionary Value	ng/L	3,000	N/A	68
Europe	Germany	N/A	PFHxA	Health-Related Indication Value	ng/L	6,000	N/A	68
Europe	Germany	N/A	PFHxS	Health-Related Indication Value	ng/L	100	N/A	68
Europe	Germany	N/A	PFHpA	Health Precautionary Value	ng/L	300	N/A	68
Europe	Germany	N/A	PFHpS	Health Precautionary Value	ng/L	300	N/A	68

(*Continued*)

Region	Country	State or Jurisdiction (if applicable)	Compound	Type	Units	Value	Additional PFASs Included in Value	Link
Europe	Germany	N/A	PFOA	Health-Related Indication Value	ng/L	100	N/A	68
Europe	Germany	N/A	PFOS	Health-Related Indication Value	ng/L	100	N/A	68
Europe	Germany	N/A	PFOSA	Health Precautionary Value	ng/L	100	N/A	68
Europe	Germany	N/A	PFNA	Health-Related Indication Value	ng/L	60	N/A	68
Europe	Germany	N/A	PFDA	Health Precautionary Value	ng/L	100	N/A	68
Europe	Italy	N/A	Sum of PFASs	Health Based Value	ng/L	500	PFBA, PFPeA, PFBS, PFHxA, PFHpA, PFHxS, PFNA, PFDA, PFUA, PFDoA	69
Europe	Italy	N/A	PFOA	Health Based Value	ng/L	500	N/A	69
Europe	Italy	N/A	PFOS	Health Based Value	ng/L	30	N/A	69
Europe	Netherlands	N/A	PFOA	Maximum Permissible Concentration	ng/L	87.5	N/A	70
Europe	Netherlands	N/A	PFOS	Maximum Permissible Concentration	ng/L	530	N/A	71
Europe	Netherlands	N/A	GenX	Maximum Permissible Concentration	ng/L	150	N/A	72
Europe	Sweden	N/A	Sum of PFASs	Maximum Tolerable Level	ng/L	90	PFOS, PFHxS, PFBS, PFOA, PFHpA, PFHxA, PFPeA, PFBA, PFNA, PFDA, 6:2 FtS	73
Europe	Switzerland	N/A	PFHxS	Maximum Tolerable Level	ng/L	300	N/A	74
Europe	Switzerland	N/A	PFOA	Maximum Tolerable Level	ng/L	500	N/A	74
Europe	Switzerland	N/A	PFOS	Maximum Tolerable Level	ng/L	300	N/A	74

PFAS Standards: Drinking Water

Region	Country	State or Jurisdiction (if applicable)	Compound	Type	Units	Value	Additional PFASs Included in Value	Link
Europe	UK	N/A	PFOA	Tier 2, 3, and 4 Values	ng/L	2=300; 3=5,000; 4=45,000	N/A	75
Europe	UK	N/A	PFOS	Tier 2, 3, and 4 Values	ng/L	2=300; 3=1,000; 4=9,000	N/A	75
N. Am	Canada	N/A	PFBA	Screening Values	ng/L	30,000	N/A	76
N. Am	Canada	N/A	PFBS	Screening Values	ng/L	15,000	N/A	76
N. Am	Canada	N/A	PFPeA	Screening Values	ng/L	200	N/A	76
N. Am	Canada	N/A	PFHxA	Screening Values	ng/L	200	N/A	76
N. Am	Canada	N/A	PFHxS	Screening Values	ng/L	600	N/A	76
N. Am	Canada	N/A	PFHpA	Screening Values	ng/L	200	N/A	76
N. Am	Canada	N/A	PFOA	Screening Values	ng/L	200	N/A	76
N. Am	Canada	N/A	PFOS	Screening Values	ng/L	600	N/A	76
N. Am	Canada	N/A	PFNA	Screening Values	ng/L	20	N/A	94
N. Am	USA	USEPA	PFOA and PFOS	USEPA Health Advisory Level	ng/L	70	PFOA and PFOS	77
N. Am	USA	USEPA	PFBS	Risk-Based Screening Level	ng/L	401,000(res)	N/A	20
N. Am	USA	USEPA	PFOA	Risk-Based Screening Level	ng/L	401(res)	N/A	20
N. Am	USA	USEPA	PFOS	Risk-Based Screening Level	ng/L	401(res)	N/A	20
N. Am	USA	Connecticut	Sum of PFASs	Drinking Water Action Level	ng/L	70	PFOA, PFOS, PFNA, PFHxS, and PFHpA	78
N. Am	USA	Indiana	PFBS	Screening Levels - Tap	ng/L	400,000	N/A	23

(Continued)

Region	Country	State or Jurisdiction (if applicable)	Compound	Type	Units	Value	Additional PFASs Included in Value	Link
N. Am	USA	Maine	PFOA and PFOS	Maximum Exposure Guideline	ng/L	70	PFOA and PFOS	79
N. Am	USA	Massachusetts	Multiple PFASs	Office of Research and Standards Guideline	ng/L	70	PFOA, PFOS, PFNA, PFHxS, and PFHpA	80
N. Am	USA	Massachusetts	PFBS	Office of Research and Standards Guideline	ng/L	2,000	N/A	96
N. Am	USA	Michigan	PFOA and PFOS	Drinking Water Criteria	ng/L	70(res); 70(i)	PFOA and PFOS	81
N. Am	USA	Minnesota	PFBA	Health Risk Limit	ng/L	7,000	N/A	82
N. Am	USA	Minnesota	PFBS	Health Risk Limit	ng/L	7,000	N/A	82
N. Am	USA	Minnesota	PFOA	Health Based Value; Health Risk Limit	ng/L	35; 300	N/A	82
N. Am	USA	Minnesota	PFOS	Health Based Value; Health Risk Limit	ng/L	27; 300	N/A	82
N. Am	USA	Nevada	PFBS	Basic Comparison Level	ng/L	66,700	N/A	28
N. Am	USA	Nevada	PFOA	Basic Comparison Level	ng/L	667	N/A	28
N. Am	USA	Nevada	PFOS	Basic Comparison Level	ng/L	667	N/A	28
N. Am	USA	New Jersey	PFOA	State MCL	ng/L	14	N/A	83
N. Am	USA	New Jersey	PFOS	State MCL	ng/L	13	N/A	84
N. Am	USA	New Jersey	PFNA	State MCL	ng/L	13	N/A	85
N. Am	USA	New Mexico	PFOA and PFOS	Screening Level	ng/L	70	PFOA and PFOS	86
N. Am	USA	New York	PFOA and PFOS	Health Advisory	ng/L	70	PFOA and PFOS	87
N. Am	USA	North Carolina	PFOA	Interim Maximum Allowable Concentration	ng/L	2,000	N/A	88

PFAS Standards: Drinking Water

Region	Country	State or Jurisdiction (if applicable)	Compound	Type	Units	Value	Additional PFASs Included in Value	Link
N. Am	USA	North Carolina	GenX	Health Goal	ng/L	140	N/A	89
N. Am	USA	Oregon	PFHpA	Initiation Level	ng/L	300,000	N/A	90
N. Am	USA	Oregon	PFOA	Initiation Level	ng/L	24,000	N/A	90
N. Am	USA	Oregon	PFOS	Initiation Level	ng/L	300,000	N/A	90
N. Am	USA	Oregon	PFOSA	Initiation Level	ng/L	200	N/A	90
N. Am	USA	Oregon	PFNA	Initiation Level	ng/L	1,000	N/A	90
N. Am	USA	Pennsylvania	PFOA and PFOS	Health Advisory	ng/L	70	N/A	91
N. Am	USA	Vermont	PFOA and PFOS	Health Based Value	ng/L	20	PFOA and PFOS	92
N. Am	USA	Washington	PFOA and PFOS	Drinking Water Limit	ng/L	70	PFOA and PFOS	93
N. Am	USA	West Virginia	PFOA	Screening Value	ng/L	150,000	N/A	34

Appendix E.5: Notes

NOTES:

1. PFAS acronyms can be found in Chapter 3.

ACRONYMS

CRC Care	Cooperative Research Centre for Contamination Assessment and Remediation of the Environment
DEE	Department of Environment and Energy (Australia)
Dept	department
DoD	Department of Defence (Australia)
EISL	Ecological Interim Screening Level
ESL	Ecological Screening Level
HEPA	Heads of EPAs Australia and New Zealand
HISL	Human Health Interim Screening Level
HH	Human Health
HSL	Health Screening Level
(i)	industrial/commercial
MCL	Maximum Contaminant Level
mg/kg	milligrams per kilogram
mg/L	milligrams per liter
N. Am	North America
N/A	not applicable
ng/L	nanograms per liter
(o)	other
PNEC	Predicted No Effect Concentration
(rec)	recreational
(res)	residential
USEPA	United States Environmental Protection Agency
USA	United States of America
µg/kg	micrograms per kilogram
µg/L	micrograms per liter

LINKS:

1. www.aph.gov.au/DocumentStore.ashx?id=54ea0797-405b-42d0-a348-cc5884207f15&subId=408775
2. www.crccare.com/files/dmfile/CRCCARETechReport38Part3_AssessmentmanagementandremediationforPFOSandPFOA_ESLs2.pdf
3. www.defence.gov.au/estatemanagement/governance/Policy/Environment/PFAS/docs/DCD8V2ScreeningGuidelines.pdf

4. www.environment.gov.au/system/files/pages/dfb876c5-581e-48b7-868c-242fe69dad68/files/draft-environmental-mgt-guidance-pfos-pfoa.pdf
5. www.epa.vic.gov.au/your-environment/land-and-groundwater/pfas-in-victoria/~/media/Files/Your%20environment/Land%20and%20groundwater/PFAS%20in%20Victoria/PFAS%20NEMP/FINAL_PFAS-NEMP-20180110.pdf
6. www.epa.vic.gov.au/~/media/Publications/1669%201.pdf
7. www.der.wa.gov.au/images/documents/your-environment/contaminated-sites/guidelines/Guideline-on-Assessment-and-Management-of-PFAS-.pdf
8. http://mst.dk/media/92446/pfoa-pfos-pfosa-datablad-final-27-april-2015.pdf
9. www.retsinformation.dk/Forms/R0710.aspx?id=180241
10. https://rp.baden-wuerttemberg.de/rpk/Abt5/Ref541/PFC/Documents/pfc_um_erlass_141222.pdf
11. No link available: Istituto Superiore di Sanità, 2015, Oggetto: valori CSC Bonifiche suoli e acque sotterranee-PFAS, June 23
12. https://onlinehaarlemmermeer.nl/bis/Raadsplein_2018_inclusief_stukken/Raadsplein_19_april_2018/Agenda_en_stukken/Raadsvergadering/Opening_mededelingen_brieven/2_Vaststelling_van/c_Lijst_van_ingekomen_brieven_en_stukken_ter_afhandeling/Stukken/Actualisatie_Beleidsregel_PFOS_en_PFOA_gemeente_Haarlemmermeer
13. https://zoek.officielebekendmakingen.nl/prb-2017-3228.html
14. http://rivm.nl/Documenten_en_publicaties/Wetenschappelijk/Rapporten/2017/mei/Risicogrenzen_PFOA_voor_grond_en_grondwater_Uitwerking_ten_behoeve_van_generiek_en_gebiedsspecifiek_beleid
15. www.rivm.nl/Documenten_en_publicaties/Wetenschappelijk/Rapporten/2016/februari/Milieukwaliteitswaarden_voor_PFOS_Uitwerking_van_generieke_en_gebiedsspecifieke_waarden_voor_het_gebied_rond_Schiphol
16. www.miljodirektoratet.no/en/Legislation1/Regulations/Product-Regulations/Chapter-4/
17. www.swedgeo.se/globalassets/publikationer/sgi-publikation/sgi-p21.pdf
18. No link available: Health Canada 2017, Updates to Health Canada Soil Screening Values for Perfluoroalkylated Substances (PFAS), January
19. www.ec.gc.ca/ese-ees/default.asp?lang=En&n=38E6993C-1
20. https://epa-prgs.ornl.gov/cgi-bin/chemicals/csl_search
21. See "Formal Proposed Amendments to 18 AAC 75" at http://dec.alaska.gov/spar/regulation-projects/cleanup-level-amendments/
22. www.dnrec.delaware.gov/dwhs/sirb/Documents/Screening%20Level%20Table.pdf
23. www.in.gov/idem/cleanups/files/risc_screening_table_2018_a6.pdf
24. https://programs.iowadnr.gov/riskcalc/Home/statewidestandards
25. www.maine.gov/dep/spills/publications/guidance/rags/ME-RAGS-Revised-Final_020516.pdf
26. www.michigan.gov/deq/0,4561,7-135-3311_4109_9846-251790-,00.html
27. www.pca.state.mn.us/waste/risk-based-site-evaluation-guidance
28. https://ndep.nv.gov/resources/risk-assessment-and-toxicology-basic-comparison-levels
29. www4.des.state.nh.us/nh-pfas-investigation/wp-content/uploads/2017/09/pfoa-soil-standard.pdf
30. www4.des.state.nh.us/nh-pfas-investigation/wp-content/uploads/2017/09/pfoa-dcrb-value.pdf
31. https://files.nc.gov/ncdeq/Waste%20Management/DWM/SF/RiskBasedRemediation/20171023_PSRGsandAppA.PDF
32. www.tceq.texas.gov/assets/public/remediation/trrp/2017%20PCL%20Tables%20March31.pdfh
33. http://dec.vermont.gov/sites/dec/files/wmp/Sites/07.11.2017.Adopted.Rule_.for_.SOS_.filing.pdf
34. http://dep.wv.gov/WWE/watershed/wqmonitoring/Documents/C-8/C-8_FINAL_CATT_REPORT_8-02.pdf
35. http://dnr.wi.gov/topic/brownfields/professionals.html
36. https://ntepa.nt.gov.au/__data/assets/pdf_file/0007/372148/decision_tree_prioritising_sites_potentially_contaminated_with_pfas.PDF
37. https://rp.baden-wuerttemberg.de/rpk/Abt5/Ref541/PFC/Documents/pfc_boden_umer-lass_150617.pdf

38. www.lfu.bayern.de/analytik_stoffe/doc/leitlinien_vorlaufbewertung_pfc_verunreinigungen.pdf
39. www.bafu.admin.ch/dam/bafu/de/dokumente/altlasten/fachinfo-daten/konzentrationswerte.pdf. download.pdf/konzentrationswerte.pdf
40. www.wfduk.org/sites/default/files/Media/UKTAG_Technical%20report_GW_Haz-Subs_ForWebfinal.pdf
41. www.colorado.gov/pacific/sites/default/files/WQ_GWStandard_PFOA_100417%20FINAL.pdf
42. https://portal.ct.gov/-/media/Departments-and-Agencies/DPH/dph/environmental_health/eoha/Groundwater_well_contamination/110916CTActionLevelListNov2016Updatepdf.pdf?la=en
43. www4.des.state.nh.us/nh-pfas-investigation/wp-content/uploads/2017/08/5-18-17-PFAS-Sampling-Letter.pdf
44. www.nj.gov/dep/rules/adoptions/adopt_20180116c.pdf
45. https://ncdenr.s3.amazonaws.com/s3fs-public/Waste%20Management/DWM/UST/Law-Rules-Memos/2L%20Rules%20040113.pdf
46. www.dem.ri.gov/programs/benviron/water/quality/pdf/pfoa.pdf
47. http://dec.vermont.gov/sites/dec/files/documents/gwprsAdoptedDec12_2016.pdf
48. https://health.gov.au/internet/main/publishing.nsf/Content/2200FE086D480353CA2580C900817CDC/$File/fs-Health-Based-Guidance-Values.pdf
49. www2.health.vic.gov.au/about/publications/factsheets/enhealth-interim-national-guidance-human-health-reference-values-per-poly-fluoroalkyl-substances
50. www.ris.bka.gv.at/Dokumente/BgblAuth/BGBLA_2016_II_363/BGBLA_2016_II_363.pdf
51. No link available: Federale Overheidsdienst Volksgezondheid, Veiligheid van de Voedselketen en Leefmilieu 2016, 15 FEBRUARI 2016. - Koninklijk besluit tot wijziging van het koninklijk besluit van 23 juni 2010 betreffende de vaststelling van een kader voor het bereiken van een goede oppervlaktewatertoestand, March 30
52. https://narodne-novine.nn.hr/clanci/sluzbeni/2015_07_78_1504.html
53. www.zakonyprolidi.cz/cs/2015-401
54. www.legifrance.gouv.fr/eli/arrete/2015/7/27/DEVL1513989A/jo/texte
55. www.gesetze-im-internet.de/ogewv_2016/OGewV.pdf
56. www.elinyae.gr/el/lib_file_upload/69b_2016.1453799648187.pdf
57. www.kozlonyok.hu/nkonline/MKPDF/hiteles/MK15125.pdf
58. www.gazzettaufficiale.it/eli/id/2015/10/27/15G00186/sg
59. www.rivm.nl/Documenten_en_publicaties/Wetenschappelijk/Rapporten/2017/september/Water_quality_standards_for_PFOA_A_proposal_in_accordance_with_the_methodology_of_the_Water_Framework_Directive
60. https://zoek.officielebekendmakingen.nl/stb-2015-394.pdf
61. http://prawo.sejm.gov.pl/isap.nsf/download.xsp/WDU20160001187/O/D20161187.pdf
62. https://dre.pt/application/file/a/70476114
63. https://legeaz.net/monitorul-oficial-633-2016/hg-570-2016-program-eliminare-evacuari-emisii-pierderi-substante-periculoase-poluanti
64. www.epi.sk/zz/2015-167
65. www.boe.es/boe/dias/2015/09/12/pdfs/BOE-A-2015-9806.pdf
66. www.mevzuat.gov.tr/Metin.Aspx?MevzuatKod=7.5.16806&MevzuatIliski=0&sourceXmlSearch=yer%C3%BCst%C3%BC
67. https://eur-lex.europa.eu/resource.html?uri=cellar:8c5065b2-074f-11e8-b8f5-01aa75ed71a1.0016.02/DOC_1&format=PDF
68. www.umweltbundesamt.de/sites/default/files/medien/374/dokumente/fortschreibung_der_uba-pfc-bewertungen_bundesgesundheitsbl_2017-60_s_350-352.pdf
69. https://sian.aulss9.veneto.it/docs/Sian/IgieneNutrizione/Acque/Pfas/Pareri/Nota_ISS_prot_1584_del_16_01_2014.pdf
70. No link available: Rijks Instituut voor Volksgezondheid en Milieu (RIVM) 2016, Afleiding richtwaarde voor PFOA in drinkwater voor levenslange blootstelling, March 24
71. www.rivm.nl/bibliotheek/rapporten/601714013.pdf

Notes 429

72. www.rivm.nl/Onderwerpen/G/GenX/Downloaden/148_2016_M_V_bijlage_afleiding_richtwaarde_FRD903_in_drinkwater_Definitief_revisie_jan_2017.org
73. www.livsmedelsverket.se/livsmedel-och-innehall/oonskade-amnen/miljogifter/pfas-poly-och-perfluorerade-alkylsubstanser/riskhantering-pfaa-i-dricksvatten
74. www.admin.ch/opc/de/classified-compilation/20143396/201705010000/817.022.11.pdf
75. www.dwi.gov.uk/stakeholders/information-letters/2009/10_2009annex.pdf
76. No link available: Health Canada 2017, Health Canada's Drinking Water Screening Values for Perfluoroalkylated Substances, January
77. www.epa.gov/sites/production/files/2016-05/documents/pfoa_health_advisory_final_508.pdf
78. https://portal.ct.gov/-/media/Departments-and-Agencies/DPH/dph/environmental_health/eoha/Groundwater_well_contamination/DrinkingWaterActionLevelPerfluorinatedAlkylSubstances-PFAS.pdf?la=en
79. www.maine.gov/dhhs/mecdc/environmental-health/eohp/wells/documents/megtable2016.pdf
80. www.mass.gov/eea/agencies/massdep/water/drinking/standards/standards-and-guidelines-for-drinking-water-contaminants.html
81. www.michigan.gov/documents/deq/deq-rrd-UpdatedGroundwaterCleanupCrieriaTableWith-FootnotesPFOSPFOA1-25-2017_610379_7.pdf
82. www.health.state.mn.us/divs/eh/risk/guidance/gw/table.html#top
83. www.nj.gov/dep/watersupply/pdf/pfoa-recommend.pdf
84. www.theintell.com/news/20171121/njdep-panel-recommends-lower-limit-for-pfos/1
85. www.state.nj.us/dep/watersupply/pdf/pfna-health-effects.pdf
86. www.env.nm.gov/wp-content/uploads/2016/11/NMED_SSG_VOL-I_-March_2017_revised.pdf
87. www.dec.ny.gov/chemical/108831.html and http://www.villageofhoosickfalls.com/Water/Documents/NYSDOH-letter-lab-reports-12012016.pdf
88. No link available: North Carolina Department of Health and Human Services 2013, Title 15A Department of Environment and Natural Resources Division of Water Quality, Subchapter 2L, Section .0100, .0200, .0300; Classifications and Water Quality Standards Applicable to the Groundwaters of North Carolina, April 1
89. https://ncdenr.s3.amazonaws.com/s3fs-public/GenX/NC%20DHHS%20Risk%20Assessment%20FAQ%20Final%20Clean%20071417%20PM.pdf
90. www.oregon.gov/deq/EQCdocs/20170913ItemD.pdf
91. www.dep.pa.gov/Citizens/My-Water/drinking_water/Perfluorinated%20Chemicals%20%e2%80%93PFOA%20and%20PFOS%20%e2%80%93%20in%20Pennsylvania/Pages/PFOA%20and%20PFOS%20Health%20Advisories%20in%20Pennsylvania.aspx
92. www.healthvermont.gov/health-environment/drinking-water/perfluorooctanoic-acid-pfoa
93. www.doh.wa.gov/CommunityandEnvironment/Contaminants/PFAS
94. No link available: Health Canada, 2018, Drinking Water Screening Value for PFNA, July 5
95. www.researchgate.net/profile/Yasuyuki_Zushi/publication/227110295_Progress_and_perspective_of_perfluorinated_compound_risk_assessment_and_management_in_various_countries_and_institutes/links/02e7e51a8e65aa6106000000.pdf
96. www.mass.gov/files/documents/2018/06/11/pfas-ors-ucmr3-recs_0.pdf
97. https://zoek.officielebekendmakingen.nl/gmb-2018-16583.html
98. www.ozhz.nl/fileadmin/uploads/bodeminformatie/PFOA_in_bodem/Handreiking_hergebruik_grond_PFOA_ZHZ_herzien_-_13_juni_2018.pdf
99. www.rivm.nl/Documenten_en_publicaties/Wetenschappelijk/Rapporten/2018/Juni/Risicogrenzen_PFOA_voor_grond_en_grondwater_Uitwerking_voor_generiek_en_gebiedsspecifiek_beleid_herziene_versie

Index

Note: *Italicized* page numbers refer to figures, **bold** page numbers refer to tables

A

action list (United States Department of Defense), 6–7
activated carbon. See granular activated carbon (GAC)
Activated Carbon Fiber (ACF20), **201**, **202**
adsorbable organofluorine (AOF) analysis, 137–138
adsorbents, 73–75, *189*, **190**, 212–213, 338–343
advanced oxidation processes (AOPs), 69–71, *189*, **190**, 213–214, 343–344. *See also* water treatment
   comparison of, **70**
   1,4-dioxane, **67**, 69–71
   per- and polyfluoroalkyl substances, 213–214
   1,2,3-trichloropropane, 343–344
advanced reduction processes (ARPs), *189*, 214
aerobic cometabolic degradation
   1,4-dioxane, 54
   1,2,3-trichloropropane, 331
aerobic mineralization (1,2,3-trichloropropane), 331–332
Agency for Toxic Substances and Disease Registry (ASTDR), 329, 333
agricultural chemicals, 315
Agrobacterium radiobacter, 332
air stripping, **67**, 68, 344
Alaska, TCP regulation in, 324
aldehyde dehydrogenase (ALDH), 48
allyl alcohol, **391**
Amb IRA-400, **206**
Amb XAD-400, **206**
Amberlite, **206**
Amberlite XAD-7HP, **203**
AMBERSORB™, 73–75
ammonium 3H-perfluoro-3-[(3-methoxypropoxy)propanoic acid (ADONA), 97
ammonium pentadecafluorooctanoate, **386**
ammonium perchlorate, **391**
ammonium perfluorooctanoate (APFO), 96
anaerobic dechlorination (1,2,3-trichloropropane), 332–333
analytical methods
   1,4-dioxane, 41–42, **43–47**
   general challenges, 364–366
   hexavalent chromium, 274–277, **275–276**
   per- and polyfluoroalkyl substances, 128–139
   1,2,3-trichloropropane, 325–327, **326**

anion-exchange resins, 204–208, **205–207**.
   See also water treatment
APTWater, **70**
atmospheric deposition, 166
Australian National Environment Protection Council, 3, 10–11

B

ball milling, 197
bamboo-derived GAC, **201**, **202**
benzene, **30**, **318**
benzidine, **391**
bioaccumulation, 140–142, **141**
bioaugmentation, **56**
biodegradation
   defined, 86
   1,4-dioxane, 54–59, 71–73
   hexavalent chromium, 284–288, 300–304
   per- and polyfluoroalkyl substances, 112–114, 115–119, 191
   1,2,3-trichloropropane, 331–333
biological treatment
   challenges in, 367
   1,4-dioxane, 65, 72–73
   hexavalent chromium, 301–302
   per- and polyfluoroalkyl substances, 191
   1,2,3-trichloropropane, 331–333
bioreactors
   1,4-dioxane, 71–73
   hexavalent chromium, 300–304
   per- and polyfluoroalkyl substances, 215
   1,2,3-trichloropropane, 332
bioremediation, 14. *See also* groundwater remediation/treatment
   1,4-dioxane, 49, 50, 54–55, **56**, 57–58, 64
   per- and polyfluoroalkyl substances, 191
biosolids, 182–183
biostimulation
   1,4-dioxane, 55, **56**
   1,2,3-trichloropropane, 333
bis(2-chloroethyl)ether, **391**
bisphenol A, **390**
bromochloroacetic acid, **392**
bromodichloroacetic acid, **392**
bromomethane, **390**
1,3-butadiene, **390**
1-butanol, **391**

Index

C

calcium polysulfide, **281**
Calgon Filtrasorb 300 (F300), **201**
Calgon Filtrasorb 400 (F400), **200**, **202**
Calgon Filtrasorb 600 (F600), **200**, **201**, **202**
Calgon URV-MOD 1, **202**
CalgonCarbon®, **70**, **202**
California, TCP regulation in, 322–323
California Department of Health Services (CDHS), 269
California Department of Public Health (CDPH), 269
California Recycled Water Policy, 8
California State Water Resources Control Board (CSWRCB), 7, 269–270, 316
candidate contaminants. *See also* emerging contaminants
    REACH, **377–383**
    USEPA, **373–376**
carcinogenicity
    1,4-dioxane, 34–35
    hexavalent chromium, 267–268
    per- and polyfluoroalkyl substances, 146–150
    perfluorooctane sulfonic acid (PFOS), 146–150
    perfluorooctanoic acid (PFOA), 146–150
    1,2,3-trichloropropane, 319–320, **320**
Chemical and Material Risk Management Program (United States Department of Defense), 6
chemical properties
    1,4-dioxane, **30**, **318**, **386**
    emerging contaminants, **385–392**
    hexavalent chromium, 255–259, **386**
    per- and polyfluoroalkyl substances, 110–119, **386–388**
    1,2,3-trichloropropane, 307–309, **310**, **318**, **386**
chlorinated organophosphates, **389**
chlorinated solvents, 86
chlorodibromoacetic acid, **392**
chlorodifluoromethane, **391**
chromate salts, 265
chromium
    geochemistry of, 263
    isotopes, 277
    primary forms of, 263–264
Class B firefighting foams, 86, 107–110, 181–186
coagulation, 211–212
cobalt, **388**
cometabolic biodegradation, of 1,4-dioxane, 54–57, 63, 64, 65, 72–73
compound specific isotope analysis (CSIA), 49–50, 64, 277
conceptual site models (CSMs), 181–188
    for fire training areas and Class B fire response areas, 181–186
    for industrial facilities, 181–188
    for land fills, 187–188
    for wastewater treatment plans, 182–183, *183*
Contaminant Candidate List (CCL), 4–5, **373–376**
    1,2,3-trichloropropane, 321–322
Cooperative Research Centre for Contamination Assessment and Remediation of the Environment (CRC CARE), 10–11
Council on Environmental Quality, 13
cyanotoxins, **358**, **391**
cyclonite, **391**
cyclotrimethylenetrinitramine, **391**

D

Dehalococcoides, 332
Dehalogenimonas, 327, 332–333
deoxyribonucleic acid (DNA), 48
di-2-ethylhexyl phthalate, **390**
dichlorodiphenyltrichloroethane, **390**
1,1-dichloroethane, **390**
1,1-dichloroethene (11DCE), **30**, 31–32
dichloropenes, 315
difluoromonochloromethane, **391**
di-iron monooxygenases, 49
1,1-dimethylhydrazine, **391**
1,4-dioxane, 5, 15, 27–77
    advanced investigation techniques, 48–50
    advanced oxidation processes (AOPs), 69–71
    analytical methods, 41–42, **43–47**
    basic information, 27–32
    bioremediation, 54–58, **56**, 71–73
    carcinogenicity of, 34–35, **35**
    chemical oxidation efficacy on, **53**
    chemical properties, **30**, **318**, **386**
    chemical structure of, 27
    in drinking water, 32
    drinking water and wastewater treatment, 65–66
    drinking water treatment technologies, 66–76
    electrochemical oxidation, 75–76
    ex situ groundwater remediation, 66–76
    federal guidelines and health standards for, **36**
    granular activated carbon and other sorbent media, 73–75
    groundwater analytical methods, **43–47**
    groundwater and drinking standards and guidelines, by state:, 37
    groundwater treatment, 51–64
    investigation approaches, 39–40
    natural attenuation, 62–64
    physical properties, **30**, **318**, **386**
    point-of-use and point-of-entry treatment, 66

regulation of, 31–32
regulatory status, 35–38
risk assessment, 32–35
site characterization, 38–50
in situ biodegradation of, 57–58
soil treatment, 50–51
sources of, **28–29**
as stabilizing agent to TCA, 31
toxicity of, 32–35, **33**
visual representation of high resolution investigation, *42*
1,4-dioxane/tetrahydrofuran monooxygenase (DXMO/THFMO), 48
disinfection byproducts, **358**, **392**
dithionite, **281**
Dow L493, **206**
Dow V493, **206**
drinking water
1,4-dioxane in, 32, 65–66, 66–76, **67–68**
hexavalent chromium in, 272–273, 292–294
per- and polyfluoroalkyl substances in, 86
perfluorooctane sulfonic acid (PFOS) in, 86
perfluorooctanoic acid (PFOA) in, 86
1,2,3-trichloropropane in, 316–317, **317**, **318**, 338
drinking water treatment
1,4-dioxane, 65–66, **67–68**
hexavalent chromium, 292–294
per- and polyfluoroalkyl substances, 198–217
1,2,3-trichloropropane, 338–345
dynamic groundwater circulation
1,4-dioxane, 61–62
hexavalent chromium, 291

E

eBeam, 196–197
electrochemical fluorination (ECF), 91, 98–99, 119–121, *121*
electrochemical fluorination (ECF) foams, **108**
electrochemical oxidation, 75–76
1,4-dioxane, **67**, 75–76
per- and polyfluoroalkyl substances, *189*, **190**
electrochemical treatment, 215
electrochemical treatment, of PFASs, 215
electrocoagulation, *189*
elimination half-lives, of PFASs, 142–143
emerging contaminants
candidate contaminant lists, **373–383**
categorizing, 355–356
challenges associated with, 15–17
characterization and analytical challenges, 364–366
characterization of, 370
classes of, **357–358**
definitions of, 2, 2–3
environmental release challenges, 359–360

exposure, 368
future of, 367–370
identification of, 3–4
life cycle of, 11–15, *12*
management challenges, 356–367
management strategies, **377**
persistent, mobile, and toxic (PMT) criteria, **393–399**
physical and chemical properties, **385–392**
regulation of, 368–370
regulatory challenges, 363–364
terminology, 1–2
toxicological risk challenges, 360–363
treatment, 370
treatment challenges, 366–367
very persistent and very mobile (vPvM) criteria, **393–399**
vulnerability, 368
empty bed contact time (EBCT), 75
emulsified vegetable oil (EVO), 286–287
endocrine disrupting compounds (EDCs), 356, **390**
environmental standards, PFAS
drinking water, **419–425**
groundwater, **408–414**
soil, **401–407**
surface water, **415–418**
erythromycin, **390**
EtFOSE-based phosphate, 98
EtFOSE-based phosphate esters, 98
European Chemicals Agency (ECHA), 9, 100, 361
European Union, 9–10
ex situ groundwater remediation, 66–76. *See also* groundwater remediation/treatment
bioreactors, 300–304
1,4-dioxane, 61–62
for hexavalent chromium, 294–305
ion exchange, 297–300
per- and polyfluoroalkyl substances, 198–217
reduction/coagulation/filtration (RCF) with ferrous iron, 294–297
reduction/filtration via stannous chloride (RF-Sn[II]), 304–305
reverse osmosis, 300
1,2,3-trichloropropane, 337, 338–345
ex situ treatment, 61–62, 290–291
experimental adsorbents, 213
extractable organofluorine (EOF) analysis, 138
extreme soil vapor extraction (XSVE), 51

F

ferrous iron, 282–283
ferrous sulfate, **281**
ferrous sulfide, 283
flame retardants, 13, **357**, **389**

Index

flocculation, *189*, 207–208, 211–212
Florida, TCP regulation in, 324
Fluka 72343 PAC, **201**, **202**
fluoropolymers, side-chain, 107
fluorotelomer alcohols (FTOHs), 100, **101**, 166, 167
fluorotelomer betaines (FTBs), 100, **102**
fluorotelomer carboxylic acids (FTCAs), 100, **101**
fluorotelomer iodide, 100
fluorotelomer sulfonamide betaine (FTSBs), **102**
fluorotelomer sulfonic acids (FTSs), 100, **101**
fluorotelomer thioether amido sulfonic acids, **102**
fluorotelomerization, *121*, 122
fluorotelomers, 100, **101–102**
    biotransformation pathways for, *117–118*
    Class B firefighting foams, **109**
fluortelomerphosphatediesters, **388**
fluortelomerphosphatemonoesters, **388**
Food and Drug Administration (FDA), 360–361
fuel hydrocarbons, 86

G

GenX, 97
German Environment Agency ("Umweltbundesamt"), 3, 148, 316
germanium, **389**
granular activated carbon (GAC)
    1,4-dioxane, 31, **67**, 73–75
    per- and polyfluoroalkyl substances, *189*, **190**, 198–204, **200–202**
    1,2,3-trichloropropane, 338–343
groundwater remediation/treatment
    1,4-dioxane, 51–64
    hexavalent chromium, 279–292
    per- and polyfluoroalkyl substances, 198–217
    1,2,3-trichloropropane, 328–337
groundwater treatment, challenges with, 366–367

H

half-lives, 142–143
Hawaii, TCP regulation in, 322–323
health advisory levels (HALs), 6, 18
Henry's constant, 31, 344
heptabromodiphenyl ethers, **389**
hexabromobiphenyl, **389**
hexabromodiphenyl ethers, **389**
hexaflouropropylene oxide-dimer acid (HFPO-DA), 97
hexafluoropropylene oxide trimer acid (HFPO-TA), 97
hexavalent chromium (Cr[VI]), 263–305
    analytical methods, 274–277, **275–276**
    chemical properties, **386**
    common species of in groundwater, 264–265
    in drinking water, 272–273
    drinking water treatment, 292–293, 294–305
    ex situ groundwater remediation, 294–305
    fate and transport of, 265–267
    geochemistry of, 263–265
    in groundwater, **271**, 273–274
    groundwater treatment, 278–292
    mineralogical analyses, 277, **278**
    occurrence of, 271–273
    physical properties, **386**
    regulatory status, **268**, 268–273
    risk assessment, 267–268
    site characterization, 273–277
    in situ reduction, 279–290
    solubility of salts, 265
    sources of, 267
    toxicity of, 267–268
    treatment technologies for, 294–305, **295–296**
high-energy electron beam, 196–197
HiPOx®, **70**
Hoosick Falls, New York, 16

I

ibuprofen, **390**
in situ biological reduction, *284*, 284–287, 284–288, **285**
in situ biological treatment
    1,4-dioxane, 54–58
    hexavalent chromium, 284–288
    per- and polyfluoroalkyl substances, 191
    1,2,3-trichloropropane, 331–333
in situ chemical oxidation (ISCO)
    1,4-dioxane, 51–54
    1,2,3-trichloropropane, 336–337
in situ chemical reduction
    hexavalent chromium, 280–284, **281**
    1,2,3-trichloropropane, 333–336
in situ hydrolysis, of 1,2,3-trichloropropane, 328–331
in situ treatment
    1,4-dioxane, 51–61
    hexavalent chromium, 279–290
    per- and polyfluoroalkyl substances, 191, 193–194, 212–213
    1,2,3-trichloropropane, 328–337
incineration, *189*, **190**, 193–194, 209–210
injectable particulate carbon, 208–209, 212–213. *See also* water treatment
ion exchange, **67**, 297–300
    per- and polyfluoroalkyl substances, *189*, **190**, 204–208, **205**
    strong base anion resins, **293**, 298–300
    weak base anion resins, **293**, 297–298
IRA67, **207**
IRA958, **207**
iron media, 304

L

land fills, as sources of PFASs, 187–188
leaching, of PFASs, 156–160
life cycle of emerging contaminants, 11–15, *12*
 awareness, 11–15
 characterization, 12–13
 management, 13
 regulation, 14
long-chain PFASs, 85–86, 87, 105–106, 111
lowest observed adverse effect level (LOAEL), 146, **147–148**

M

manganese, **389**
Mann-Kendall statistical trend analysis, 291
maximum contaminant levels (MCLs), 18, 35
MeadWestvaco Co. BioNuchar, **201**, **202**
MeadWestvaco Co. WVB 14x35, **201**, **202**
mechanochemical destruction, of PFASs, 197
membrane bioreactors, for hexavalent chromium treatment, 302
mesoporous organosilica (MPOS), for PFAS treatment, 210
messenger ribonucleic acid (mRNA), 48
metabolic biodegradation, of 1,4-dioxane, 48, 50, 54, **56**, 72
metals, **357**
methyl bromide, **390**
methyl tertiary butyl ether (MTBE), 5, 14
Methylosinus trichosporium OB3b, 331
Microbial Insights, 327
microcystin-LR, **391**
microfiltration, for 1,4-dioxane treatment, **68**
microplastics, **358**
Minnesota, TCP regulation in, 324
Missouri, TCP regulation in, 324
mode of action (MOA), of 1,4-dioxane, 34–35
molecular biology tools (MBTs), 48
molybdenum, **389**
monitored natural attenuation (MNA), 62–64, 291

N

N-alkyl perfluoroalkane sulfonamides, **103**
N-alkyl perfluoroalkane sulfonamidoacetic acids, **103**
nanofiltration, for FPAS treatment, *189*, **190**, 208–209
nanomaterials, **358**
National Drinking Water Advisory Council, 4–5
National Environment Protection Measures (NEPMs), 10–11
natural attenuation, 62–64, 291–292. *See also* groundwater remediation/treatment

tier I, 291
tier II, 292
tier III, 292
tier IV, 292
N-ethyl perfluorooctane perfluorooctane sulfonamide (NEtFOSA), 98
N-ethyl perfluorooctane sulfonamido ethanol-based phosphate diester (SAmPAP diester), **104**
N-ethyl perflurorooctane sulfonamidoacetate (NEtFOSAA), 98
N-ethyl-perfluorooctane sulfonamide, **387**
N-ethyl-perfluorooctane sulfonamidoethanol, **387**
New Jersey, TCP regulation in, 323
New York, TCP regulation in, 324
Nitrosomas europaea, 331
N-methyl perfluorooctane sulfonamide (NMeFOSA), 98, **387**
N-methyl perfluorooctane sulfonamidoacetate (NMeFOSAA), 98
N-methyl perfluorooctane sulfonamidoethanol, **387**
N-nitrosodimethylamine, **391**
no observed adverse effect level (NOAEL), 146, **147–148**
Norit 1240W PAC, **202**
Norwegian Geotechnical Institute, 10
nZVI, **281**

O

oligomerization, *121*, 122–123
Organization for Economic Cooperation and Development (OECD), 361
ozofractionation, for FPASs, 211. *See also* water treatment
ozofractionation, for PFASs, *189*, **190**, 211
Ozofractionative Catalyzed Reagent Addition (OCRA), for PFASs, 211

P

Pace Energy Services, 327
particle-induced gamma emission (PIGE) spectroscopy, 138
Pengcheng Activated Carbon Co. GAC, **201**, **202**
Pengcheng Activated Carbon Co. PAC, **201**, **202**
pentabromodiphenyl ethers, **389**
pentadecafluorooctanoic acid ammonium salt, **386**
per- and polyfluoroalkyl substances (PFASs), 85–217
 acceptable guidance concentrations, 86
 acid dissociation constants, 112
 acute toxicity of, 144
 atmospheric deposition, 166
 Australian regulation of, 153

Index

background concentrations of, 167–180, **168–178**
bioaccumulation, 140–142, **141**
biological activity towards, 112–114
biological properties, 110–119
biological treatment of, 191
carcinogenicity of, 146–150
chemical analysis methods, 128–139
chemical properties, 110–119
chemistry of, 88–110
chronic toxicity of, 144
in Class B firefighting foams, 107–110, **108–109**
conceptual site models, 181–188
detection in drinking water, 86
difference from common contaminants, 86–87
distribution in environmental matrices, 155–167
distribution in tissue, 139–140
environmental standards, **401–425**
epidemiological studies, 144–145
European regulation of, 153
exposure routes, 139, *140*
families and subgroups, *89*
fate and transport of, 154–188
firefighting, 179–180
global treaties and conventions on, 150–152
in groundwater, 160–165
ionic state, 91
leaching, 156–160
linear and branched isomers, 91
long-chain, 85–86, 87, 105–106
main groups, 85
nonbiodegradiblity of, 86
perfluoroalkyl substances, 92–97
physical properties, 110–119
polyfluoroalkyl substances, 98–100
polymeric, 106
production of, 119–123, *121*
reference doses/tolerable daily intakes, 146, **147–148**
regulation of, 150–154
replacement, 107
sampling of, 125–128
in sediments, 165–166
short-chain, 85–86, 105–106
sites of concern, 180–188
in soil, 156
soil and sediment sampling, 125
soils and sediment treatment of, 191–197
sources of, 86
as surface coatings, 123
in surface water, 165–166
surface water and groundwater sampling, 125–128
as surfactants, 123

treatment technologies, 187–217
U.S. regulation of, 152–153
uses of, 86, 123–125, *124*
vapor migration, 166
water solubility of, 111
water treatment, 198–217
perchlorate, 5, **391**
perchloric acid, **391**
perfluordecylethanol, **387**
perfluordodecylethanol, **387**
perfluorethylethanol, **387**
perfluorhexylethanol, **387**
perfluorinated compounds (PFCs), 92
perfluormethylethanol, **387**
perfluoroalkane sulfonamides, 98
perfluoroalkane sulfonamidoacetic acids (FASAAs), **103**
perfluoroalkanesulfonamidoethanols, 98
perfluoroalkyl acids (PFAAs), 85, **93–94**
  acceptable guidance concentration, 86
  background concentrations of, **168–178**
  background levels of, 179
  degradation rates of precursors, **116**
  in rain and surface waters, 179
perfluoroalkyl carboxylates (PFCAs), naming convention, 92
perfluoroalkyl carboxylic acids, 96
perfluoroalkyl ether carboxylates (PFECAS), 97, 179
perfluoroalkyl ether sulfonates, 97
perfluoroalkyl ether sulfonic acids (PFESAs), 92, 179
perfluoroalkyl ethers, regulation of, 154
perfluoroalkyl phosphonic acids (PFPAs), 96–97
perfluoroalkyl substances, 92–97
  chemical properties, **386–388**
  perfluoroalkyl carboxylic acids, 96
  perfluoroalkyl ether carboxylates, 97
  perfluoroalkyl phosphonic acids (PFPAs), 96–97
  perfluoroalkyl sulfonic acids, 95
  physical properties, **386–388**
perfluoroalkyl sulfonamide amines, 98
perfluoroalkyl sulfonamide amino carboxylates, **104**
perfluoroalkyl sulfonamido amines, **103**
perfluoroalkyl sulfonamido substances, **103–104**
perfluoroalkyl sulfonates (PFSAs), 92
perfluoroalkyl sulfonic acids, 95
perfluorobutane sulfonate (PFBS), 95, **386**
perfluorobutanesulfonic acid, **387**
perfluorobutanoic acid (PFBA), **386**
perfluorobutyl phosphonic acid, **387**
perfluorocylethanol, **387**
perfluorodecane sulfonate, **386**
perfluorodecanesulfonic acid, **387**
perfluorodecanoic acid, **386**

perfluorodecyl phosphonic acid, **387**
perfluorododecanoic acid, **386**
perfluoroethane sulfonate (PFEtS), 95
perfluoroethanoic acid (PFEtA), 96
perfluoroheptane sulfonate, **386**
perfluoroheptanoic acid, **386**
perfluorohexane sulfonate, **386**
perfluorohexane sulfonic acid (PFHxS), 85
perfluorohexanesulfonic acid, **387**
perfluorohexanoic acid, **386**
perfluorohexyl phosphonic acid, **387**
perfluorononanoic acid, **386**
perfluorooctane sulfonamide (FOSA), 98, **387**
perfluorooctane sulfonamidoacetate (FOSAA), 98
perfluorooctane sulfonamidoethanol, **387**
perfluorooctane sulfonate, **386**
perfluorooctane sulfonic acid (PFOS), 85
    background levels of, 178–179
    bioaccumulation, **141**
    carcinogenicity of, 146–150
    chemical structure of, *95*
    detection in drinking water, 86
    elimination half-lives of, 142–143
    linear and branched isomers, 91
    reference doses/tolerable daily intakes, 146, **147–148**
    structure of, *90*
perfluorooctanesulfonic acid, **387**
perfluorooctano phosphonic acid (PFOPA), 97
perfluorooctanoic acid (PFOA), 85, **386**
    acceptable guidance concentration, 86
    bioaccumulation, **141**
    carcinogenicity of, 146–150
    chemical structure of, *96*
    detection in drinking water, 86
    elimination half-lives of, 142–143
    reference doses/tolerable daily intakes, 146, **147–148**
perfluorooctyl phosphonic acid, **387**
perfluoropentadecanoic acid, **386**
perfluoropentanoic acid, **386**
8:8 perfluorophosphonic acid (8:8 PFPiA), 97
perfluoropolyethers (PFPEs), 107
perfluoropropane sulfonate (PFPrS), 95
perfluorotetradecanesulfonic acid, **387**
perfluorotetradecanoic acid, **386**
perfluorotridecanoic acid, **386**
perfluoroundecanesulfonic acid, **387**
perfluoroundecanoic acid, **386**
permeable reactive barriers, for hexavalent chromium treatment, 288–289
permethrin, **390**
persistent organic pollutants (POPs), 8–9, 64, 86–87, 150, 167
persistent, bioaccumulative, and toxic (PBT) criteria, 10

persistent, mobile, and toxic (PMT) criteria, 10, **393–399**
personal care products (PCPs), 357, **390**
pesticides, **358**
PFAS environmental standards
    drinking water, **419–425**
    groundwater, **408–414**
    soil, **401–407**
    surface water, **415–418**
PFASs. See per- and polyfluoroalkyl substances (PFASs)
pharmaceuticals, 357, **390**
Phase II assessments, 7
phenol hydroxylase, 48
phosphinic acids, 96–97
PhotoCat®, **70**
photolysis, for PFASs, *189*, **190**, 216–217. *See also* water treatment
photovolatilization, 60
physical properties
    1,4-dioxane, **30**, **318**, **386**
    emerging contaminants, **385–392**
    hexavalent chromium, **386**
    per- and polyfluoroalkyl substances, 110–119
    perfluoroalkyl substances, **386–388**
    1,2,3-trichloropropane, **318**, **386**
phytodegradation, 60
phytoextraction, 58
phytoremediation, 212. *See also* water treatment
    1,4-dioxane, 59–60
    per- and polyfluoroalkyl substances, 211–212
phytostabilization
    1,4-dioxane, 58
    hexavalent chromium, **296**, 302–303
point-of-entry (POE) treatment. *See also* water treatment
    1,4-dioxane, 66
    hexavalent chromium, 293–294
point-of-use (POU) treatment. *See also* water treatment
    1,4-dioxane, 66
    hexavalent chromium, 293–294
polybrominated biphenyls (PBBs), 13, **389**
polybrominated diphenyl ethers (PBDEs), 13, **389**
polychlorinated biphenyls (PCBs), 86–87
polydiallyldimethylammonium chloride (PDM), 190
polyfluoroalkane sulfonamides (FASAs), 98
polyfluoroalkyl phosphates (PAPs), 100, **102**
polyfluoroalkyl substances (PFASs), 5, 13, 15, 16, 98–100, **101–104**
    abiotic transformation of, 115
    biotic transformation of, 115–119, **119**
    ECF-derived, 98–99
    elimination half-lives of, 142–143
    fluorotelomerization-derived, 100
    perfluoroalkyl ether sulfonates, 97

Index 437

toxicity of, 145
transformation of, 114–119
polyfluoroalkyl sulfonamides, biotransformation
   pathways for, *120*
polymeric PFASs, 106–107
polysulfide, **281**
potassium perchlorate, **391**
propane biosparging, for 1,4-dioxane treatment,
   57, 57–58, *58*, *59*
propane monooxygenase (PPO), 48
2-propen-1-ol, **391**
Pseudonocardia dioxanivorans CB1190, 55
public sensitivity, 16
public water systems (PWSs), 315–316
pug mills, 189
pump and treat strategy, 290–291
Purifics, **70**
Purolite, **205–207**

Q

quadrupole time of flight-mass spectroscopy
   (QTOF-MS), 139
QuantArray®, 327
quantitative polymerase chain reaction (qPCR),
   48, 327

R

RayOx®, **70**
reduced sulfur, 283
reduction/coagulation/filtration (RCF) with
   ferrous iron, 294–297
reduction/coagulation/filtration (RCF), for
   hexavalent chromium, 294–297, **295**
reduction/filtration via stannous chloride
   (RF-Sn[II]), 304–305
reduction/filtration via stannous chloride
   (RF-Sn[II]), for hexavalent chromium,
   304–305
reference concentration, 319
reference dose, 319
Regional Screening Levels (RSLs), 38
Registration, Evaluation, Authorisation, and
   Restriction of Chemicals (REACH),
   4, 9
   candidate contaminant list, **377–383**
regulatory status
   1,4-dioxane, 35–38
   hexavalent chromium, 268–273
   per- and polyfluoroalkyl substances,
     150–154
   1,2,3-trichloropropane, 321–324
replacement PFASs, 107
reverse osmosis, 208–209, 300
   1,4-dioxane, **68**
   hexavalent chromium, **296**, 300

per- and polyfluoroalkyl substances, *189*, **190**,
   208–209
1,2,3-trichloropropane, 345
rhizodegradation, 58
rhizofiltration, 58
Rhodococcus ruber ENV425, 55
Rhodococcus sp., 332
ring-hydroxylating toluene monooxygenases, 48
risk assessment
   1,4-dioxane, 32–35
   hexavalent chromium, 267–268
   per- and polyfluoroalkyl substances, 138–147
   1,2,3-trichloropropane, 317–321
Royal Demolition Explosive (RDX), 7

S

Safe Drinking Water Act (SDWA), 4–5, 293
SAmPAP diester, **104**
SAmPAP monoester, **104**
SAmPAP triester, **104**
sediment treatment, for PFASs, **190**, 191–197
   incineration, 192–193
   mechanochemical destruction, 197
   vapor energy generator, 194–195
   washing, 196
sediments, PFASs in, 165–166
Sen's slope estimate, 291
short-chain PFASs, 85–86, 105–106
side-chain fluoropolymers, 107
Siemens Inc. AquaCarb 1240C, **201**, **202**
SiREM, 327
site characterization
   1,4-dioxane, 38–50
   hexavalent chromium, 273–277
   per- and polyfluoroalkyl substances, 125–139,
     180–188
   1,2,3-trichloropropane, 325–327
sodium perchlorate, **391**
soil
   1,4-dioxane in, 50–51
   per- and polyfluoroalkyl substances in, 125,
     156, 187–193
   PFAS environmental standards, **401–407**
soil treatment, 191–197
   1,4-dioxane, 50–51
   high-energy electron beam, 196–197
   incineration, 192–193
   mechanochemical destruction, 197
   per- and polyfluoroalkyl substances, 192–197
   stabilization/solidification, 193–194
   vapor energy generator, 194–195
   washing, 196
solidification, for PFASs, 189–190
soluble methane monooxygenase (sMMO), 48,
   54–55, 64, 327
sonolysis, 215–216

sonolysis, for PFASs, **190**, 215–216
sorbent media, 73–75
Sphingobium japonicum UT26, 332
stabilization/solidification, for PFASs, 193–194
stable isotope probing (SIP), for 1,4-dioxane, 49, 57–58, *59*
state agencies, 7–8
Stockholm Convention on Persistent Organic Pollutants, 8–9, 150
Strategic Environmental Research and Development Program (SERDP), 51
strong base anion ion exchange (SBA-IX), for hexavalent chromium treatment, 293, **295**, 298–300
strontium, **388**
submerged packed bed reactors, 301–302
sulfolane, **390**
sulfonamido ethanol-based phosphate esters (SAmPAP), 98
sulfur hexafluoride, **390**
Superfund, 291
surface coatings, PFAS use as, 123
surface water
    per- and polyfluoroalkyl substances in, 165–166
    PFAAs in, 179
    PFAS environmental standards, **415–418, 419–425**
surfactants, PFAS use as, 123

T

Telone, 315
tetrabromodiphenyl ethers, **389**
2,3,7,8-tetrachlorodibenzo-para-dioxin, **390**
thermal in situ sustainable remediation (TISR™), for TCP, 329
thermal in-situ treatment, for 1,4-dioxane treatment, 60–61
titanium dioxide, **390**
tolerable daily intakes (TDIs), for PFASs, 146, **149**
total oxidizable precursor (TOP) assay, for PFAS analysis, 129–137
    conditions for, 132–134
    data quality objectives, 134–136
    prediction of terminal products with, 136–137
toxicity
    1,4-dioxane, 32–35, **33**
    hexavalent chromium, 267–268
    per- and polyfluoroalkyl substances, 144
    polyfluoroalkyl substances, 145
    1,2,3-trichloropropane, 317–321
TreeWell® System, for 1,4-dioxane treatment, 60
triboplasmas, 193
tribromoacetic acid, **392**
tribufos, **390**
1,1,1-trichloroethane (TCA), 30, 31–32, **318**
trichloroethene (TCE), 14, **30**, 31–32, **318**
1,2,3-trichloropropane (TCP), 5, 315–345
    agriculture-related sources of, 315
    analytical methods for, 325–327, **326**
    cancer slope factors and inhalation risks, **320**
    chemical properties, **318**, **386**
    in drinking water, 316–317, **317**
    drinking water treatment, 338–345
    ex situ groundwater remediation, 338–345
    groundwater remediation for, 328–337
    international guidance for, 324
    physical properties, **318**, **386**
    regulation of, 321–324
    risk assessment, 317–321
    site characterization, 325–327
    in situ treatment, 328–337
    sources of, 315
    structure of, *316*
    toxicity of, 317–321
    toxicological reference values, **319**
    treatment technologies for, 338–345
    U.S. federal regulations, 321–322
    U.S. state regulations, 322–324
    water treatment, 338
triclosan, **390**
trifluoromethane sulfonate (TFMS), 95
1,3,5-trinitroperhydro-1,3,5-triazine, 7
trivalent chromium (Cr[III]), 263
    reoxidation of, 290
TrojanUV®, **70**
TrojanUVPhox™, **70**

U

U.S. Coast Guard Support Center, 283–284
ultrafiltration, for 1,4-dioxane treatment, **68**
uncertainty, 16–17
Uncontaminated Monitoring Rules (UCMRs), 315
United States Department of Defense (DoD), 6–7
United States Environmental Protection Agency (USEPA), 4–6, 32, 360–361
    candidate contaminants, **373–376**
    Third Unregulated Contaminant Monitoring Rule (UCMR3), 152
United States Geologic Survey (USGS), 7
Unregulated Compounds Monitoring Rule (UCMR), 4–5
US Filter A-399, **206**
US Filter A-714, **206**, 207

Index

V

vanadium, **389**
vapor energy generator, for PFAS treatment, 194–195
vapor migration, PFASs and, 166
very persistent and very mobile (vPvM) criteria, 10, **393–399**
very persistent, very bioaccumulative (vPvBs), 10
visibility, 16

W

waste water treatment
    conceptual site models for (PFASs), 182–183, *183*
    1,4-dioxane, 65–66
    per- and polyfluoroalkyl substances, 182–183, *183*
watch list (United States Department of Defense), 6–7
Water Resources Control Board (California), 7
water treatment
    1,4-dioxane, 65–66, **67–68**
    hexavalent chromium, 292–294
    per- and polyfluoroalkyl substances, 198–217
    1,2,3-trichloropropane, 338–345
weak base anion ion exchange (WBA-IX), for hexavalent chromium treatment, 293, **295**, 297–298
Wedeco® MiPro™ eco$_3$, **70**
Wedeco® MiPro™ eco$_3$ plus, **70**
Wedeco® MiPro™ photo, **70**
Wedeco® Pro mix®, **70**

X

Xinhua Activated Carbon Co. PAC, **201**
Xylem, **70**

Z

zero-valent iron (ZVI)
    hexavalent chromium, **281**, 289
    1,2,3-trichloropropane, 333–336
zero-valent zinc (ZVZ), for TCP treatment, 333–336, 345